Descriptive Geometry and Geometric Modeling

A Basis for Design

About the Cover

The cover photograph, produced on a PS330 calligraphic refresh display and used by permission of Evans and Sutherland Inc., illustrates how three-dimensional information is lost when a single, two-dimensional representation of an object is used. A designer must continually maintain a spatial awareness of objects under consideration so that three-dimensional information can be recovered. This can be done either manually from a series of two-dimensional drawings, or computationally from a computer database. This text is written to help satisfy this design need for spatial awareness.

Descriptive Geometry and Geometric Modeling

A Basis for Design

J. ALAN ADAMS *United States Naval Academy*

LEON M. BILLOW *Late of the United States Naval Academy*

HOLT, RINEHART AND WINSTON, INC.
New York Chicago San Francisco
Philadelphia Montreal Toronto
London Sydney Tokyo

Publisher/Acquisitions Editor Ted Buchholz
Senior Project Editors Suzanne Magida, Herman Makler
Production Manager Paul Nardi
Design Supervisor Bob Kopelman
Text Design York Production Services
Illustrations J & R Services

Library of Congress Cataloging-in-Publication Data

Adams, J. Alan (James Alan), 1936–
Descriptive geometry and geometric modeling.

Includes index.
1. Geometry, Descriptive. 2. Geometry—Data processing. 3. Computer graphics. I. Billow, Leon M. II. Title.
QA501.A223 1988 516 87-12006

ISBN 0-03-009514-X

Printed in the United States of America

8 9 0 1 090 9 8 7 6 5 4 3 2 1

Holt, Rinehart and Winston, Inc.
The Dryden Press
Saunders College Publishing

One month before the final manuscript was submitted for publication, Leon M. Billow was killed in a car accident. The last three years of his life were primarily spent working on this text. He was concerned that students be properly taught to think and reason in a spatial context. Leon demonstrated a dedicated, patient, disciplined approach to his profession. His work was fastidious in every detail. He loved his students and his peers, and he was proud of the effort he had made on this book. The book is dedicated to his memory, and to his lifelong friend and wife, Kathryn S. Billow.

Preface

The standard of living we enjoy today is the result of inventive skills that depend on imagination and creativity. Knowledge and understanding of geometric principles by which ideas and concepts are developed and recorded provide the bridge from abstract thinking to physical reality. Thus, the study of geometry is fundamental to the creative process and is an important part of engineering and technical education.

The combination of descriptive geometry with geometric modeling provides new possibilities for creative design. Descriptive geometry is a graphical method of formulation and analysis that gives solutions to design problems while working in a two-dimensional plane. Its basic principles are crucial for geometric communication of ideas and for interpretation of three-dimensional information presented on drawings or computer terminals. Geometric modeling provides the basis for a mathematical formulation of shape information, and computational techniques for obtaining design solutions based upon geometric objects stored in computer memory. This approach is rapidly growing in importance due to the popularity of microbased computer-aided design (CAD) systems and engineering computer work stations.

For many years, descriptive geometry was the first quantitative course required of all beginning engineering students. Recently, the need to introduce computer programming and other introductory courses has resulted in a decline in the coverage of graphical solution techniques, a limited use of geometric and spatial reasoning, and less emphasis on graphical communication skills. The authors feel that a modern, integrated approach to descriptive geometry and geometric modeling can present fundamental principles and concepts which will restore geometry as the cornerstone of an engineering education. Associated skills in programming, graphics, and use of commercial software can be obtained from a variety of educational experiences to supplement this foundation.

The mathematical techniques used to complement descriptive geometry help provide a spatial awareness necessary when dealing with traditional mechanical drawing and design, or with computer-aided design and manufacture (CAD/CAM). This awareness is required for those who wish to make intelligent use of CAD/CAM software systems available today. It also prepares one for later study of advanced geometric modeling and computational geometry which includes mathematical and numerical techniques for curve, surface, shape, and solid model definitions.

This text integrates the traditional geometric reasoning of descriptive geometry with an introduction to mathematical formulations and solutions. Sample computer programs are given in Appendix 1 to serve as a guide for students who wish to create similar programs. ANSI BASIC is used because of its readability, but the programs and subroutines can easily be adapted to other software languages due to their modular nature. The greatest benefit will be obtained from this text if the reader has access to a computer. However, the computations required can be implemented on a variety of hardware, including small, programmable calculators.

In this text, the term "design" is used to mean conceptual, preliminary, or geometric design. Before simplifying, engineering assumption can be made, before equations or free-body diagrams can be applied, before any quantitative analysis can be carried out, a device, configuration, shape, prototype, or mental image must exist. This conceptual design can be documented using manual, graphical techniques or interactive, computer techniques. Regardless of how it is done, conceptual design documentation must be the foundation for subsequent engineering study. For this reason, the mathematical level of this text is limited to be suitable for an early undergraduate course; mathematical methods that are common in subsequent engineering courses are emphasized.

The mathematical prerequisites for this introductory text are algebra, analytical geometry, and trigonometry. Elementary concepts from vector algebra and matrix algebra are presented as needed. Vectors are used to represent points, lines, and planes in space. Operations such as vector cross product, scalar dot product, and scalar triple product are used to describe geometric attributes needed for solving geometry problems. It is assumed that the student has been introduced to computer programming, using one of the popular, high-level languages.

Descriptive geometry forces the student to think. It stimulates the imagination to form mental pictures, overcome complications, and formalize solutions to spatial problems. The computer expands this capability through its ability to rapidly analyze large amounts of data, improve accuracy, and solve complex problems.

A first course in geometry at the college level should meet the following objectives:

a. Use both classical and modern techniques to develop a spatial awareness and a three-dimensional reasoning ability.

b. Develop a logical and orderly method of problem solving based on fundamental principles.

c. Make use of elementary mathematical methods, such as vector algebra, which provide a computational approach to geometry.

d. Provide an opportunity for use and development of computer software for specific three-dimensional geometric applications.

e. Provide many examples to illustrate various applications of descriptive geometry and geometric modeling.

The reader will find several unusual approaches in the presentation of the textual material. New technology requires new combinations of academic material and departure from some established traditions. For example, there is no mention of a 45 degree transfer line or a reference line since this implies a mechanical, routine way of taking measurements from a drawing. The *folding line* or *reference plane* is used to impress upon the student the fact that the edge of a plane is used, and that measurements are not taken from just a line. The definition of geometric concepts such as points, lines, and planes are discussed from both a graphical and mathematical point of view within the same chapters. Thus, the student has the option to approach a problem using traditional descriptive geometry or a computational method.

The example solutions presented in the text are designed to encourage the student to develop a methodical and systematic approach to all problems, based upon general principles. Self-study is encouraged in some cases where the student must develop the step-by-step procedure to agree with the illustrated solution. Computer graphics are used to present solutions to some of the geometric modeling examples. Commercial software can also be used to produce graphical output if the student has access to this option. However, students must first learn to cope with the three-dimensionality of a design problem, especially when a computer does all of the graphics display work. This text is designed to help students meet this goal.

We would like to thank the following reviewers whose constructive comments and suggestions helped us to improve the final manuscript: Jack C. Brown, University of Alabama; Chung Ha Suh, University of Colorado, Boulder; John T. Demel, Ohio State University; Jon M. Duff, Purdue University; Robert J. Foster, Pennsylvania State University; Garland K. Hilliard, North Carolina State University; Richard C. Latimer, University of Louisville; James A. Leach, Auburn University; Bruce C. Rogers, University of Northern Iowa; and William H. Sanders, Kansas Technical Institute.

J.A.A./L.M.B.

To the Student

Gaspard Monge (1746–1818) introduced the principles of descriptive geometry in the eighteenth century while he was a military student in France. Since that time, this graphical method of problem solving has developed from a French military secret to a subject that has been offered in almost every engineering college in the world in order to teach students to analyze problems through visualization and reasoning.

In addition to the objectives set forth in the preface, this text is concerned with the geometric fundamentals needed to produce engineering drawings and to solve engineering problems using both manual and computer techniques. The book is not a text on engineering drawing per se, but deals with descriptive geometry and geometric modeling as a means of reaching technical objectives. For information on the many details of engineering graphics such as line construction, fasteners, dimensioning, and the like, the reader should consult the available literature. Students should also be aware of the available computer software packages that can be used as an aid in the rapid and accurate production of engineering drawings, as well as for solutions to geometry problems.

Suggested student goals, in order of importance, are as follows:

1. Develop a spatial awareness and reasoning capability.
2. Create a graphical and mathematical foundation for solving geometric problems.
3. Learn to use the computer as a tool for analysis and design.

Contents

Preface vii

To the Student xi

1 ORTHOGRAPHIC PROJECTION *1*

1-1 Projection—Perspective and Orthographic 1
1-2 Third Angle Projection 3
1-3 Objects in Space 4
1-4 Visualization 4
1-5 Principal Views 6
1-6 Folding Lines 7
1-7 Points in Space 9
1-8 Lines in Space 11
1-9 Planes in Space 14
1-10 Solids in Space 16
1-11 Visibility 17
1-12 Vector Representation of a Point 19
1-13 Wire Frame Projections 21
1-14 Vector Addition 28
1-15 Unit Vectors 32

2 PRIMARY AUXILIARIES—PLANES 38

2-1 Normal Views 38
2-2 Normal View of a Plane—Frontal Auxiliary 40
2-3 Normal View of a Plane—Elevation Auxiliary 42
2-4 Normal View of a Plane—Profile Auxiliary 43
2-5 Reference Planes 44
2-6 Normal View of an Irregular Plane 45
2-7 Normal and Edge Views—A Summary and Preview 47
2-8 Vector Cross Product 48
2-9 Eye Positions for Orthographic Views 54
2-10 Area of Plane Objects 60
2-11 Polygon Files 65

3 PRIMARY AUXILIARIES—LINES 73

3-1 Primary Auxiliaries—Lines in True Length 73
3-2 True Length Principal Lines 75
3-3 True Length Line—Frontal Auxiliary 75
3-4 True Length Line—Horizontal Auxiliary 76
3-5 True Length Line—Profile Auxiliary 77
3-6 Point View of Inclined Line 78
3-7 Points on a Line 80
3-8 Intersecting and Nonintersecting Lines 82
3-9 True Length of Line by Revolution 84
3-10 Bearing, Slope, and Grade of a Line 86
3-11 Vector Equation of a Line 91
3-12 True Length of a Three-Dimensional Line 92
3-13 Direction Cosines 95
3-14 Intersecting Two-Dimensional Lines 97
3-15 Two-Dimensional Line Construction 99
3-16 Intersecting Three-Dimensional Lines 100

4 LINES AND PLANES 105

4-1 Edge View of a Plane 105
4-2 Planes—Graphical Representation 108
4-3 Points in Planes 111
4-4 Lines in Planes 115

4-5 Vector Dot Product 117
4-6 Vector Equation of a Plane 120
4-7 Point in a Plane 124
4-8 Parametric Equation of a Plane 126

5 SUCCESSIVE AUXILIARY VIEWS *138*

5-1 Second Auxiliary Views 138
5-2 Successive Auxiliaries 140
5-3 Point View of Oblique Line 142
5-4 Normal View of an Oblique Plane 143
5-5 Normal View of an Irregular Oblique Plane 145
5-6 Orthographic Views with Computer Graphics 147
5-7 Special Orthographic Views 150

6 PARALLELISM *158*

6-1 Parallel Lines 158
6-2 Line Parallel to a Plane 161
6-3 Plane Parallel to a Line 162
6-4 Parallel Planes 164
6-5 Geometric Modeling for Parallel Lines 166
6-6 Geometric Modeling for Parallel Planes 167
6-7 Geometric Modeling for Parallel Line and Plane 168

7 PERPENDICULARITY *175*

7-1 Perpendicular Lines 175
7-2 Shortest Distance from a Point to a Line 178
7-3 Common Perpendicular (Shortest Distance) Between Two Skew Lines 181
7-4 Shortest Horizontal Distance Between Two Skew Lines 183
7-5 Shortest Line of Specified Grade Between Two Skew Lines 184
7-6 Line Perpendicular to a Plane 185
7-7 Plane Perpendicular to a Line 188
7-8 Plane Perpendicular to a Plane 189
7-9 Geometric Modeling for Perpendicularity 191

8 INTERSECTING LINES AND PLANES 202

8-1 Intersecting Line and Plane-Piercing Points 202
8-2 Projection of a Line on a Plane 205
8-3 Line of Intersection: Intersecting Plane Segments Given 208
8-4 Line of Intersection: Nonintersecting Plane Segments Given 210
8-5 Modeling the Intersection of a Line and Plane 212
8-6 Computational Descriptive Geometry 216
8-7 Piercing Point Using Parametric Equation for Plane 220
8-8 Modeling the Projection of a Line 222
8-9 Shortest Distance Between Two Skew Lines 224

9 ANGLES 234

9-1 Angle Between Two Intersecting Lines 234
9-2 Angle Between Two Nonintersecting Skew Lines 235
9-3 Angle Between an Oblique Line and Principal Plane 236
9-4 Angle Between an Oblique Plane and Principal Plane 237
9-5 Angle Between a Line and an Oblique Plane 238
9-6 Dihedral Angle: Angle Between Two Planes 242
9-7 Analytical Calculation of Angles 246

10 REVOLUTION 254

10-1 Revolution of a Point 254
10-2 Revolution of a Line 256
10-3 Edge View of a Plane by Revolution 259
10-4 Normal View of a Plane by Revolution 260
10-5 Revolution of a Solid 261
10-6 Angle Between Oblique Line and Oblique Plane by Revolution 264
10-7 Dihedral Angle by Revolution 265
10-8 Line Making Specified Angles with Two Perpendicular Planes 266
10-9 Modeling Surfaces of Revolution 268
10-10 Recapitulation 272

11 INTERSECTIONS—LINE AND SOLID 277

11-1 Intersection—Line and Prism 277
11-2 Intersection—Line and Cylinder 280
11-3 Intersection—Line and Pyramid 283
11-4 Intersection—Line and Cone 284
11-5 Two-Dimensional Clipping 285
11-6 Three-Dimensional Clipping 288

12 TANGENCIES 294

12-1 Line Tangent to Circles 294
12-2 Plane Tangent to a Cone 294
12-3 Plane Tangent to a Cylinder 298
12-4 Plane Tangent to a Sphere 301
12-5 Plane Making a Specified Angle with a Principal Plane and Containing a Given Line 302

13 INTERSECTIONS—PLANES AND SOLIDS 309

13-1 Plane and Prism 310
13-2 Plane and Pyramid 313
13-3 Plane and Cylinder 314
13-4 Plane and Cone 319
13-5 Plane and Double-Curved Surface (Sphere) 322
13-6 Fair Surfaces 323
13-7 Intersecting Planes 327
13-8 Intersecting Plane and Cone 329

14 INTERSECTIONS—SOLIDS 336

14-1 Prisms (Edge-View Method and Cutting-Plane Method) 336
14-2 Prism and Pyramid (Cutting-Plane Method) 338
14-3 Cylinders (Cutting-Plane Method) 339
14-4 Cylinder and Cone 343
14-5 Prism and Cone 348
14-6 Cones 349
14-7 Cylinder and Torus 349

14-8 Geometric Modeling for Intersecting Surfaces 352
14-9 Geometric Modeling for Prism-Cylinder Intersections 355
14-10 Solid Modeling 355

15 SURFACE DEVELOPMENTS *361*

15-1 Right Circular Cylinder—Parallel Line Development 362
15-2 Right Prism—Parallel Line Development 364
15-3 Right Pyramid—Radial Line Development 365
15-4 Right Circular Cone—Radial Line Development 367
15-5 Oblique Prism—Parallel Line Development 368
15-6 Oblique Cylinder—Parallel Line Development 369
15-7 Oblique Cone—Radial Line Development 370
15-8 Transition Piece—Triangulation 372
15-9 Intersection and Development of Prism and Pyramid 373
15-10 Analytical Development of a Cone 374

References 388

Appendices

1 **COMPUTER PROGRAMS 389**

2 **MATHEMATICAL CONCEPTS 437**

3 **GEOMETRIC FIGURES 449**

4 **GEOMETRIC CONSTRUCTION 454**

5 **DRAFTING TECHNIQUES 457**

6 **VECTORS 461**

Index 471

Descriptive Geometry and Geometric Modeling

A Basis for Design

1

Orthographic Projection

The student needs to be concerned with the exact interpretation of ideas and must be able to transfer them by an exchange of thought which is, in itself, exact. This chapter is concerned with orthographic projection, an accurate and precise method of communication; one that is understood and recognized throughout the world. Using this "universal language" an engineer can make a drawing that will show the true size, shape, and details of any object.

1-1 PROJECTION—PERSPECTIVE AND ORTHOGRAPHIC

A. Perspective Projection

The pictorial form of drawing referred to here is used by the architect and artist to produce illustrations that represent objects as they appear to the eye, Figure 1-1. This picture-type drawing is familiar to us but is not exact enough for most engineering work. The inaccuracy of perspective projection is caused by the foreshortening of lines which varies according to the changing positions of the eye and object with relation to the picture plane.

Figure 1-2 shows the eye of an observer located at a finite distance from the picture plane. The lines of sight (L.O.S.) from the observer to the tops and bottoms of objects *x*, *y*, and *z*, intersect the picture plane at different places. Every point on the object is projected onto the picture plane at the point where the line of sight intersects the plane. This means that the images of *X*, *Y*, and *Z*, when projected onto the plane, will produce three pictures which will vary in size according to (a) the distance between

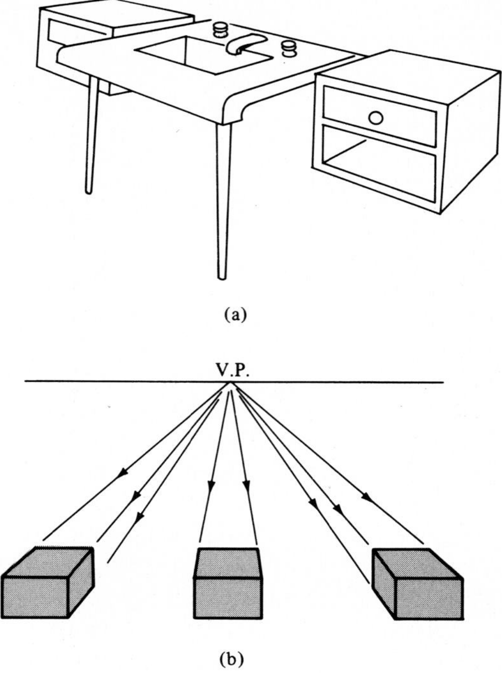

Figure 1-1
Perspective drawing.

the observer and the picture plane, and (b) the distance between the object and the plane. These relationships, noted in Figure 1-2, show why an object seems to grow smaller as it recedes and accounts for seeing objects as we normally do.

B. Orthographic Projection

Webster defines orthographic projection as ". . . a projection in which the projection lines are perpendicular to the plane of projection." This tech-

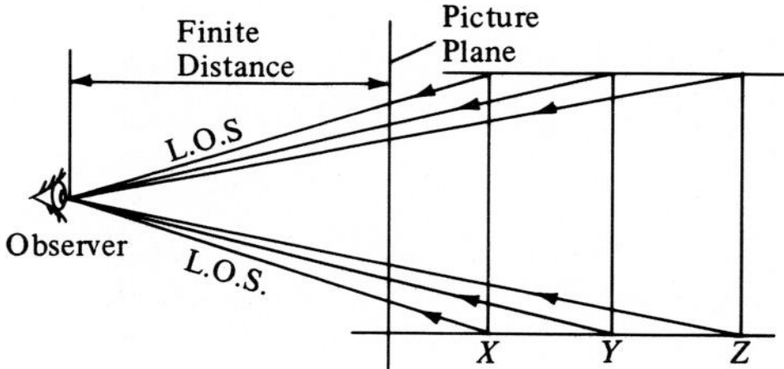

Figure 1-2
Perspective theory.

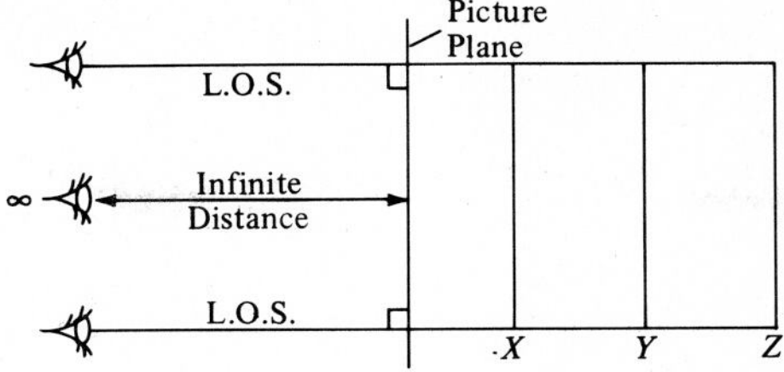

Figure 1-3
Orthographic theory.

nique is also referred to as *orthographic drawing, multiview drawing, engineering drawing, multiplanar drawing,* and *orthogonal drawing.* In any of the above the observer, by definition, views the object so that the line of sight to any point on the object is perpendicular to the plane through which the object is viewed. This plane is known as the "picture plane" or the "plane of projection."

Figure 1-3 shows the projection when the observer moves back farther from the picture plane. As the viewer retreats, the lines of sight tend to become parallel until at infinity they *are* parallel and pierce the projection plane at right angles. Every point on the object is projected along the line of sight until it intersects the picture plane and creates a visible point. A series of projected points are connected to form a picture, or view, of the object. This is *orthographic projection,* the method which is used to describe the details of any object (see Figure 1-4).

1-2 THIRD ANGLE PROJECTION

Figure 1-5 shows that four quadrants are formed by the right angle intersection of the vertical and horizontal picture planes. In first-angle projec-

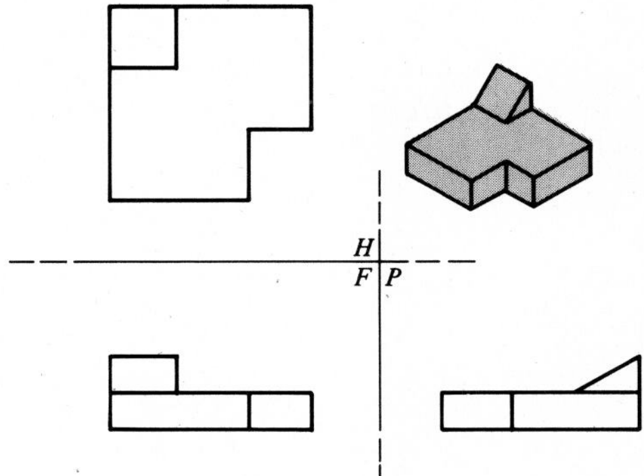

Figure 1-4
Orthographic projection.

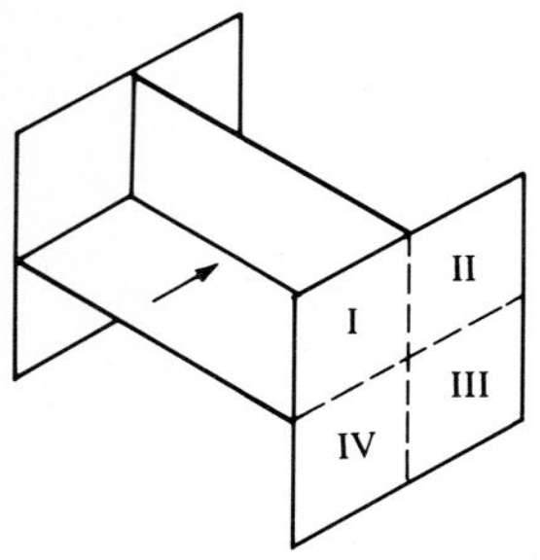

Figure 1-5
Projection planes.

tion, which is used in some countries, the object is placed in the first quadrant and the object is projected onto the plane. This method which places the object between the plane and the observer, is seldom used in the United States. It is mentioned here so that the student will be familiar with the procedure if it is found in research or in specialized drawings.

This text will deal *only* with third-angle projection since it is used almost exclusively in the United States. A study of Figure 1-6 will show that third-angle projection provides for a logical arrangement of planes. The object is placed in the third quadrant as shown in Figure 1-7 so that the planes of projection are between the object and the observer. This concept is the basis for future discussion of the principal views of objects in space.

1-3 OBJECTS IN SPACE

All material objects occupy space and therefore have three dimensions: *width* (*W*), the horizontal dimension from left to right; *height* (*H*), the vertical distance from top to bottom; and *depth* (*D*), the horizontal distance from front to back, as shown in Figure 1-8. The fact that these three distances are mutually perpendicular is very important and is one of the fundamental principles of orthographic projection.

1-4 VISUALIZATION

Consider a method by which the surfaces of a three-dimensional object may be reproduced on a two-dimensional medium such as a sheet of drawing paper or a computer terminal screen.

1. Visualize a solid object in space such as the one shown in Figure 1-9.
2. Imagine that planes are folded in such a way that they envelop the entire object. The planes, which must be parallel to the three principal sides of the object, are mutually perpendicular and form a rectangular box which occupies the three dimensions shown in Figure 1-10.
3. Visualize the box as glass or clear plastic and identify each plane as a picture plane with reference to the object within as shown in Figure 1-11. The vertical plane, which is parallel to the front of the object, is identified as the *frontal* (*F*) plane. It is through this plane that we see the front of the object and can determine its height and width. Perpendicular to the frontal plane and parallel to the top of the object is the *horizontal* (*H*) plane. The top of the object is viewed through this plane and we can determine the width and depth dimensions. The vertical plane, which is located at the side of the object and which is perpendicular to both the horizontal plane and the frontal plane, is

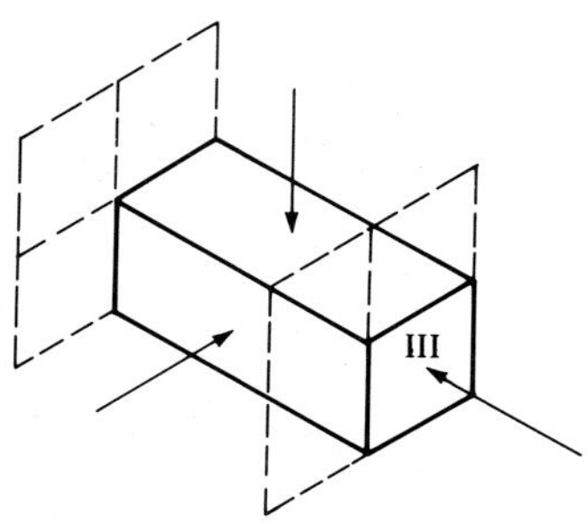

Figure 1-6
Third-angle projection.

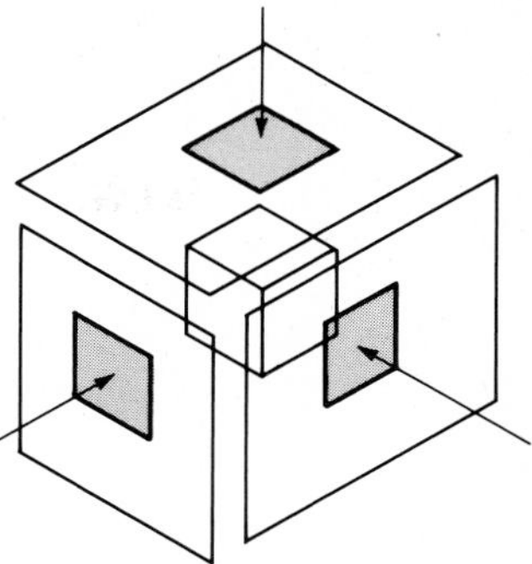

Figure 1-7
Exploded view: third-angle projection.

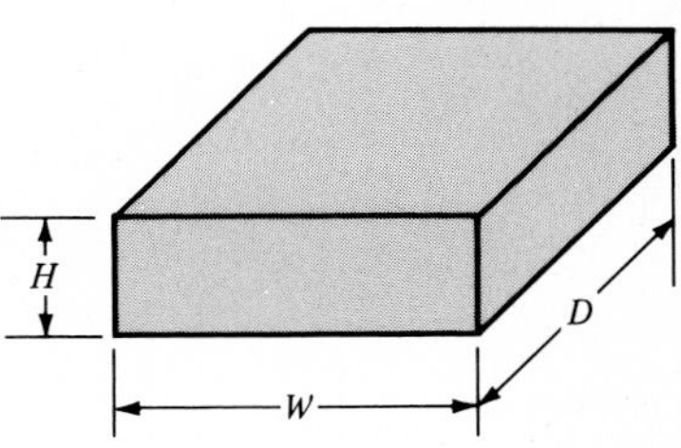

Figure 1-8
Dimensioning.

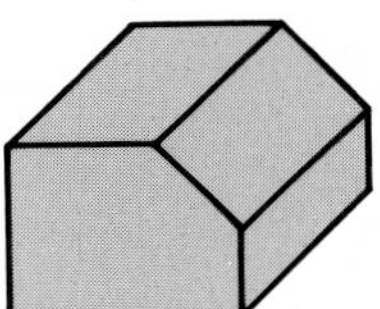

Figure 1-9
Object in space.

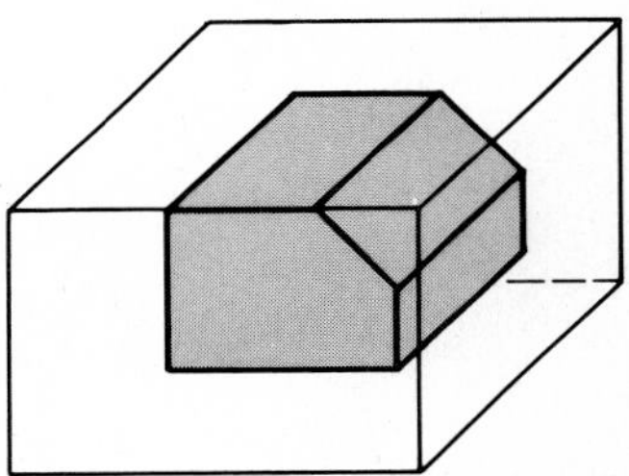

Figure 1-10
Object in space.

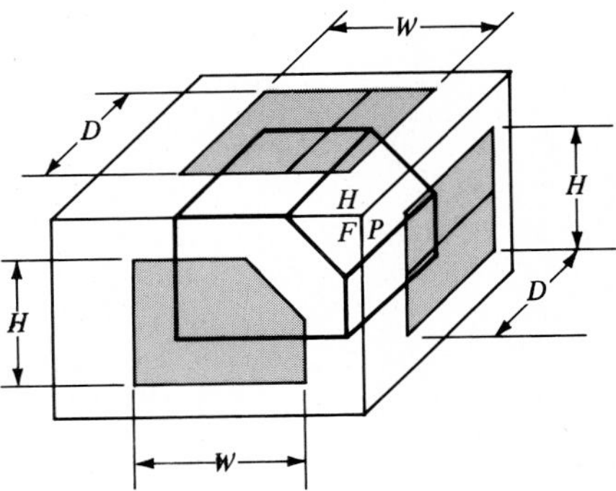

Figure 1-11
Viewing planes.

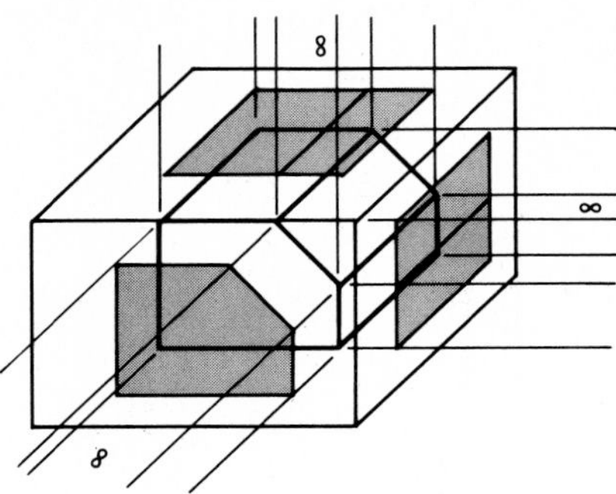

Figure 1-12
Orthographic lines of sight.

known as the *profile* (*P*) plane. Through this plane we see the side of the object and the height and depth dimensions can be determined.

4. We know now that in orthographic projection each point of the object is projected along the lines of sight until it intersects the frontal, horizontal, and profile picture planes at an angle of 90 deg. The points which have been projected onto the picture planes or onto your drawing paper show the front, top, and right-side views of the enclosed object as in Figure 1-12.

1-5 PRINCIPAL VIEWS

Six orthographic (principal) views of an object are possible, as indicated by the arrows in Figure 1-13. These planes can be rotated into the plane

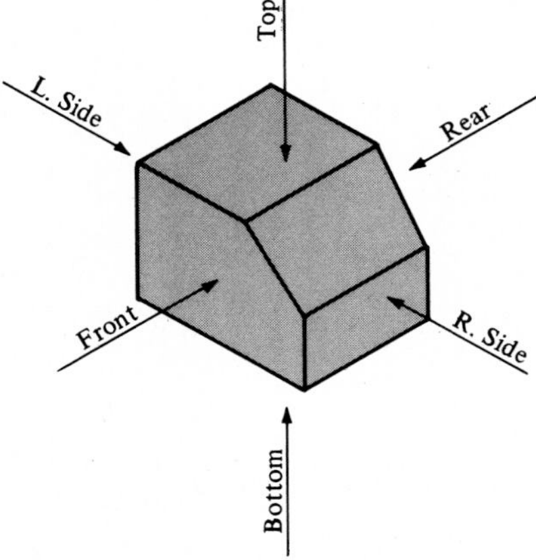

Figure 1-13
Principal views.

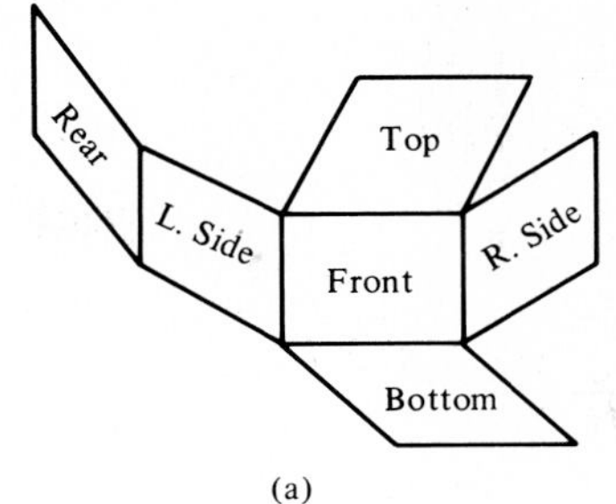

(a)

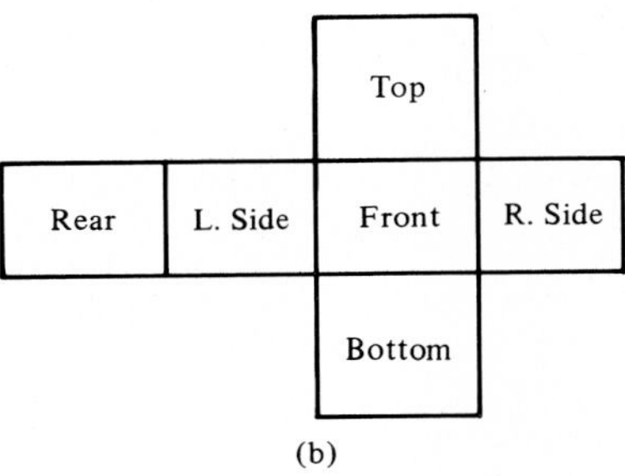

(b)

Figure 1-14
Principal view representation.

of the paper as illustrated in Figure 1-14(a) and 1-14(b). Of the six possible views, the combination that is used most often is that of *top, front,* and *right side.* Since the concept of descriptive geometry can be explained using only this given combination, we will confine our studies of principal views in this book to the projection on the following planes, as shown in Figure 1-15:

a. *Frontal plane:* front view, front elevation, vertical projection.
b. *Horizontal plane:* top view, plan view, horizontal projection.
c. *Right profile plane:* right-side view, right-side elevation, profile projection.

When the concept of orthographic drawing and descriptive geometry is understood, the student will have no difficulty using any of the six views shown in Figure 1-14.

1-6 FOLDING LINES

The projection (picture) planes are bounded by their intersections with the planes that are perpendicular to them. The lines about which the planes hinge when they are rotated into the plane of the paper will, in this text,

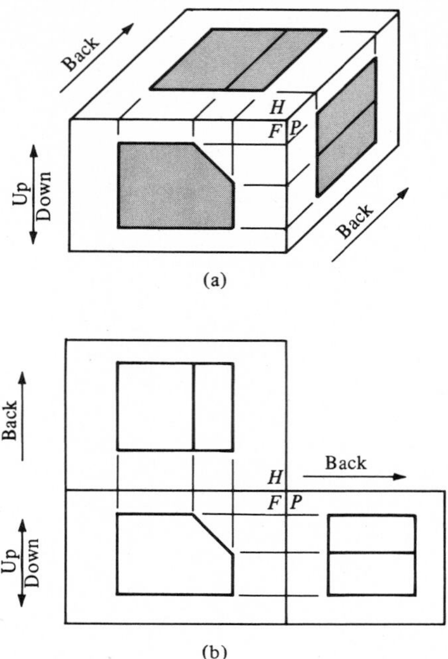

Figure 1-15
Front, top, and right side views.

be called *folding lines,* although they are known by other names (i.e., reference lines, reference planes, or hinge lines). As shown in Figure 1-15(a), the folding line between the horizontal plane and the frontal plane is labeled *H/F*. The one between the frontal plane and profile plane is labeled *F/P*. When the horizontal and profile views are rotated into the plane of the paper the top view is *always* aligned directly above the front view and the side view is *always* aligned horizontally with the front view. This orientation of views is indicated in Figure 1-15(b). When the top, front, and side views are drawn in the plane of the paper, the folding lines *H/F* and *H/P* will be perpendicular to each other as shown in Figure 1-16. The space directions as they would appear on a two-dimensional surface are also given.

Folding lines are generally removed from finished drawings in industry since the end-product and its manufacture are of prime importance. It is strongly suggested, however, that all work done in connection with this text show folding lines and/or reference planes as solid lines which have two dashes at each end and are labeled according to the planes represented, that is, $--\frac{H}{F}--$. A folding line designated in this manner will not be confused with other lines, and will assist in the orientation, solving, and checking of the required work.

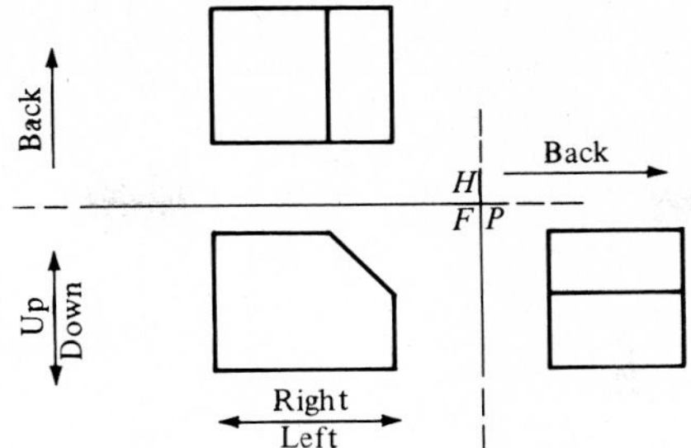

Figure 1-16
Coordinate system and folding lines.

1-7 POINTS IN SPACE

A point indicates a definite position in space, but it has no dimensions. For the discussion in this book, however, a point will be represented by a dot or a symbol + and will be considered an infinitesimally small part of a solid object. The method used in this text will be valid not only for one point, but for any and all points or parts of an object.

In Figure 1-17 point "O" has been surrounded by a three-dimensional box. Three of these planes, the frontal, horizontal, and profile, will be used. Until the point has been located on our drawing it has not been fixed in space and can be moved in any direction.

To "fix" point "O" in the vertical direction the following steps are used:

1. Place point "O_F" an arbitrary distance "y" below the horizontal plane and project the point onto the frontal plane on a line of sight which is perpendicular to the frontal plane, Figure 1-18(a). Since the edge of the horizontal plane is represented in the front view of our drawing by the *H*/*F* folding line, the point "O_F" is located "y" distance below the folding line as shown on Figure 1-18(b). Point "O" is now fixed vertically and cannot be moved up or down. All other views showing point "O" and the edge of the horizontal plane *must* show the point located "y" distance below the H folding line.
2. Figure 1-19(a) shows the projection of point "O" on the *H*-plane. Since point "O" has not been fixed in the horizontal direction, its location can be arbitrary with reference to the frontal plane and profile plane. The projection of the point on the *H*-plane will be at the 90 deg intersection of the plane and the line of sight to the point. The location of the point projection has been chosen as "z" distance back of the frontal plane and "x" distance from the profile plane, (Figure 1-19(a)). Because each line of sight is projected onto the picture plane at a 90 deg angle, the top and front views of that point will lie on lines

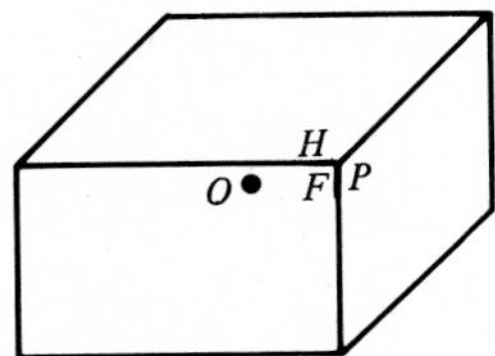

Figure 1-17
Point in space.

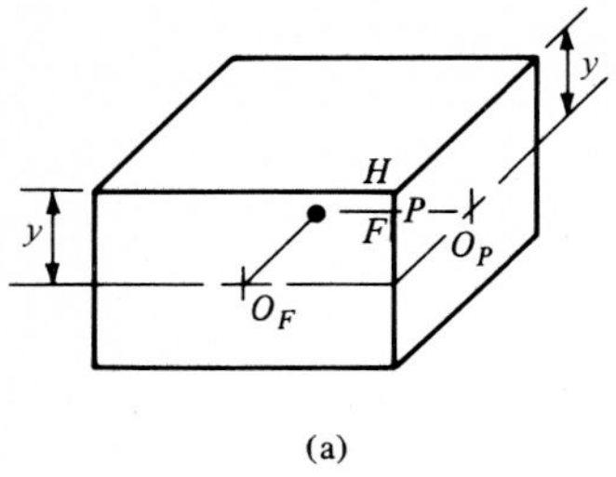

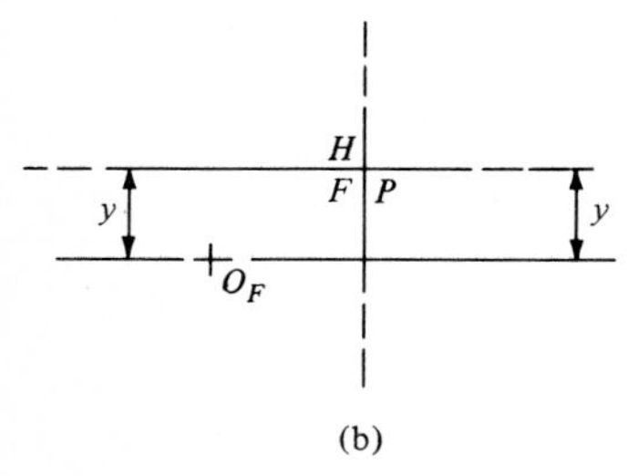

Figure 1-18
Projection of a point.

that are perpendicular to the folding line *H/F*. The location of "O_H" will be as shown in Figure 1-19(b).

3. Since point "*O*" is now fixed in space, it cannot be moved in any direction. The projection of the point on the profile plane lies on a horizontal line (parallel to the *H/F* folding line) extending from "O_F" into the side view. "O_P" will be located on this line. When looking through the profile plane, the viewer sees folding line *F/P* as the edge of the frontal plane and can locate the point by determining its distance "*z*" back of that plane (Figure 1-20(a) and 1-20(b)).

Because the preceding discussion is the foundation of orthographic projection and descriptive geometry, we will emphasize the following points:

1. The projection of any point is shown on the picture plane at the point where the line of sight intersects the picture plane at right angles.
2. a. When the observer looks through the horizontal plane the *H/F* folding line becomes the edge of the frontal plane.
   b. When looking through the profile plane, the *F/P* folding line becomes the edge of the frontal plane.
   c. When the viewer looks through the frontal plane, the *H/F* fold-

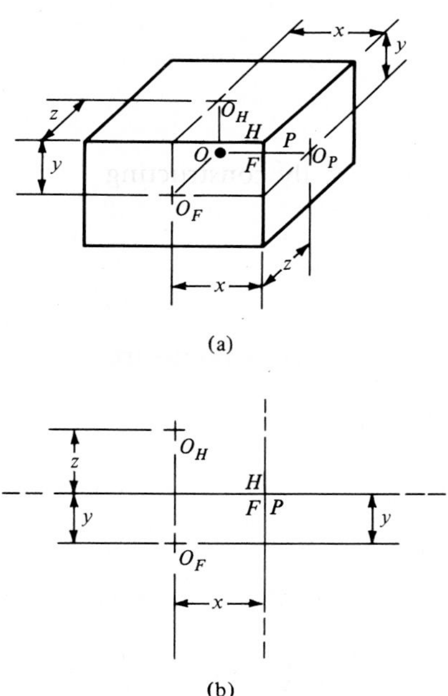

Figure 1-19
Fixing a point in space: front and top views.

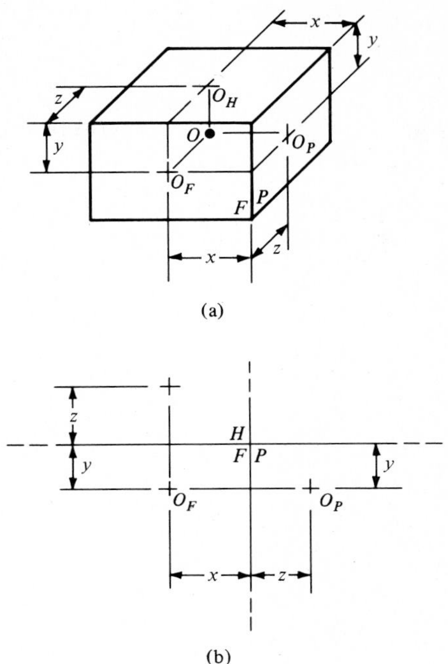

Figure 1-20
Fixing a point in space: front, top and side views.

ing line becomes the edge of the horizontal plane, while the *F*/*P* folding line represents the edge of the profile plane.

3. When graphically constructing the required views, dividers are used to transfer distances from one to another using the folding line as a base.
4. Points that are projected from one view to another are projected on lines which *must* be perpendicular to the folding lines. Since this method of projecting points from one view to another is valid, we can use the same process in future work for any view. Points are connected to obtain lines and lines are connected to obtain surfaces.

1-8 LINES IN SPACE

In order to intelligently discuss orthographic projection and descriptive geometry one should have a knowledge of lines, especially their use and their nomenclature. A *line* is defined as a locus of two or more points. Theoretically, it has infinite length, but neither breadth nor thickness. We will, for the purpose of illustration, deviate from the strict definition and discuss the types, location, and visibility of lines.

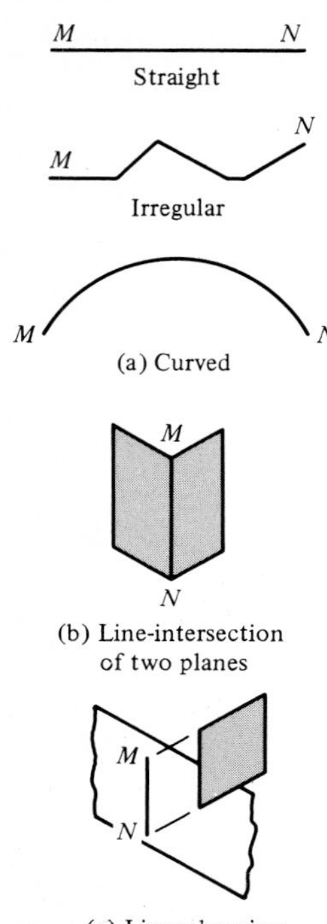

Figure 1-21
Lines in space.

A. Line Definition

In general use, a line is defined, but not limited, by two points. For ease in problem solution, the *line segment* will be given or defined in future discussions. The student, however, should feel free to extend the concept of the line segment whenever and wherever necessary.

1. A single line segment may represent a pipe, rod, wire, etc. and is terminated by end points (Figure 1-21 (a)).
2. A line may also represent the intersection of two surfaces as shown in Figure 1-21(b).
3. The edge view of a plane surface is shown by a line. For this to be true, the plane surface must be perpendicular to the projection plane as shown by the line *MN* in Figure 1-21(c).

B. Line Classification

Lines may be classified according to their position in space and their relation to the principal planes as follows:

1. A *Horizontal* line is a line that is parallel to the horizontal plane. It appears true length (T.L.) in the top view (Figures 1-22(a) and (b)).
2. A *Frontal* line is a line which is parallel to a frontal plane. It appears true length in the front view (Figures 1-23(a) and (b)).
3. A *Profile* line is parallel to the profile plane. It appears true length in the side view (Figures 1-24(a) and (b)).
4. An *Inclined* line is a line which is parallel to one principal plane and forms an angle with the other two (Figure 1-25). It may be a frontal, horizontal, or profile line.
5. An *Oblique* line is not parallel to any principal plane (Figure 1-26).
6. A *Vertical* line is perpendicular to the horizontal plane and parallel to the frontal and profile planes (Figure 1-27).

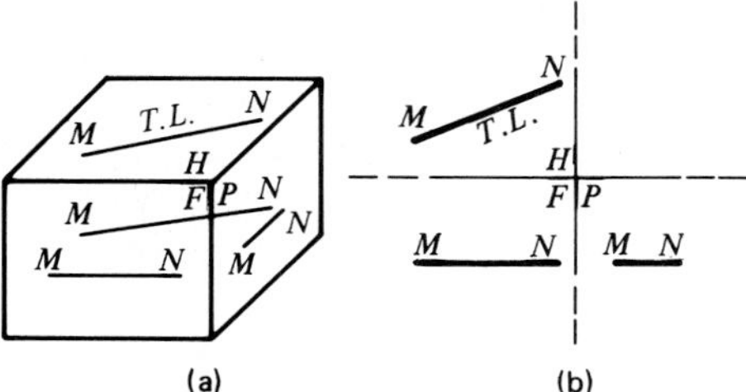

Figure 1-22
Horizontal line.

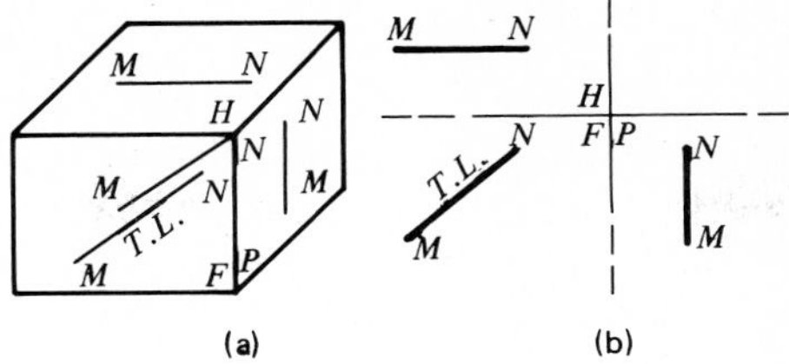

Figure 1-23
Frontal line.

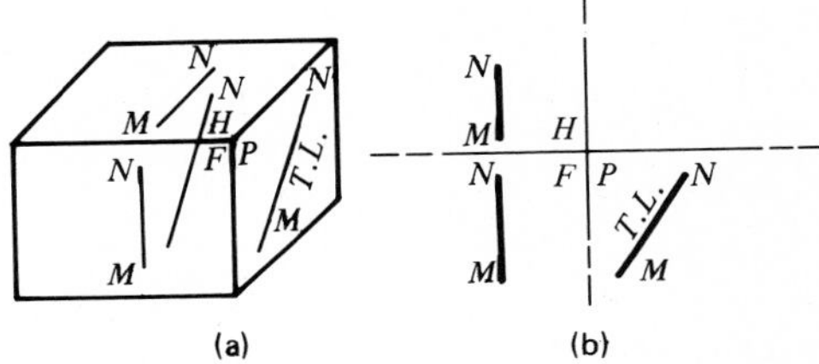

Figure 1-24
Profile line.

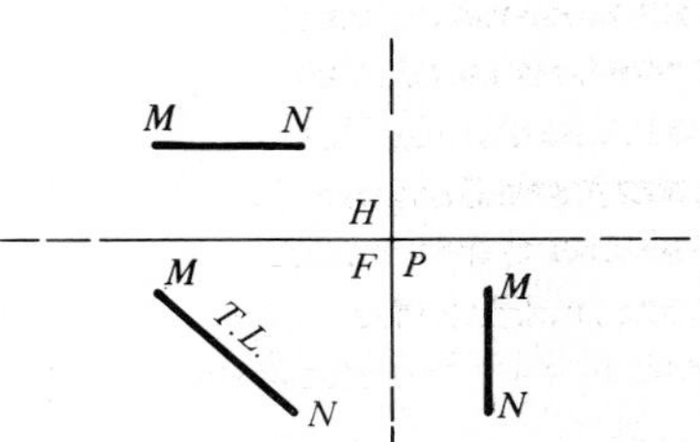

Figure 1-25
Inclined line.

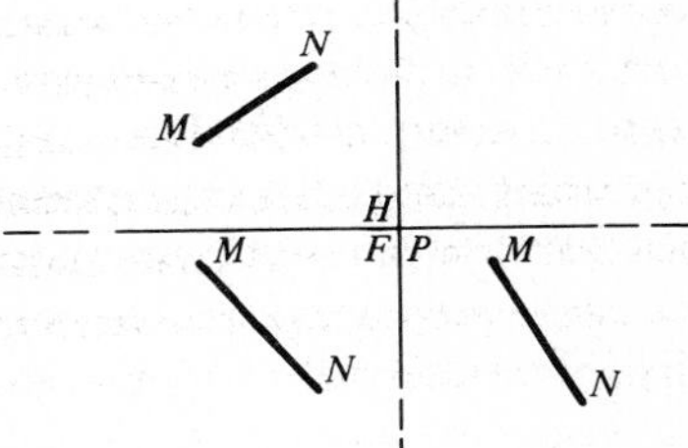

Figure 1-26
Oblique line.

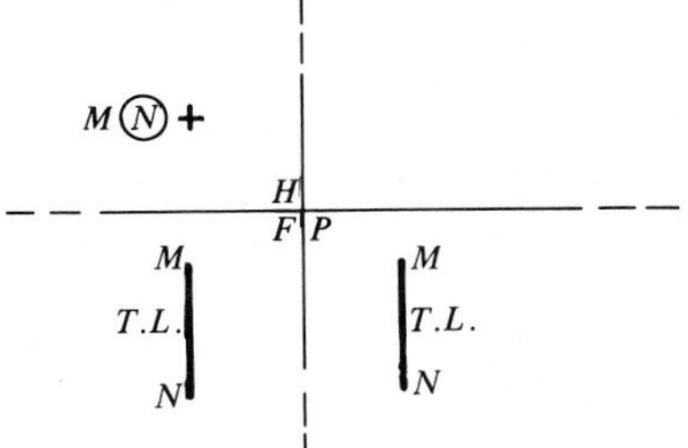

Figure 1-27
Vertical line.

1-9 PLANES IN SPACE

A plane is defined as a flat surface in which any two points may be connected by a straight line which lies completely in that surface. Planes are considered to have no thickness, but have infinite length and breadth. In practical use, however, the area of a *plane segment* will generally be defined, while the thickness of the plane will be shown by a line.

A. Plane Definition

A plane surface may be defined by:

1. two intersecting lines
2. a point and a line
3. two parallel lines
4. three noncollinear points.

More consideration will be given to the above in Chapter 4.

B. Plane Classification

Planes can be classified as follows:

1. A *Normal* plane is parallel to one of the principal planes and perpendicular to the other two (Figures 1-28(a) and (b)).
2. An *Inclined* plane is perpendicular to one principal plane and forms an angle with the other two (Figures 1-29(a) and (b)).
3. An *Oblique* plane forms an angle with the three principal planes (Figures 1-30(a) and (b)).
4. A *Vertical* plane forms an angle of 90 deg with the horizontal plane. It can be either parallel, perpendicular, or inclined to the other two (Figures 1-31(a) and (b)).

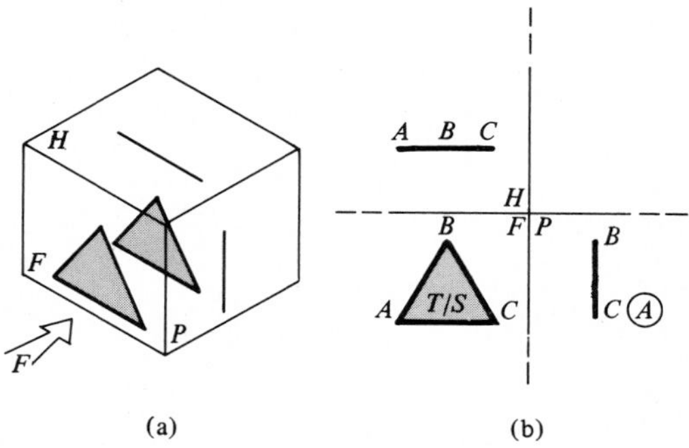

Figure 1-28
Normal plane.

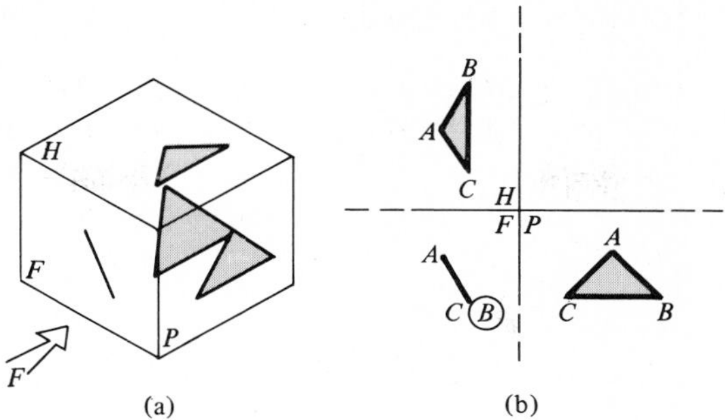

Figure 1-29
Inclined plane.

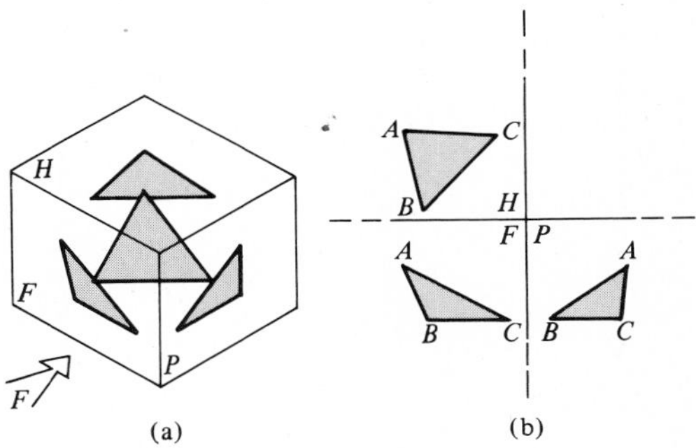

Figure 1-30
Oblique plane.

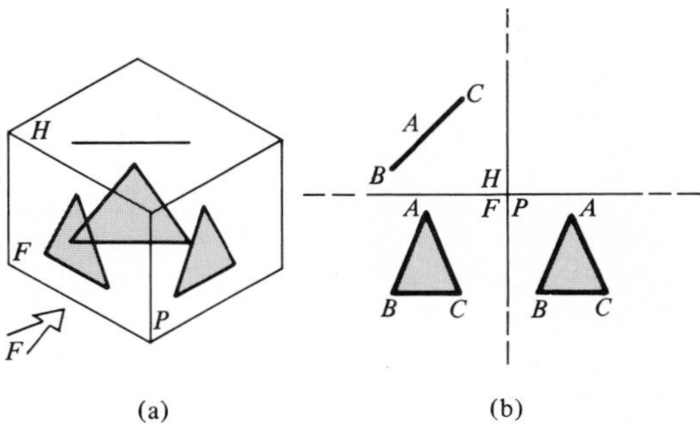

Figure 1-31
Vertical plane.

1-10 SOLIDS IN SPACE

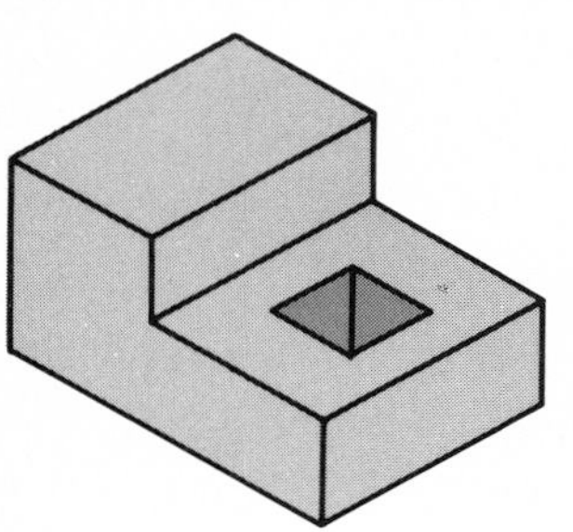

Figure 1-32
Mutilated block.

We can now show points, lines, and planes on a three-view orthographic drawing. Since solids are made up of planes, the same procedures may be used when describing the shape and details of three-dimensional (3-D) objects. For a review, it is suggested that the reader follow the steps of Example 1-1, which shows the procedure used in making a three-view drawing of the simple block given in Figure 1-32.

Example 1-1

Given: Front and top views of a mutilated block.
Task: Complete the three-view orthographic drawing of the block
Solution:

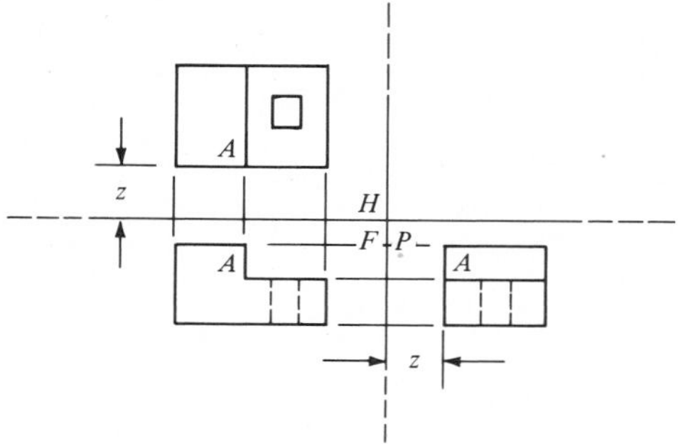

Example 1-1
Orthographic drawing.

1. Project lines of sight of given points horizontally from the front view into the side view.
2. Using dividers, determine the distance "z" from point "A_H" to the frontal plane in the top view.
3. On the line which projects from the front view of point A into the side view, and using the F/P folding line as a reference, lay off point A so that it is "z" distance behind the frontal plane in the side view.
4. Employing the same method as above, transfer all other points from the top view into the side view.
5. Connect all points with lines to complete the drawing of the solid as shown in Example 1-1. ◀

The folding line method of projection used in the preceding discussion is fundamental in constructing two-dimensional representations of three-dimensional objects. The short introduction presented here gives the necessary knowledge to allow the reader to benefit from the following material on graphical descriptive geometry. As mentioned in the Preface, traditional descriptive geometry projects three-dimensional space onto two-dimensional viewing planes before solutions to the three-dimensional problems are obtained.

1-11 VISIBILITY

Even though all, or part, of an object is located between the observer and a line, the line must be shown on an orthographic drawing. When part of the line is invisible, that part is represented by a dashed line. If the line is completely hidden, then the entire line is dashed.

Since visibility is a very important part of orthographic projection and since many of the illustrations given later deal with hidden lines, we now explore the subject further. Two points to be remembered are:

1. The outline of an object is *always* visible.
2. Generally speaking, we can say that all lines terminating at an invisible point will be invisible either for their entire length, or as they approach the invisible point.

Figure 1-33(a) gives the top and front view of a tetrahedron. The front view shows that point *D* is the lowest point on the object, which means that in the top view point *D* will be invisible. Therefore, all lines which terminate at point *D* will be invisible in the top view.

By the same reasoning used above, we can see that in Figure 1-33(b), the side view will show that *A* is an invisible point since it is the farthest point from the profile plane and hence, from the observer who is looking through that plane.

Before visibility of nonintersecting rods or pipes is discussed, let us take a look at the top and front views of the two spheres shown in Figure 1-34(a). When the observer looks *down* through the horizontal plane only one sphere is visible, the one that is closer to the horizontal plane. The front view, by inspection, shows us that the sphere seen in the top view would be the black one since it is higher than the white.

In Figure 1-34(b), by looking through the horizontal plane, we see that the white sphere has been placed in front of the black one (closer to the frontal plane). The front view will show only the white sphere.

The foregoing method may be used to determine the visibility of two skew (nonintersecting) linear objects as shown by the two cylinders in Figure 1-35. Inspection of the given views proves that the cylinders do *not* intersect since the *apparent points of intersection* of the centerlines are not aligned vertically. Because the cylinders do not intersect, one must be shown crossing above the other in the top view, and one crossing in front of the other in the front view.

Use only the centerlines of the cylinders and concentrate on a small

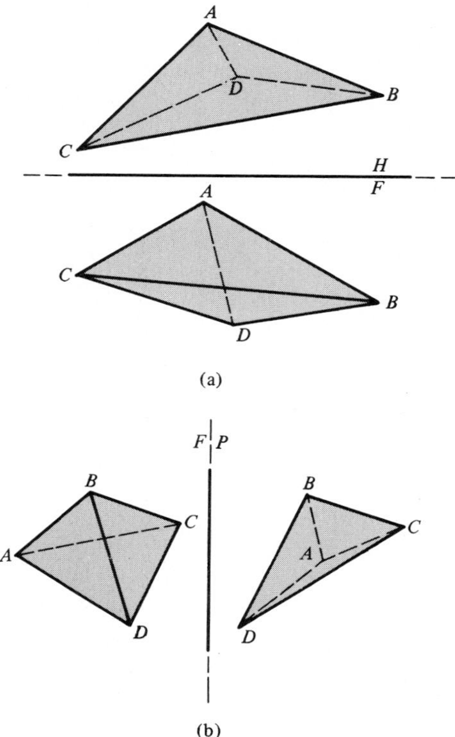

Figure 1-33
Visibility.

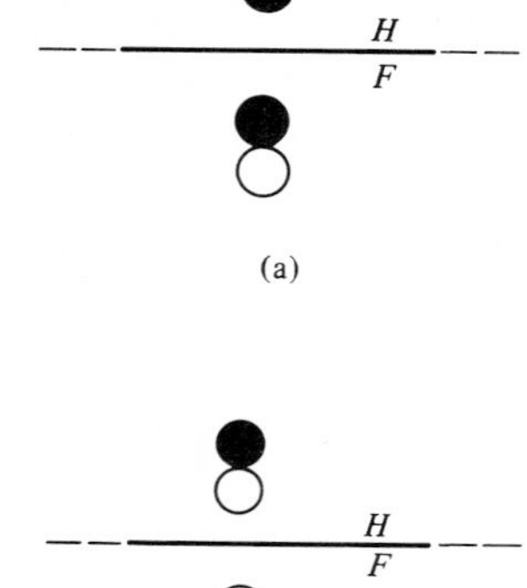

Figure 1-34
Two spheres.

spherical section of the "cross over" in the top view. Project this small area into the front view where the cylinder *MN* is shown *above* cylinder *AB*. In the top view, cylinder *MN* will be shown with solid lines throughout, while cylinder *AB* will be shown with hidden (i.e., dashed) lines where it passes under *MN*.

The front view may be completed by again focusing your attention on a small area where centerlines of the cylinders cross. Project this small area into the top view which shows cylinder *AB* closer to the frontal plane at that location. The front view will show cylinder *AB* with solid lines throughout since that cylinder is in front. *MN* will be shown in dashed lines where it passes behind *AB*.

Note: Always use centerlines when working with cylindrical objects such as pipes, rods, tubing, and similar objects.

This method of determining visibility can be used in *any two adjacent* views. Figure 1-36 shows the visibility of a tetrahedron using the front and right side views. It is suggested that this figure be used to check the reader's comprehension of the problem. In your study, think of the plane edges as lines and follow the step-by-step procedure shown above. A thor-

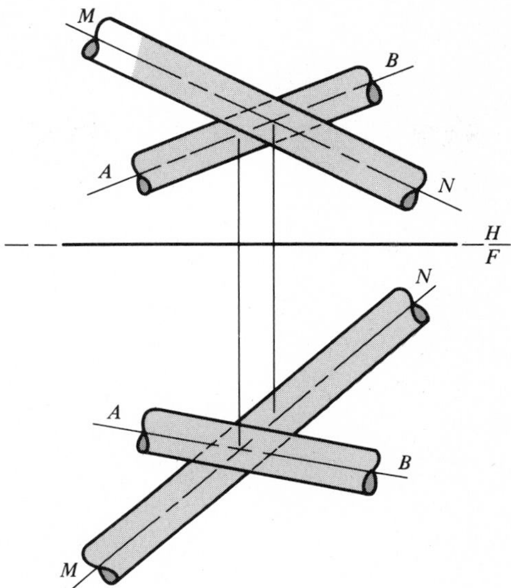

Figure 1-35
Visibility of cylinders.

ough understanding of this process will prove useful in future studies of descriptive geometry.

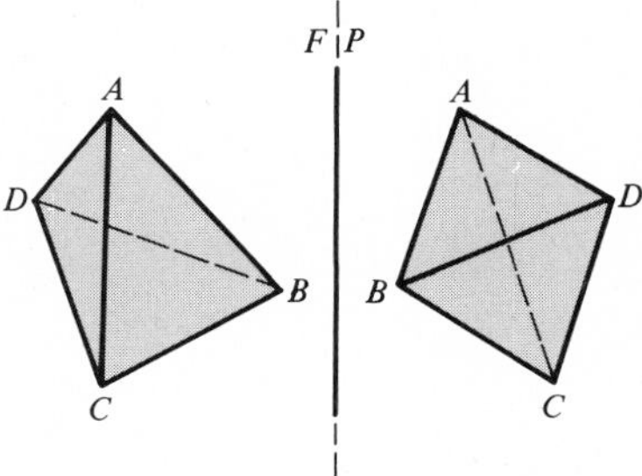

Figure 1-36
Visibility of a pyramid.

1-12 VECTOR REPRESENTATION OF A POINT

A *vector* is a quantity with both magnitude and direction. It is represented by a directed line segment whose length represents the magnitude and whose line of action indicates direction. An arrow is used to indicate the sense of the vector along the line of action, either forward or backward.

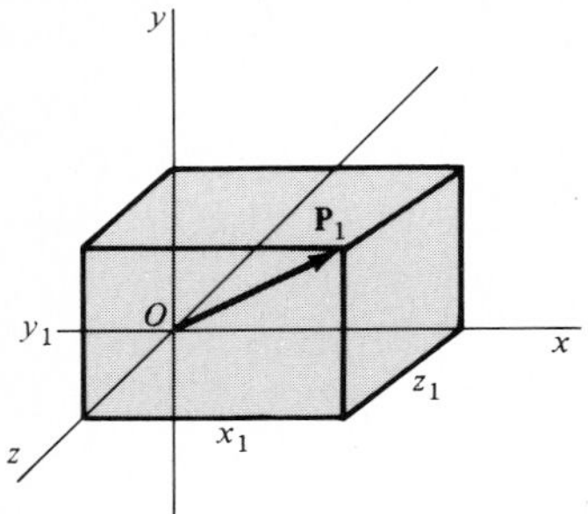

Figure 1-37
Point in space.

A point is not a vector. It has neither magnitude nor direction. However, a point in space can be *represented* by a vector directed from the origin of a three-dimensional coordinate system to the point. Such a vector is called a *base, fixed,* or *localized* vector. This is in contrast to a *free* vector which represents direction and magnitude only, and can be moved to other locations as long as it remains parallel to its original line of action in the same sense.

Figure 1-37 shows a point in space represented by the vector $\mathbf{P}_1$. The three coordinates of the point are the three components of the vector along the principal x,y,z axes. These three components of vector $\mathbf{P}_1$ can be written as $P_{1,x}$, $P_{1,y}$, and $P_{1,z}$ or simply x_1, y_1, z_1.

The three-dimensional coordinate system used in Figure 1-37 is a right-hand coordinate (R.H.C.) system. Notice that positive y is upward in Figure 1-37. Positive z points toward the observer.

When orthographic projections are used to represent a point, two principal views are required to fix a point in space. When vector representation is used, the three components are specified together and often written as a matrix. For example, $\mathbf{P}_1 = [x_1\ y_1\ z_1]$ is a row matrix with one row and three columns. Alternately, one can use a column matrix of three rows and one column and write

$$\mathbf{P}_1 = \begin{bmatrix} x_1 \\ y_1 \\ z_1 \end{bmatrix}$$

Each point on a solid in space can be defined by using the vector representation of a point. Consider Figure 1-38 with numbered vertices as shown. Twenty vertices can be identified, each represented by a vector drawn from the origin. Each of these vectors can be placed in a *row* matrix (20 rows and 3 columns) or a *column* matrix (3 rows and 20 columns.) In row matrix form:

$$[\mathbf{R}] = \begin{bmatrix} x_1 & y_1 & z_1 \\ x_2 & y_2 & z_2 \\ x_3 & y_3 & z_3 \\ \cdot & \cdot & \cdot \\ \cdot & \cdot & \cdot \\ x_{20} & y_{20} & z_{20} \end{bmatrix}$$

In column matrix form:

$$[\mathbf{C}] = \begin{bmatrix} x_1 & x_2 & x_3 & \cdot & \cdot & x_{20} \\ y_1 & y_2 & y_3 & \cdot & \cdot & y_{20} \\ z_1 & z_2 & z_3 & \cdot & \cdot & z_{20} \end{bmatrix}$$

Each row in the $[\mathbf{R}]$ matrix is a vector, and each column in the $[\mathbf{C}]$ matrix is a vector. A matrix, then, is a compact way to represent a series of points

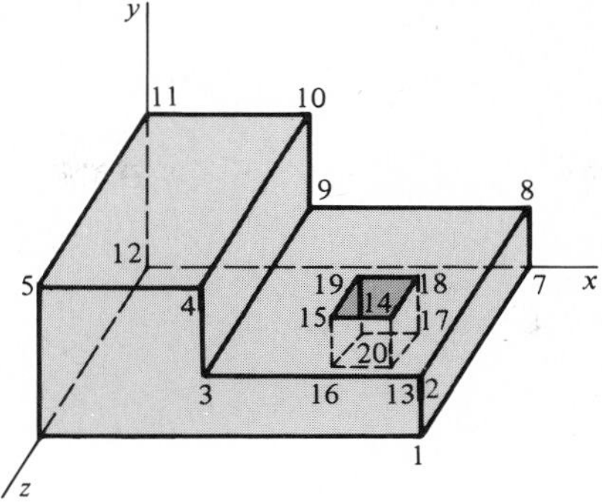

Figure 1-38
Numbered vertices.

in space. When used in this manner it is called a *three-dimensional data matrix.*

Vectors are used not only to represent points, but also lines and planes. These techniques are discussed throughout this text. Descriptive geometry problems are solved graphically by projecting three-dimensional objects onto two-dimensional surfaces and then obtaining solutions by making measurements in these planes. Computer solutions will be obtained directly using geometric information contained in a three-dimensional data matrix coupled with vector algebra and computer programming.

1-13 WIRE FRAME PROJECTIONS

We now consider how to write computer programs and use graphics to create principal views of a three-dimensional object. The programs which are listed in ANSI BASIC can be adapted to generate similar output on personal computer terminals or larger work stations.

Consider the 20 vertices of the object shown in Figure 1-38. Arrays *X*( ), *Y*( ), *Z*( ) are used in the program to contain the vertices. Many ways can be used to define the elements in these arrays within the program. We will use an external data file which we choose to call CORNER as shown in Table 1-1 to contain the components of each vertex. These vertices are then read into the elements of the *X*( ), *Y*( ), *Z*( ) arrays. Program VIEWS listed in Appendix 1 gives an example of reading this data file.

If the principal views are to be drawn automatically, then a line file is needed to identify which vertices are connected by straight lines, in addition to the vertex file that defines the vertices. The line file can be

Table 1-1
Vertices

CORNER	
10,0,3	0,4,0
10,2,3	0,0,0
4,2,3	8,0,2
4,4,3	8,2,2
0,4,3	6,2,2
0,0,3	6,0,2
10,0,0	8,0,1
10,2,0	8,2,1
4,2,0	6,2,1
4,4,0	6,0,1

used to produce a *wire frame* drawing where all lines are shown solid, whether they are actually visible or not. Advanced techniques exist in the computer graphics literature which determine computationally the visibility of lines and surfaces. Those techniques are beyond the scope of this text.

A line file called CONNECT is created for the object shown in Figure 1-38 as given in Table 1-2. Each value is a vertex number for the object as it appears in the file CORNER. The line file is used to connect the given vertex pairs with solid lines. Read the file CONNECT as follows: connect vertex 1 with 2, vertex 2 with 3, . . . and vertex 16 with 20. The numbering scheme is shown in Figure 1-38.

The user must define the dimensions of a three-dimensional "ortho-box" which encloses the object and whose edges form the folding lines. The screen coordinates x_s, y_s which are needed for a computer display are calculated as follows, assuming that the origin is in the lower left corner of the screen.

$$\text{Horizontal View: } x_s = x; \qquad y_s = \text{HEIGHT} + \text{DEPTH} - z$$

$$\text{Frontal View: } x_s = x; \qquad y_s = y \tag{1-1}$$

$$\text{Right Profile View: } x_s = \text{WIDTH} + \text{DEPTH} - z; \qquad y_s = y$$

where x,y,z are the coordinates of each vertex. Figure 1-39 illustrates this idea.

Subroutine PVIEWS (Appendix 1) implements these three equations to calculate screen coordinates and produce graphical output using PLOT commands. The variables passed to this subroutine are the dimensions of the "ortho-box," the number of lines in the line file along with the double

Table 1-2
Lines

CONNECT	
1,2	4,10
2,3	5,11
3,4	6,12
4,5	13,14
5,6	14,15
6,1	15,16
7,8	16,13
8,9	17,18
9,10	18,19
10,11	19,20
11,12	20,17
12,7	13,17
1,7	14,18
2,8	15,19
3,9	16,20

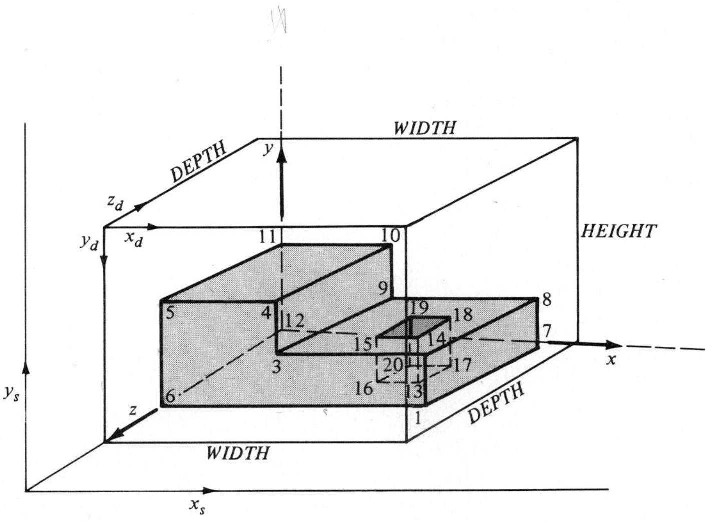

Figure 1-39
Object with ortho-box.

subscripted variable $L(,)$ which contains the drawing information from the line file, and the three-dimensional vectors stored in data arrays $X(\ )$, $Y(\ )$, $Z(\ )$.

The main program VIEWS given in Appendix 1 reads the data points from the vertex file and the vertex pairs from the line file, and then calls subroutine PVIEWS which draws the three principal wire frame views. These views are shown in Figure 1-40. Notice that the hidden lines are shown as solid lines in a wire frame rendering. The same program VIEWS can be used to display any three-dimensional object in a similar manner once the vertex and line files are created. The names of these two files are requested as input within the main program. Additional examples are given in Figures 1-41 through 1-45.

Figure 1-41 shows the three principal views of a cut block using a wire frame representation. The reader should determine which lines are hid-

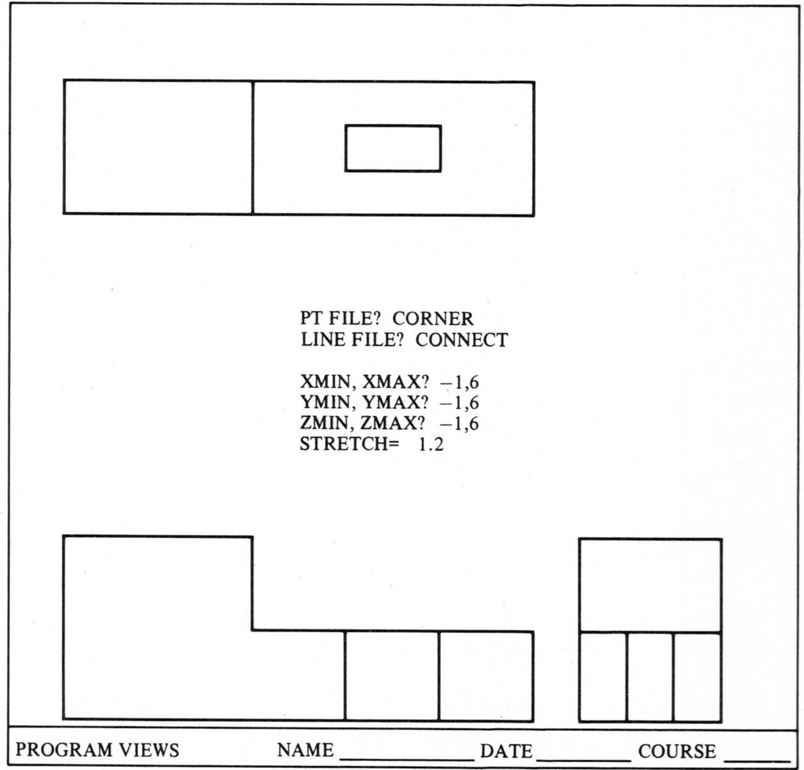

Figure 1-40
Machine part.

den in the frontal and profile views. Figure 1-42 shows principal views of two three-dimensional skew lines, with the shortest horizontal line and the shortest perpendicular line between the skew lines drawn in each view. The reader should identify each line in each view. Methods for determining these shortest distances are given later in the text.

Figure 1-43 shows the intersection between a four-sided solid and a plane. The intersection points are calculated as the piercing points between the four edges of the solid and the plane. Methods for determining piercing points are discussed in Chapter 8.

Figure 1-44 gives the principal views for three simple objects; a line, triangle, and rectangle. The views show that no object touches another, and no object is seen in true length or size in any of the principal views. These concepts will be discussed in more detail later in the text.

Another test of spatial awareness is given in Figure 1-45. Here, a wire

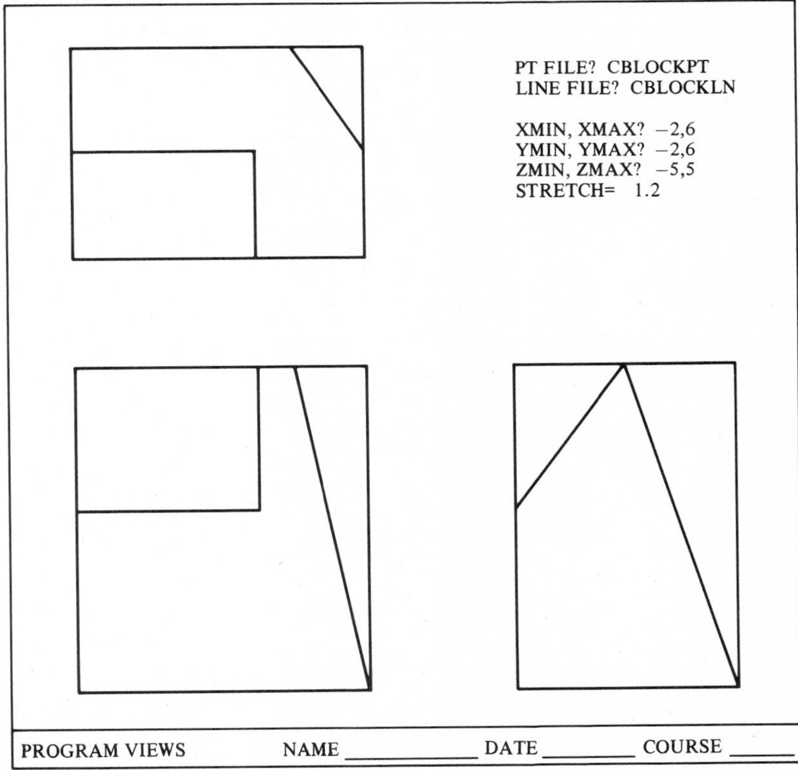

Figure 1-41
Cut block.

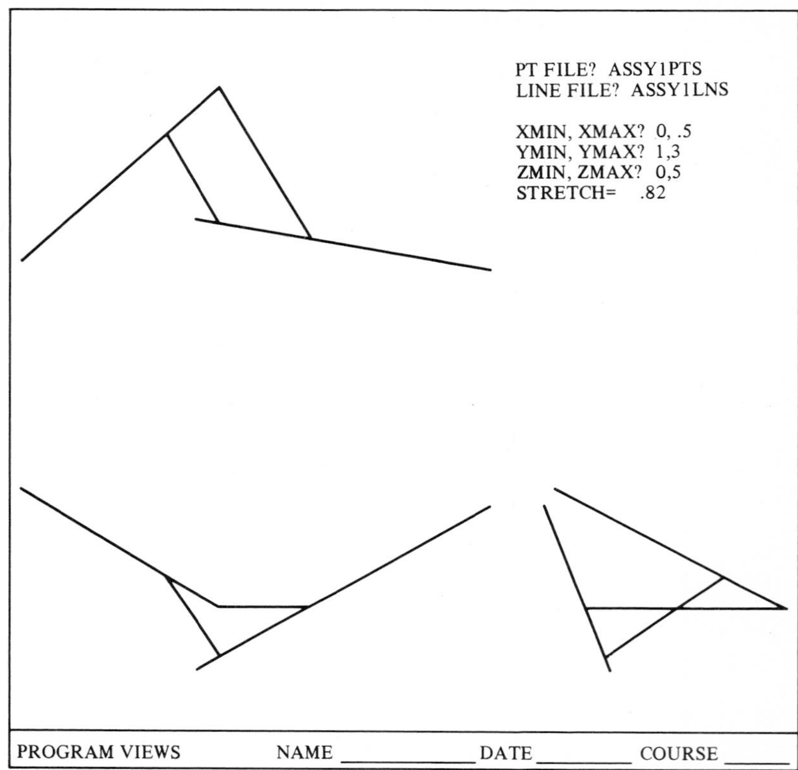

Figure 1-42
Skew lines.

frame representation of two intersecting solids is given. Often an orthographic projection made perpendicular to another line of sight will make the three-dimensional interpretation easier. This is illustrated in the next chapter.

Notice that program VIEWS also calls subroutine TITLE (given in Appendix 1) which creates the title block around the output. If the reader has access to a computer graphics capability, it is suggested that a program similar to VIEWS be used and that vertex and line files be created to generate wire frame, principal orthographic views. If color capability is available, then the wire frame surfaces can be filled in with color to create the appearance of solid surfaces. Techniques for doing this will be found in the user's manual for the available system. Commercial programs such as Autocad® or Cadkey® can be used if available.

It is important to again emphasize the two right-hand coordinate sys-

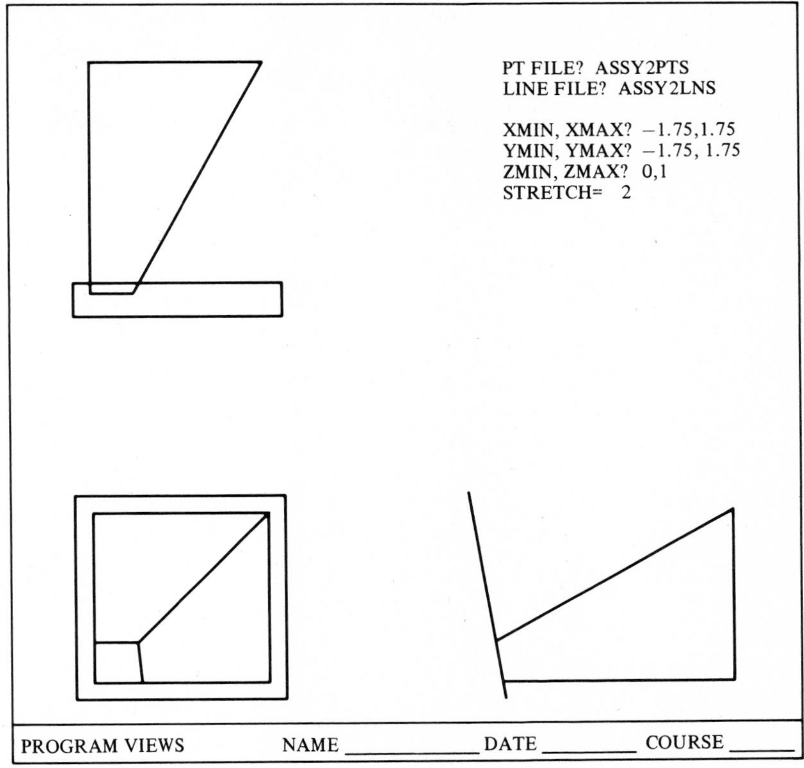

Figure 1-43
Piercing points.

tems used in this text. These are indicated on the "ortho-box" shown in Figure 1-46. When drawing principal views in descriptive geometry, the x_d, y_d, z_d system is used. When defining objects for geometric modeling the x,y,z system is chosen. This latter system was used to define the objects displayed by the program VIEWS and given in Figures 1-40 through 1-45.

When principal views of a three-dimensional object are shown in a two-dimensional plane, always remember which coordinate system applies. This is necessary to properly interpret the graphical results. It is easy to convert from one system to the other as follows (see Figure 1-46):

$$x = x_d$$

$$y = \text{HEIGHT} - y_d \tag{1-2}$$

$$z = \text{DEPTH} - z_d$$

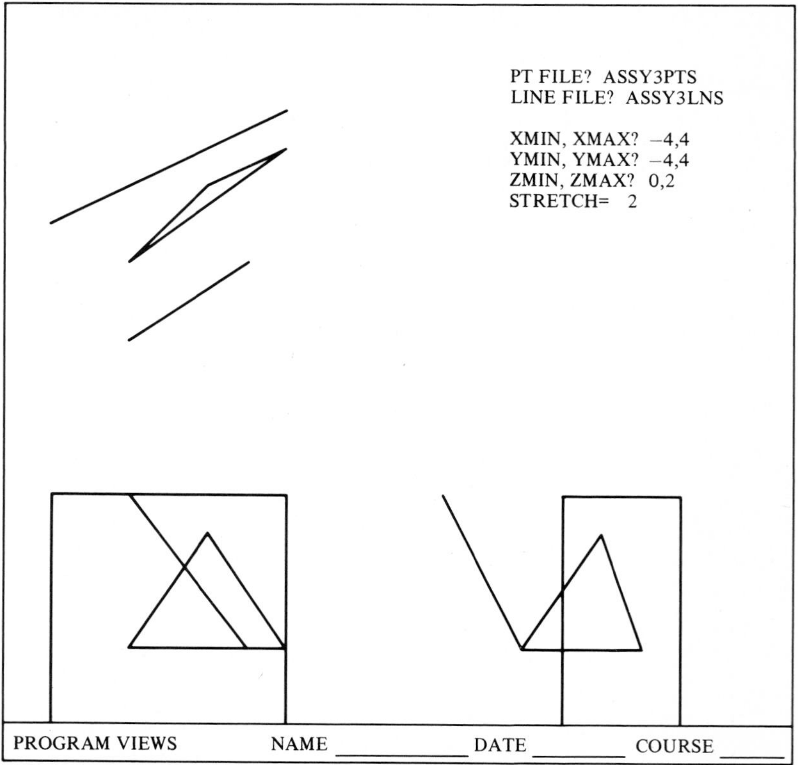

Figure 1-44
Simple shapes.

1-14 VECTOR ADDITION

Vectors can be added using the rules of vector algebra. The sum of two vectors is obtained by adding the corresponding components together. For example, in matrix form let $\mathbf{A} = [x_1\ y_1\ z_1]$ and $\mathbf{B} = [x_2\ y_2\ z_2]$. Then the sum $\mathbf{R}$ is given by

$$\mathbf{R} = \mathbf{A} + \mathbf{B} = [(x_1 + x_2) \quad (y_1 + y_2) \quad (z_1 + z_2)] \tag{1-3}$$

The subtraction of two vectors is similar to addition except that the corresponding scalar components are subtracted. Multiplication of a vector by a scalar is obtained by multiplying each component by the scalar value. Matrix multiplication of two vectors is discussed later and summarized in Appendix 2.

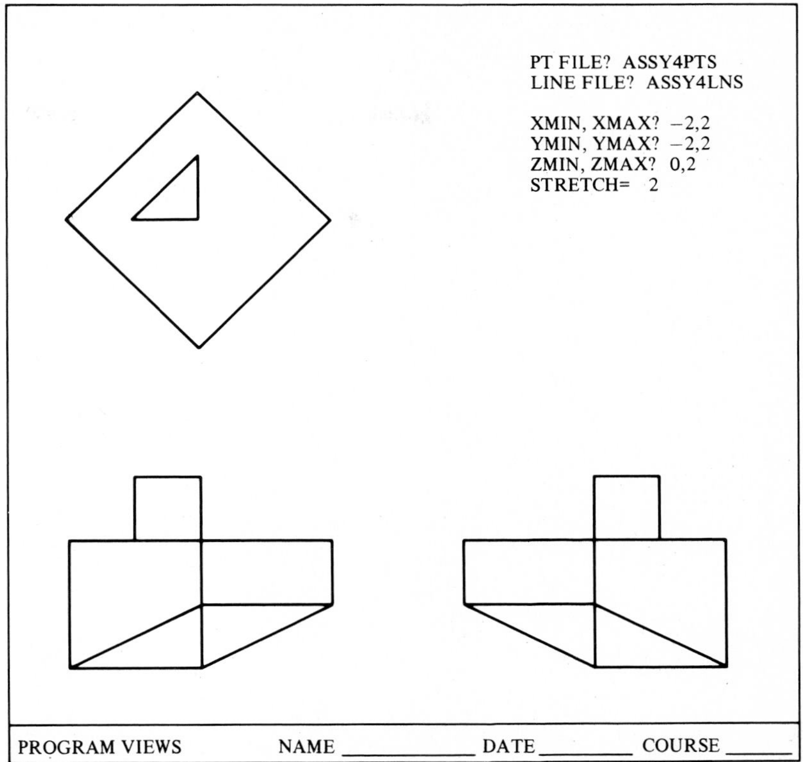

Figure 1-45
Two solids.

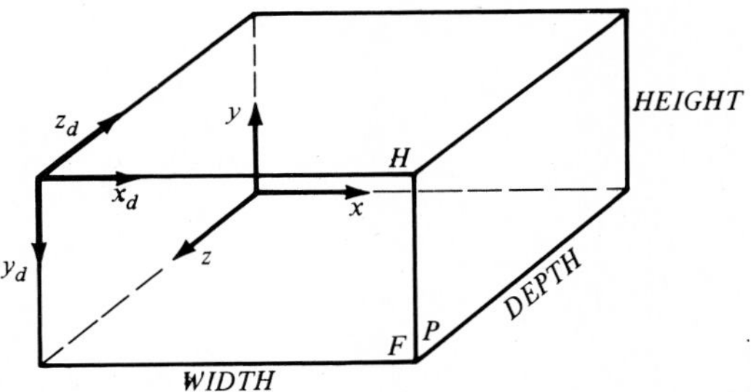

Figure 1-46
Two R.H.C. systems.

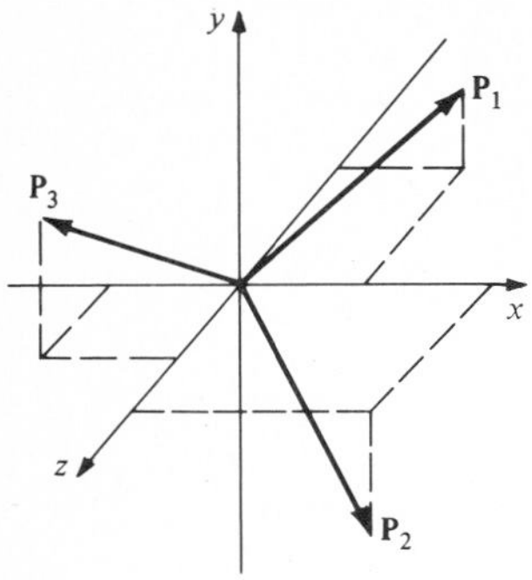

Figure 1-47
Base position vectors.

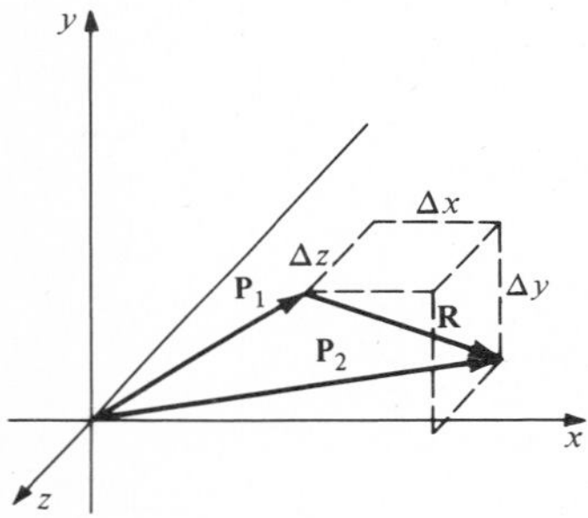

Figure 1-48
Base and relative vectors.

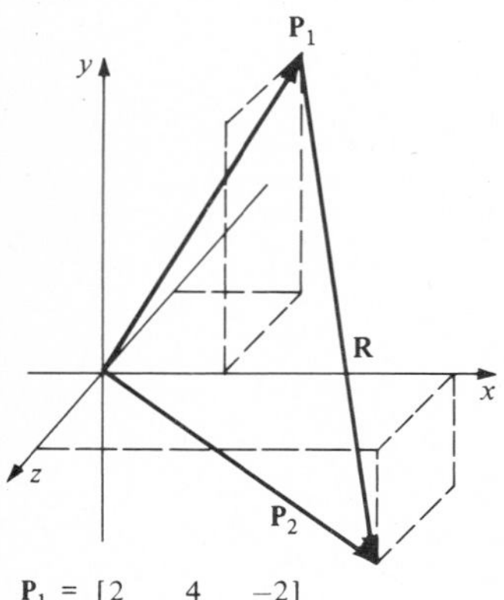

Figure 1-49
Vector addition example.

Example 1-2

Given: $\mathbf{A} = [3\ 1\ -2]$, $\mathbf{B} = [2\ 4\ 1]$, $\mathbf{C} = [1\ -3\ 2]$
Task: Find $\mathbf{R} = \mathbf{A} + \mathbf{B} + \mathbf{C}$, $\mathbf{S} = \mathbf{A} - \mathbf{B} - \mathbf{C}$, and $\mathbf{M} = 3\mathbf{R}$.
Solution:

$$\mathbf{R} = [(3 + 2 + 1) \quad (1 + 4 - 3) \quad (-2 + 1 + 2)] = [6\ 2\ 1]$$

$$\mathbf{S} = [(3 - 2 - 1) \quad (1 - 4 + 3) \quad (-2 - 1 - 2)] = [0\ 0\ -5]$$

$$\mathbf{M} = [(3)(6) \quad (3)(2) \quad (3)(1)] = [18\ 6\ 3]$$ ◄

Consider the three points in space indicated in Figure 1-47. Each point is represented by a vector. Each vector has three scalar components measured relative to the origin along the axes of an R.H.C. system. The three components of each point are scalar numbers which are placed in a matrix of one row and three columns, giving

$$\mathbf{P}_1 = [x_1\ y_1\ z_1], \qquad \mathbf{P}_2 = [x_2\ y_2\ z_2] \quad \text{and} \quad \mathbf{P}_3 = [x_3\ y_3\ z_3]$$

The base point of a vector can begin at any point in space. If one begins at base point $\mathbf{P}_1 = [x_1\ y_1\ z_1]$ and moves to point $\mathbf{P}_2 = [x_1 + \Delta x \quad y_1 + \Delta y \quad z_1 + \Delta z]$, then vector **R** can be defined as shown in Figure 1-48. Vector $\mathbf{P}_1$ is referred to as the *base vector* for vector **R**. Vector **R** can be written in matrix form as $\mathbf{R} = [\ \Delta x \quad \Delta y \quad \Delta z]$. Vectors such as **R** which do not begin or end at the origin of a coordinate system are called *relative* position vectors or *directional* vectors.

The vector **R** by itself represents **any** three-dimensional *free* vector in space with components given by the same values of Δx, Δy, and Δz. The vector becomes a unique vector with a particular point of application when a *base vector* is given for the vector **R**.

Vector $\mathbf{P}_2$ in Figure 1-48 is a base vector beginning at the origin. It is written as $\mathbf{P}_2 = [x_1 + \Delta x \quad y_1 + \Delta y \quad z_1 + \Delta z] = [x_2\ y_2\ z_2]$, where $x_2 = x_1 + \Delta x$, $y_2 = y_1 + \Delta y$, $z_2 = z_1 + \Delta z$. This is another example of vector addition. One can write $\mathbf{P}_2 = \mathbf{P}_1 + \mathbf{R} = [x_1 + \Delta x \quad y_1 + \Delta y \quad z_1 + \Delta z]$. If $\mathbf{P}_2 = [x_2\ y_2\ z_2]$, then $\Delta x = x_2 - x_1$, $\Delta y = y_2 - y_1$, $\Delta z = z_2 - z_1$.

Another example is shown in Figure 1-49. The vector **R** is the difference between $\mathbf{P}_2$ and $\mathbf{P}_1$. The base vector for vector **R** is $\mathbf{P}_1$. Vector $\mathbf{P}_2$ is the sum of vectors $\mathbf{P}_1$ and **R**. The scalar components for each vector are given in matrix form in Figure 1-49.

In Figure 1-50 vector **A** begins at the origin. Thus, the scalar components of this base vector are the three space coordinates x_1, y_1, z_1 of point **A**. In matrix form, the components of **A** are written as a single row matrix in one of two ways:

$$\mathbf{A} = [A_x\ A_y\ A_z] = [x_1\ y_1\ z_1] \tag{1-4}$$

The magnitude of **A** is given in terms of its scalar components by use of the Pythagorean Theorem.

$$|\mathbf{A}| = \{A_x^2 + A_y^2 + A_z^2\}^{1/2} \tag{1-5}$$

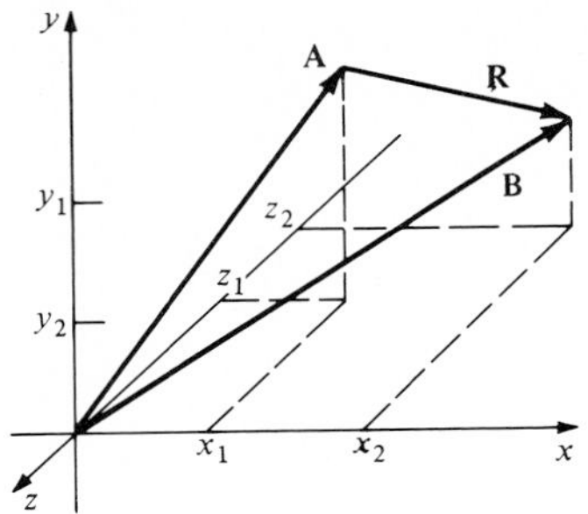

Figure 1-50
Scalar components.

The magnitude of **A** is equal to the true length of the line from the origin to point $[A]$. Figure 1-51 shows the geometric interpretation of Equation (1-5).

Vector **B** in Figure 1-50 is a base vector. In matrix notation

$$\mathbf{B} = [B_x \; B_y \; B_z] = [x_2 \; y_2 \; z_2] \tag{1-6}$$

Vector $\mathbf{R} = [R_x \; R_y \; R_z]$ is a directed line segment from point **A** to **B**. It can be seen in Figure 1-50 that

$$\mathbf{A} + \mathbf{R} = \mathbf{B} \tag{1-7}$$

Solving for **R** gives

$$\mathbf{R} = \mathbf{B} - \mathbf{A} = [x_2 \; y_2 \; z_2] - [x_1 \; y_1 \; z_1] \tag{1-8}$$

Using the rule of matrix subtraction Equation (1-7) becomes

$$\mathbf{R} = [(x_2 - x_1) \quad (y_2 - y_1) \quad (z_2 - z_1)] \tag{1-9}$$

Hence, the scalar components of **R** are $R_x = (x_2 - x_1)$, $R_y = (y_2 - y_1)$, and $R_z = (z_2 - z_1)$. The true length of vector **R** is given by

$$|\mathbf{R}| = \{(x_2 - x_1)^2 + (y_2 - y_1)^2 + (z_2 - z_1)^2\}^{1/2} \tag{1-10}$$

As discussed in Section 1-12, corners of solids with flat faces, such as the polyhedron shown in Figure 1-52, can be thought of as points or base vectors. In addition, boundaries of the solid can be represented by directional vectors. These directional vectors can be expressed as the difference between two position vectors. In Figure 1-52 one can see that $\mathbf{A} = \mathbf{P}_2 - \mathbf{P}_1$; $\mathbf{B} = \mathbf{P}_3 - \mathbf{P}_2$; and $\mathbf{C} = \mathbf{P}_4 - \mathbf{P}_2$. Vectors can also be used to represent the surface normals to each facet as will be shown later.

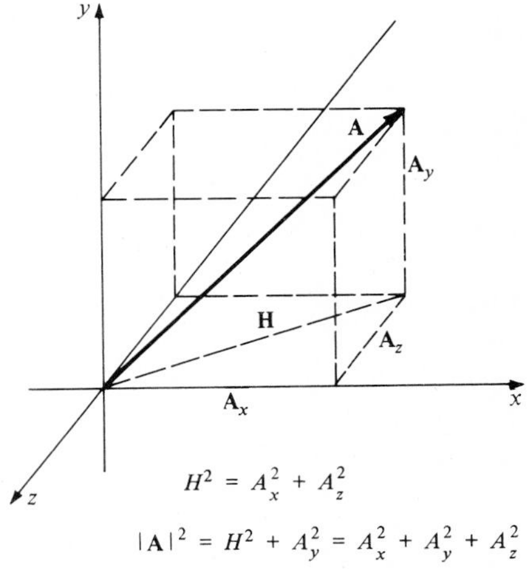

Figure 1-51
Magnitude of base vector.

1-15 UNIT VECTORS

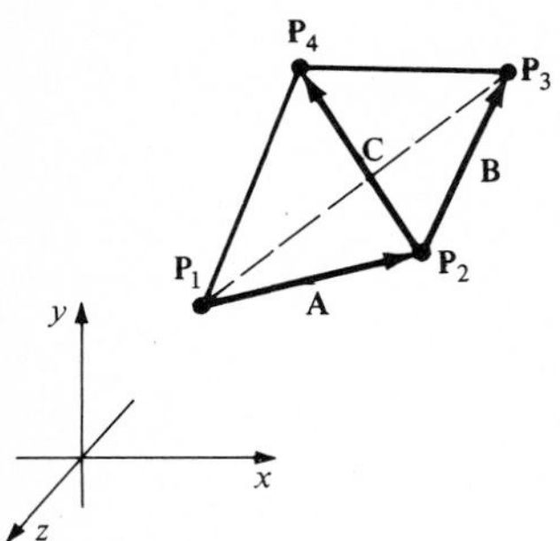

Figure 1-52
Polyhedron surface.

A *unit vector* is a vector with magnitude equal to 1.0. Normally, the symbols **i, j, k** are used to indicate positive unit vectors along the principal *x, y, z* axes of a rectangular coordinate system. Unit vectors for a right-hand coordinate system are shown in Figure 1-53. Each unit vector has only one nonzero scalar component along a single principal axis. Also, each unit vector is perpendicular to the other two.

In matrix notation, using single row matrices, the unit vectors are expressed as

$$\mathbf{i} = [i_x\ i_y\ i_z] = [1\ 0\ 0]$$

$$\mathbf{j} = [j_x\ j_y\ j_z] = [0\ 1\ 0]$$

$$\mathbf{k} = [k_x\ k_y\ k_z] = [0\ 0\ 1]$$

The vector **A** can also be expressed in terms of its scalar components in the x-, y-, and z-directions, and the associated unit vectors. The vector **A** in Figure 1-51 can be written as

$$\mathbf{A} = A_x\mathbf{i} + A_y\mathbf{j} + A_z\mathbf{k} \tag{1-11}$$

This is equivalent to the matrix representation $\mathbf{A} = [A_x\ A_y\ A_z]$.

The scalar component gives the magnitude, and the unit vector gives the direction of each vector component, thus making up the vector sum which defines the single vector **A**. The vector **R** in Figure 1-50 is expressed in terms of unit vectors as follows:

$$\mathbf{R} = R_x\mathbf{i} + R_y\mathbf{j} + R_z\mathbf{k} = (x_2 - x_1)\mathbf{i} + (y_2 - y_1)\mathbf{j} + (z_2 - z_1)\mathbf{k} \tag{1-12}$$

The reader should learn to visualize object representations, such as the orthographic projections given in Figures 1-41 through 1-45, in terms of their vector definitions.

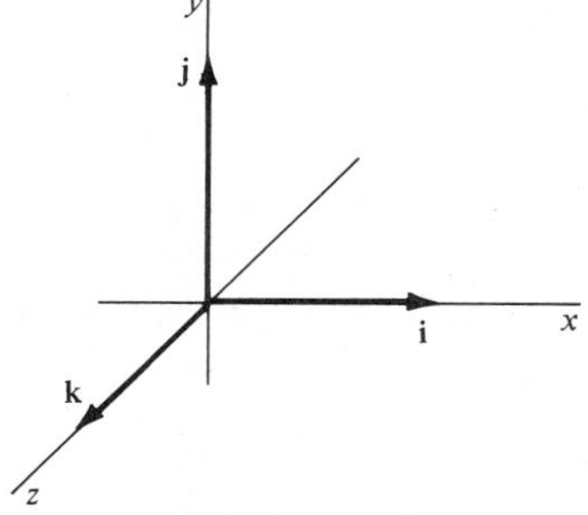

Figure 1-53
Unit vectors.

PROBLEMS: CHAPTER 1

1-1 *Given:* Top and front views of point **A** = [4 4 3].
Task: Find the profile view of point **A**.

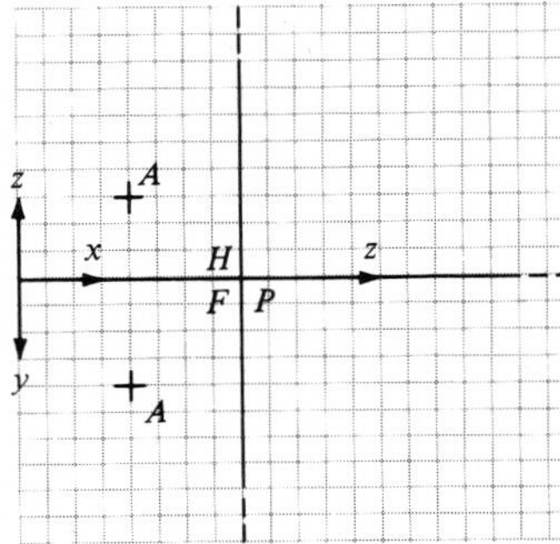

Figure 1-1P

1-2 *Given:* Top and side views of line *MN*, where

M = [2 6 4] and **N** = [6 2 1]

Task: Find the front view of line *MN*.

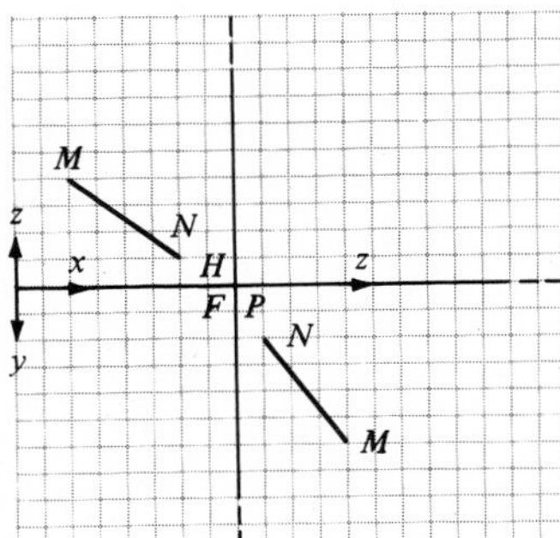

Figure 1-2P

1-3 *Given:* (open-ended problem)
Task: Draw top, front and profile views of any horizontal line *MN*.

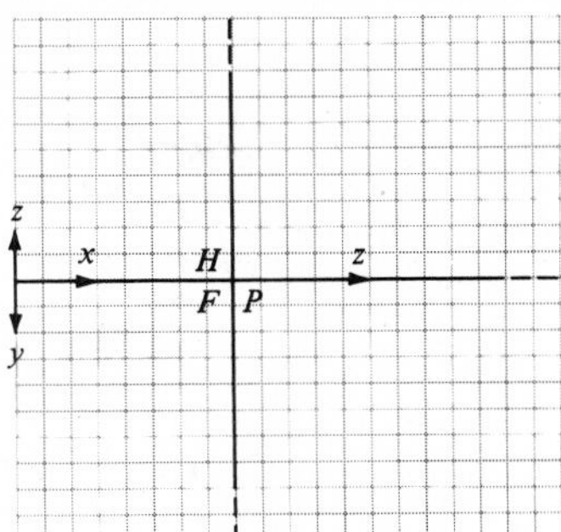

Figure 1-3P

1-4 *Given:* Top and front views of triangle *ABC*, where **A** = [1 6 4], **B** = [6 4 5] and **C** = [3 2 1].
Task: Draw the profile view of the given triangle.

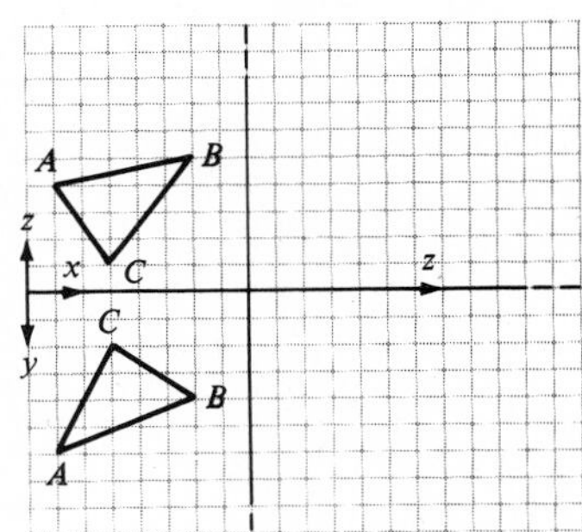

Figure 1-4P

1-5 *Given:* Top and profile views of a mutilated block, where **A** = [6 5 1].
Task: Draw the front view of the given block.

1-6 *Given:* Incomplete top and front views of a tetrahedron with **A** = [2 1 4], **B** = [5 2 6], **C** = [7 5 3], and **D** = [3 6 2].
Task: Show the correct visibility in both views.

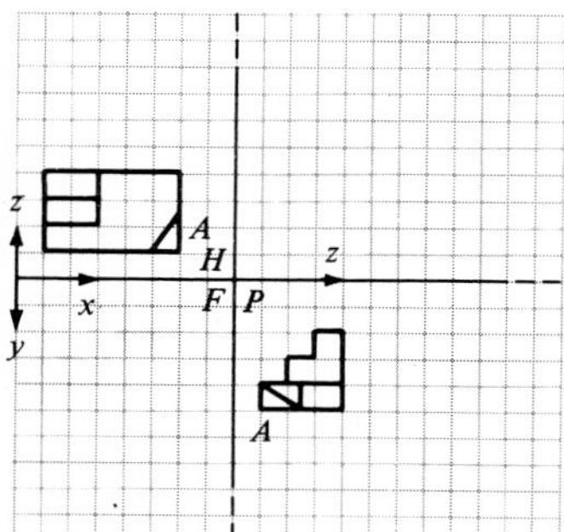

Figure 1-5P

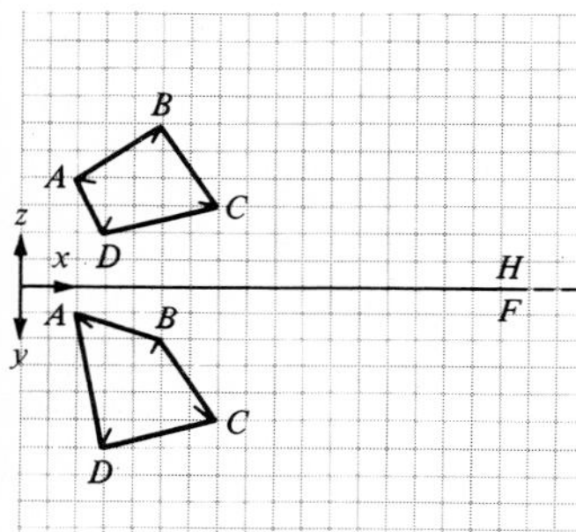

Figure 1-6P

1-7 *Given:* Incomplete top and front views of two nonintersecting pipes with endpoints **A** = [2 6.5 5], **B** = [7 1 2.5], **M** = [2 2.5 1.5] and **N** = [6 4.5 7].
Task: Show correct visibility.

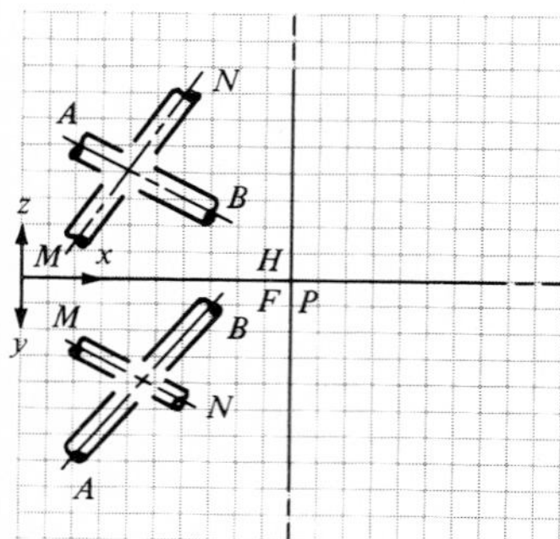

Figure 1-7P

1-8 *Given:* Top and front views of a tetrahedron with **A** = [1 5 2], **B** = [5 6 5], **C** = [6 3 0] and **D** = [3 8 2].
Task: Show correct visibility.

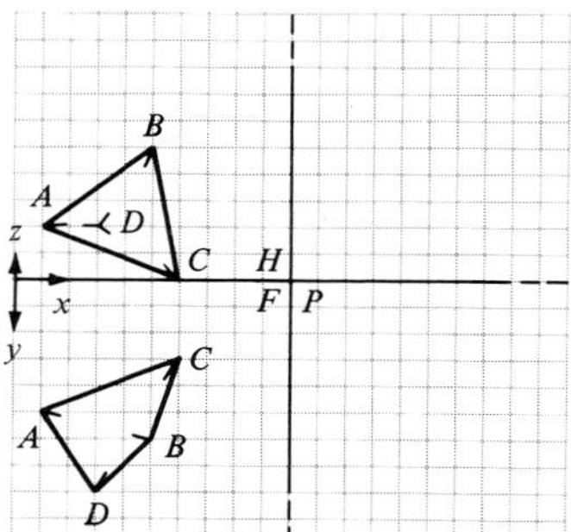

Figure 1-8P

1-9 *Given:* Incomplete top and front views of three nonintersecting pipes. **A** = [5 9.5 5.5], **B** = [30 4.5 10.5].
Task: Show correct visibility in both views.

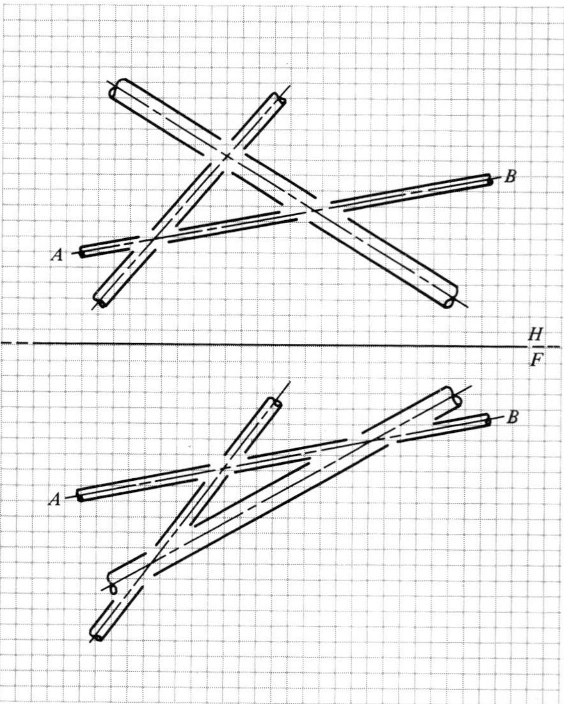

Figure 1-9P

1-10 *Given:* Top and right side views. **A** = [2 16 2]
Task: Draw the front view.

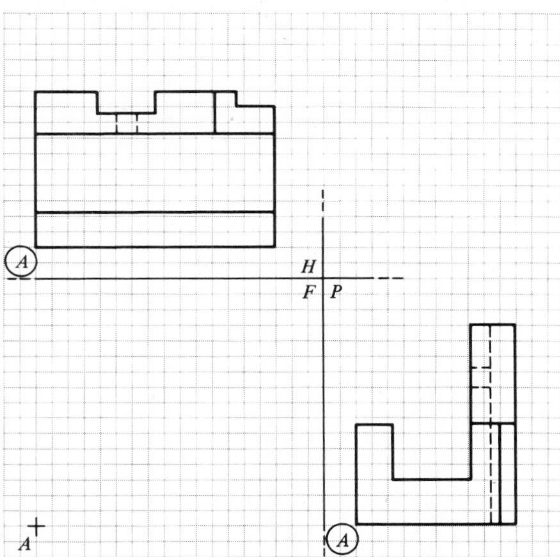

Figure 1-10P

1-11 *Given:* Front and right side views with **A** = [2 15 2].
Task: a. Draw the top view.
b. Complete visibility in both views.
c. Draw profile view.

1-12 *Given:* Figures 1-40 and 1-43.
Task: Sketch the shape of the solid objects that are represented in the given figures.

1-13 *Given:* Figures 1-40 and 1-43.
Task: Redraw the principal views and show all hidden lines dashed.

1-14 *Given:* A three-dimensional line is defined from **A** = [3 −2 1] to **B** = [4 3 −2], relative to a R.H.C. system such as the *x*, *y*, *z* axes shown in Figure 1-46.
Task: Draw the three principal views of this line.

1-15 *Given:* Two triangular facets *ABC* and *BCD* have a common edge *BC* defined relative to the *x*, *y*, *z* R.H.C. system. Vectors which define the corner points are **A** = [−1 2 2], **B** = [−1 1 −2], **C** = [1 1 −1], and **D** = [2 3 −3].
Task: Draw the three principal views for these two facets.

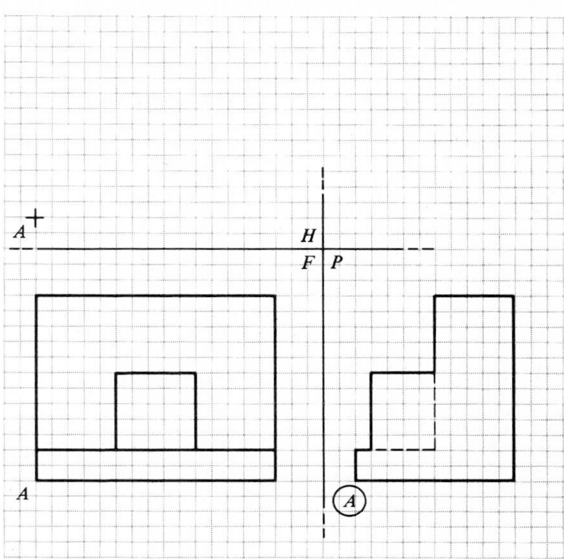

Figure 1-11P

1-16 *Given:* Object shown in Figure 1-16P. Hidden lines are not shown but can be inferred. The scale is to be defined by the reader using a R.H.C. system to define the points.
Task: a. Create a point file and a line file for a similar object.
b. Create principal views for a similar object, either by hand or by computer.

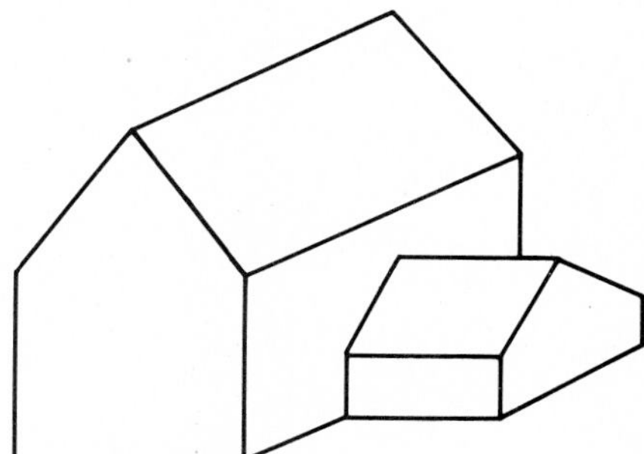

Figure 1-16P
House.

1-17 *Given:* Object shown in Figure 1-17P.
Task: a. Create a point file and a line file for a similar object using a R.H.C. system as shown.

b. Create the three principal views by hand or by computer.

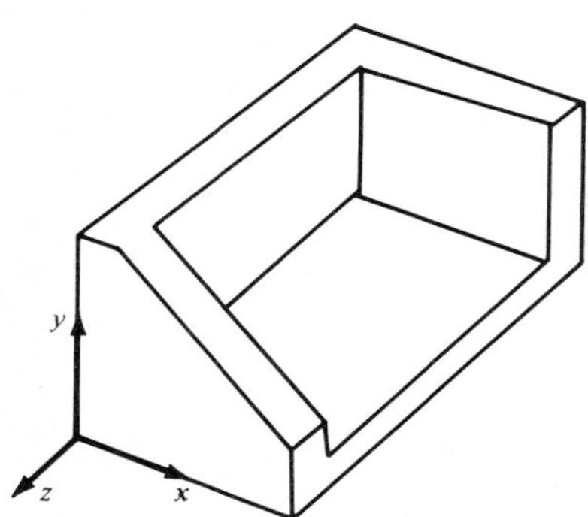

Figure 1-17P
Open box.

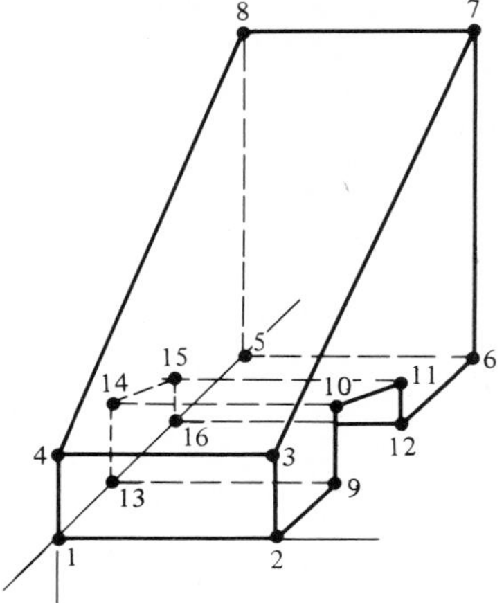

Figure 1-18P
Block with groove.

1-18[c] *Given:* Object shown in Figure 1-18P, defined by the data file below.
Task: Create a line definition file and write a program such as VIEWS which calls PVIEWS and draws a wire frame display of the three principal views.

Point	x	y	z
1	0	0	6
2	5	0	6
3	5	2	6
4	0	2	6
5	0	0	0
6	5	0	0
7	5	8	0
8	0	8	0
9	5	0	4
10	5	2	4
11	5	1	2
12	5	0	2
13	0	0	4
14	0	2	4
15	0	1	2
16	0	0	2

[c]Superscript *c* indicates that a computer solution is required.

1-19 *Given:* The line in space defined by two endpoints in Problem 1-2.
Task: a. Write the directional vector for this line in matrix form. See Equation (1-9).
b. Calculate the true length of the line using Equation (1-10).

1-20 *Given:* Triangle *ABC* defined in Problem 1-4.
Task: Calculate the length of the perimeter around the triangle.

1-21 *Given:* The tetrahedron defined in Problem 1-6.
Task: Represent each edge of the tetrahedron by a directional vector expressed in matrix form.

1-22 *Given:* Two nonintersecting pipes defined in Problem 1-7.
Task: Determine which pipe is the longest.

1-23 *Given:* The tetrahedron defined in Problem 1-8.
Task: Define unit vectors for the coordinate system shown in Figure 1-8P and express each edge of the tetrahedron as a directional vector expressed in the form of Equation (1-12).

1-24 *Given:* Three pipes defined in Problem 1-9.
Task: Define an ortho-box as shown in Figure 1-46 with width=30, height=20, and depth=20. Sketch the three pipes within this box.

1-25 *Given:* Three vectors $\mathbf{A} = 3\mathbf{i} + 4\mathbf{j} - \mathbf{k}$; $\mathbf{B} = 2\mathbf{i} - \mathbf{j} + 3\mathbf{k}$; $\mathbf{C} = \mathbf{i} + 5\mathbf{j} - 4\mathbf{k}$.

Task: a. Show that these vectors represent the edges of a triangular plane in space. Hint: Show $\mathbf{A} = \mathbf{B} + \mathbf{C}$.

b. Repeat with $\mathbf{A} = [3\ 1\ -2]$; $\mathbf{B} = [-1\ 3\ 4]$; $\mathbf{C} = [4\ -2\ -6]$.

c. Find the length of the three medians for the triangle in part a. A median is drawn from a vertex to the midpoint of the opposite side.

2

Primary Auxiliaries—Planes

Some geometric solutions and renderings which formerly were done by drafters using a drawing board can now be done using available computer systems for design and drafting. One can choose from two-dimensional or three-dimensional computer-aided design software. In a two-dimensional approach, the drawing and orthographic projections are generated using a thought process similar to that of a drafter working with a drawing board. Instead, a computer is used to generate the required graphic displays. Automatic features may include dimensioning, cross-sectioning, lettering, and template formation. In a three-dimensional system, each separate object is stored in computer memory using three-dimensional vertex, line, and surface information. General two-dimensional orthographic views may then be generated automatically. The simplest object definition scheme is a wire frame model.

Concepts of descriptive geometry are necessary to interpret the two-dimensional information contained on drawings, regardless of how they are generated. Vector analysis concepts are needed to understand and develop useful computer procedures to make use of a three-dimensional approach to object definition. This chapter will demonstrate means of producing orthographic views both manually with pencil and paper, and automatically with the help of computer graphics and a wire frame model.

2-1 NORMAL VIEWS

Our previous study dealt with orthographic projection related to three principal planes: frontal, horizontal, and profile. We know from this study that observers view an object through a picture plane so that their line of sight to each point on the object makes an angle of 90 deg with that plane. If a plane figure (*ABCDE* of Figure 2-1(a)) is viewed orthographically we

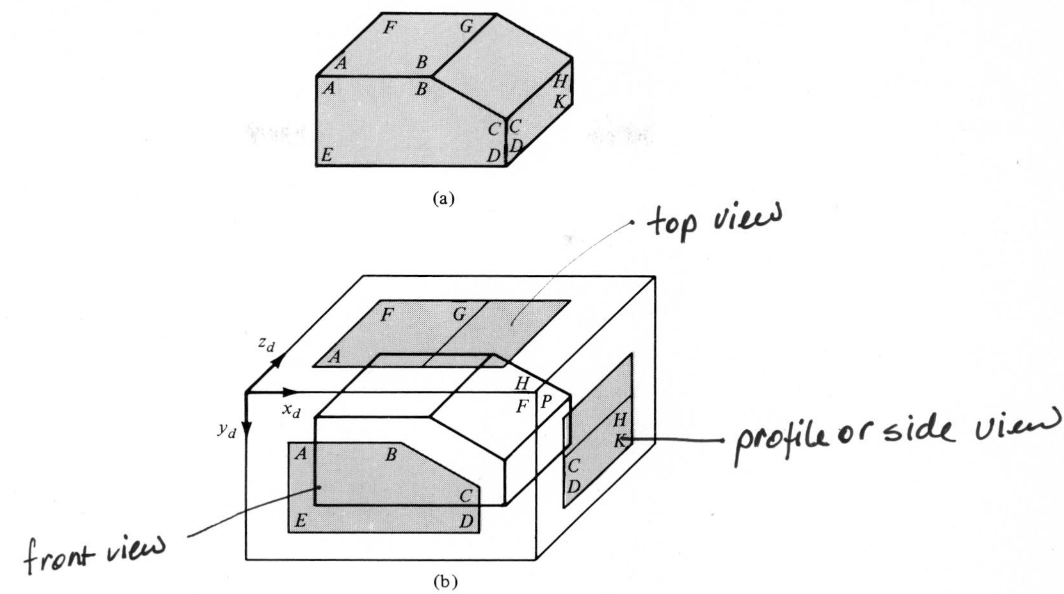

Figure 2-1
Normal view of planes.

can make the following statement: *Any plane figure which is parallel to the picture plane will project onto the picture plane so it will be seen in true size and shape* (T/S/S). This projection will also be a normal view of the plane since "normal" means "at right angles to" or "perpendicular to." In Figure 2-1(b) the *true size and shape* of *ABCDE* is seen when looking through the frontal plane (front view). *AFGB* is seen through the horizontal plane (top view), and *CHKD* through the profile plane (side view).

In Figure 2-2, plane *BGHC* does not show in true size and shape in any

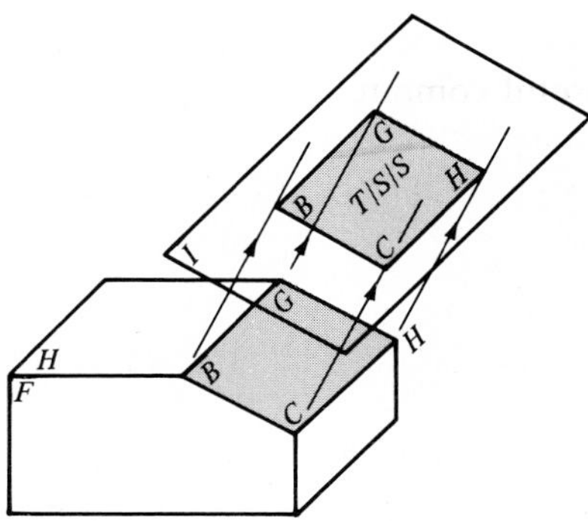

Figure 2-2
Auxiliary plane.

of the principal views because it is not parallel to the principal planes. In order to show it as a normal view, a plane must be inserted parallel to the inclined surface. The observer then views *BGHC* so that the lines of sight pass through the plane at an angle of 90 deg. The true size and shape of *BGHC* will be projected onto the plane which is, by definition, a primary auxiliary plane. In other words: *A primary auxiliary plane is perpendicular to one of the principal planes and inclined to the other two.*

2-2 NORMAL VIEW OF A PLANE—FRONTAL AUXILIARY

Example 2-1 refers to Figure 2-2 (plane *BGHC*) which projects "foreshortened" views onto all principal planes. By applying the same principles given above, the normal view of plane *BGHC* may be found using a frontal auxiliary. In this procedure, the auxiliary plane which is placed parallel to plane *BGHC* will be perpendicular to the frontal plane and inclined to the other two. The projection of *BGHC* on the primary auxiliary plane will produce the image which will be referred to as a *frontal auxiliary* or *first auxiliary*.

Figure 2-3 shows the intersection of the first auxiliary plane and the frontal plane. It is suggested that this figure be studied in conjunction with Example 2-1. Figure 2-3(a) typifies the front view in which the observer is looking through the frontal plane and sees the edge of the first auxiliary plane. This edge is represented by a folding line which is labeled *F*/1 in Example 2-1. In Figure 2-3 the auxiliary plane has been folded into the plane of the paper (auxiliary view) and the observer is looking through it at a 90 deg angle. The folding line now becomes the edge of the frontal plane and depth measurements can be taken from it in the horizontal view and plotted from it in the first auxiliary view. It can be seen, also,

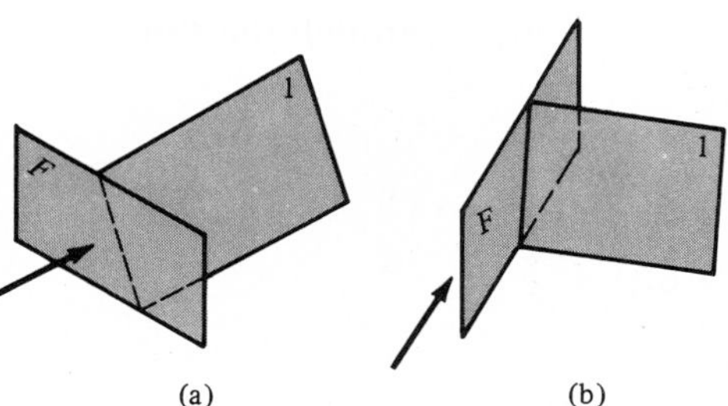

Figure 2-3
Intersection: frontal and first auxiliary planes.

that the image will *always* be true size and will not vary according to the distance between the object and the plane.

Example 2-1 shows the procedure when the first auxiliary is projected from the front view. Arrows are used to indicate auxiliary projections when space permits.

Example 2-1

Given: Top and front views of a mutilated block.

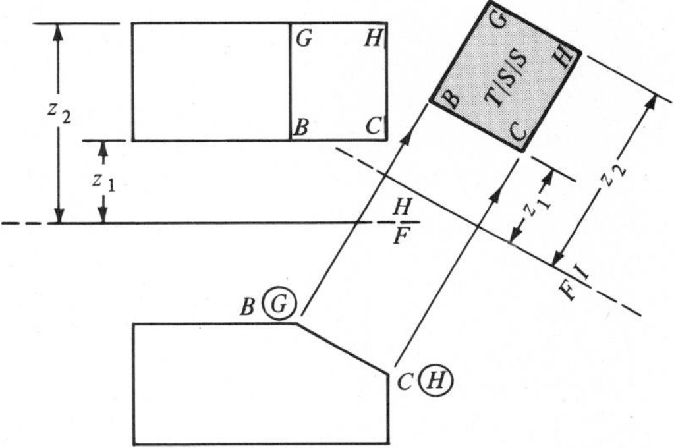

Example 2-1
True size and shape: inclined plane, frontal auxiliary.

Task: Find the true size and shape of plane *BGHC.*
Solution:

1. Find the view in which plane *BGHC* appears on edge. In this case it is the front view.
2. Pass a primary auxiliary plane parallel to the edge view. This auxiliary plane appears as a folding line which is labeled *F*/1. When looking through the *horizontal* plane we see the edge of the frontal plane; when looking through the *first auxiliary plane* we also see the edge of the frontal plane.
3. Establish lines of sight by drawing them into the first auxiliary view from each point of the plane in the front view so that they (the lines of sight) are perpendicular to the folding line.
4. Find the distances z_1 and z_2 from the folding line *F/H back* to the four points *BGHC* in the top view.
5. When looking in the auxiliary view, the *F*/1 folding line represents the edge of the frontal plane. On their respective lines of sight, and using the measurements for z_1 and z_2 obtained in the top view, mea-

sure the correct distances into the auxiliary view and lay off points *BGHC*.

6. Connect the four points to obtain the true size and shape of plane *BGHC*. ◀

2-3 NORMAL VIEW OF A PLANE—ELEVATION AUXILIARY

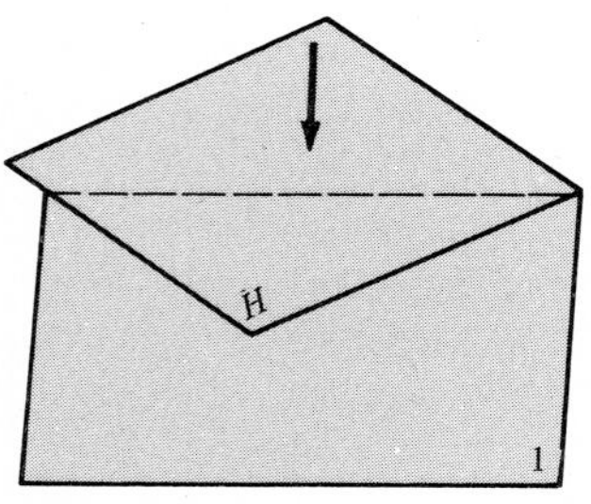

Figure 2-4
Intersection: horizontal and first auxiliary planes.

The procedure for finding the elevation auxiliary view of an object by projecting from the horizontal view is basically the same as described in Example 2-1. However, the primary auxiliary plane, when placed parallel to the plane surface in this method, will be perpendicular to the horizontal plane and inclined to the frontal and profile planes. Figure 2-4 shows the line formed by the intersection of the horizontal plane and the first auxiliary plane. Reference to Example 2-2 will show that this line is represented by the folding line *H*/1. When the elevation auxiliary is folded into the plane of the paper, the true height of *ABCDEFGH* can be seen. Dimensions can be taken from folding line *H*/*F* in the front view and plotted from *H*/1 in the first auxiliary. Both *H*/*F* and *H*/1 represent the edge of the horizontal plane.

Example 2-2

Given: Top and front views of the mutilated block shown in Figure 2-5.
Task: Find the true size and shape of surface *ABCDEFGH*.
Solution:

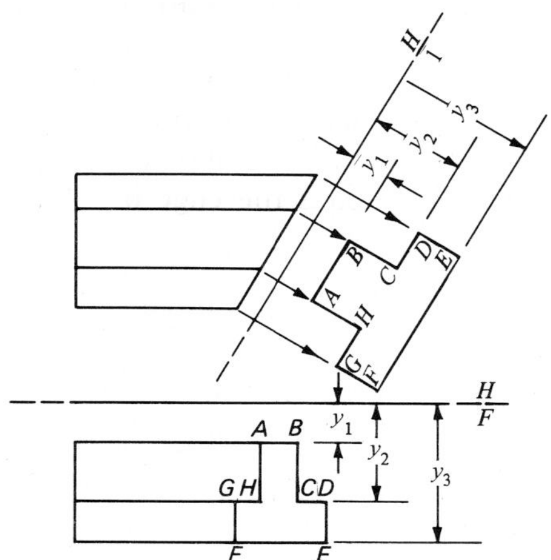

Example 2-2
True size and shape: inclined plane, elevation auxiliary.

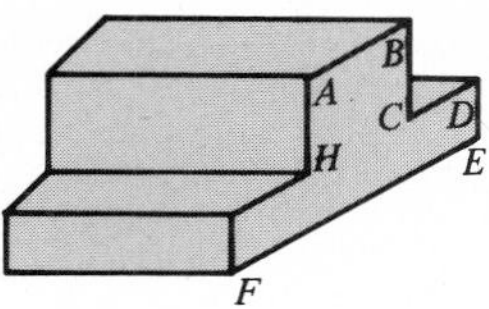

Figure 2-5
Mutilated block.

1. Find the view in which plane *ABCDEFGH* appears as an edge. In this example it is the top view.
2. Pass a primary auxiliary plane parallel to the edge view. We know this plane is a first auxiliary plane since it will be perpendicular to the horizontal plane and will hinge to it. A mental image of this intersection, represented by folding line *H*/1, is then formed. When viewed through the frontal plane, the folding line *H*/*F* represents the edge of the horizontal plane. In this view we can see that points *A, B, C, D, E, F, G,* and *H* lie at distances y_1, y_2, and y_3 *below* the horizontal plane. When looking through the first auxiliary plane, the folding line *H*/1 represents the edge of the horizontal plane. We know from prior inspection that points *A* through *H* lie at distances y_1, y_2, and y_3 *below* the horizontal plane.
3. Establish lines of sight from each point of the plane into the auxiliary view. These lines of sight must be perpendicular to folding line *H*/1.
4. On the lines of sight in the first auxiliary view, locate the points at their respective distances y_1, y_2, and y_3 below the folding line *H*/1.
5. Connect the eight points to find the true size and shape of the plane figure *A,B,C,D,E,F,G,H.* ◀

2-4 NORMAL VIEW OF A PLANE—PROFILE AUXILIARY

Example 2-3 shows, graphically, the operations necessary to find the normal view of a plane by projecting from the profile view. Here again, the same fundamental procedures are followed as those employed in finding frontal and elevation auxiliaries. Figure 2-6 shows that in this case the primary auxiliary plane makes a 90 deg intersection with the profile plane. The line formed by this intersection is portrayed in Example 2-3 by folding line 1/*P*. When the observer looks through the profile plane the folding line 1/*P* depicts the edge of the primary auxiliary plane. When looking through the auxiliary plane the folding line represents the edge of the profile plane.

Examples 2-1 and 2-2 set forth the step-by-step procedures for finding a frontal auxiliary and an elevation auxiliary. The student should now be able to develop the solution for Example 2-3 and list the operations necessary to obtain a profile auxiliary. (Note that the front view is not needed to solve this problem.)

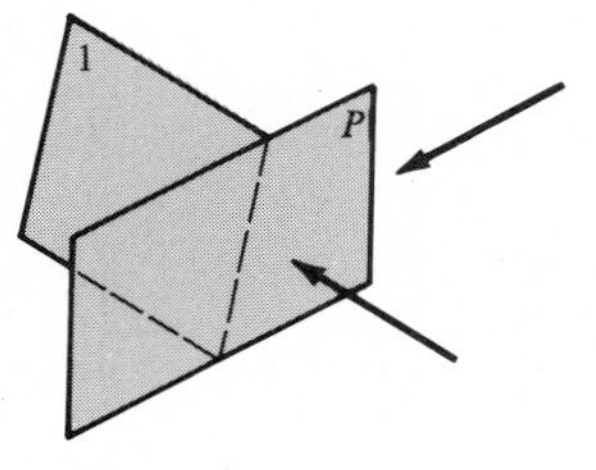

Figure 2-6
Intersection: profile and first auxiliary planes.

Example 2-3

Given: Top and side views of the mutilated block shown in Figure 2-7.
Task: Find the true size and shape of the inclined surface *ABCD.* ◀

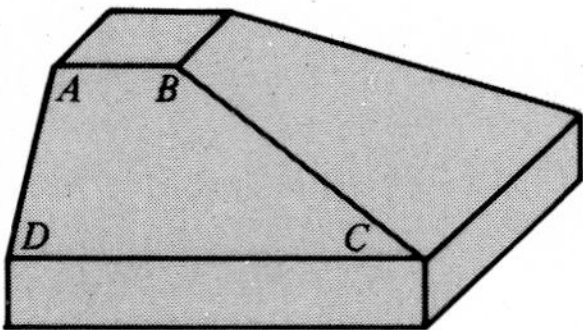

Figure 2-7
Mutilated block.

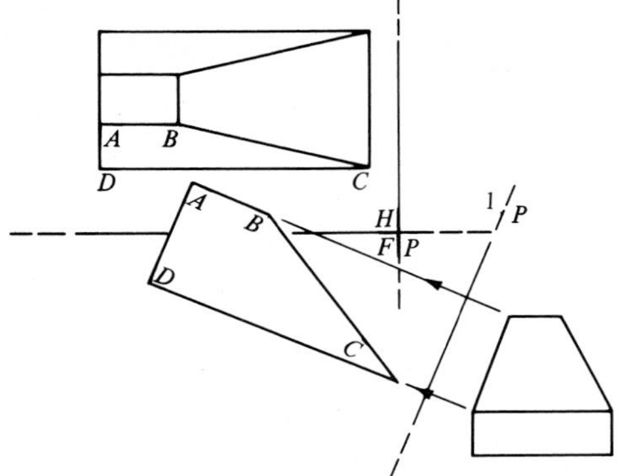

Example 2-3
True size and shape: inclined plane, profile auxiliary.

2-5 REFERENCE PLANES

In previous work the problems and solutions were presented so that the folding lines represented hinges between two planes. Because of limited space, symmetrical parts, and subject orientation, a special case may arise where it is advantageous to move the folding lines. Figure 2-8(a) shows that, because of symmetry, the *H/F* folding line has been moved so it passes through points *F* and *C* in the top view and coincides with the centerline (℄). When points *F* and *C* are plotted in the side view, they will fall on the centerline, the new location of the folding line. By moving the folding line so that the points are equidistant from it the points can be plotted in less time than they could be by the previous method.

Moving the folding lines may cause confusion if the student tries to visualize them as hinge points which are located between planes. So that a logical procedure may be followed, the reference planes should be shown in sets and so labeled. The notation "*RP*" will be used as shown in Figure 2-8(a). A convenient rule that can be used is as follows: *A folding line or reference plane may be moved to a more convenient location if it remains perpendicular to the lines of sight.*

Figure 2-8(b) shows the reference plane (*RP* #1) placed at the *base* of the object *ABCD*. The distance is still taken from the *reference plane* to various points. In this figure, the reference plane is shown in the elevation auxiliary and the object is placed *above* it. Generally, the dimensions would be taken *away* from the top view in both the front view and the auxiliary view. In this case, however, the dimensions are taken *toward* the top view in both the front and auxiliary views. *ABCD* is then shown *above* *RP* #1 in the elevation auxiliary view and agrees with its location in the front view.

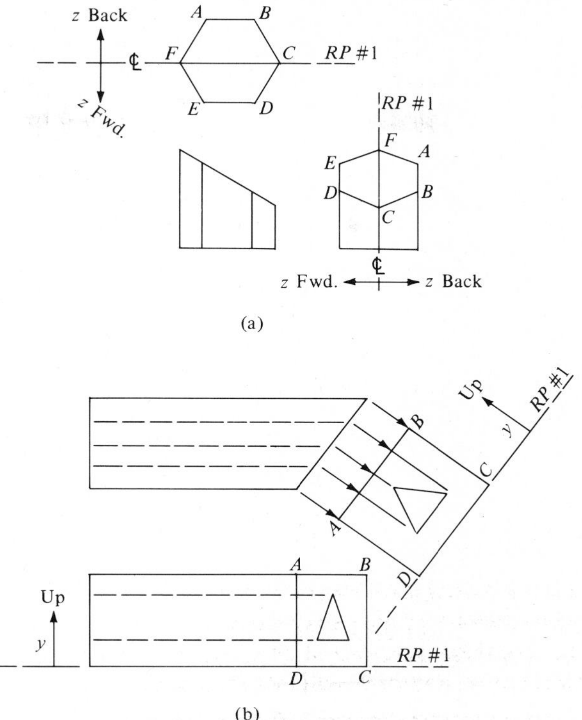

Figure 2-8
Reference plane.

2-6 NORMAL VIEW OF AN IRREGULAR PLANE

Not all planes are limited by straight lines but are bound by definite outlines that form geometric curves or irregular curves. The plane which does not have straight line boundaries is designated in this article as an *irregular plane.* A plane surface which has this type boundary either as an interior or exterior limit must be drawn graphically in order to obtain its normal view. It is only after finding its true size and shape that a mating part can be made.

The top and front views of a truncated cylinder are given in Figure 2-9. This shows a *circle* in the top view. The required normal view of the truncated surface could easily be plotted as a frontal auxiliary if enough points on the surface could be transposed. However, the top and front views show only four points which can be projected (i.e., the points where the centerlines intersect the surface boundary). Inspection shows that more than four points must be plotted in the first auxiliary in order to

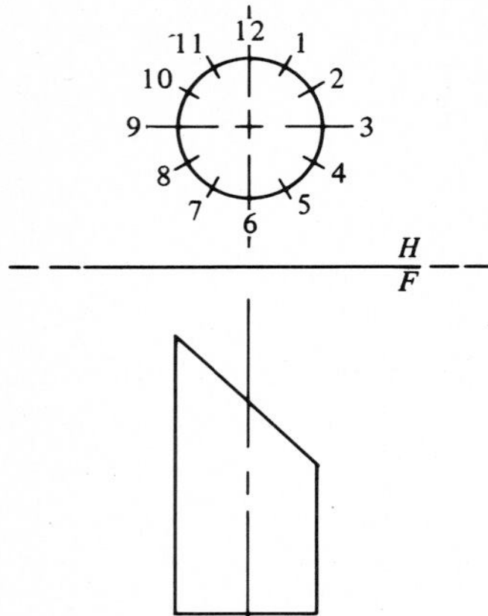

Figure 2-9
Points located on truncated cylinder.

obtain a normal view of the elliptical surface. It is recommended that at least 12 points be used when a circle or ellipse is plotted. For noncircular or irregular curves a sufficient number of points should be used so that a smooth curve can be delineated. A sharp curve would require more points than a flat curve.

Note that the procedure given in Section 2-5 has been followed in Example 2-4 and the folding line moved to the centerline so that it passes through numbers 3 and 9 and is labeled "*RP* #1." This is particularly convenient when circles are involved since four points can be plotted and transferred to both sides of the auxiliary "*RP*" at the same time.

EXAMPLE 2-4

Given: Top and front views of the truncated cylinder.

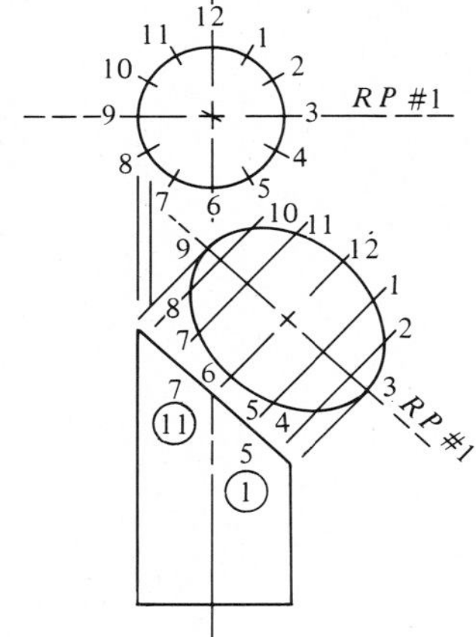

Example 2-4
Normal view: truncated surface.

Task: Show the normal view of the truncated surface.
Solution:

1. Plot at least 12 points at 30 deg intervals on the circumference of the circle projection of the cylinder in the top view.
2. It is recommended that the points be numbered to avoid confusion when plotting. (The clock method has been used here.)
3. Project the 12 points into the front view.
4. Place the folding line in the top view so that it coincides with the centerline (℄) and passes through points 3 and 9. Label this line *RP* # 1 since it is called a reference plane. (See Section 2-5.)

5. Draw the edge of the primary auxiliary plane, *RP* #1, parallel to the edge view of the truncated surface.
6. Project the lines of sight from the front view and perpendicular to *RP* #1 into the first auxiliary view.
7. Take the dimensions of the 12 points from the front and rear of *RP*# 1 in the top view. Transfer these dimensions into the first auxiliary and plot them on their respective lines of sight so they are located to the front and rear of *RP* #1.
8. Connect the points with an irregular curve to complete the elliptical boundary.
9. Draw the centerlines. ◄

2-7 NORMAL AND EDGE VIEWS—A SUMMARY AND PREVIEW

Studies in descriptive geometry are based largely on the theories discussed in this chapter. It is advisable, at this point, to summarize some of the preceding discussions and preview some mathematical ideas:

A. Figure 2-10 is an orthographic drawing consisting of three principal views and a frontal auxiliary. Some of the plane surfaces which make up this object are shown as follows:
   1. Normal View
      a. *AFGB*—Top view
      b. *CHKD*—Side view
      c. *ABCDE*—Front view
      d. *BGHC*—First auxiliary view
   2. Edge Views
      a. *ABCDE*—Top, side, and auxiliary views
      b. *AFGB*—Front and side views
      c. *CHKD*—Top and front views
B. When an auxiliary view is projected from the front view, it shows the true depth dimension.
C. An elevation auxiliary which is projected from the top view will give the true height dimension.
D. The true width dimension will be shown in a profile auxiliary which is projected from the side view.
E. When viewed orthographically any plane figure which is perpendicular to a picture plane will appear on that plane as an edge (line). It will be seen in true size and shape (normal view) if a picture plane is placed parallel to it.

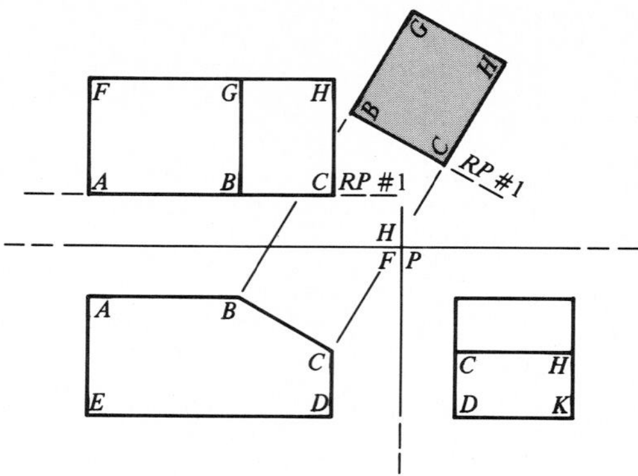

Figure 2-10
Normal and edge views.

A mathematical way to define a normal to a plane is to use the *vector cross product.* If two nonparallel edges of a plane are treated as vectors, then the vector cross product can be used to define a vector which is normal to the plane. This technique is used in the next section.

Another product which is used later in the text is the *scalar* (or *inner*) *product* of two vectors. Unlike the vector cross product which produces a vector, the result of a scalar product is a single, scalar number. The rules of vector algebra needed to produce these products are given when the operations are first used.

A final product which will be introduced and used throughout the text is the *matrix product.* Matrix multiplication obeys a set of rules which are quite different from normal multiplication. A three-dimensional data matrix can be multiplied by a transformation matrix to create solid body rotations of the object about a principal axis. This is necessary to create an orthographic projection using computational techniques. The theory of matrix multiplication is reviewed in Appendix 2.

2-8 VECTOR CROSS PRODUCT

A plane can be defined by specifying three noncollinear points in space. These three points can be considered as position vectors from the origin of the coordinate system as shown in Figure 2-11. In matrix form these three vectors are written

$$
\begin{aligned}
\mathbf{P}_1 &= [x_1\ y_1\ z_1] \\
\mathbf{P}_2 &= [x_2\ y_2\ z_2] \\
\mathbf{P}_3 &= [x_3\ y_3\ z_3]
\end{aligned}
\tag{2-1}
$$

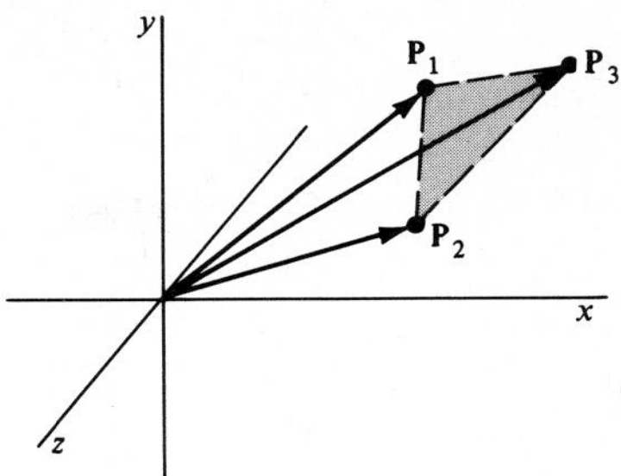

Figure 2-11
Three points in a plane.

The sides of the triangular facet shown in Figure 2-11 can also be represented by vectors $\mathbf{S}_1$, $\mathbf{S}_2$ and $\mathbf{S}_3$, indicated in Figure 2-12. These boundary vectors can be defined by vector subtraction. Using the sense of each vector as shown in Figure 2-12, and using matrix notation to represent the components gives

$$
\begin{aligned}
\mathbf{S}_1 &= \mathbf{P}_2 - \mathbf{P}_1 = [(x_2 - x_1) \quad (y_2 - y_1) \quad (z_2 - z_1)] \\
\mathbf{S}_2 &= \mathbf{P}_3 - \mathbf{P}_2 = [(x_3 - x_2) \quad (y_3 - y_2) \quad (z_3 - z_2)] \qquad (2\text{-}2) \\
\mathbf{S}_3 &= \mathbf{P}_3 - \mathbf{P}_1 = [(x_3 - x_1) \quad (y_3 - y_1) \quad (z_3 - z_1)]
\end{aligned}
$$

Alternatively, one can use vector addition and write $\mathbf{P}_3 = \mathbf{P}_2 + \mathbf{S}_2$, $\mathbf{P}_2 = \mathbf{P}_1 + \mathbf{S}_1$, and so on. Remember that in vector addition or subtraction one must add or subtract the corresponding components. Matrix notation provides an easy way to accomplish this task. The traditional graphical method of combining vectors is discussed in Appendix 6.

Example 2-5

Given: $\mathbf{A} = [\ 3\ 6\ -1]$; $\mathbf{B} = [1\ 3\ 2]$; $\mathbf{C} = [-1\ 1\ 0]$
Task: Find $\mathbf{A} + \mathbf{B}$; $\mathbf{A} - \mathbf{B}$; $\mathbf{A} + \mathbf{B} + \mathbf{C}$; $\mathbf{A} - \mathbf{B} + \mathbf{C}$
Solution:

$$\mathbf{R}_1 = \mathbf{A} + \mathbf{B} = [4\ 9\ 1]$$

$$\mathbf{R}_2 = \mathbf{A} - \mathbf{B} = [\ 2\ 3\ -3]$$

$$\mathbf{R}_3 = \mathbf{A} + \mathbf{B} + \mathbf{C} = [3\ 10\ 1]$$

$$\mathbf{R}_4 = \mathbf{A} - \mathbf{B} + \mathbf{C} = [1\ 4\ -3]$$

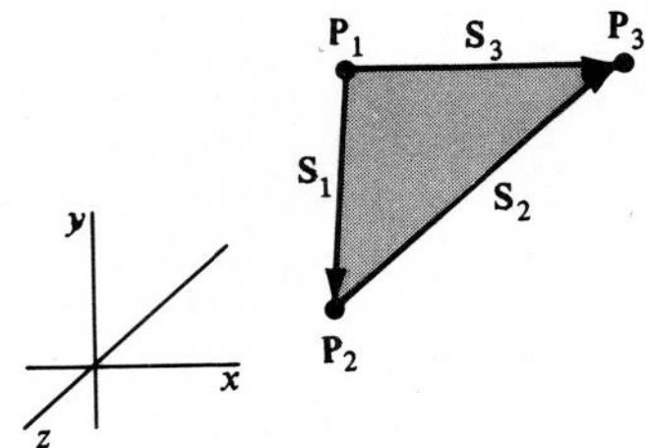

Figure 2-12
Triangular facet with boundary vectors.

It is suggested that the reader sketch the given vectors and the four resultant vectors on a three-dimensional coordinate system. ◀

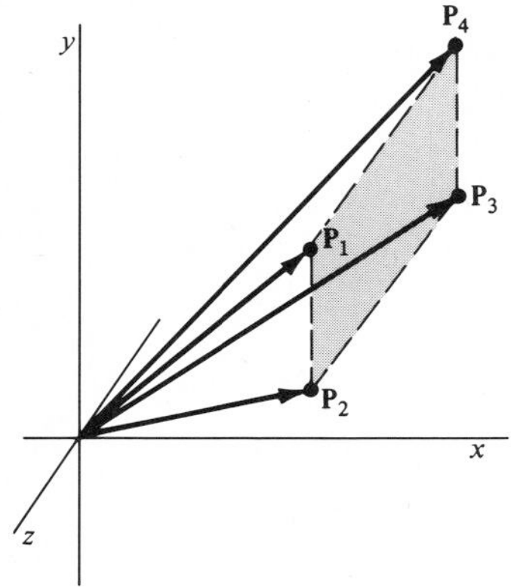

Figure 2-13
Rectangular facet.

If a four-sided rectangular plane facet is to be defined, then the fourth vertex can be calculated as $\mathbf{P}_4 = [x_4\ y_4\ z_4]$, where $\mathbf{P}_4 = \mathbf{P}_1 + \mathbf{P}_3 - \mathbf{P}_2$. Hence, the scalar components of point $\mathbf{P}_4$ are

$$\begin{aligned} x_4 &= x_1 + x_3 - x_2 \\ y_4 &= y_1 + y_3 - y_2 \\ z_4 &= z_1 + z_3 - z_2 \end{aligned} \tag{2-3}$$

Figure 2-13 shows a four-sided plane facet. Even though a plane can be defined by three noncollinear points within the plane, it must be realized that mathematically, a plane is infinite in extent. The properties of vectors will be used later to formulate equations for a plane which can be used to lead to computational solutions of descriptive geometry problems.

When vectors are defined relative to an orthogonal, rectangular coordinate system, the cross product is most easily evaluated by expanding a determinant as discussed in Appendix 2. The cross product $\mathbf{A} \times \mathbf{B}$ is given by

$$\mathbf{C} = \mathbf{A} \times \mathbf{B} = \det \begin{vmatrix} \mathbf{i} & \mathbf{j} & \mathbf{k} \\ A_x & A_y & A_z \\ B_x & B_y & B_z \end{vmatrix}$$

Expanding the determinant gives

$$\mathbf{C} = (A_yB_z - A_zB_y)\mathbf{i} - (A_xB_z - A_zB_x)\mathbf{j} + (A_xB_y - A_yB_x)\mathbf{k}$$

The result of the cross product $\mathbf{A} \times \mathbf{B}$ is another vector $\mathbf{C}$ which is perpendicular to the plane containing $\mathbf{A}$ and $\mathbf{B}$. The cross product is also called the *vector product.*

Let the two sides of a rectangular, or triangular, plane facet be given by the vectors $\mathbf{A}$ and $\mathbf{B}$ where

$$\begin{aligned} \mathbf{A} &= \mathbf{P}_1 - \mathbf{P}_2 \\ \mathbf{B} &= \mathbf{P}_3 - \mathbf{P}_2 \end{aligned} \tag{2-4}$$

Notice that the base of each boundary vector $\mathbf{A}$ and $\mathbf{B}$ is at $\mathbf{P}_2$. Figure 2-14 shows a rectangular, plane facet.

The vector cross product $\mathbf{A} \times \mathbf{B}$ produces a vector $\mathbf{N}$ which is normal to the plane containing $\mathbf{A}$ and $\mathbf{B}$ as shown in Figure 2-14. The direction of $\mathbf{N}$ in a right-hand coordinate (R.H.C.) system is toward the observer who sees the rotation from $\mathbf{A}$ to $\mathbf{B}$ as counterclockwise (CCW) for $0 \leq \theta \leq 180$ deg. The correct sense for $\mathbf{N}$ is shown in Figure 2-14 for the facet represented. Using the definition of a vector cross product in rectangular coordinates gives

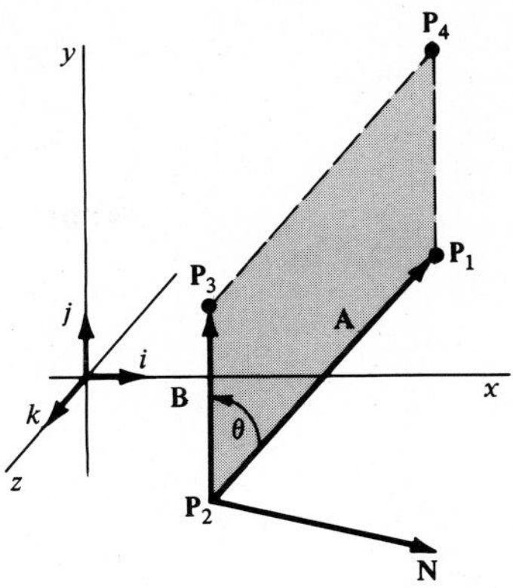

Figure 2-14
Normal vector to a plane.

$$\mathbf{N} = \mathbf{A} \times \mathbf{B} = (\mathbf{P}_1 - \mathbf{P}_2) \times (\mathbf{P}_3 - \mathbf{P}_2) \tag{2-5}$$

$$= \det \begin{vmatrix} \mathbf{i} & \mathbf{j} & \mathbf{k} \\ (x_1 - x_2) & (y_1 - y_2) & (z_1 - z_2) \\ (x_3 - x_2) & (y_3 - y_2) & (z_3 - z_2) \end{vmatrix}$$

If **N** has components indicated by $[N_1\ N_2\ N_3]$, then evaluation of the determinant in Equation (2-5) gives

$$N_1 = (y_1 - y_2)(z_3 - z_2) - (y_3 - y_2)(z_1 - z_2)$$

$$N_2 = (x_3 - x_2)(z_1 - z_2) - (x_1 - x_2)(z_3 - z_2) \tag{2-6}$$

$$N_3 = (x_1 - x_2)(y_3 - y_2) - (x_3 - x_2)(y_1 - y_2)$$

(See Examples 2-6 and 2-7.) Unit vectors in the x,y,z directions are **i, j, k** as shown in Figure 2-14.

The magnitude of the vector which is the cross product **A** × **B** is known to be given by

$$|\mathbf{N}| = |\mathbf{A}||\mathbf{B}| \sin\theta; \quad 0 \le \theta \le \pi \tag{2-7}$$

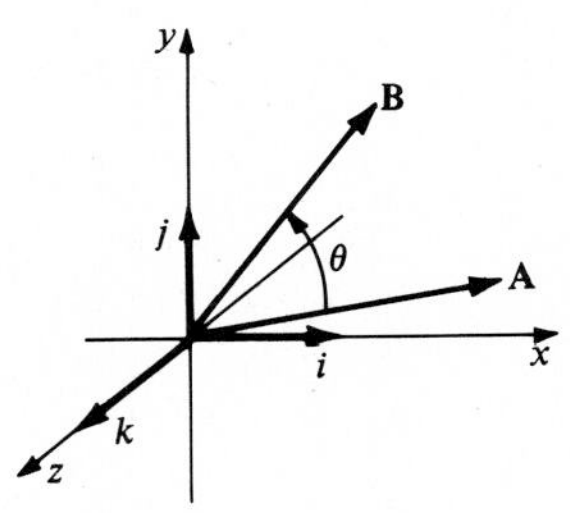

Figure 2-15
Vector cross product.

where θ is the angle between **A** and **B** indicated in Figure 2-15. The vertical bars have been used to denote the magnitude of a vector. If $\mathbf{A} = [x_1\ y_1\ z_1]$ and $\mathbf{B} = [x_2\ y_2\ z_2]$, then the magnitude of **A** is $|\mathbf{A}| = \{x_1^2 + y_1^2 + z_1^2\}^{1/2}$ and $|\mathbf{B}| = \{x_2^2 + y_2^2 + z_2^2\}^{1/2}$. The magnitude of a vector is a scalar quantity.

Notice that if $\theta = 0$, then **A** and **B** are collinear and the cross product is zero since $\theta = 0$. If $\theta = \pi/2$, then $\sin\theta = 1.0$ and the magnitude of the cross product is the product $|\mathbf{A}||\mathbf{B}|$. In summary, the cross product of two vectors is another vector with both magnitude and direction, the direction being perpendicular to the plane containing the original two vectors.

Example 2-6

Given: Vectors $\mathbf{A} = [1\ 2\ 3]$ and $\mathbf{B} = [3\ 2\ 1]$.
Task: Find $\mathbf{A} \times \mathbf{B}$ and $\mathbf{B} \times \mathbf{A}$.
Solution:

$$\mathbf{R}_1 = \mathbf{A} \times \mathbf{B} = \det \begin{vmatrix} \mathbf{i} & \mathbf{j} & \mathbf{k} \\ 1 & 2 & 3 \\ 3 & 2 & 1 \end{vmatrix}$$

Expanding the determinant gives

$$\mathbf{R}_1 = \mathbf{i}(2 - 6) + \mathbf{j}(9 - 1) + \mathbf{k}(2 - 6)$$

or

$$\mathbf{R}_1 = -4\mathbf{i} + 8\mathbf{j} - 4\mathbf{k} = [-4\ 8\ -4]$$

$$\mathbf{R}_2 = \mathbf{B} \times \mathbf{A} = \det \begin{vmatrix} \mathbf{i} & \mathbf{j} & \mathbf{k} \\ 3 & 2 & 1 \\ 1 & 2 & 3 \end{vmatrix}$$

$$\mathbf{R}_2 = \mathbf{i}(6 - 2) + \mathbf{j}(1 - 9) + \mathbf{k}(6 - 2)$$

$$\mathbf{R}_2 = 4\mathbf{i} - 8\mathbf{j} + 4\mathbf{k} = [4\ -8\ 4]$$

Notes:

a. Notice that the matrix containing the components of a vector represents the same vector as defined by use of the unit vectors **i, j, k** along the three coordinates.

b. Notice that $\mathbf{A} \times \mathbf{B}$ is not equal to $\mathbf{B} \times \mathbf{A}$ since one is the negative of the other. Hence, the vectors $\mathbf{R}_1$ and $\mathbf{R}_2$ point in opposite directions. ◀

Example 2-7

Given: Three points in space $\mathbf{P}_1 = [3\ 2\ 5]$, $\mathbf{P}_2 = [-2\ -3\ 2]$, and $\mathbf{P}_3 = [1\ 4\ -4]$.
Task: Calculate the components of the normal vector **N** to the plane containing the three points.
Solution:

$$\text{Let } \mathbf{A} = \mathbf{P}_1 - \mathbf{P}_2 = [5\ 5\ 3]$$

$$\mathbf{B} = \mathbf{P}_3 - \mathbf{P}_2 = [3\ 7\ -6]$$

$$\mathbf{N} = \mathbf{A} \times \mathbf{B} = \det \begin{vmatrix} \mathbf{i} & \mathbf{j} & \mathbf{k} \\ 5 & 5 & 3 \\ 3 & 7 & -6 \end{vmatrix}$$

$$\mathbf{N} = \mathbf{i}(-30 - 21) + \mathbf{j}(9 + 30) + \mathbf{k}(35 - 15)$$

$$\mathbf{N} = -51\mathbf{i} + 39\mathbf{j} + 20\mathbf{k} = [-51\ 39\ 20]$$ ◀

The subroutine VECPROD listed in Appendix 1 will accept the components of two vectors A(1), A(2), A(3) and B(1), B(2), B(3) and return the cross product components R(1), R(2), R(3), where $\mathbf{R} = \mathbf{A} \times \mathbf{B}$. Many examples of main programs which call the subroutine VECPROD can be found in Appendix 1.

The vector cross product of the two unit vectors **i** and **j** in rectangular coordinates is given by

$$\mathbf{i} \times \mathbf{j} = \det \begin{vmatrix} \mathbf{i} & \mathbf{j} & \mathbf{k} \\ 1 & 0 & 0 \\ 0 & 1 & 0 \end{vmatrix} = \mathbf{i}(0 - 0) - \mathbf{j}(0 - 0) + \mathbf{k}(1 - 0) = \mathbf{k}$$

On the other hand:

$$\mathbf{i} \times \mathbf{i} = \det \begin{vmatrix} \mathbf{i} & \mathbf{j} & \mathbf{k} \\ 1 & 0 & 0 \\ 1 & 0 & 0 \end{vmatrix} = \mathbf{i}(0 - 0) - \mathbf{j}(0 - 0) + \mathbf{k}(0 - 0) = 0$$

The reader can verify that $\mathbf{j} \times \mathbf{k} = \mathbf{i}$; $\mathbf{k} \times \mathbf{i} = \mathbf{j}$; $\mathbf{i} \times \mathbf{j} = \mathbf{k}$; $\mathbf{j} \times \mathbf{i} = -\mathbf{k}$; $\mathbf{j} \times \mathbf{j} = 0$, and so on. Remember that the result of the cross product of any two nonparallel vectors is a third vector which is perpendicular to the plane of the initial two vectors.

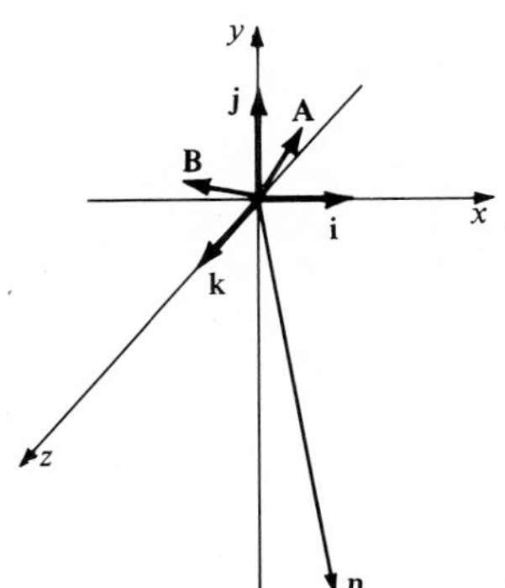

Figure 2-16
Unit normal vector.

Sometimes a normal vector with a magnitude (length) equal to 1.0 is useful. This is called a *unit normal vector* and is indicated by a boldface, lower case letter **n** (see Figure 2-16). The unit vectors **i, j,** and **k** are special unit vectors which point in the x, y, and z directions, respectively, in a rectangular coordinate system. They are used in the calculation of a vector cross product. If the components of the regular normal vector $\mathbf{N} = \mathbf{A} \times \mathbf{B}$ in the x, y, z directions are indicated by N_1, N_2, and N_3 respectively, then the magnitude $|\mathbf{N}|$ is given by

$$|\mathbf{N}| = \{N_1^2 + N_2^2 + N_3^2\}^{1/2} \tag{2-8}$$

The three components of the unit normal vector **n** (in the direction **N**) are

$$n_1 = \frac{N_1}{|\mathbf{N}|}$$

$$n_2 = \frac{N_2}{|\mathbf{N}|} \tag{2-9}$$

$$n_3 = \frac{N_3}{|\mathbf{N}|}$$

and the matrix representation of this unit normal vector is

$$\mathbf{n} = [n_1 \; n_2 \; n_3] \tag{2-10}$$

If we let **n** be the unit vector in the direction of the vector cross product $\mathbf{N} = \mathbf{A} \times \mathbf{B}$, then we can use an alternate way to write the vector **N**. The vector will have a direction indicated by the components of **n** and a magnitude given by $|\mathbf{A}| \; |\mathbf{B}| \sin \theta$. Thus we write:

$$\mathbf{N} = \{|\mathbf{A}| \; |\mathbf{B}| \sin \theta\}\mathbf{n} \tag{2-11}$$

This represents the same vector given by expanding the determinant to obtain the cross product **N**. Any vector, such as **L**, can be written

$$\mathbf{L} = |\mathbf{L}|\mathbf{n} \tag{2-12}$$

if **n** is defined as the unit vector in the direction of **L**, and $|\mathbf{L}|$ is the scalar magnitude of **L**.

Example 2-8

Task: Find the components of the unit normal vector to the plane defined in the previous example.
Solution:

$$|\mathbf{N}| = \{N_1^2 + N_2^2 + N_3^2\}^{1/2} = \{(51)^2 + (39)^2 + (20)^2\}^{1/2} = 67.25$$

$$n_1 = \frac{51}{67.25} = 0.758$$

$$n_2 = \frac{39}{67.25} = 0.580$$

$$n_3 = \frac{20}{67.25} = 0.297$$

$$\mathbf{n} = [0.758 \quad 0.580 \quad 0.297]$$

Notice that $n_1^2 + n_2^2 + n_3^2 = 1.0$. This will always be true. ◀

2-9 EYE POSITIONS FOR ORTHOGRAPHIC VIEWS

The components of a vector in space determine the direction of the vector as shown in Figure 2-17. In fact, the components are often called *direction*

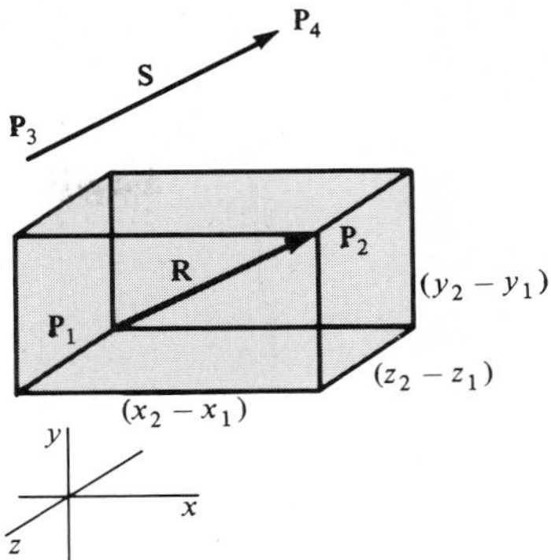

Figure 2-17
Parallel lines in space.

components. If two lines in space are parallel, then their components will be proportional. For example, let

$$\mathbf{R} = [(x_2 - x_1) \quad (y_2 - y_1) \quad (z_2 - z_1)]$$

and (2-13)

$$\mathbf{S} = [(x_4 - x_3) \quad (y_4 - y_3) \quad (z_4 - z_3)]$$

If **R** and **S** are parallel, then the direction ratios will be equal. That is:

$$\frac{(x_2 - x_1)}{(x_4 - x_3)} = \frac{(y_2 - y_1)}{(y_4 - y_3)} = \frac{(z_2 - z_1)}{(z_4 - z_3)} \tag{2-14}$$

An orthographic view can be observed from any point in space. The line of sight is along a line from the origin of the coordinate system to the eye position $\mathbf{E} = [E_1 \ E_2 \ E_3]$. The orthographic projection will be onto a picture plane which is perpendicular to the line of sight as shown in Figure 2-18. Recall that since points are projected along parallel lines, the

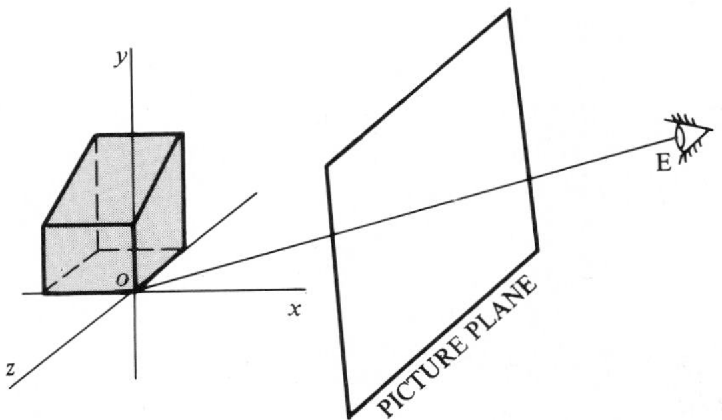

Figure 2-18
Picture plane normal to line of sight.

distance from the eye to the origin on a given line of sight does not affect the results. Also, the line of sight can be parallel to the line of sight shown in Figure 2-18 without affecting the results. The projection will change when the direction of the line of sight changes.

If the line of sight is parallel to a line on an object, then that line will project as a point. If the line of sight is perpendicular to a line, that line will appear in true length. If the line of sight is parallel to the normal vector to a plane, the plane will project in true size. The method for creating these types of projections using computer graphics is to transform the data by a series of rotations to align the line of sight with the z-axis and then plot the transformed x, y components of the object. This creates an orthographic projection onto the x-y plane.

The eye position is given by $\mathbf{E} = [E_1\ E_2\ E_3]$ as shown in Figure 2-19. Two rotations of the line of sight shown in Figure 2-19 will place it on the z-axis. These are a clockwise rotation about the y-axis by θ, followed by a counterclockwise rotation about the x-axis by ϕ. In Figure 2-19 it can be seen that

$$\theta = \tan^{-1}\left(\frac{E_1}{E_3}\right)$$
$$\phi = \tan^{-1}\left(\frac{E_2}{\sqrt{E_1^2 + E_3^2}}\right) \qquad (2\text{-}15)$$

If a three-dimensional data matrix is transformed in the same manner, then a plot of x,y values gives the desired orthographic projection of the object represented by the three-dimensional data matrix.

The remainder of this section is for those readers who have access to a computer with computer graphics capability. It may be omitted without loss in continuity of material presentation. The transformation matrices needed to create an orthographic projection from any eye position are given in Appendix 2. Consult this appendix for the mathematical details.

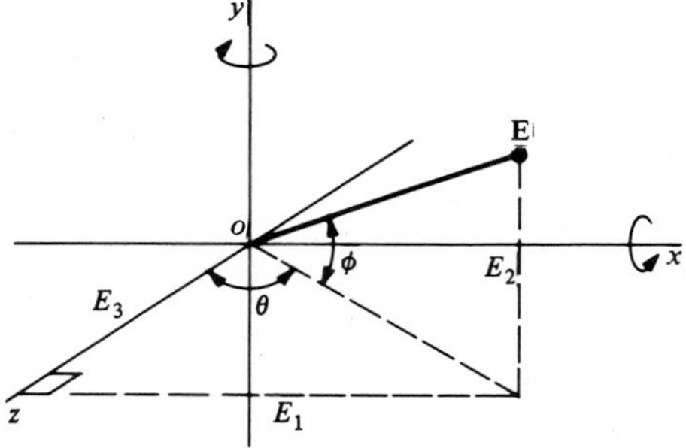

Figure 2-19
Rotations of eye position.

A series of matrix multiplications are used to produce a single transformation matrix. This single matrix will produce the required two rotations about the y-axis and x-axis when operating on a three-dimensional data matrix according to the rules of matrix multiplication as discussed in Appendix 2.

A program called ORTHOEYE and its associated subroutines are given in Appendix 1 which implement the matrix transformations and produce computer graphics output. The details of the matrix operations do not need to be fully understood at this point. [Examples of general orthographic views from specified eye positions are shown in Figures 2-20 through 2-23.] Students with access to, and experience with, a computer may want to write and implement a similar program, or use commercially available software to accomplish the same thing.

When the eye position is changed by the user, the WINDOW limits usually require modification. The modified limits are input by the user from the keyboard in the program ORTHOEYE. These WINDOW limits

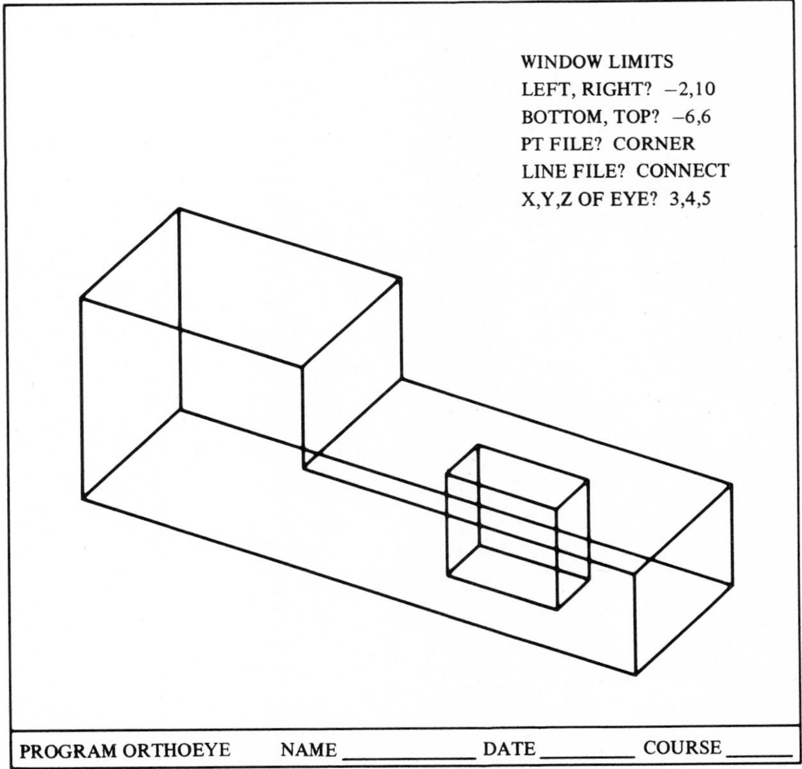

Figure 2-20
Machine part.

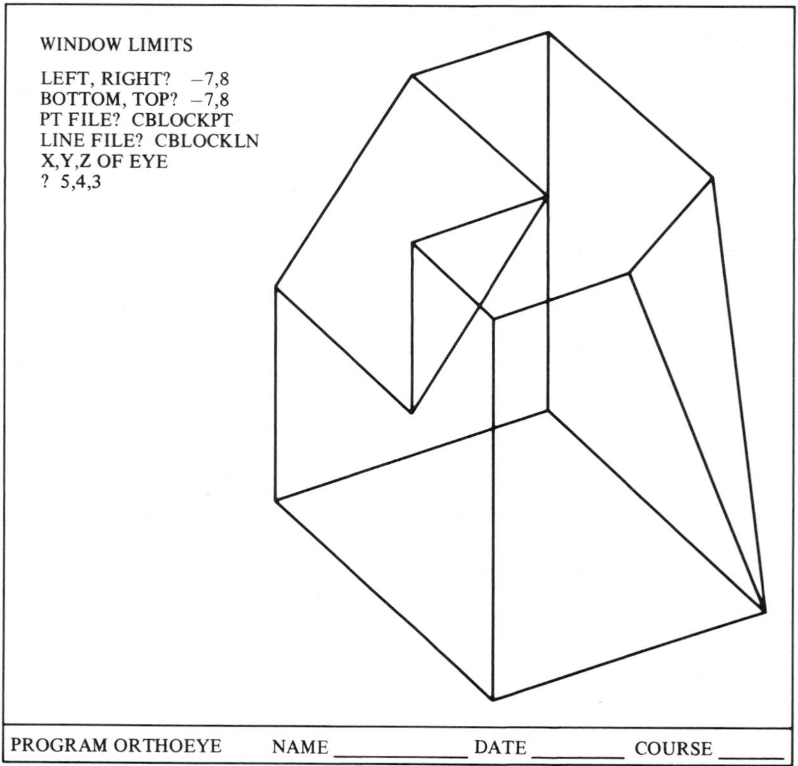

Figure 2-21
Cut block.

define the two-dimensional coordinate system onto which the data will be displayed, after the transformations and orthographic projections are made. The correct limits are a function of the three-dimensional data matrix and the direction components of the line of sight. Usually trial and error are necessary to determine the best choice for WINDOW limits.

In the program ORTHOEYE the subroutine EYETOZ creates the single 3×3 transformation matrix **[R]** which places the line of sight defined by $\mathbf{E} = [E_1\ E_2\ E_3]$ along the z-axis. Matrix multiplication of the data in $X(\,)$, $Y(\,)$, $Z(\,)$ with the matrix **[R]** occurs in lines 500–530. The orthographic projection is then created by plotting the transformed x, y values which are stored in arrays $A(\,)$ and $B(\,)$.

Figure 2-20 gives a view of the part shown in Figure 1-40 as seen from $\mathbf{E} = [3\ 4\ 5]$. Figure 2-21 shows a wire frame orthographic view of the cut block in Figure 1-41, with eye position $E = [5\ 4\ 3]$. The three-dimensional shape is more obvious in this view. The plane-solid intersection of Figure 1-43 is shown in Figure 2-22, and the two intersecting solids of Figure 1-45 are given in Figure 2-23, as seen from $\mathbf{E} = [3\ 3\ 5]$. A wire frame pro-

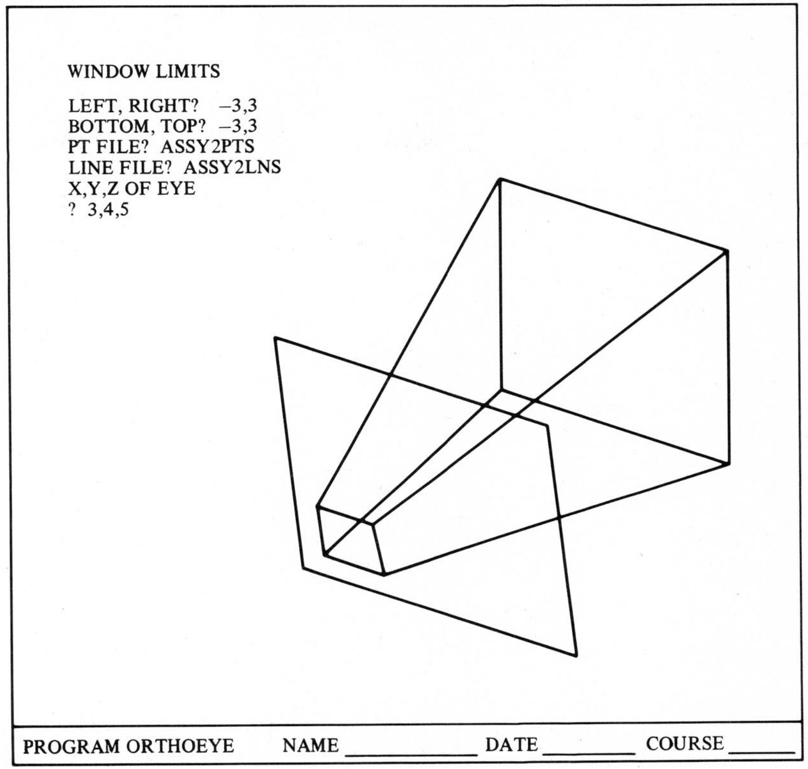

Figure 2-22
Piercing points.

jection of an auto safety design simulation is given in Figure 2-24, and a missile design application appears in Figure 2-25. These types of projections are very useful in computer-aided design.

Another example of controlling the eye position is the main program PLANEDGE given in Appendix 1. This program allows the user to define the coordinates of three corners of a plane. The components of the normal vector are then calculated and printed. The user can choose a line of sight direction which is defined by letting the three eye components equal the three normal vector components. Figure 2-26 gives one example from this program. Here the three points input to the program are $\mathbf{P}_1 = [2\ 2\ 2]$, $\mathbf{P}_2 = [2\ 4\ -3]$, and $\mathbf{P}_3 = [-2\ 8\ -2]$. The components of the normal vector obtained from the vector cross product are $\mathbf{N} = [22\ 20\ 8]$. The eye coordinates are then chosen to be equal to these normal components. Hence, the line of sight is parallel to the normal vector and the triangular plane will appear in true size and shape.

In Figure 2-27, the eye coordinates are chosen to be equal to the direction components of the line from $\mathbf{P}_1$ to $\mathbf{P}_2$. These are given by [(2 − 2)

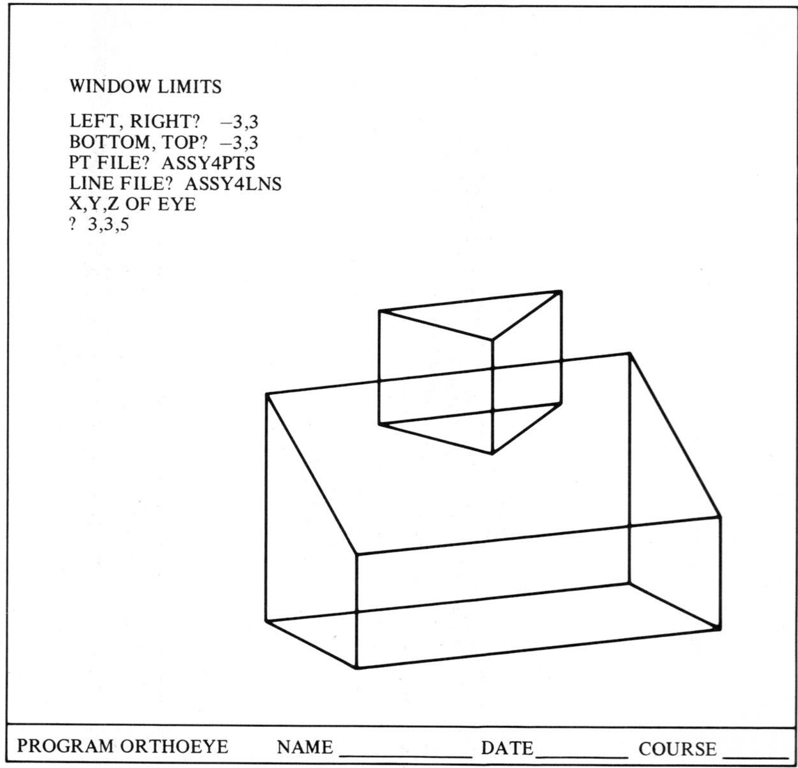

Figure 2-23
Intersecting solids.

$(4 - 2)\ (-3 - 2)] = [0\ 2\ -5]$. This then shows the plane on edge with the line of sight parallel to the line between $\mathbf{P}_1$ and $\mathbf{P}_2$. Thus, this line will appear as a point. If any line in a plane appears as a point, the plane will project as a line as shown in Figure 2-27. Other eye components which will show the plane as a line are the direction components of the line between $\mathbf{P}_1$ and $\mathbf{P}_3$ given by $[-4\ 6\ -4]$ and the components of the line between $\mathbf{P}_2$ and $\mathbf{P}_3$ given by $[-4\ 4\ 1]$.

2-10 AREA OF PLANE OBJECTS

When a plane facet of any shape is shown in true size, then the x, y vertices which create the projection can be used to calculate the area enclosed by the plane facet. Assume that such a facet consists of n vertices in a two-dimensional plane, defined in a counterclockwise manner. The enclosed area can be calculated using Equation (2-16).

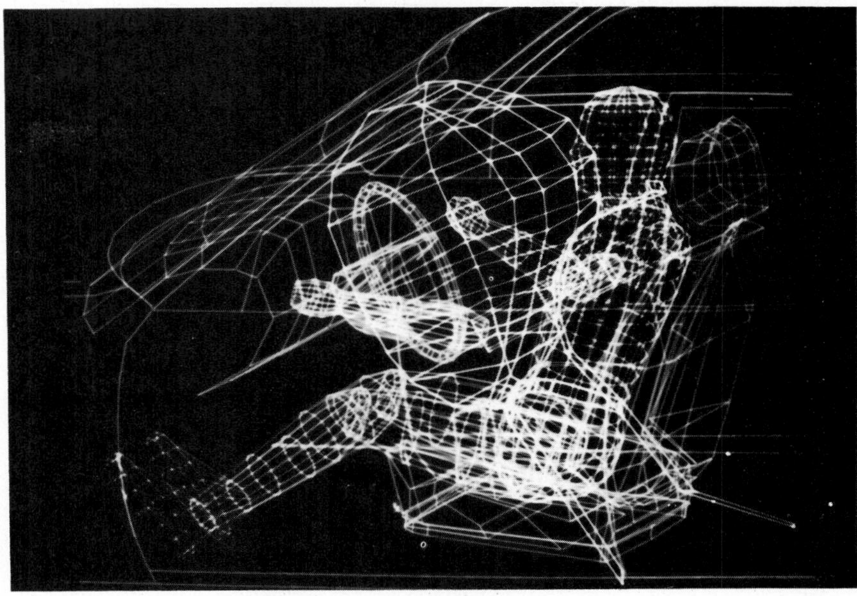

Figure 2-24
An example of a wire frame model used in the design of auto safety equipment. Courtesy of Evans and Sutherland Inc.

$$\text{Area} = -\frac{1}{2}[(x_2 - x_1)(y_1 + y_2) + (x_3 - x_2)(y_2 + y_3) + \cdots\cdots + (x_1 - x_n)(y_n + y_1)] \quad (2\text{-}16)$$

For a triangular facet, the area is simply

$$\text{Area} = -\frac{1}{2}[(x_2 - x_1)(y_1 + y_2) + (x_3 - x_2)(y_2 + y_3) + (x_1 - x_3)(y_3 + y_1)] \quad (2\text{-}17)$$

Notice that the area is closed by assigning the last point to be equal to the first point. The area calculation for the triangle in the previous two figures made use of the subroutine AREA discussed in the following example.

Example 2-9

Given: A 16-sided facet is symmetrical about the y-axis. The true size of the right half of the facet is given by the following x,y values for the vertices.

Vertex	x	y
1	0	−1.5
2	1.5	−2
3	1	−1
4	2	−1.5
5	1.5	0
6	1	0.5
7	1	1
8	0.5	0.5
9	0	1.5

Task: Write a subroutine that will evaluate the area of any n-sided facet. Then write a main program that will calculate the complete area of the facet.

Solution: The subroutine AREA implements Equation (2-16) and the main program FINDAREA defines the facet by use of DATA statements and then

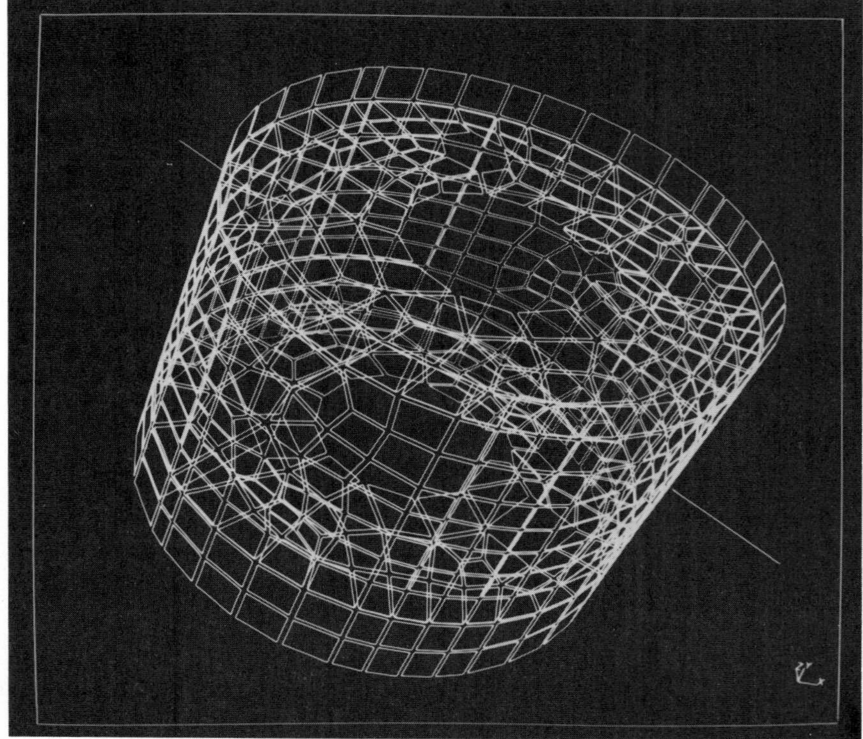

Figure 2-25
Wire frame display of missile housing. Courtesy of Evans and Sutherland Inc.

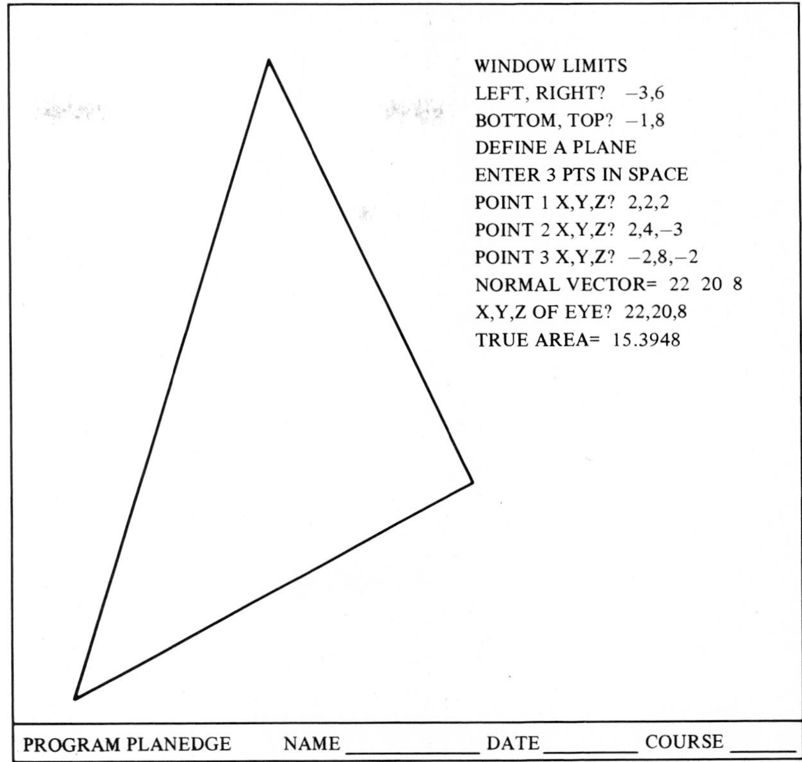

Figure 2-26
True size of plane.

uses subroutine AREA to calculate the half area. The total area of the facet = 7.5 units^2. ◀

```
70 REM
80 REM
90 REM*****AREA*****
100 EXTERNAL
110 SUB AREA(N,X( ),Y( ),AREA)
120 REM SUBROUTINE TO CALCULATE AREA OF 2-D POLYGON
130 REM N IS THE NUMBER OF POLYGON VERTICES
140 REM X( ) and Y( ) CONTAIN THE VERTICES, DEFINED IN THE MAIN PROGRAM
150 REM *** VERTICES MUST BE DEFINED IN CCW MANNER FOR POSITIVE AREA***
160 REM AREA IS RETURNED TO MAIN PROGRAM
170 OPTION NOLET
180 REM CLOSE AREA BY DEFINING LAST VERTEX=FIRST VERTEX
190 LET X(N+1)=X(1)
200 LET Y(N+1)=Y(1)
```

```
210 REM INITIALIZE AREA
220 LET AREA=0
230 REM CALCULATE ENCLOSED AREA
240 FOR I=1 TO N
250 AREA=AREA+(X(I+1)-X(I))*(Y(I)+Y(I+1))
260 NEXT I
270 REM CALCULATE ACTUAL AREA
280 AREA=-AREA/2
290 END SUB
```

```
100 !*****FINDAREA*****
110 !
120 !CALCULATE AREA OF 2-D REGION DEFINED BY N VERTICES
130 !VERTICES ARE DEFINED USING DATA STATEMENTS
140 LIBRARY"AREA"
150 DIM X(10),Y(10)
160 LET N=9 !NUMBER OF VERTICES IN DATA
```

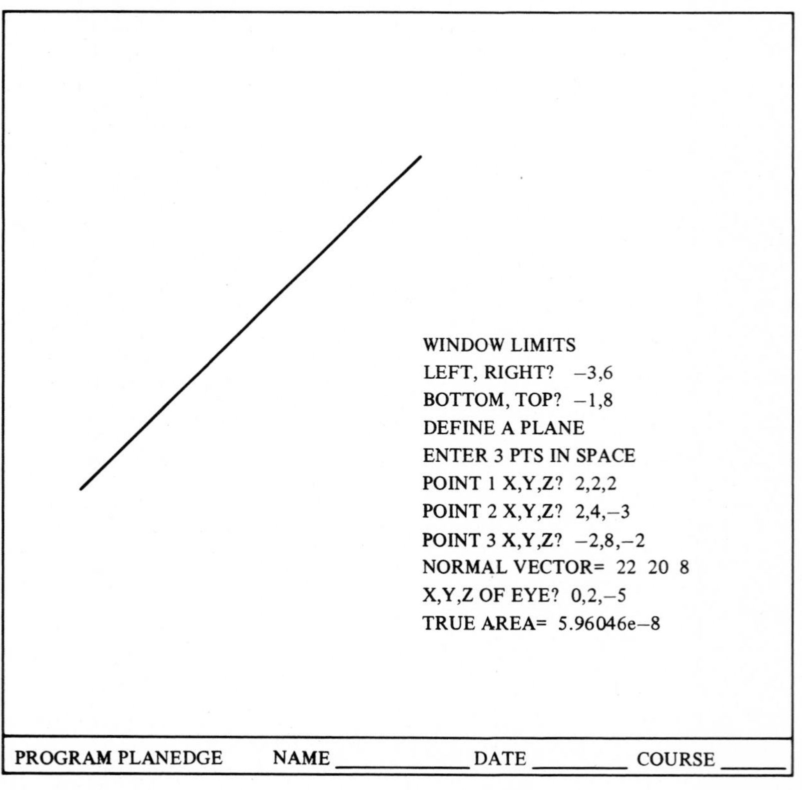

Figure 2-27
Edge view of plane.

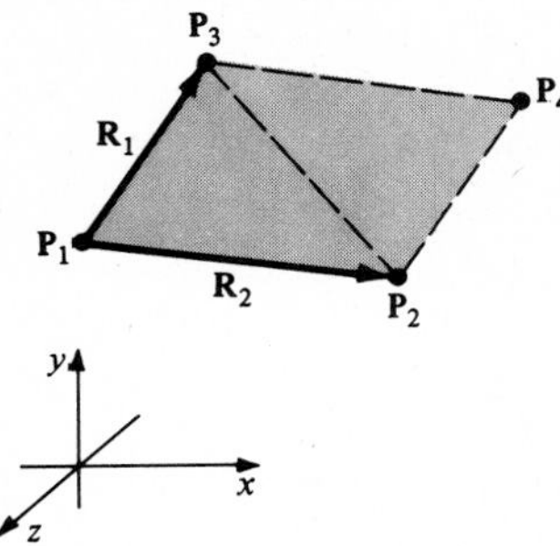

Figure 2-28
Plane facets.

```
170 ! PLACE VERTICES IN X( ),Y( )
180 FOR I=1 TO N
190 READ X(I),Y(I)
200 NEXT I
210 ! USE SUBROUTINE 'AREA' TO FIND HALF-AREA
220 CALL AREA(N,X,Y,HALF)
230 ! PRINT OUT FINAL ANSWER
240 PRINT"AREA =";2*HALF
250 DATA 0,-1.5,1.5,-2,1,-1,2,-1.5,1.5,0
260 DATA 1,0.5,1,1,0.5,0.5,0,1.5
270 END
```

Often plane facets are simply triangles or parallelograms. The area of these two types of facets can be found directly from the vector cross product. Let $\mathbf{R}_1$ and $\mathbf{R}_2$ be two adjacent sides of a parallelogram in Figure 2-28. The area is given by the magnitude of the cross product.

$$A_P = |\mathbf{R}_1 \times \mathbf{R}_2| \tag{2-18}$$

The triangular area created by a diagonal line from $\mathbf{P}_2$ to $\mathbf{P}_3$ is simply

$$A_T = \frac{1}{2}|\mathbf{R}_1 \times \mathbf{R}_2| \tag{2-19}$$

Notice that $\mathbf{R}_1 = \mathbf{P}_3 - \mathbf{P}_1$ and $\mathbf{R}_2 = \mathbf{P}_2 - \mathbf{P}_1$ when the vertices are defined as shown in Figure 2-28.

Any plane area can be divided into a series of triangular areas and Equation (2-19) applied to each triangular section. The accumulated results give the total area. This procedure will produce the same area as obtained by use of the subroutine AREA, based upon Equation (2-17). A main program such as FINDAREA which calls AREA can be used for any planar shape by simply modifying the DATA statements in lines 250–260, and changing the value of N in line 160 to equal the number of two-dimensional vertices defined in the DATA. Variables X and Y must be dimensioned to N + 1 in line 150.

2-11 POLYGON FILES

Commercial, three-dimensional packages are available that are based upon the ideas given in this chapter. Some provide the ability to rotate and project lines and surfaces by simply choosing an item from a menu. Thus, it becomes possible to train students to obtain solutions to descriptive geometry problems without requiring a knowledge of the vector and matrix methods upon which the solutions are based. These packages are more suitable for an industrial environment and should be used with care by beginning students in an academic environment.

Use of available software for generating computer graphics displays requires that objects be defined according to a chosen format. This definition can be done manually or automatically with properly designed software. Once this is done then various auxiliary, orthographic views can be generated and displayed very quickly. One example of a simple file format was discussed in Chapter 1. Another example, called a *polygon file,* is discussed below. This file format is used to produce both perspective and orthographic projections in both wire frame and shaded rendering, when used in conjunction with software written by Steven G. Satterfield for computer graphics equipment at the United States Naval Academy. Similar software exists for other facilities.

A polygon file contains both vertex and surface facet definitions. It begins with a string identifier followed by two numbers (v, f) which indicate the number of vertices used to define the object and the number of surface facets that make up the total surface. Then the three-dimensional components for (v) vertices are listed, row by row, in a chosen order. Finally, (f) rows of data are used to define each facet.

Each surface facet can contain three or more vertices. The first number in each row of a facet definition gives the number of vertices used to define that facet. Then the vertex numbers that make up the facet are listed. The numbers correspond to the order of the defined three-dimensional vertex components. For certain applications where surface normals are computed to support surface rendering schemes, it is necessary to limit the number of vertices for each facet and list the vertex numbers in a certain order, such as counterclockwise as seen from an external observer.

A polygon file for the block shown in Figure 2-29 is given below. The

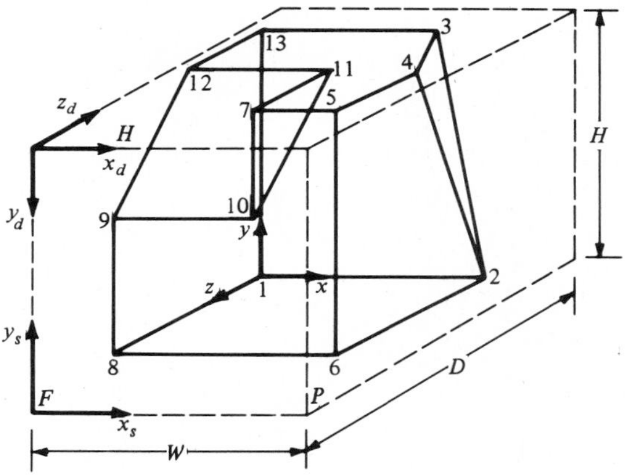

Figure 2-29
Orthographic projection to screen coordinates.

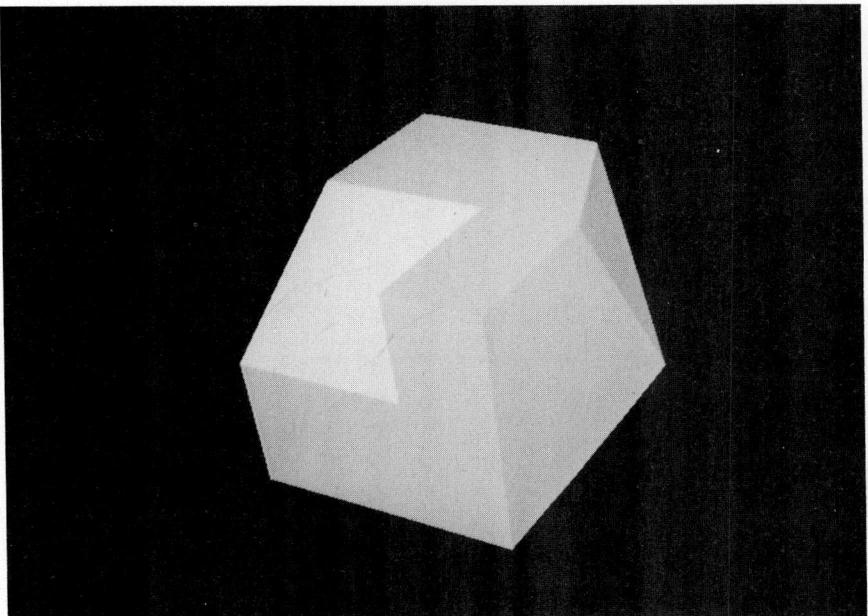

Figure 2-30
Image of CUTBLOCK. Produced on a PS/340 in the Anderson Lab at the Center for Computer Assisted Engineering at Colorado State University using software written by Kris English. Courtesy of CCAE, Colorado State University.

Table 2-1
Polygon File for CBLOCK

Cblock	
13,9	0,9,3
0,0,0	0,9,0
8,0,0	4,6,8,1,2
6,9,0	6,6,5,7,10,9,8
8,9,3	4,6,2,4,5
8,9,6	3,2,3,4
8,0,6	4,1,13,3,2
5,9,6	5,1,8,9,12,13
0,0,6	4,9,10,11,12
0,5,6	3,10,7,11
5,5,6	7,5,4,3,13,12,11,7
5,9,3	

block consists of 13 vertices and 9 facets. Notice that all facets are defined in a counterclockwise manner. Figure 1-41 shows the three principle views with hidden lines not indicated. Figure 2-30 is a shaded rendering of the same object. When software exists to produce this type of output, the user does not have to write programs such as VIEWS, PVIEWS, and ORTHOEYE.

PROBLEMS: CHAPTER 2

2-1 *Given:* Top and front views of a solid. Vectors which represent the vertices on the inclined face are **A** = [4 4 4], **B** = [7 6 3], **C** = [7 6 1], **D** = [4 4 0].

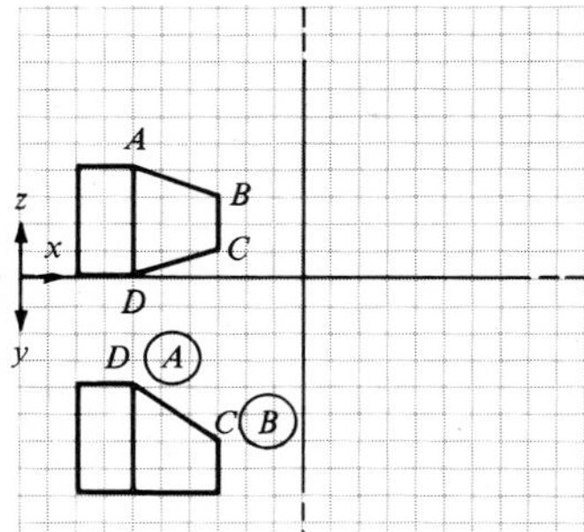

Figure 2-1P

Figure 2-2P

Task: Find true size and shape of plane *ABCD*.

2-2 *Given:* Top and front views of block shown. The vectors which represent vertices are **A** = [3 0 5], **B** = [8 0 10], **C** = [8 5 10], **D** = [3 5 5].
Task: Show the true size and shape, and the features of plane *ABCD*.

2-3 *Given:* Front and right side views of mutilated block shown. Vectors **A**, **B**, **C**, and **D** are given by **A** = [1 5 4], **B** = [2 2 6], **C** = [6 2 6], **D** = [7 5 4].
Task: Find the normal view of *ABCD*, and top view.

2-4 *Given:* Front and top views of a pyramid as shown. The defining vectors are **A** = [1 6 3], **B** = [7 6 4], **C** = [5 6 0], and **O** = [4 3 2].

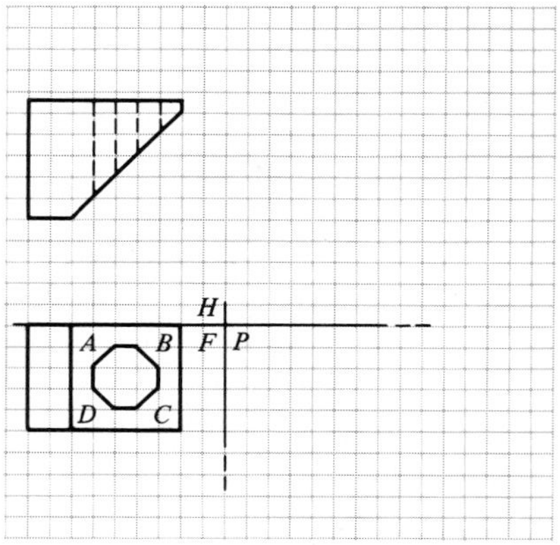

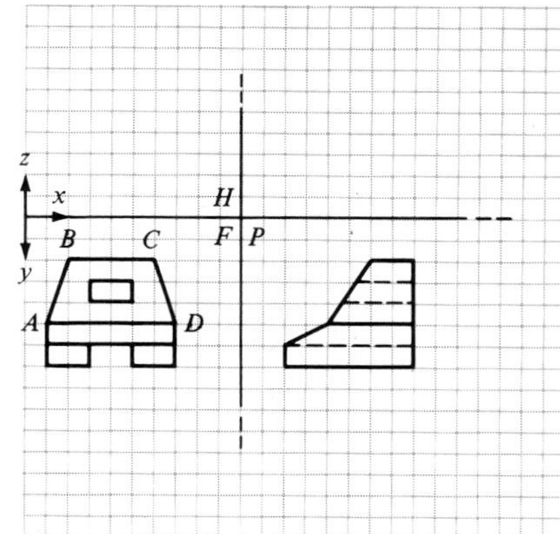

Figure 2-3P

Task: Label all given lines which appear in true length. Then add a complete auxiliary view of the pyramid in which line *OC* appears in true length. Calculate the true length of *OC*.

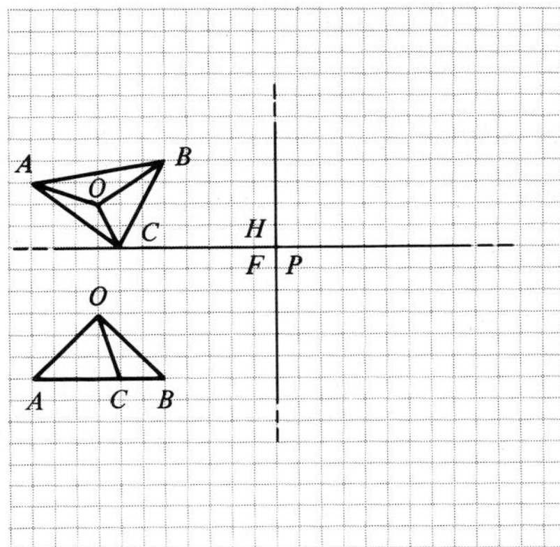

Figure 2-4P

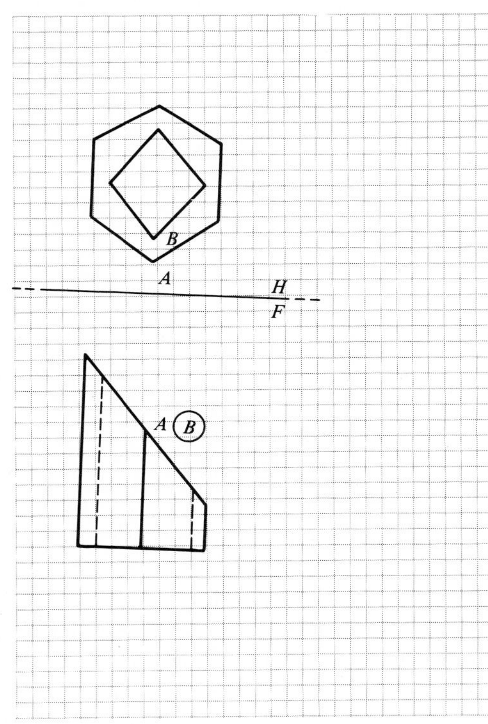

Figure 2-5P

2-5 *Given:* Front and top views of a truncated hollow hexagonal prism.

A = [9 9 2] **B** = [9 9 3.5]

Task: Draw a first auxiliary view showing the truncated surface in true size and shape. Show all features of the surface.

2-6 *Given:* Front and top views of a truncated hexagonal prism.

A = [8 14 5]

Task: Find the normal view of the inclined surface.

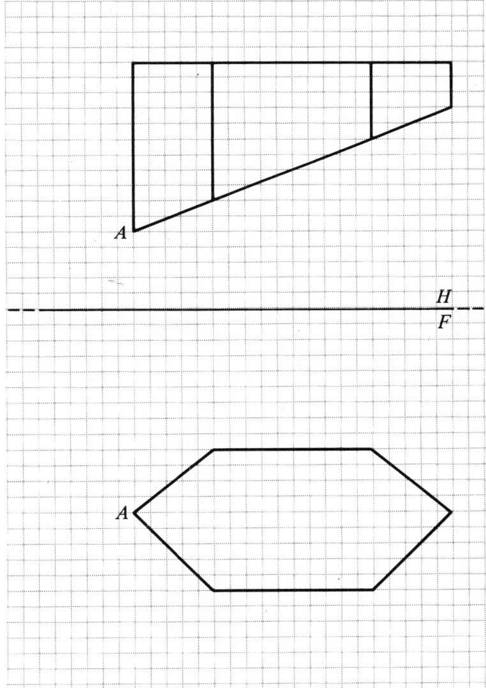

Figure 2-6P

2-7 *Given:* Front and top views of a truncated cylinder.

P = [10 11 8]

Task: Show the normal view of the truncated surface.

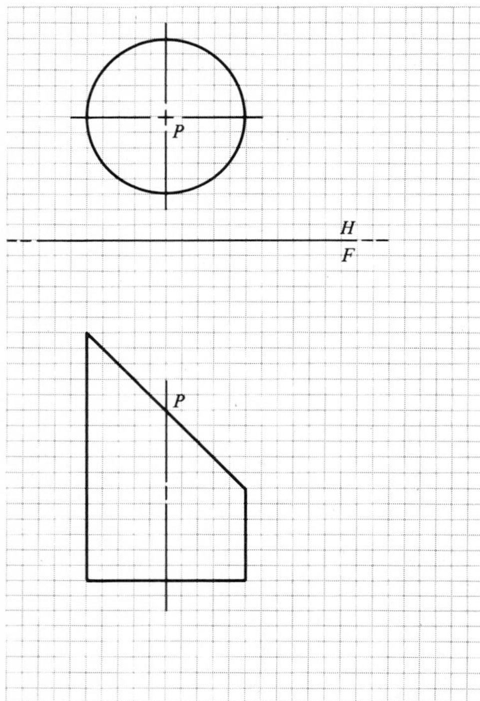

Figure 2-7P

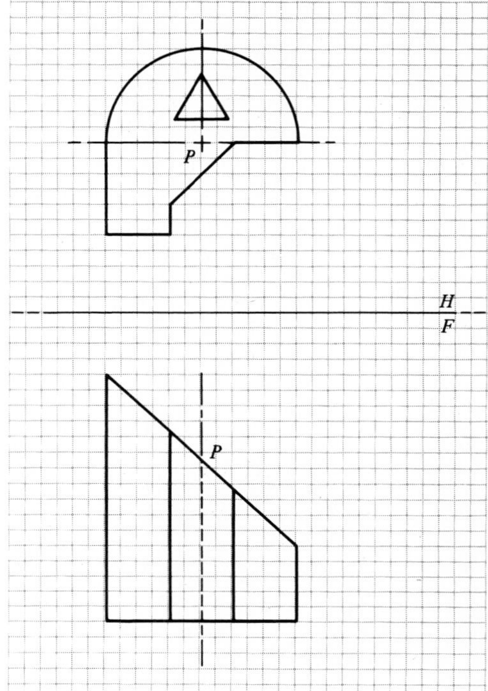

Figure 2-8P

2-8 *Given:* Front and top views.

$$\mathbf{P} = [12\ 9.5\ 11]$$

Task: Show the normal view of the truncated surface.

2-9 *Given:* Front and top views of a multilated block.

$$\mathbf{A} = [22\ 11\ 11]$$

Task: Find the normal views of the front and rear inclined surfaces. Use the given folding lines.

2-10 *Given:* Front and top views.
Task: Show the normal view of the upper inclined surface and all features on that surface.

2-11 *Given:* Two vectors $\mathbf{A} = [3\ -1\ 2]$ and $\mathbf{B} = [2\ -1\ 3]$.
Task: Perform the following vector calculations:

a. $\mathbf{R} = \mathbf{B} - \mathbf{A}$

b. $\mathbf{S} = \mathbf{B} + \mathbf{A}$

c. magnitudes $|\mathbf{R}|$ and $|\mathbf{S}|$

2-12 *Given:* Three vectors $\mathbf{A} = [-1\ 2\ 1]$, $\mathbf{B} = [3\ -2\ 2]$, and $\mathbf{C} = [3\ 1\ -2]$.
Task: Perform the following:

a. Calculate $\mathbf{R} = \mathbf{A} + \mathbf{B} + \mathbf{C}$

b. Calculate $\mathbf{S} = \mathbf{A} + \mathbf{B} - \mathbf{C}$

c. Plot **A**, **B**, **R**, **S** on a R.H.C. three-dimensional coordinate system.

2-13 *Given:* Two vectors $\mathbf{A} = [8\ 5\ 2]$ and $\mathbf{B} = [4\ 6\ -3]$.
Task: Perform the following:

a. Calculate $\mathbf{A} \times \mathbf{B}$

b. Calculate $\mathbf{B} \times \mathbf{A}$

c. Calculate $\mathbf{A} - \mathbf{B}$

d. Calculate $\mathbf{A} + \mathbf{B}$

2-14 *Given:* A line in space passes from $\mathbf{A} = [-3\ -2\ 1]$ to $\mathbf{B} = [3\ 2\ -1]$.

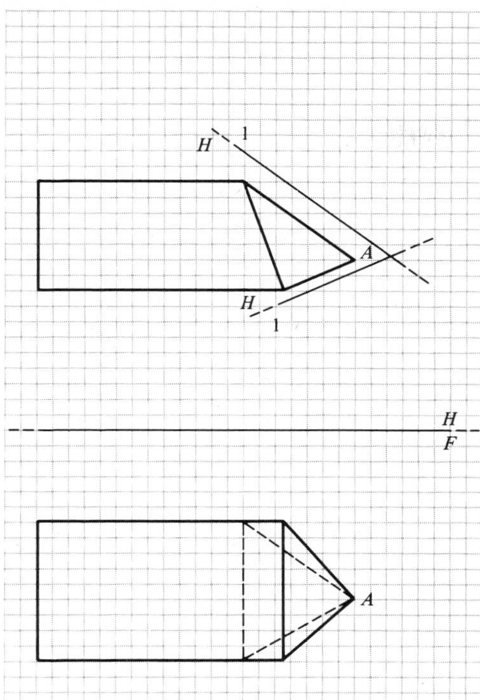

Figure 2-9P

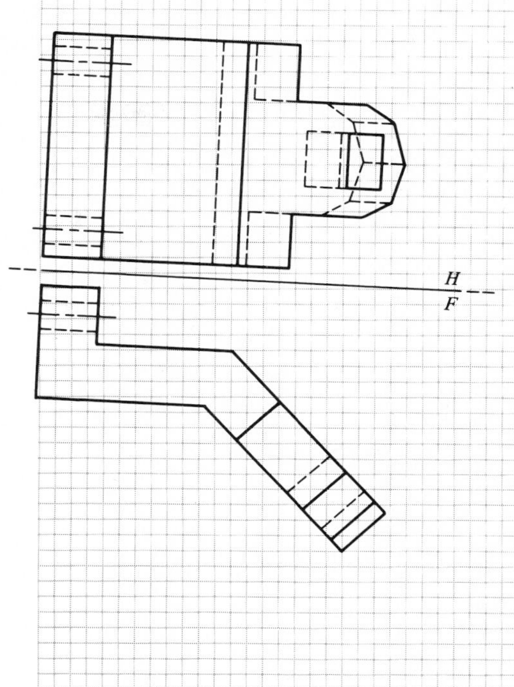

Figure 2-10P

Task: a. Calculate the components of the unit vector **n** along this line.

b. If **R** = **B** − **A**, express the vector **R** in the form **R** = |**R**|**n**.

2-15 *Given:* The following pairs of vectors:

a. $\mathbf{A}_1 = [10\ 0\ 5]$; $\mathbf{B}_1 = [12\ 2\ 5]$

b. $\mathbf{A}_2 = [5\ 0\ 2]$; $\mathbf{B}_2 = [5\ 5\ 5]$

c. $\mathbf{A}_3 = [-3\ 3\ 3]$; $\mathbf{B}_3 = [3\ 3\ -3]$

d. $\mathbf{A}_4 = [4\ 0\ 0]$; $\mathbf{B}_4 = [0\ 5\ 0]$

Task: Determine which pair of vectors has the largest magnitude of the vector cross product $\mathbf{A}_i \times \mathbf{B}_i$ for i = 1, 2, 3, 4.

2-16 *Given:* The eye position is located at **E** = [1 3 4] and the line of sight passes from **E** to the origin of a R.H.C. coordinate system as shown in Figure 2-19.
Task: Calculate the two angles ϕ and θ which can be used to align the line of sight with the positive z-axis as discussed in Section 2-9.

2-17 *Given:* The triangular plane facet as defined in Problem 1-4.
Task: a. Calculate the coordinates of a fourth point to create a rectangular plane facet.

b. Calculate the components of the unit normal vector to the plane.

2-18 *Given:* A plane triangle in the x-y plane with vertices **A** = [1 1], **B** = [5 1], **C** = [3 6].
Task: a. Calculate the area using the formula $A = ½bh$.

b. Calculate the area using Equation (2-16).

c. Calculate the area using Equation (2-19).

2-19[c] *Given:* Program ORTHOEYE and subroutines AXISROT, EYETOZ, and TITLE given in Appendix 1.
Task: Write a program that will produce a general orthographic view from any eye position for the object defined in Problem 1-18.[c]

[c]Superscript *c* indicates that a computer solution is required.

2-20^c *Given:* Same as Problem 2-19.^{c}
Task: Write a program and create point and line files which will generate orthographic views of objects similar to those defined in Problems 2-1 and 2-4.

2-21^c *Given:* Subroutine AREA in Appendix 1.
Task: Write a program such as FINDAREA (Section 2-10) which calls subroutine AREA and duplicate the results given in Exercise 2-9. (If subroutines are not supported on your computer system, combine these two algorithms into a single main program.)

2-22^c *Given:* A plane surface in the *x-y* plane is defined by ten vertices as given in the following column matrix:

$$\begin{bmatrix} 3 & 2 & 0 & 0 & -1 & -1 & -2 & 0 & 2 & 1 \\ 0 & 1 & 1 & 2 & 0 & -1 & -1 & -2 & -2 & -1 \end{bmatrix}$$

Task: Use the programs in Section 2-10 as a guide and develop software which calculates the area of the plane object. Check the answer by manually summing the areas of the triangles and rectangles that make up the object.

2-23^c *Given:* Same as 2-22^c except that a square hole of one unit on a side exists at the origin of the plane object.
Task: Same as 2-22^c. (Hint: to account for the hole, define the outer vertices in a counterclockwise manner and the inner vertices in a clockwise manner.

2-24 *Given:* Problem 2-1 and Figure 2-1P.
Task: Calculate the area of the plane surface ABCD. Choose a scale.

2-25^c *Given:* Problem 2-3 and Figure 2-3P.
Task: Calculate the area of the plane surface ABCD. Account for the rectangular hole. Choose a scale.

2-26 *Given:* Problem 2-6 and Figure 2-6P.
Task: Calculate the eye coordinate **E** for the line of sight EO which will show the true size of the inclined surface.

2-27 *Given:* The triangular plane defined in Problem 1–4.
Task: Calculate the area of the triangle.

2-28 *Given:* Definition of a polygon file in Section 2-11.
Task: Generate polygon files for the solids in Figure 2-1P, Figure 2-2P, Figure 2-4P, and Figure 2-6P.

2-29 *Given:* Definition of a polygon file in Section 2-11.
Task: Generate polygon files for the solids in Figures 1-16P, 1-17P, and 1-18P.

3

Primary Auxiliaries—Lines

The line is one of the basic constituents in the study of geometry. For this reason, some of the definitions and classifications from Chapter 1 are repeated here to emphasize their importance. The mathematical definition of a line is given as a vector, parametric equation to support geometric modeling techniques.

Although a line was defined as being a locus of two or more points with infinite length, it is generally used in the form of a line segment. A straight line segment is the shortest distance between two points and when shown graphically can be seen in three configurations:

1. As a point—when the line of sight is parallel to it.
2. Foreshortened—when it is shown to be less than true length because it is not parallel to the picture plane.
3. True length—when it is parallel to the picture plane.

A finite line segment is defined mathematically by limiting the range of the parameter in the defining equation. The true length of a line segment can be easily calculated when the three components of the two endpoints are known. Direction cosines also help define a line in space. This chapter concludes with some geometric modeling ideas applied to two-dimensional lines in a plane.

3-1 PRIMARY AUXILIARIES—LINES IN TRUE LENGTH

A plane is made up of an infinite number of straight lines. Therefore, all of these lines will be shown in true length when the normal view of a

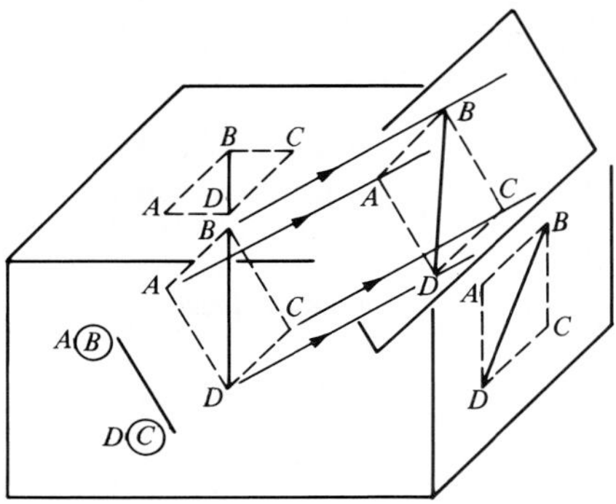

Figure 3-1
True length of line *BD.*

plane is obtained. Figure 3-1 shows a dotted inclined plane, *ABCD,* placed in a glass box so that an edge view of the plane is shown in the front view while a foreshortened view appears in the top and side views. A line has been drawn in the plane from corner *B* to corner *D*. This figure along with Example 3-1 will show the similarity in finding the normal view of a plane and finding the true length of a line.

Example 3-1

Given: Top and front views of line *BD* in the plane *ABCD*.

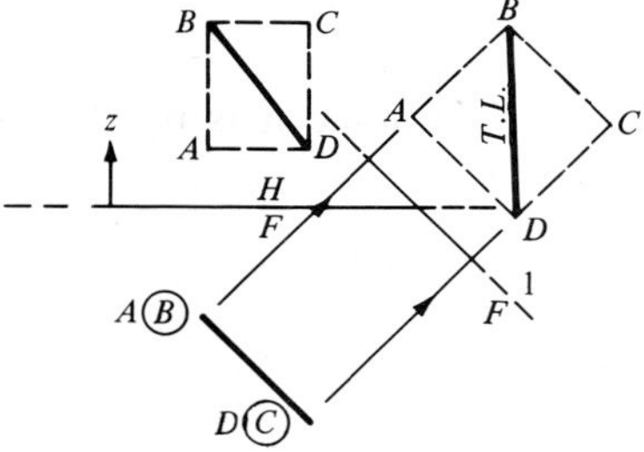

Example 3-1
True length of line *BD.*

Task: Find the normal view of plane *ABCD* and consequently show the true length of line *BD*.
Solution:

1. Draw a folding line *F*/1 parallel to the plane on edge. *This is also parallel to line BD which lies in the plane.* The observer is now looking normal to the line; hence, the lines of sight will be perpendicular to the *F*/1 folding line.

2. Obtain measurements of points in the top view (z-values) and lay off points in the auxiliary view.
3. The true size and shape of plane *ABCD* is then shown.
4. The true length of line *BD* is shown in the same auxiliary.

The projection of a straight line onto a plane which has been placed parallel to it will appear as a true length line. ◀

3-2 TRUE LENGTH PRINCIPAL LINES

Principal lines are defined as those which lie in, or parallel to, a principal plane. This category would include all horizontal, frontal, and profile lines. Figures 3-2(a), 3-2(b), and 3-2(c) show how the true lengths of principal lines may be found by inspection when the three views are given. It is well to inspect the problem that is presented in order to ascertain whether it qualifies as one of the "special cases" which allows a short-cut method to be used in its solution.

The line *MN* in Figures 1-22 and 3-2(a) is parallel to the horizontal folding line *H/F* and therefore parallel to the horizontal plane. When the observer looks through the plane he sees the true length of the line *MN* in the top view.

Figures 1-23 and 3-2(b) show frontal line *MN* parallel to the *H/F* folding line and therefore parallel to the frontal plane. This relationship of line and plane determines that true length of line *MN* is shown on the frontal plane.

The profile line *MN* is projected as a true length line onto the profile plane in Figure 3-2(c) since it is parallel to the profile plane. (Also see Figure 1-24.)

3-3 TRUE LENGTH LINE—FRONTAL AUXILIARY

Prior studies have established the fact that the true length of a line will be projected onto a first auxiliary plane which has been passed parallel to it. Example 3-2 shows the procedures involved in obtaining the true length of a line by projecting from the front view. Lines of projection are always perpendicular to the folding line.

Example 3-2

Given: Front and top views of oblique line *MN*.
Task: Find the true length of line *MN* by projecting from the front view.

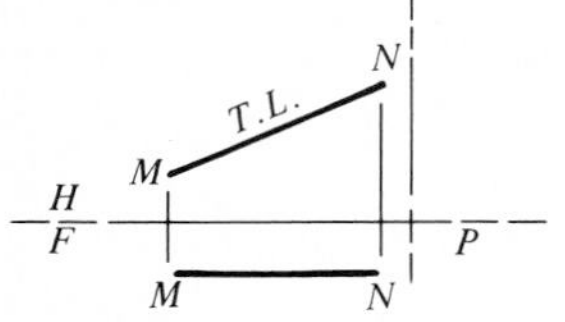

(a) Horizontal line

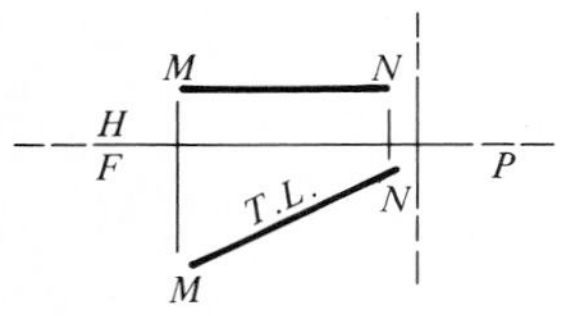

(b) Frontal line

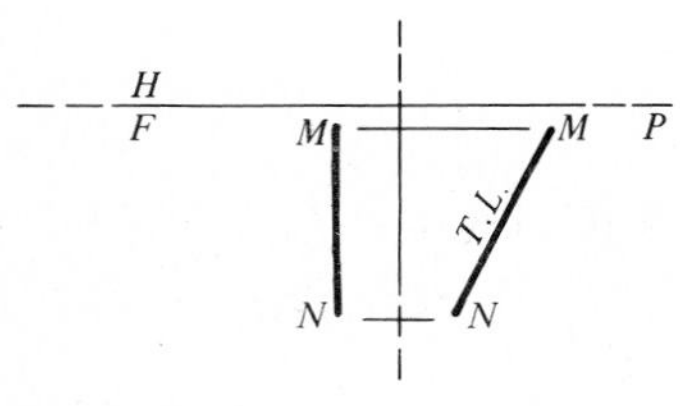

(c) Profile line

Figure 3-2
True length of line.

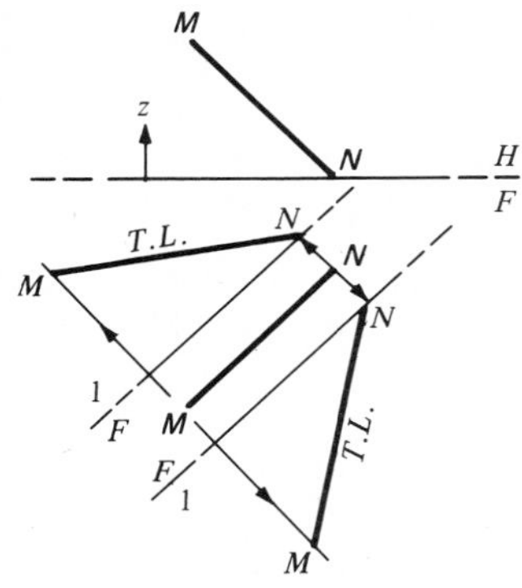

Example 3-2
True length line: frontal auxiliary solution.

Solution:

1. In the front view, draw folding line *F*/1 parallel to line *MN*. (The first auxiliary plane may be placed on either side of line *MN* as shown.)
2. Draw folding line *H*/*F* between the top and front views. Let *H*/*F* pass through point *N* in the top view of *MN* to eliminate one step in obtaining dimensions. Take the dimensions from the edge of the frontal plane (*H*/*F* folding line) in the *z*-direction.
3. Plot these dimensions on their respective lines of sight from the edge of the frontal plane (*F*/1 folding line) into the first auxiliary view to obtain the true length of *MN*. ◀

The drawing shows that the true length line will be obtained regardless of the viewing direction in the first auxiliary.

3-4 TRUE LENGTH LINE—HORIZONTAL AUXILIARY

The following example shows the steps to be followed in finding the true length of a line by using an elevation auxiliary projected out of the horizontal view.

Example 3-3

Given: Top and front views of oblique line *MN*.
Task: Find the true length of line *MN*.
Solution:

1. Draw folding line *H*/1 parallel to line *MN* in the top view.
2. Draw folding line *H*/*F* between the top and front views so that it passes through the front view of point *N*.

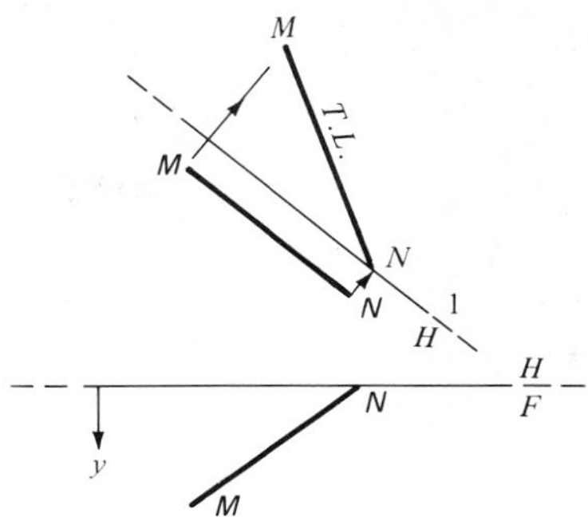

Example 3-3
True length line: elevation auxiliary.

3. Take dimensions from folding line *H/F* into the front view. ("*y*" direction).
4. Plot the dimensions of their respective lines of sight from folding line *H*/1 into the first auxiliary to obtain the true length of line *MN*. ◀

3-5 TRUE LENGTH LINE—PROFILE AUXILIARY

The preceding sections discussed situations in which a frontal auxiliary or an elevation auxiliary was used to acquire the true length of a line. The student can use the same reasoning to analyze the solution shown in Example 3-4 and set down the steps used in obtaining the true length of line *MN*.

Example 3-4

Given: Front and profile views of oblique line *MN*.

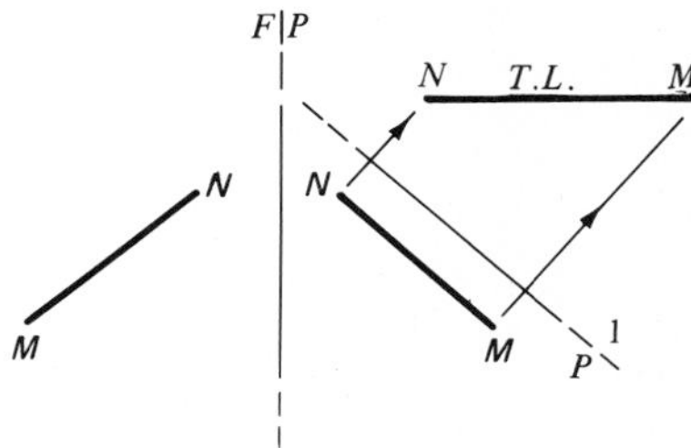

Example 3-4
True length line: profile auxiliary.

Task: Use a profile auxiliary to find the true length of line *MN*. ◀

3-6 POINT VIEW OF INCLINED LINE

When a line shows as a horizontal line in the front view and a frontal line in the top view it will be true length in both those views and will appear as a point in the profile view (Figure 3-3(a)).

A frontal-profile line will appear true length in the front and side views and as a point in the top view (Figure 3-3(b)).

A horizontal-profile line will appear true length in the horizontal and profile views and as a point in the frontal view (Figure 3-3(c)).

In all the preceding cases, a picture plane has been placed perpendicular to the true length projection of line *AB*. This is the same as placing it perpendicular to the line itself. As the observer looks through the picture plane the lines of sight are at right angles to the plane and are therefore

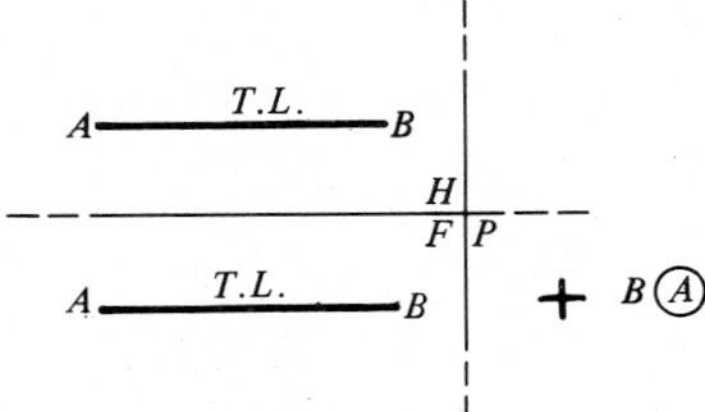

(a) Horizontal-frontal line

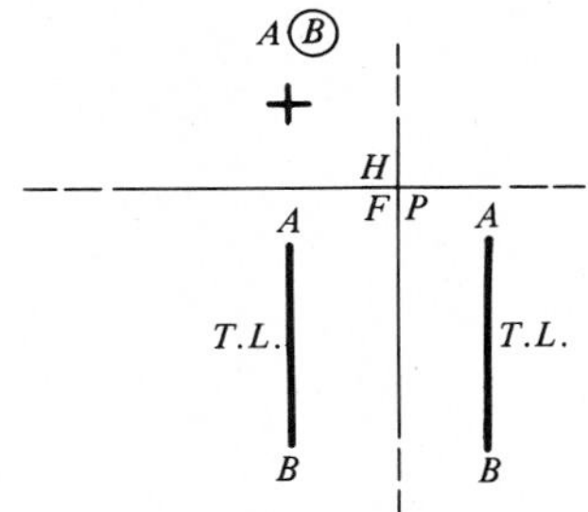

(b) Frontal-profile line

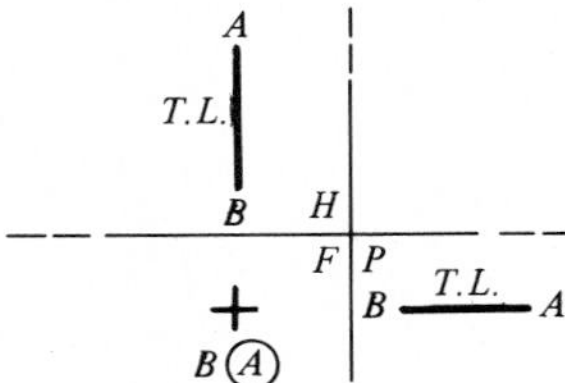

(c) Horizontal-profile line

Figure 3-3
Point view of line.

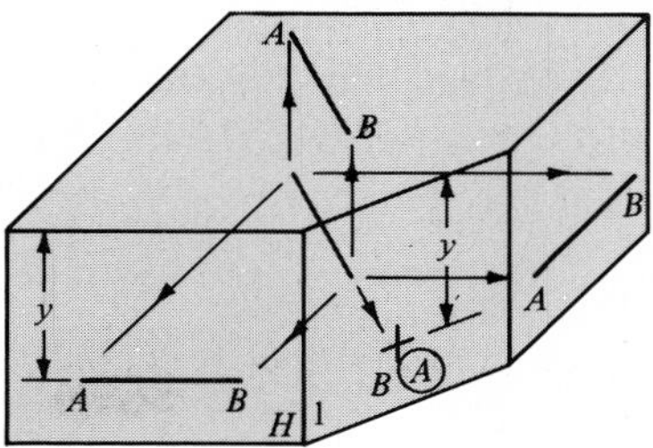

Figure 3-4
Point view: horizontal line.

parallel to the true length line. The point view of the line *AB* is then projected onto the plane as shown in the accompanying figures.

Figure 3-4 shows a glass box in which a horizontal line *AB* appears horizontal in the front view and true length in the top. The point view appears on the first auxiliary plane which has been placed perpendicular to the true length line.

Example 3-5 illustrates the steps in finding the point view of a line using a primary auxiliary projected out of the top view. A similar method would be used if the true length were shown in either the front or side view.

Example 3-5

Given: Front and top views of horizontal line *AB*.

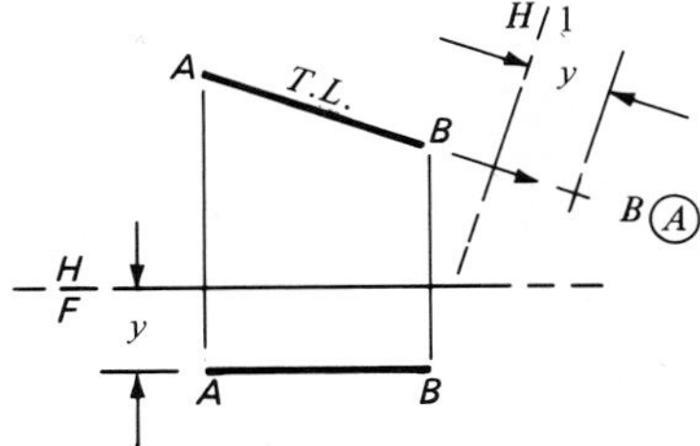

Example 3-5
Point view: inclined line.

Task: Find the point view of line *AB*.
Solution:

1. We can see, by inspection, that *AB* is true length in the top view.
2. Draw a folding line *H*/1 perpendicular to the true length line.
3. Project the lines of sight from points *A* and *B*, perpendicular to folding line *H*/1, into the first auxiliary view.

4. Measure distance "y" in the front view, and lay off the same distance in the first auxiliary below folding line $H/1$. ◀

3-7 POINTS ON A LINE

It has been stated that a line is made up of a series of points. This section, however, recognizes the general use of the line symbol and deals with the relationship of a *specific* point and a *given* line.

When a line is divided into parts or segments, the ratio of the segments will be the same in all views. In Figure 3-5(a) it is seen that "X" which is located at the midpoint of line MN in the top view will, when projected orthographically, be located at the midpoint of MN in the front and side views.

Figure 3-5(b) shows point "X" located one-third of the distance from M to N in the top view. Orthographic projection of the point will show that the line is divided in the same ratio in the other views.

It must be remembered, however, that *actual* measurements cannot be taken unless the true length of the line is shown.

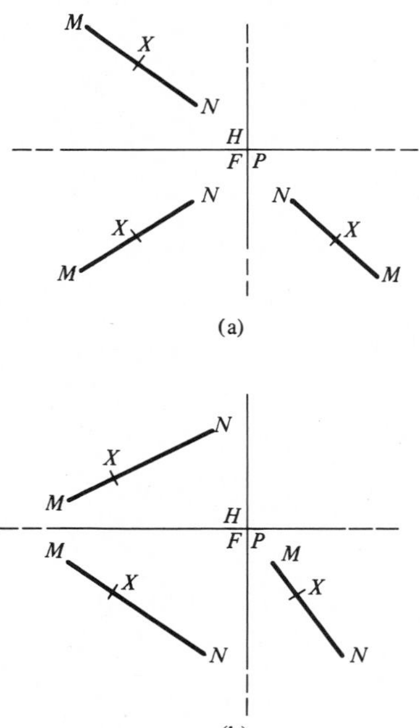

Figure 3-5
Line ratio.

A point that is on a line segment will project perpendicular to each folding line and will appear on the given line segment in all views.

Example 3-6

Given: Top and front views of line *MN* and top view of point *X*.

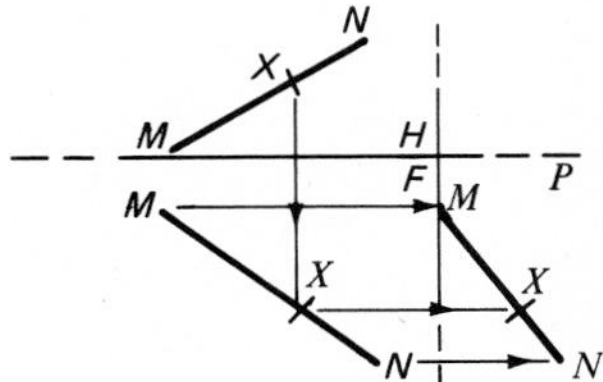

Example 3-6
Point on a line.

Task: Place point *X* on all views of line *MN*.
Solution:

1. Project point *X* until it falls on line *MN* in the front view.
2. Project line *MN* into the side view. Point *X* will fall on line *MN*. This view is proof that point *X* is on line *MN* and it is not necessary to fix point *X*. ◀

The general condition where a point can be shown on an oblique line in two views is shown in Example 3-6. A special case is shown in Example 3-7 where the projection of the line is perpendicular to the folding line. A third view is necessary here since direct projection cannot solve the problem.

Example 3-7

Given: Top and front views of line *MN* and point *X*.

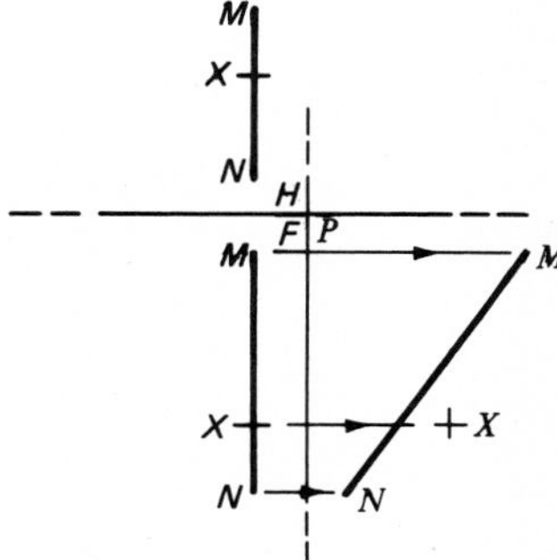

Example 3-7
Point and line.

Task: Determine whether point *X* is on line *MN*.
Solution:

1. Obtain the distances that points *M* and *N* lie back of the frontal plane in the top view. Plot these dimensions on their respective lines of sight into the side view and draw line *MN*.
2. Find the distance point *X* lies behind the frontal plane and transfer this distance into the side view on its line of sight. Point *X* will not fall on line *MN*. ◀

3-8 INTERSECTING AND NONINTERSECTING LINES

A. Intersecting Lines

Intersecting lines are those which are joined at a point that lies in, and is common to, both of the given lines. This point may be projected perpendicular to the folding line into any view. Example 3-8 shows the solution which proves that two lines intersect.

Example 3-8

Given: Top and front views of two lines *AB* and *MN*.

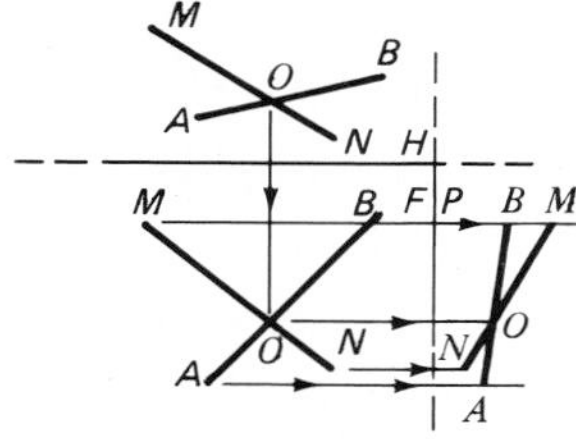

Example 3-8
Intersecting lines.

Task: Determine whether the given line segments intersect.
Solution:

1. By inspection, it can be seen that the apparent intersection at *O* in the top view can be projected as a point into the front view. For this reason we can say that point *O* is common to both line segments and is, therefore, an intersection.
2. Although a third view is not necessary, it has been added to prove the solution. ◀

B. Nonintersecting Lines

Conversely, we note in Example 3-9 that the given lines do not intersect.

Example 3-9

Given: Top and front views of line segments *AB* and *MN*.

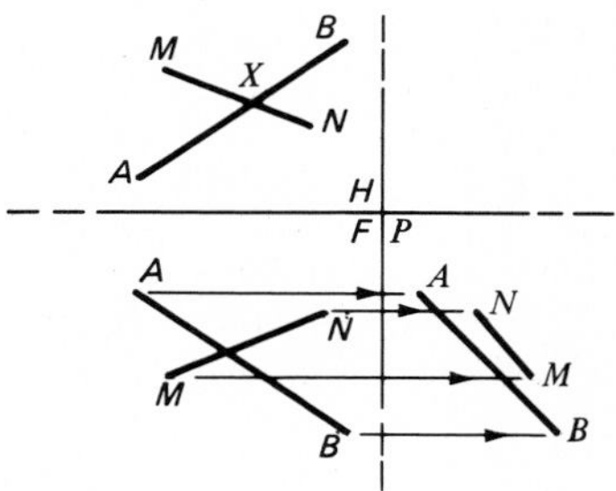

Example 3-9
Nonintersecting lines.

Task: Determine whether the given lines intersect.
Solution:

1. By inspection, we can see that the *apparent* point of crossing at *X* in the top view will not project as a point into the front view. Therefore, we know that *X* is not a point common to both lines, and the lines do not intersect.
2. As in Example 3-8, the side view is not necessary, but is included to verify the solution. ◀

C. Special Case—One Line Parallel to a Principal Plane

Consider the particular instance of intersecting or nonintersecting lines where one line is parallel to a principal plane, as shown in Example 3-10.

Example 3-10

Given: Top and front views of line *AB* and line *MN*.
Task: Determine whether the given line segments intersect.

1. Line *AB* is parallel to the profile plane in both given views. For this reason, the apparent crossing cannot be projected to an exact location without drawing another view.
2. In this example, a side view shows that the lines do not intersect. ◀

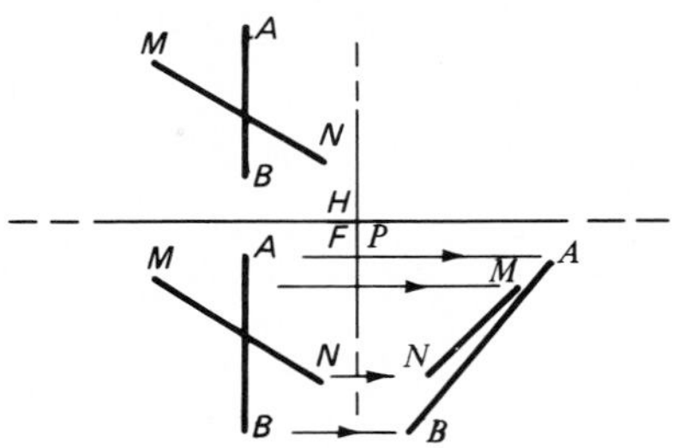

Example 3-10
Nonintersecting lines.

3-9 TRUE LENGTH OF LINE BY REVOLUTION

The true length of a line segment is the *actual* distance of that line measured between its designated end points. The true distance can be measured when the line segment lies in a plane which is perpendicular to the viewer's line of sight. It is for this reason that we can solve some problems with the techniques used in Figure 3-6(a) and (b).

An oblique line is not in any principal plane. However, if it can be placed in, or parallel to, a folding line (principal plane), we can see its true length without constructing an auxiliary view. We know that a line which revolves about an axis does not change in length because all points on that line revolve through the same number of degrees. Example 3-11 illustrates a solution procedure based upon this fact.

Example 3-11

Given: Front and top views of line *AB*.

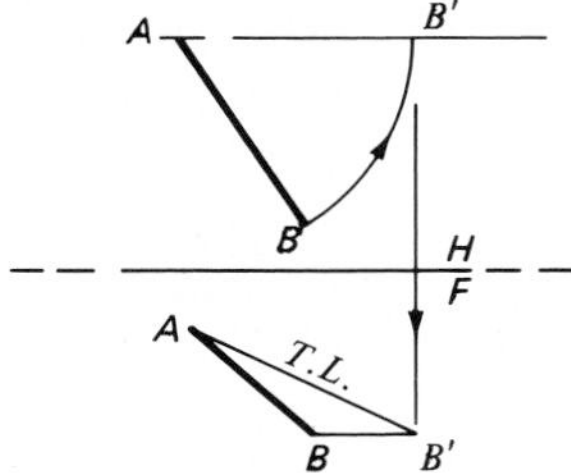

Example 3-11
True length of line by revolution.

Task: Determine the true length of line *AB*.
Solution:

1. Visualize line *AB* as an element of a cone, as shown in Figure 3-6(a). The axis of the cone passes through *A* and is perpendicular to the horizontal plane.

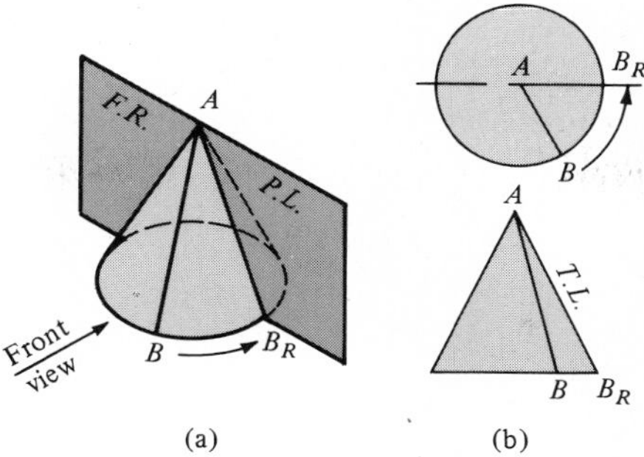

Figure 3-6
Revolution.

2. Using *A* as a center and line *AB* as a radius, swing point *B* until line *AB* lies in a frontal plane. Line *AB* will then be parallel to folding line *H/F*.
3. Project the relocated point *B′* into the front view.
4. Connect point *A* to the newly projected point *B′*.
5. The new segment *AB′* will be the true length of *AB*. ◀

The axis of revolution may be perpendicular to the frontal plane and the true length of *AB* found according to the following:

Example 3-12

Given: Top and front views of line *AB*.

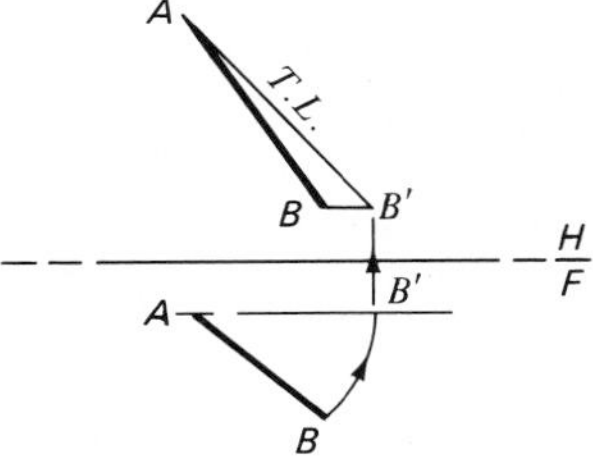

Example 3-12
True length of line by revolution.

Task: Find the true length of line *AB* using the method of revolution.
Solution:

1. Consider line *AB* as an element of a cone whose axis passes through *A* and is perpendicular to the frontal plane.
2. Working in the front view, use *A* as a center and line *AB* as a radius. Swing point *B* until line *AB* lies parallel to folding line *H/F*. Line *AB* will then lie in the horizontal plane.

3. Project the relocated point B' into the top view until it meets a frontal line projected from point B.
4. Connect point A to B' in the top view.
5. The new line segment AB' is the true length of line AB. ◀

The following example obtains the true length of line AB by using the revolution procedure where the axis of the imaginary cone is perpendicular to the profile plane. It is recommended that the student study Example 3-13 and list the step-by-step procedure for the solution required.

Example 3-13

Given: Front and side views of line AB.

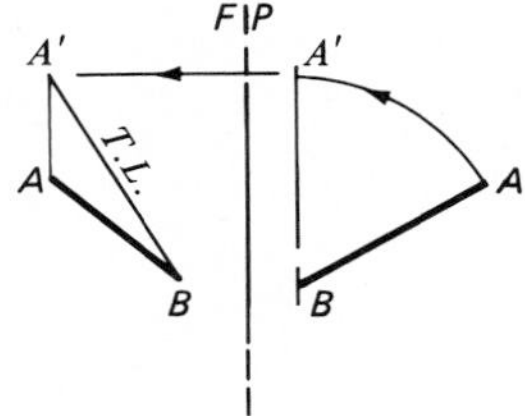

Example 3-13
True length of line by revolution.

Task: Using an imaginary cone with its axis perpendicular to the profile plane, find the true length of line AB by revolution. ◀

3-10 BEARING, SLOPE, AND GRADE OF A LINE

The engineer will often find that the position of a line in space will be defined by its bearing, slope, and grade. The following definitions apply:

A. Bearing of a Line

A line's *bearing* gives its direction relative to the earth's surface. The bearing angle is measured relative to the projection of that line on a *horizontal* plane. It is measured from true north or true south. In general, and if not noted otherwise, true north will be assumed to be at the top of the drawing as shown in Figure 3-7(a). The bearing of a line is *always* noted in the top view or horizontal view, as in map reading, and is read according to the direction of the line. Line AB (A toward B—Figure 3-7(a)) has a bearing of N 40 deg E. The same line could be read BA (B toward A), but would then have a bearing of S 40 deg W.

Directions are measured from a north-south line called a *meridian,* and by convention, are taken from north to either east or west, or from south

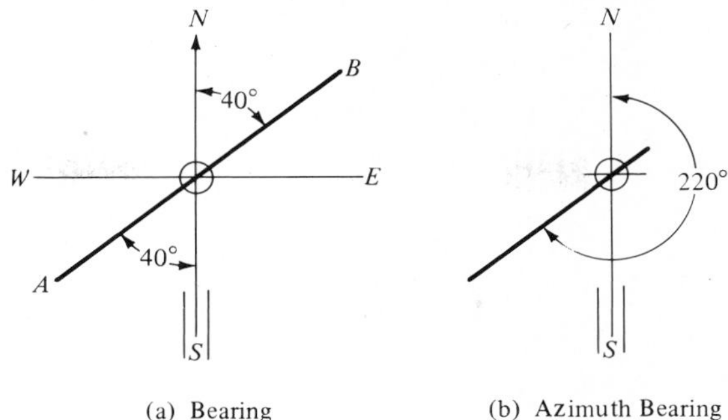

Figure 3-7
Direction of a line in space.

to either east or west. All directions should be specified with the bearing angles given where they show less than 90 deg.

Some forms of engineering, navigation, and military activities use a form of designation known as *azimuth bearing.* In this case, the required angle is the clockwise departure of the line from true north. In navigation, the direction of the line shown in Figure 3-7(b) would be given as an azimuth bearing of N 220 deg or 220 deg True.

B. Slope of a Line

The *slope* of a line is, by definition, the angle between a line and a horizontal plane. In order to measure the slope the line must be shown in true length, and the horizontal plane must be shown on edge, as indicated in Figure 3-8.

Two methods of finding the slope are now available to the student. Example 3-14 shows the first method, where the requirements of the line and plane are found by revolution as noted in Example 3-11.

Example 3-14

Given: Top and front views of a line *AB*.

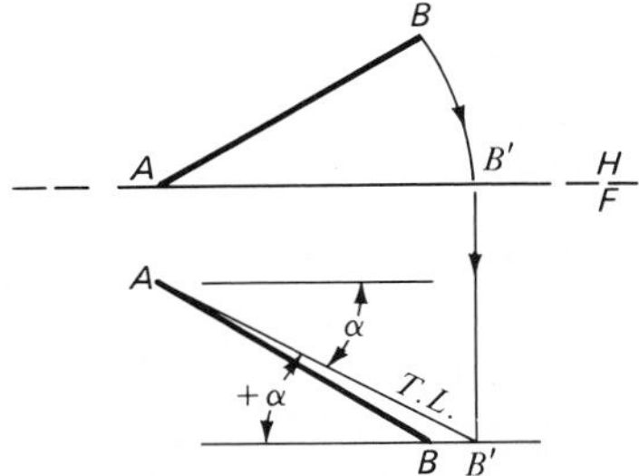

Example 3-14
Slope of a line by revolution.

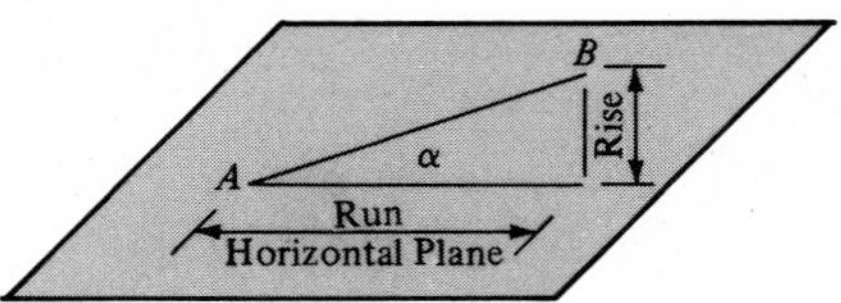

Figure 3-8
Slope (grade) of line.

Task: Find the slope of line *AB*. Use the revolution method to find the true length of the line.
Solution:

1. With "*A*" as a center, revolve line *AB* so it lies in a frontal plane. Point *B* will move to the location of point *B′*.
2. Project *B′* into the front view. The horizontal plane is on edge in the front view.
3. Draw the true length line from *A* to *B′* in the front view.
4. Measure the angle of declension (negative slope) between line *AB′* and the horizontal plane.
5. The angle between line *B′A* and the horizontal plane would show the positive slope. ◀

The second procedure, solution by the auxiliary method, is shown in Example 3-15.

Example 3-15

Given: Top and front views of line *AB*.

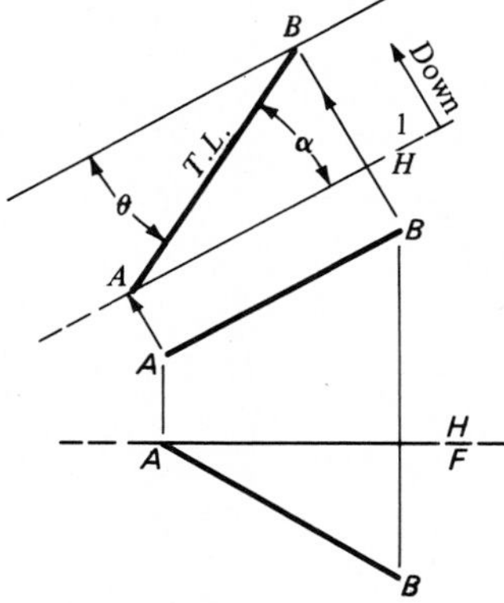

Example 3-15
Slope of a line by auxiliary method.

Task: Find the slope of line *AB* using the auxiliary method.

Solution:

1. In order to find the edge view of the horizontal plane and the true length of *AB* in the same auxiliary, the folding line *H*/1 must be placed parallel to *AB* in the top view.
2. Project line *AB* into the first auxiliary view and measure the angle.
3. We find that line *AB* has a negative slope (α), and line *BA* has a positive slope (θ). ◀

C. Grade of a Line

In engineering, or in any occupation where maps are used, grade lines are established for highways, railroads, sewers, and so on. The grade, as well as the slope, is the deviation of the line from the horizontal. It is, however, given in percentage and is designated as plus (+) when the slope is upward and minus (−) when the slope is downward. As with the slope, the grade of a line is *generally* found in a view where the line is in true length and the horizontal plane is on edge. The grade of a line is the ratio of the vertical rise to the horizontal run. It can be expressed as a percentage as follows (see Figure 3–8):

$$\text{percent grade} = \frac{\text{vertical rise}}{\text{horizontal run}} \times 100$$

It is convenient to use an engineer's scale when graphically calculating grade. The horizontal run is set off parallel to the horizontal plane on edge, while the rise is set off perpendicular to the horizontal plane (see Figure 3-9). By using a scale, it becomes a simple problem to find the number of units of vertical rise or fall per 100 units of horizontal distance. The slope of a line can be found in any auxiliary that shows the line in true length and the horizontal plane on edge.

Example 3-16

Given: Front and top views of line *AB*.
Task: Find the percent grade of line *AB*.
Solution:

1. Use an *H*/1 folding line parallel to line *AB* in the top view and project the line into an elevation auxiliary. This first auxiliary will show the horizontal plane on edge and the line *AB* in true length.
2. Lay off 100 units from "*B*" parallel to the horizontal plane.
3. At this point, lay off a vertical distance to line *AB* and measure the distance.

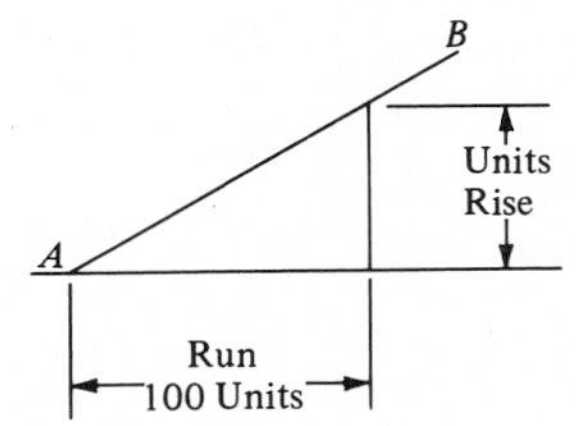

Figure 3-9
Grade of a line.

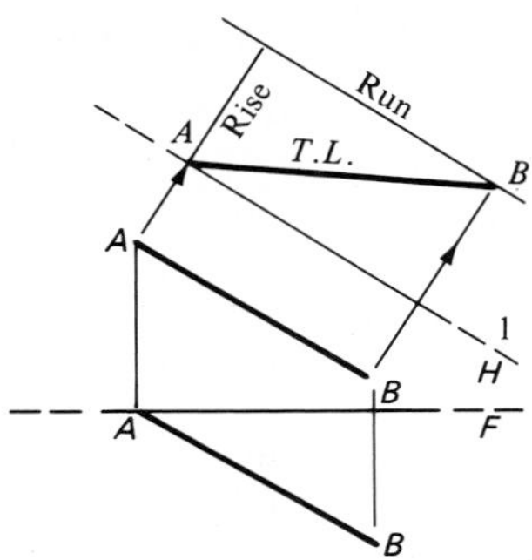

Example 3-16
Percent grade.

4. Calculate percent grade of line *AB*.
5. Since line *AB* is downhill, the grade is minus. ◀

Unless otherwise required, it is not necessary to find the true length of a line in order to calculate percent grade. The necessary information can be obtained from the front and top views only. Again referring to the formula for grade, we see that the information needed is the rise and the run of the line in question.

Example 3-17

Given: Top and front views of line *AB*.

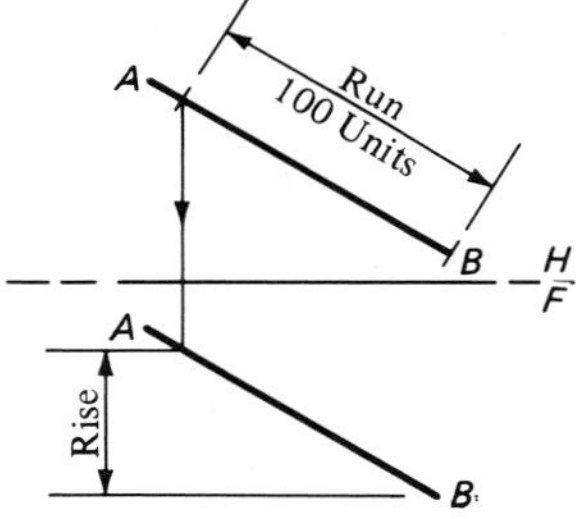

Example 3-17
Percent grade.

Task: Determine the percent grade of oblique line *AB*.
Solution:

1. Calculate the horizontal run of line *AB* in the top view (measure the length of 100 units on the projection of *AB* on the horizontal plane).
2. Calculate the vertical rise of "*A*" in the front view.
3. Calculate the grade of line *AB* using the formula.
4. Since *AB* slopes downward, the grade is minus. ◀

In the previous descriptive geometry solutions, three-dimensional lines were first represented on two-dimensional planes and then three-dimensional solutions obtained by using principal or auxiliary views. Geometric modeling solutions, on the other hand, are normally obtained in three-dimensional space which are then projected onto a two-dimensional plane for viewing and analysis. This approach requires that lines in space be represented mathematically. The computational approach is feasible because of the storage, accuracy, and speed of digital computers used for computer-aided layout and design.

3-11 VECTOR EQUATION OF A LINE

An object that is stored in computer memory may consist of hundreds of lines, defined between pairs of points in space. In order to determine geometric attributes of these lines relative to other points, lines, and planes by computations within a computer, it is best to define these lines mathematically. In this section, we introduce the parametric, vector equation for a three-dimensional line in space.

Let point **Q** be any point on the vector **R** = **B** − **A** between points **A** and **B** as shown in Figure 3-10. An additional vector can be defined as **L** = **Q** − **A**. A scalar parameter t is then defined as the ratio of the magnitudes of these two vectors. **n** is the unit vector along **R** and **L**.

$$t = \frac{|\mathbf{L}|}{|\mathbf{R}|} = \frac{|\mathbf{L}|\mathbf{n}}{|\mathbf{R}|\mathbf{n}} = \frac{\mathbf{L}}{\mathbf{R}} \tag{3-1}$$

Using the above definitions of **R** and **L** gives

$$\frac{\mathbf{L}}{\mathbf{R}} = \frac{(\mathbf{Q} - \mathbf{A})}{(\mathbf{B} - \mathbf{A})} = \frac{\mathbf{Q}}{(\mathbf{B} - \mathbf{A})} - \frac{\mathbf{A}}{(\mathbf{B} - \mathbf{A})} = t \tag{3-2}$$

Solving for vector **Q** gives

$$\mathbf{Q} = \mathbf{A} + t\,(\mathbf{B} - \mathbf{A}) = \mathbf{A} + t\,\mathbf{R} \tag{3-3}$$

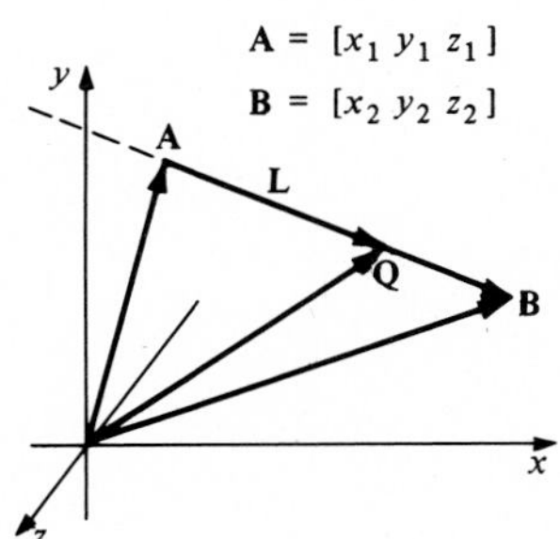

Figure 3-10
Vector equation of a line.

Equation (3-3) is called the *base vector form* of the vector, parametric equation for a three-dimensional line in space. The absolute vector **A** is called the **base vector.** The relative vector **R** = (**B** − **A**) is called the *directional vector.* As the scalar parameter t varies between $0 \leq t \leq 1.0$, the point **Q** falls on the vector **R** between **A** and **B**. However, Equation (3-3) actually defines an infinite line in space when $-\infty < t < \infty$, with **A** and **B** being two points (position vectors) on the line. When **A** and **B** are treated as endpoints of a finite line, then parameter values between $0 \leq t \leq 1.0$ describe the entire line segment. Further, when t =. 0.5, **Q** is halfway between **A** and **B**. When $t = 0.75$, $|\mathbf{L}| = 0.75\ |\mathbf{R}|$, and so on.

Any vector can be written in terms of its three scalar components. If a point **Q** on line **R** has coordinates $[x\ y\ z]$ then the components of the vector equation for the line become

$$x = x_1 + t(x_2 - x_1) = x_1(1 - t) + tx_2$$
$$y = y_1 + t(y_2 - y_1) = y_1(1 - t) + ty_2 \quad (3\text{-}4)$$
$$z = z_1 + t(z_2 - z_1) = z_1(1 - t) + tz_2$$

where

$$\mathbf{A} = [x_1\ y_1\ z_1] \quad \text{and} \quad \mathbf{B} = [x_2\ y_2\ z_2]$$

The BASIC subroutine BLINE (Appendix 1) can be used to generate the components of the base vector **B** and the directional vector **D** = **R**. First the two endpoints **A** and **B** are stored in arrays X(1), Y(1), Z(1) and X(2), Y(2), Z(2) in the main program. Then a call to BLINE will return base vector components B(1), B(2), B(3) and directional vector components D(1), D(2), D(3). This subroutine will be used in later programs.

3-12 TRUE LENGTH OF A THREE-DIMENSIONAL LINE

Graphical methods for obtaining the true length of a line were given earlier in this chapter. If the endpoint vectors of a three-dimensional line segment are known, then the directional components can be used to calculate the true length of the line segment. The true length is simply the magnitude of the directional vector between the endpoint vectors. If the directional vector is **R** = **B** − **A**, where $\mathbf{A} = [x_1\ y_1\ z_1]$ and $\mathbf{B} = [x_2\ y_2\ z_2]$, then the directional components of **R** are:

$$R_1 = x_2 - x_1; \quad R_2 = y_2 - y_1; \quad R_3 = z_2 - z_1 \quad (3\text{-}5)$$

Using the notation in Figure 3-11, one can write

$$d^2 = R_1^2 + R_3^2$$
$$|R|^2 = d^2 + R_2^2 = R_1^2 + R_3^2 + R_2^2$$

and then the true length of the line between **A** and **B** is

$$|R| = [R_1^2 + R_3^2 + R_2^2]^{1/2} \quad (3\text{-}6)$$

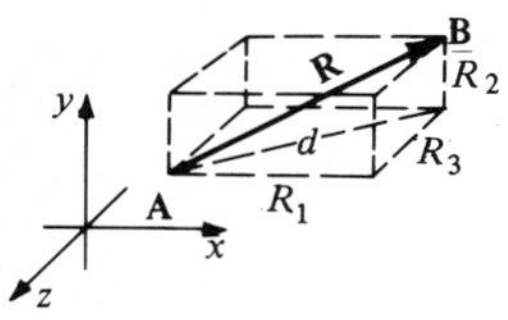

Figure 3-11
Directional components.

This computation was performed earlier in Chapters 1 and 2.

If the parameter value is known at each end of a line segment which lies on the line through vectors **A** and **B**, t_i and t_f, then

$$\mathbf{Q}_i = [x_i\ y_i\ z_i] = \mathbf{A} + t_i\mathbf{R}$$
$$\mathbf{Q}_f = [x_f\ y_f\ z_f] = \mathbf{A} + t_f\mathbf{R} \quad (3\text{-}7)$$

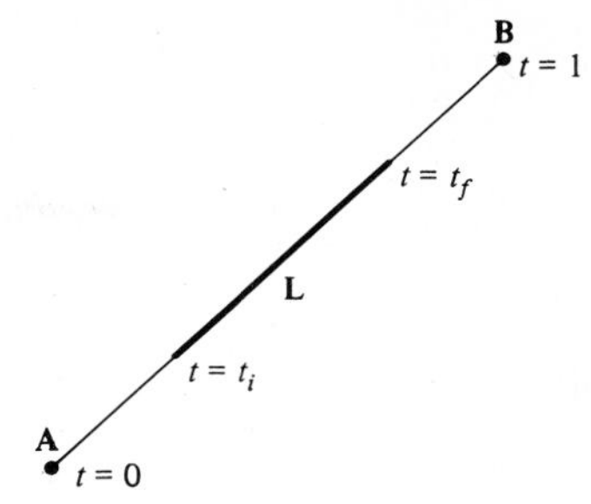

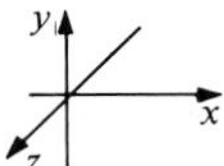

Figure 3-12
Line segment.

The directional components of the line segment are then

$$L_1 = (x_f - x_i); \quad L_2 = (y_f - y_i); \quad L_3 = (z_f - z_i) \tag{3-8}$$

where

$$x_i = x_1 + t_iR_1; \quad y_i = y_1 + t_iR_2; \quad z_i = z_1 + t_iR_3 \tag{3-9a}$$

and

$$x_f = x_1 + t_fR_1; \quad y_f = y_1 + t_fR_2; \quad z_f = z_1 + t_fR_3 \tag{3-9b}$$

with $\mathbf{R} = \mathbf{B} - \mathbf{A}$.

The true length of the line segment is

$$|\mathbf{L}| = [L_1^2 + L_3^2 + L_2^2]^{1/2} \tag{3-10}$$

Figure 3-12 gives an example of a line segment **L**.

Example 3-18

Given: A line passes through the points $\mathbf{A} = [1\ 1\ 1]$ and $\mathbf{B} = [3\ 4\ 2]$.
Task: Determine the parametric, vector equation for the line segment between **A** and **B**, and the coordinates of the midpoint **M**. Then find the length of the line between $-1 \leq t \leq 2$.
Solution:

$$\mathbf{Q} = \mathbf{A} + t(\mathbf{B} - \mathbf{A})$$

In matrix form this equation is

$$[x\ y\ z] = [1\ 1\ 1] + t[2\ 3\ 1]$$

For $t = \frac{1}{2}$

$$x = 1 + \tfrac{1}{2}(2) = 2$$
$$y = 1 + \tfrac{1}{2}(3) = 2.5$$
$$z = 1 + \tfrac{1}{2}(1) = 1.5$$

Thus:

$$\mathbf{M} = [2 \quad 2.5 \quad 1.5]$$

At $t = -1$

$$x_i = 1 - 2 = -1; \quad y_i = 1 - 3 = -2; \quad z_i = 1 - 1 = 0$$

At $t = 2$

$$x_f = 1 + 4 = 5; \quad y_f = 1 + 6 = 7; \quad z_f = 1 + 2 = 3$$

The true length of the line between $-1 \leq t \leq 2$ is then

$$|\mathbf{L}| = \{(5 + 1)^2 + (7 + 2)^2 + (3 - 0)^2\}^{1/2} = 11.225$$ ◀

Example 3-19

Given: Two views of an oblique line *AB* as shown in Figure 3-13. The spatial coordinates are **A** = [2 6 4] and **B** = [9 2 1].
Task: Calculate the coordinates of a point on the line 3.0 units from end **B**.
Solution: (a) A *graphical solution* is shown in Figure 3-13. First the true length of line *AB* is obtained in the first auxiliary. Then distance B_1P_1 is measured at 3.0 units from end B_1 in this auxiliary view. Point P_1 is then

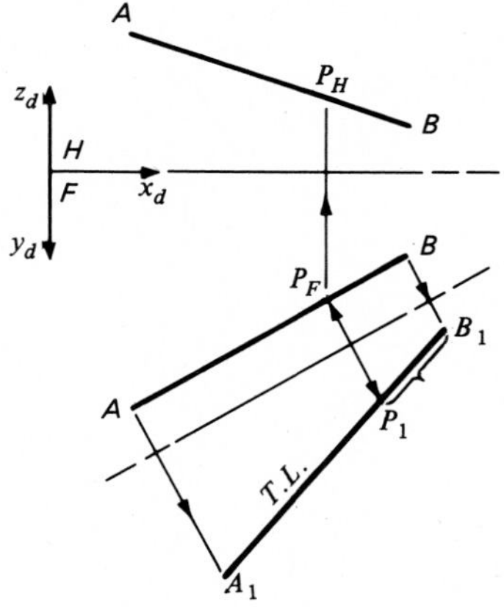

Figure 3-13
Two given views and auxiliary.

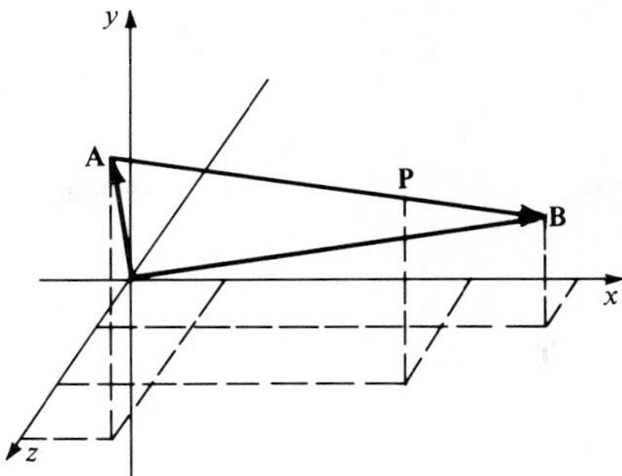

Figure 3-14
Point on a line.

projected back into the frontal and horizontal planes. The coordinates of P_F and P_H can be measured as $x_F = 6.6$, $y_F = 3.4$; $x_H = 6.6$, $z_H = 2.1$. (b) Using a *computational approach,* let $\mathbf{R} = \mathbf{B} - \mathbf{A}$. Then:

$$|\mathbf{R}| = \{(9-2)^2 + (2-6)^2 + (1-4)^2\}^{1/2} = 8.6$$

Calculate the parameter value $t = (8.6 - 3)/8.6 = 0.651$. Then the components of point **P** on line **B** − **A** are given by:

$$x = x_1 + t(x_2 - x_1) = 2 + 0.651(9 - 2) = 6.56$$

$$y = y_1 + t(y_2 - y_1) = 6 + 0.651(2 - 6) = 3.40$$

$$z = z_1 + t(z_2 - z_1) = 4 + 0.651(1 - 4) = 2.05$$

Figure 3-14 shows the point **P** in a three-dimensional right-hand coordinate system. ◀

3-13 DIRECTION COSINES

A unit vector was discussed in Section 1-15. The components of a unit vector lead to the definition of direction cosines. Let a vector **R** have directional components given by

$$\mathbf{R} = [(x_2 - x_1) \quad (y_2 - y_1) \quad (z_2 - z_1)]$$

The magnitude of **R** is given by

$$|\mathbf{R}| = \{(x_2 - x_1)^2 + (y_2 - y_1)^2 + (z_2 - z_1)^2\}^{1/2} \tag{3-11}$$

The unit vector **n** along the direction of **R** is then

$$\mathbf{n} = \frac{\mathbf{R}}{|\mathbf{R}|} = [n_1 \; n_2 \; n_3]$$

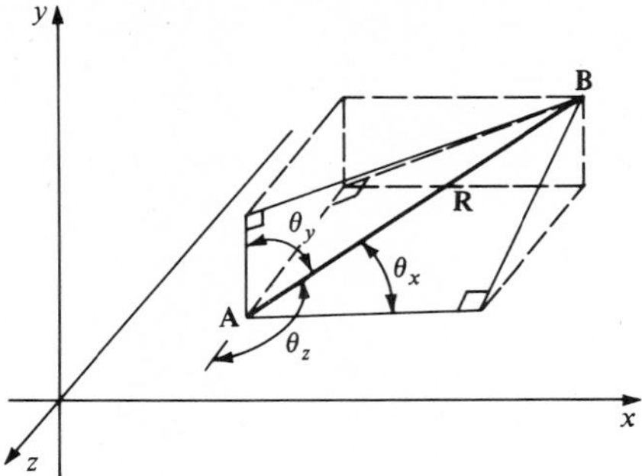

Figure 3-15
Direction cosines.

By comparing the above three equations it can be seen that the components of the unit vector **n** are

$$n_1 = \frac{(x_2 - x_1)}{|\mathbf{R}|}; \quad n_2 = \frac{(y_2 - y_1)}{|\mathbf{R}|}; \quad n_3 = \frac{(z_2 - z_1)}{|\mathbf{R}|} \tag{3-12}$$

These three scalar components of the unit vector are called the *direction-cosines* of the vector **R**. Thus:

$$\begin{aligned} n_1 &= \cos\theta_x = \frac{(x_2 - x_1)}{|\mathbf{R}|} \\ n_2 &= \cos\theta_y = \frac{(y_2 - y_1)}{|\mathbf{R}|} \\ n_3 &= \cos\theta_z = \frac{(z_2 - z_1)}{|\mathbf{R}|} \end{aligned} \tag{3-13}$$

The angles θ_x, θ_y, θ_z are shown in Figure 3-15. These angles are called the *direction angles.* They define the inclination of **R** with respect to the coordinate axes as shown in Figure 3-15. Notice that

$$\begin{aligned} 1.0 &= \cos^2\theta_x + \cos^2\theta_y + \cos^2\theta_z \\ &= \frac{(x_2 - x_1)^2}{|\mathbf{R}|^2} + \frac{(y_2 - y_1)^2}{|\mathbf{R}|^2} + \frac{(z_2 - z_1)^2}{|\mathbf{R}|^2} \end{aligned} \tag{3-14}$$

Thus, if two direction cosines are given, the third must satisfy the above equation. If a direction cosine is negative, then the supplement of the angle is used. Recall that $\cos(\pi - \theta) = -\cos\theta$. For example, if $\cos\theta = -0.66$, then the angle $\theta = 131.3$ deg.

Example 3-20

Given: Two planes intersect and form line **B** − **A**. The endpoints of this line of intersection are given by

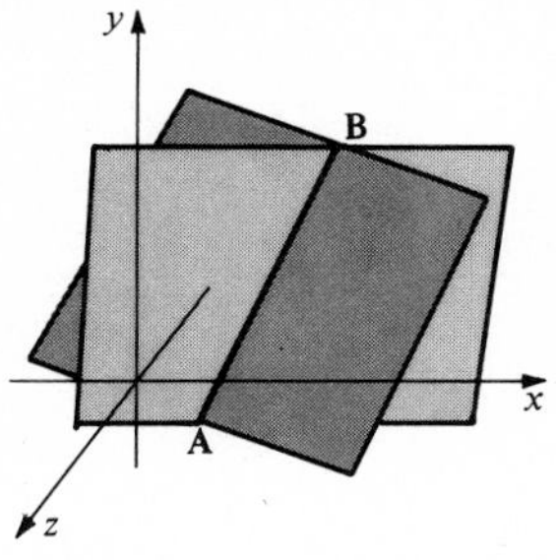

Figure 3-16
Two intersecting planes.

$$\mathbf{A} = [1 \;\; -2 \;\; 1]$$

$$\mathbf{B} = [3 \quad 4 \;\; 3]$$

Task: Find the direction cosines for the line of intersection. Figure 3-16 shows this line.

Solution:

$$\text{Let } \mathbf{R} = \mathbf{B} - \mathbf{A}$$

$$|\mathbf{R}| = \{(3-1)^2 + (4+2)^2 + (3-1)^2\}^{1/2}$$

$$|\mathbf{R}| = 6.63$$

Then:

$$\cos\theta_x = \frac{(3-1)}{6.63} = 0.301 \quad \text{or } \theta_x = 72.4 \text{ deg}$$

$$\cos\theta_y = \frac{(4+2)}{6.63} = 0.905 \quad \text{or } \theta_y = 25.2 \text{ deg}$$

$$\cos\theta_z = \frac{(3-1)}{6.63} = 0.301 \quad \text{or } \theta_z = 72.4 \text{ deg}$$ ◀

3-14 INTERSECTING TWO-DIMENSIONAL LINES

The point of intersection between two lines in a plane can be found by solving two linear, algebraic equations. Let $\mathbf{A}_1 = [x_1 \; y_1]$ and $\mathbf{A}_2 = [x_2 \; y_2]$ be two points on line 1. Let $\mathbf{B}_1 = [x_3 \; y_3]$ and $\mathbf{B}_2 = [x_4 \; y_4]$ be two points on line 2. The parametric, vector equations for these two lines in a plane are (see Figure 3-17)

$$\begin{aligned} \mathbf{Q}_1 &= \mathbf{A}_1 + t(\mathbf{A}_2 - \mathbf{A}_1) \\ \mathbf{Q}_2 &= \mathbf{B}_1 + s(\mathbf{B}_2 - \mathbf{B}_1) \end{aligned} \tag{3-15}$$

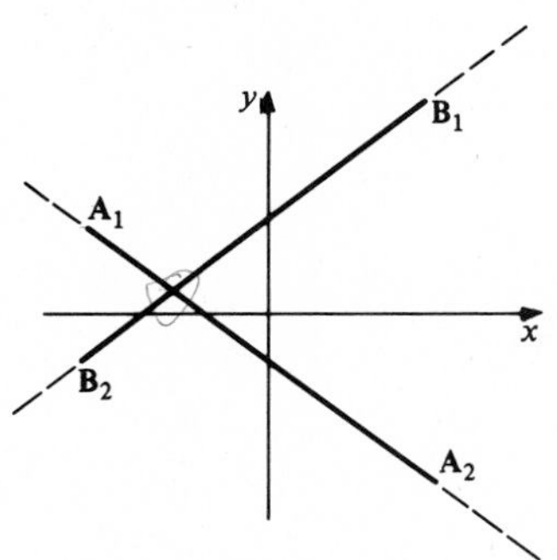

Figure 3-17
Intersecting two-dimensional lines.

Different scalar parameters (t and s) are used for each line.

Equate the two scalar components of the two vector equations for the given lines. This gives two algebraic equations in terms of the two unknown parameter values at the point of intersection.

$$\begin{aligned} x_1 + t(x_2 - x_1) &= x_3 + s(x_4 - x_3) \\ y_1 + t(y_2 - y_1) &= y_3 + s(y_4 - y_3) \end{aligned} \tag{3-16}$$

If the lines are not parallel, then Cramer's rule or substitution can be used to solve Equations (3-16) for t and s.

Using the definition of matrix multiplication as defined in Appendix 2, Equations (3-16) can be written in an equivalent matrix form given by

$$\begin{bmatrix} (x_2 - x_1) & (x_3 - x_4) \\ (y_2 - y_1) & (y_3 - y_4) \end{bmatrix} \begin{bmatrix} t \\ s \end{bmatrix} = \begin{bmatrix} (x_3 - x_1) \\ (y_3 - y_1) \end{bmatrix} \tag{3-17}$$

Let the coefficients of the 2 × 2 matrix in Equation (3-17) be indicated by a_{ij}, and the right side 2 × 1 matrix by c_{ij}. Then the general matrix form is

$$\begin{bmatrix} a_{11} & a_{12} \\ a_{21} & a_{22} \end{bmatrix} \begin{bmatrix} t \\ s \end{bmatrix} = \begin{bmatrix} c_{11} \\ c_{21} \end{bmatrix} \tag{3-18}$$

The determinant D of the a_{ij} matrix is given by

$$D = a_{11}a_{22} - a_{12}a_{21} \neq 0 \tag{3-19}$$

The solutions for t and s as given by Cramer's rule are

$$t = \frac{(a_{22}c_{11} - a_{12}c_{21})}{D}$$
$$s = \frac{(a_{11}c_{21} - a_{21}c_{11})}{D} \tag{3-20}$$

If $D = 0$ the lines are parallel and no solution is possible.

Once t or s is found, the actual components $[Q_x \; Q_y]$ of the point of intersection between the two lines can be calculated using the two scalar equations of one of the lines. For example, when the value of $t = t_i$ at the point of intersection is known, the components of the point of intersection are given by

$$Q_x = x_1 + t_i(x_2 - x_1)$$
$$Q_y = y_1 + t_i(y_2 - y_1) \tag{3-21}$$

If the calculated value of t_i does not fall between 0 and 1 then the point of intersection falls on an extended line through points $\mathbf{A}_1$ and $\mathbf{A}_2$.

Subroutine INTER2D calculates the point of intersection, if any, for two two-dimensional lines in a plane. The BASIC calling variables passed to this routine are single variables. Using the position vectors indicated in Figure 3-17 these would be

X1 = x_1; Y1 = y_1

X2 = x_2; Y2 = y_2

X3 = x_3; Y3 = y_3

X4 = x_4; Y4 = y_4

The parameter values at the point of intersection are t and s, while the point of intersection is returned as X,Y. If X and Y are set equal to 99999 in the subroutine, then the lines are parallel. Other "flags" can be used to indicate that no intersection exists.

The main program 2DLINE given in Appendix 1 calls subroutine INTER2D and then prints the intersection point. The endpoints of the two-dimensional lines are input from the keyboard in the program 2DLINE. If many pairs of lines are to be investigated then other input techniques such as reading data files should be used.

3-15 TWO-DIMENSIONAL LINE CONSTRUCTION

Graphical construction of lines with certain attributes such as perpendicular bisectors, lines through given points, normals to a line through a point, and so on is aided by drafting tools such as dividers, drawing compass, straightedge, *T*-square, and a collection of plastic triangles. Analytical construction requires a little more thought. A variety of approaches is possible. Two examples are given here for illustration. References [7] and [15] contain additional ideas.

The equation for a two-dimensional line through point $\mathbf{P} = [x_3\ y_3]$ and perpendicular to a reference line, such as a folding line, can easily be found if two points of the reference line are given. Let $\mathbf{A} = [x_1\ y_1]$ and $\mathbf{B} = [x_2\ y_2]$ lie on the reference line. The parametric equation for the line is given by

$$\begin{aligned} x &= x_1 + t(x_2 - x_1) \\ y &= y_1 + t(y_2 - y_1) \end{aligned} \tag{3-22}$$

The slope of the line, as shown in Figure 3-18, is given by

$$m = \frac{(y_2 - y_1)}{(x_2 - x_1)} \tag{3-23}$$

Analytical geometry theory proves that a line perpendicular to a line of slope m will have a slope of $-(1/m)$. Thus, the slope of the perpendicular to the reference line is

$$m' = -\frac{(x_2 - x_1)}{(y_2 - y_1)} \tag{3-24}$$

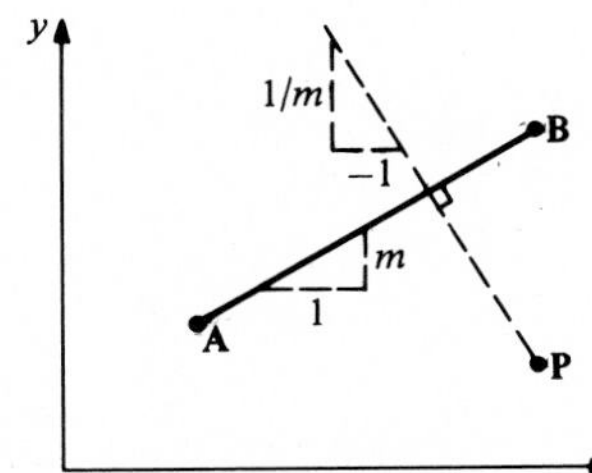

Figure 3-18
Line slopes.

One can then "construct" the parametric equation for a line through the point **P** which is perpendicular to the given line as follows:

$$\begin{aligned} x &= x_3 - s(y_2 - y_1) \\ y &= y_3 + s(x_2 - x_1) \end{aligned} \tag{3-25}$$

In matrix form, Equations (3-22) and (3-25) are given by

$$\begin{bmatrix} (x_2 - x_1) & (y_2 - y_1) \\ (y_2 - y_1) & -(x_2 - x_1) \end{bmatrix} \begin{bmatrix} t \\ s \end{bmatrix} = \begin{bmatrix} (x_3 - x_1) \\ (y_3 - y_1) \end{bmatrix} \tag{3-26}$$

Cramer's rule gives t and s which can then be used in either Equation (3-22) or (3-25) to calculate the point of intersection (see Appendix 2).

If the perpendicular bisection of line AB is required, then point **P** can be placed at its midpoint. If $\mathbf{P} = [x_3\ y_3]$ then

$$x_3 = \frac{(x_1 + x_2)}{2} \qquad y_3 = \frac{(y_1 + y_2)}{2} \tag{3-27}$$

Then, any other point of the perpendicular bisector can be calculated using Equation (3-25) for a chosen value of the parameter s;

$$\mathbf{Q} = [x_4\ y_4]$$

where

$$x_4 = x_3 - s(y_2 - y_1) \qquad y_4 = y_3 + s(x_2 - x_1) \tag{3-28}$$

3-16 INTERSECTING THREE-DIMENSIONAL LINES

If Equations (3-15) define two three-dimensional lines in space then a third scalar equation, obtained by equating the z-components, is given by

$$z_1 + t(z_2 - z_1) = z_3 + s(z_4 - z_3) \tag{3-29}$$

When this equation is added to the two in Equations (3-16) one has three equations and two unknowns. The procedure is to obtain a solution for t and s from *any* pair of the three equations as discussed above. Then, if these values of t and s satisfy the remaining third equation, the three-dimensional lines actually intersect.

The subroutine XCHECK implements this procedure. A main program 3DLINE given in Appendix 1 provides an example of calling XCHECK to find the intersection between two, three-dimensional lines in space. This subroutine uses Cramer's rule to solve for the unknown parameters. In Chapter 9 a vector approach will be given to obtain the solution for parameters t and s.

Sometimes computer graphics is the best way to observe whether two lines actually intersect. Animation of physical systems can be used to see if required clearances exist for every position of the mechanism. Figure 3-19 illustrates this idea.

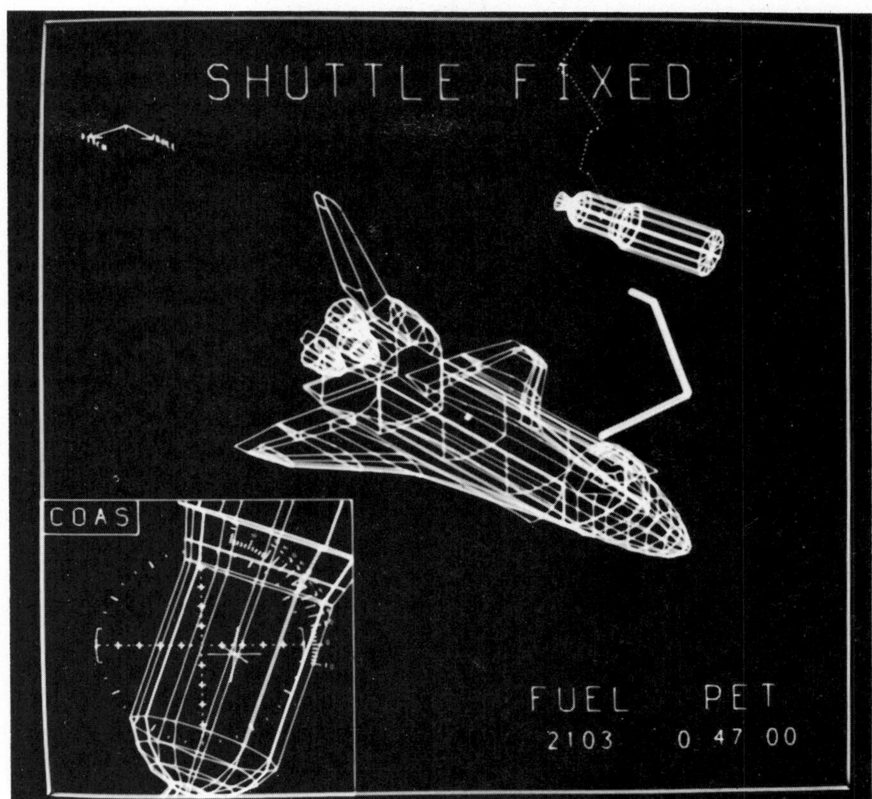

Figure 3-19
Three-dimensional lines used to define a system of components. Courtesy of Evans and Sutherland Inc.

PROBLEMS: CHAPTER 3

3-1 *Given:* Top and front views of triangle *ABC*, where **A** = [1 4 7], **B** = [7 3 7], and **C** = [3 0 3].
Task: Use both descriptive geometry and computation to find the true length of each side of the given triangle.

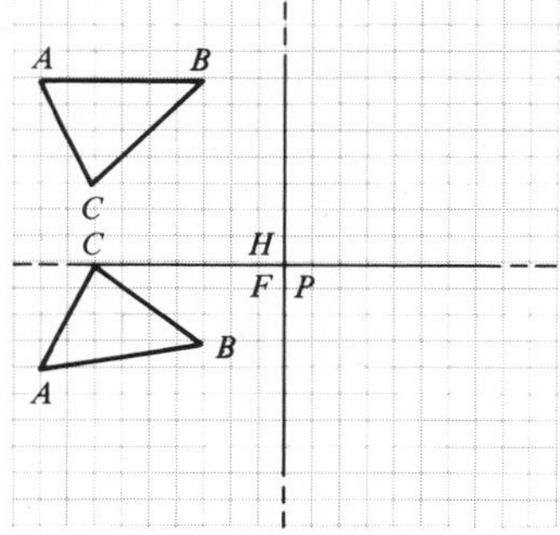

Figure 3-1P

3-2 *Given:* Top and front views of line *MN*, where **M** = [1 4 1] and **N** = [6 0 5].
Task: Point **O** is located on line *MN* three units from *N*. Show line *MN* and point *O* in the top, front, and right-side views.

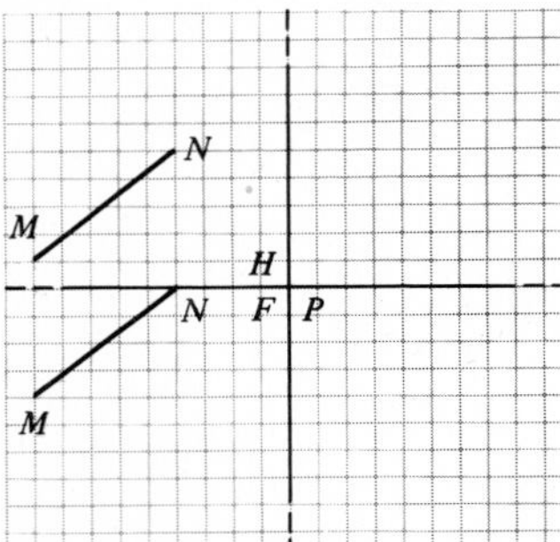

Figure 3-2P

3-3 *Given:* Top and front views of line *MN* and point *O*, where **M** = [4 7 6], **N** = [4 1 1], and **O** = [4 4 3.5].
Task: Determine whether point *O* is on line *MN*. Use both a graphical and computational technique.

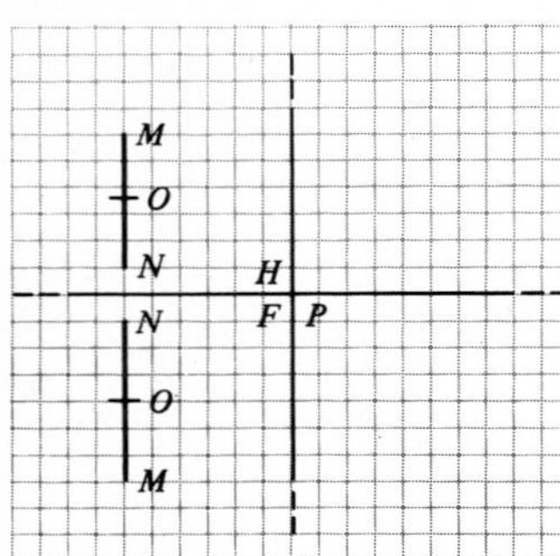

Figure 3-3P

3-4 *Given:* Side view of intersecting lines *AB* and *MN*, and front view of line *AB*. The representing vectors are given by **A** = [1 2 6], **B** = [7 7 1], **M** = [x_1 6 6], and **N** = [x_2 3 1], where x_1 and x_2 are components to be determined.
Task: Complete the front and top views of lines *AB* and *MN*.

3-5 *Given:* Front and side views of lines *AB* and *MN*, where **A** = [2 6 6], **B** = [7 2 2], **M** = [1 4 7], and **N** = [7 4 1].
Task: Determine whether the given lines intersect. Use both a graphical and computational technique.

3-6 *Given:* Top and front views of line *AB* between **A** = [2 1 5] and **B** = [5 6 1].
Task: Using revolution find the true length of line *AB*.

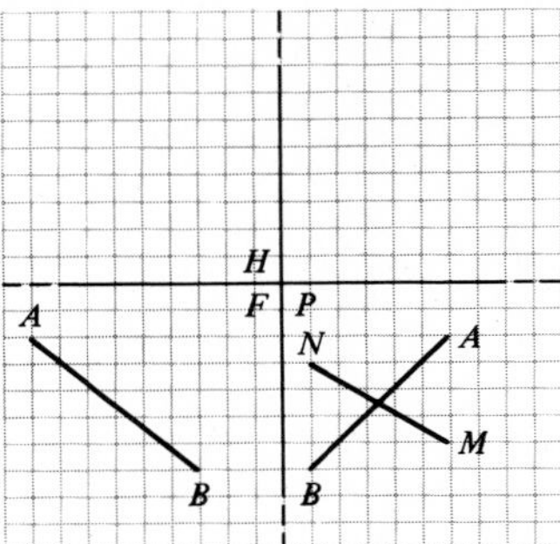

Figure 3-4P

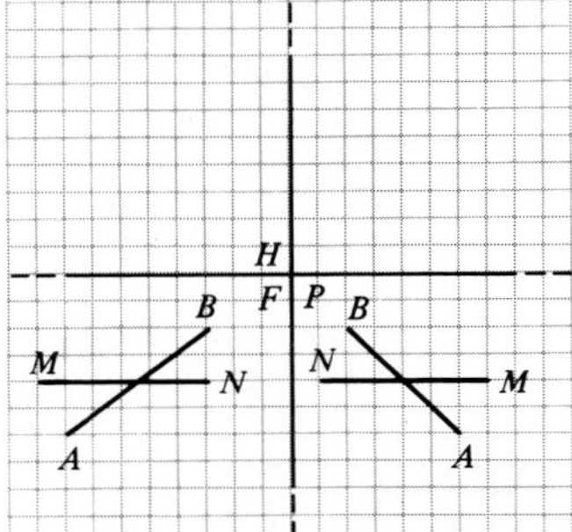

Figure 3-5P

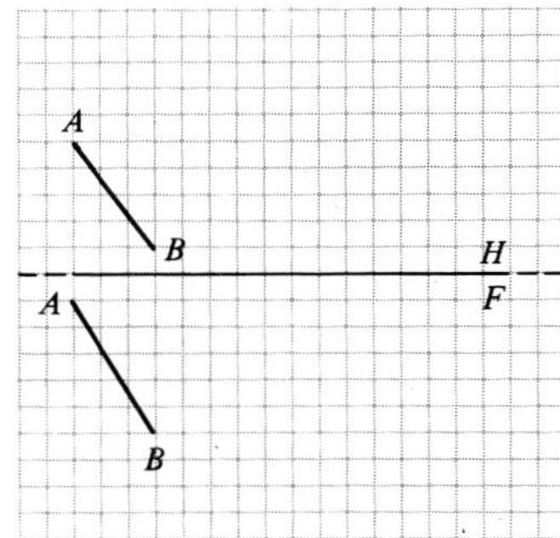

Figure 3-6P

3-7 *Given:* Top and front views of line *AB* between **A** = [1 3 4] and **B** = [7 1 7].
Task: Find bearing, slope, and grade of line *AB*.

3-8 *Given:* Top and front views of point *M*.
Task: From point *M* an 85 ft shaft having a bearing of S75 deg W is dug. Show the shaft in the front and top views for a 50° angle with the horizontal.

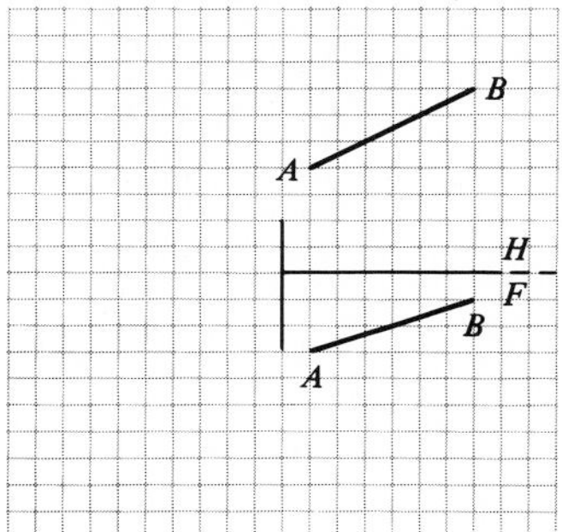

Figure 3-7P

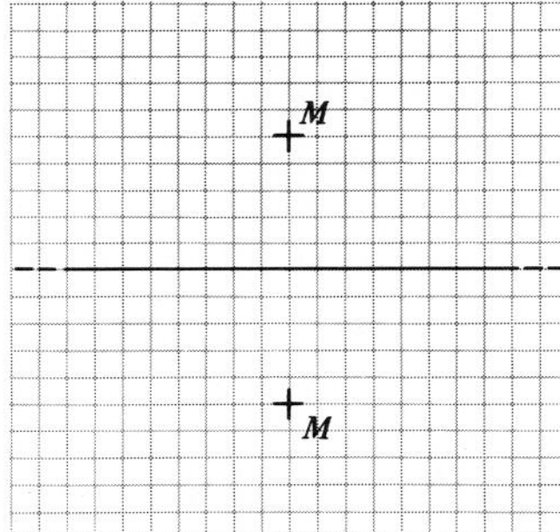

Figure 3-8P

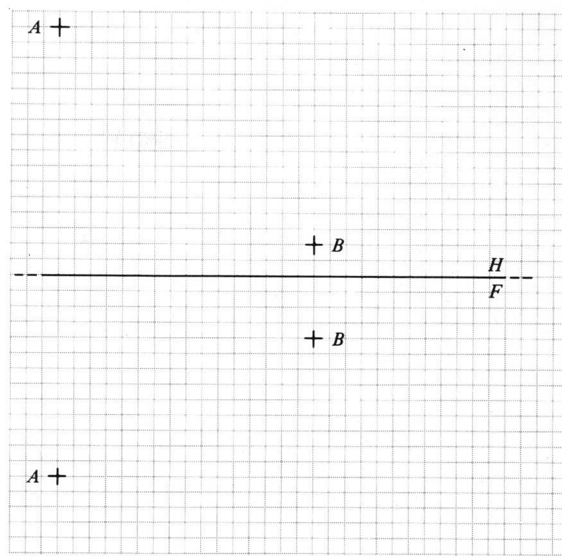

Figure 3-9P

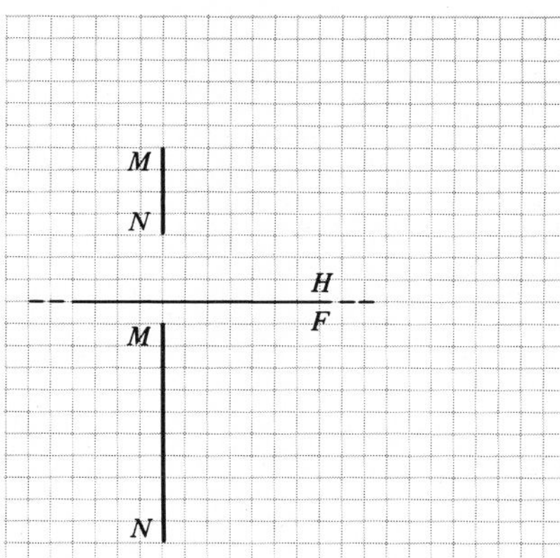

Figure 3-10P

3-9 *Given:* The front and top views of a pipeline to be installed are defined by letters *A* and *B*. **A** = [3 13 16] **B** = [19 4 2]
Task: Determine length of pipe needed. Scale: 1 unit = 4 ft.

3-10 *Given:* Front and top views of line *MN*. **M** = [7 1 7] **N** = [7 11 3].
Task: Determine the angle the line makes with the horizontal plane.

3-11 *Given:* An infinite line in space passes through **A** = [3 2 2] and **B** = [−2 1 −2].
Task: a. Calculate the magnitude of the directional vector.
b. Write the parametric, vector equation for the line segment between **A** and **B**.
c. Repeat parts a and b for each side of the plane given in Problem 3-1.
d. Repeat parts a and b for the line defined in Problem 3-6.

3-12 *Given:* The frontal and horizontal views of two lines are shown in Figure 3-12P. The endpoints are given by **A** = [4 4 4], **B** = [11 1 3], and **C** = [8 4 1], **D** = [6 7 3].

Task: a. Determine the true length of each line using auxiliary views.
b. Calculate the true lengths of lines **B** − **A** and **D** − **C**.
c. Calculate the direction cosines of each line.

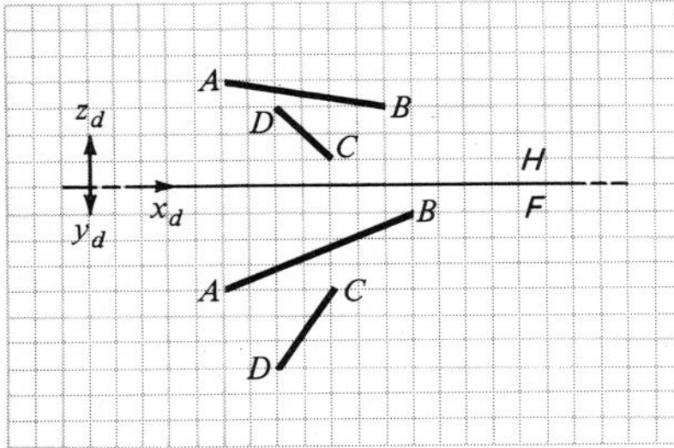

Figure 3-12P
Two lines.

3-13 *Given:* Problem 3-2
Task: Calculate the coordinates of point *O* using the vector, parametric equation for the line *MN*, where point *O* is located three units from point *N* on Line *MN*.

3-14 *Given:* The line *AB* defined in Problem 3-7.
Task: Calculate the direction cosines and compare with the slope, bearing, and grade.

3-15 *Given:* Two two-dimensional lines between **A** = [4 1] and **B** = [1 5] and between **C** = [1 1] and **D** = [5 4].

Task: Calculate the point of intersection between the two lines.

3-16[c] *Given:* The two lines in Problem 3-5.
Task: Use a main program like 3DLINE and subroutines XCHECK and BLINE to determine whether the lines actually intersect.

3-17[c] *Given:* A support cable is attached between **A** = [1 10 5] and **B** = [10 0 −2]. A second cable is to be attached to a vertical tower located at x = 10, z = 4. It originates at the origin of the coordinate system.
Task: How high can the second cable be attached on the tower without contacting the support cable? The support cable must remain on top of the second cable. Use of a computer program such as 3DLINE is optional.

3-18 *Given:* A two-dimensional line is defined between **A** = [1 5] and **B** = [6 2].
Task: Calculate the point of intersection between this line and a line drawn from point **C** = [4 4] which is perpendicular to the given line.

3-19 *Given:* A two-dimensional line between **A** = [1 5] and **B** = [6 2].
Task: Write the scalar form of the equations for a two-dimensional line which is the perpendicular bisector to the given line. What is the slope of this bisector?

3-20[c] *Given:* Two three-dimensional lines between **A** = [100 60 60] and **B** = [40 80 30] and between **C** = [80 50 40] and **D** = [−10 20 50].
Task: Generate the three principal views.

---

[c]Superscript *c* indicates that a computed solution is required.

4

Lines and Planes

In this chapter, the concept of a line is embellished to create the definition of a plane in space. Conditions are defined under which a plane will appear as a line. Various ways of defining a plane are then presented and auxiliary views of planes are created. These include both descriptive and geometric modeling methods. More vector techniques are introduced so that a plane can be defined mathematically and computations made numerically. Subroutines are suggested that can be used to implement the theory.

4-1 EDGE VIEW OF A PLANE

To understand the basic concepts of descriptive geometry and to be able to solve the problems presented it is necessary to be able to find: (1) the true length of a line, (2) the point view of a line, (3) the edge view of a plane and, (4) the normal view of a plane. Previous work explained the principles of (1), (2), and (4), but only discussed the relation between the edge view and normal view of a plane.

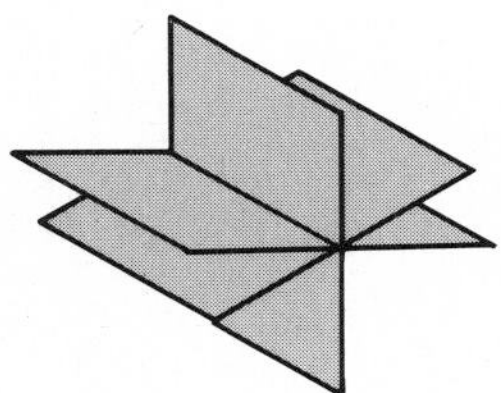

Figure 4-1
Point view of line: edge view of plane.

When a given plane *ABCD* is placed perpendicular to a picture plane, the lines of sight will travel from the eye through the picture plane and be parallel to *ABCD*. The observer will see the *edge* of *ABCD* which will appear on the picture plane as a line. If a line is placed *in plane ABCD* so that its true length is perpendicular to the picture plane, it will appear on that plane as a point. The point view will be on the edge view of *ABCD*. From the above and from Figure 4-1 it can be seen that the point view of a line placed in a plane will *always* show the plane on edge. If a normal view of the plane is needed as a final answer it could be found by placing a folding line or picture plane parallel to it. Example 4-1 illustrates the above principle by obtaining the point view of one of the edges of a plane.

Example 4-1

Given: Top and front views of triangle *ABC*. *AC* is a frontal line.

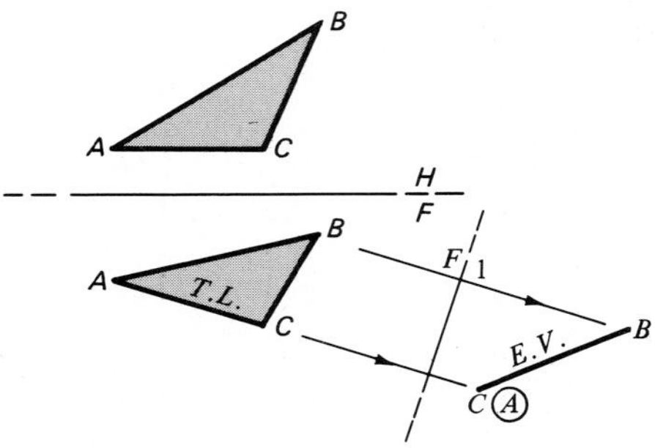

Example 4-1
Edge view of plane: true length line given.

Task: Find the edge view of plane *ABC*.
Solution:

1. Line *AC*, one of the limiting lines of triangle *ABC*, is a frontal line. For this reason, it appears as a true length line in the front view.
2. Place folding line *F*/1 perpendicular to the true length line *AC* in the front view.
3. Project lines of sight from all points into the first auxiliary and perpendicular to folding line *F*/1.
4. Obtain the distances that points *A*, *B*, and *C* lie behind the frontal plane in the top view and plot these distances into the first auxiliary on their respective lines of sight.
5. Line *AC* will appear as a point and triangle *ABC* will be seen as an edge view. ◀

The views of the triangle in Example 4-1 showed a limiting line of the plane as a frontal line. In most cases, a problem will concern an oblique triangle where none of the edges will appear as a principal line. Finding the point view of a line would require two auxiliaries; one to find the true length of the line and the second to find the point view. This being the case, let us make use of the principles noted in Section 3-2 and place a principal line *in* the plane. A point view of this line would show the edge view of the plane thereby projecting only one auxiliary instead of two. The following example places a horizontal line in plane *ABC*.

Example 4-2

Given: Top and front views of triangle *ABC*.

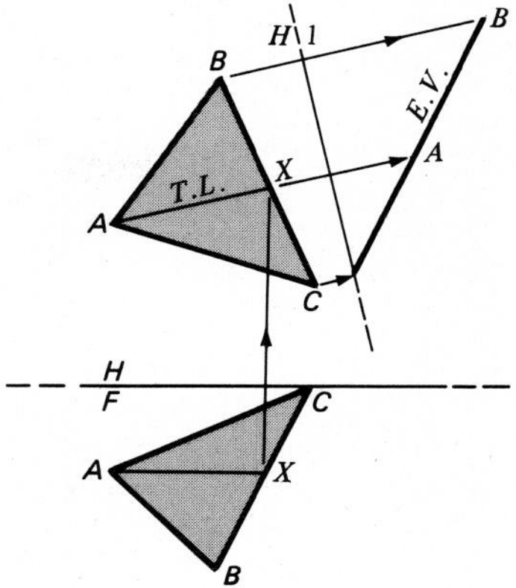

Example 4-2
Edge view of plane: horizontal line inserted.

Task: Find the edge view of *ABC*.

1. Place horizontal line *AX* in plane *ABC*.*
2. Project line *AX* into the top view. Since the line is *in* the plane, the top view of *AX* will extend from the given point *A* to intersect line *BC* at point *X* (projected from the front view).
3. Line *AX* is a horizontal line and will now show true length in the top view.
4. Place folding line *H*/1 perpendicular to the true length line *AX*.
5. Project lines of sight from points *A*, *B*, and *C* into the first auxiliary so that they are parallel to the true length line and perpendicular to the folding line *H*/1.
6. Obtain the distances that points *A*, *B*, and *C* lie below the horizontal plane in the front view. Plot these distances below folding line *H*/1 in the first auxiliary view. Points *A*, *B*, and *C* should lie in a straight line. Since construction line *AX* is in the plane and now appears as a point, the straight line *ABC* signifies the edge view of the plane. ◀

Note: The practice of plotting only two points and drawing the line between them does not provide a check on the plotting procedure. Accuracy dictates that at least three points be represented if possible.

A profile line is used in Example 4-3 to find the edge view of plane

*A reference line can be drawn parallel to the folding edge in any primary view.

ABC. The graphical solution is shown for Example 4-3 so that the student can use it as a self-study problem.

Example 4-3

Given: Front and side views of triangle *ABC*.

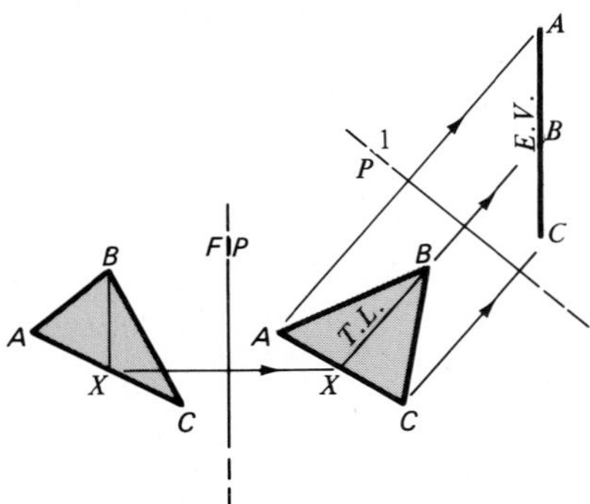

Example 4-3
Edge view of plane: profile line inserted.

Task: Use the true length of a profile line to determine the edge view of triangle *ABC*.
Solution: Reader should provide steps.

4-2 PLANES—GRAPHICAL REPRESENTATION

Planes and their relation with points and lines play an important part in the study of space geometry. As noted in Section 1-9, we will be concerned here with their graphical representation and the use of the fundamental principles that orient them in space. It must be remembered that the construction shown here does not limit the boundaries of planes. They are considered to be infinite but are limited by lines in order to show segments which can be considered in the solution of space problems.

A. Two Intersecting Lines Form a Plane

Example 4-4

Given: Top and front view of two intersecting lines *AB* and *MN*.
Task: Prove that the intersecting lines form a plane.
Solution:

1. Assume that a plane will be formed. Connect the end points of the line segments.

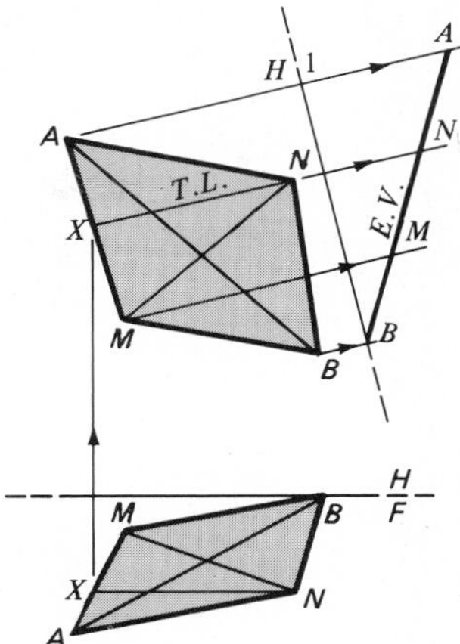

Example 4-4
Plane formed by two intersecting lines.

2. Draw a horizontal line *XN* in the front view of *AMBN*.
3. The line *XN* will be true length in the top view.
4. Place the *H*/1 folding line perpendicular to true length line *XN*. Project points on lines of sight which are parallel to the true length line and perpendicular to *H*/1.
5. A plane on edge will be formed, proving that two intersecting lines form a plane. ◀

B. A Point and a Line Form a Plane

Example 4-5

Given: Top and front views of line *MN* and point *O*.

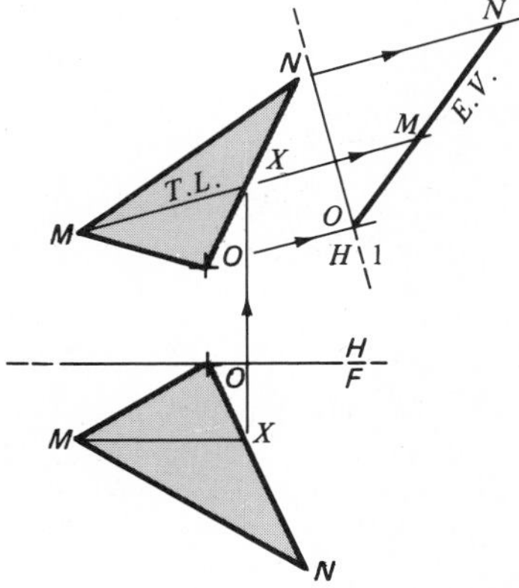

Example 4-5
Plane formed by point and line.

Task: Prove that a point and a line form a plane.
Solution:

1. Connect the ends of the line segment *MN* with point *O*.
2. Draw horizontal line *MX* in the front view of *MON*.
3. Line *MX* is true length in the top view.
4. Draw the elevation auxiliary in which *MON* appears as a plane on edge. ◀

C. Two Parallel Lines Form a Plane

Example 4-6

Given: Two parallel lines *AB* and *MN*, shown in top and front views.

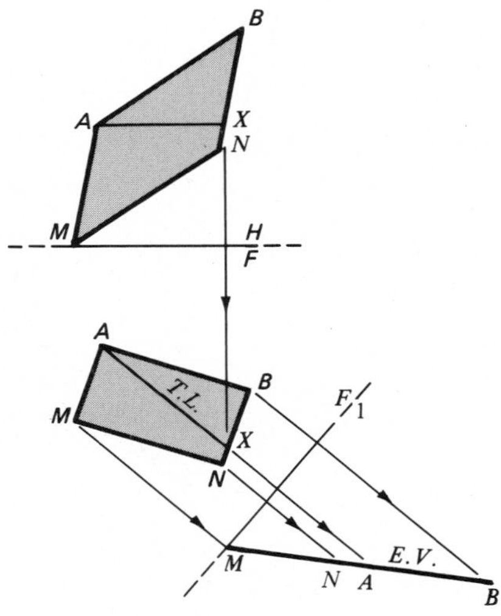

Example 4-6
Plane formed by two parallel lines.

Task: Prove that the two parallel lines form a plane.
Solution: The student should now provide the procedures needed for Example 4-6 until a plane on edge is shown in the first auxiliary. ◀

D. Three Noncollinear Points Form a Plane

Example 4-7

Given: Top and front views of three noncollinear points *A*, *B*, and *C*.

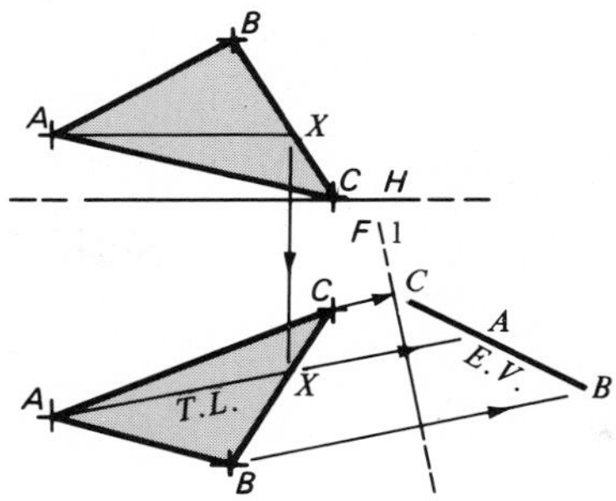

Example 4-7
Plane formed by three noncollinear points.

Task: Prove that the given points form a plane.
Solution: It is again suggested that the student provide the procedures needed for Example 4-7. This method of defining a plane is often used in the vector analysis approach to plane definition. ◀

4-3 POINTS IN PLANES

A point has been defined as a definite *position* in space with no physical dimensions. For general purposes, however, a point will be represented here by a dot (·) or two crossed lines (+) in order to show its location.

A point in a plane will be on a line which is in the plane. Since a line is made up of an infinite number of points, a plane that contains a line will contain any and all points on that line. Any point which is on a line in a plane or on an extension of that line can be projected in the usual manner.

The following examples describe two methods used to locate a point with reference to a plane. The two-view method determines whether a point lies in a plane while the auxiliary method is used to show a definite relationship between a point and a plane.

A. Two-View Method

The solution given in Example 4-8 shows the short method of finding whether a point lies in a plane.

Example 4-8

Given: Top and front views of plane *ABC* and point *O*.
Task: Determine whether point *O* lies in the plane *ABC*.
Solution:

1. In the front view draw line *BX* from point *B* through point *O* to intersect line *AC* at *X*.

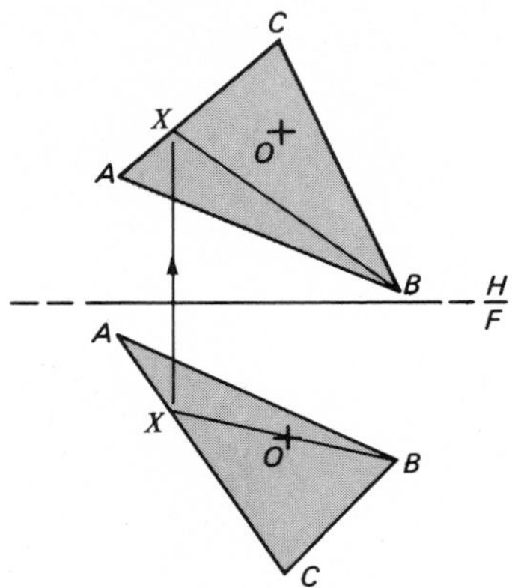

Example 4-8
Point in plane.

2. Project point *X* from the front view to intersect line *AC* in the top view. Since both point *B* and point *X* are in the plane, the line *BX* lies in the plane.
3. Since point *O* does not lie on line *BX* it does not lie in the plane *ABC*.

Example 4-9

Given: Front and top views of plane *ABCD*. Front view of point *O* which lies in plane *ABCD* (defined) and point *P* which lies in plane *ABC* (infinite).

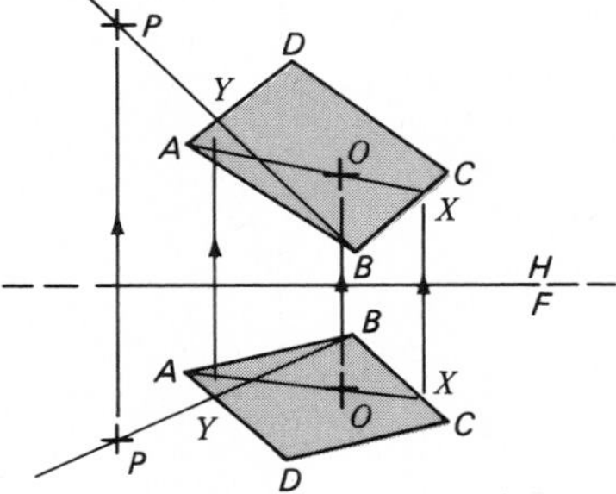

Example 4-9
Points in plane.

Task: Show the top view of points *O* and *P*.
Solution:

1. Draw a line from point *A* through point *O* to intersect line *BC* at *X* in the front view.
2. Project point *X* into the top view until it intersects line *BC*. Line *AX* is now in plane *ABCD*.

3. Project point *O* into the top view until it intersects line *AX*. Point *O* is now on line *AX* and in plane *ABCD*.
4. In the front view draw a line from point *B* through point *P*. This line will intersect *AD* at point *Y*.
5. Project point *Y* to intersect line *AD* in the top view. Extend a line drawn from *B* through *Y* in the top view. This line is now in plane *ABCD* (infinite). Project point *P* into the top view until it intersects line *BY* extended. Point *P* is now on line *BY* and in the infinite plane. ◀

B. Auxiliary Method

Example 4-10 shows the relationship between a point and a plane using the auxiliary method of solution.

Example 4-10

Given: Top and front views of plane *ABC* and point *O*.

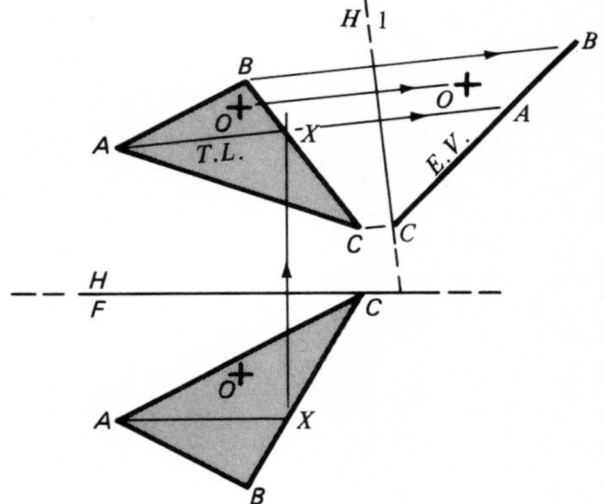

Example 4-10
Relationships: point and plane.

Task: Determine the position of point *O* relative to plane *ABC*.
Solution:

1. Construct a horizontal line *AX* in the front view. Project *AX* into the top view where it appears in true length.
2. Place folding line *H*/1 perpendicular to *AX* and project points *A*, *B*, *C*, and *O* into the first auxiliary.
3. Plane *ABC* appears as an edge in the elevation auxiliary where point *O* is shown above the plane. ◀

Measurements in space are always taken perpendicularly from a specified principal plane or from a reference plane which is parallel to it.

Points are then measured above or below a horizontal reference plane, in back or in front of a frontal plane, or to the left or right of a profile plane. A point would not have to be directly below another point to be measured but be on a horizontal plane which shows as an edge view and passes through that point. This is shown in the following example.

Example 4-11

Given: Top and front views of triangle *ABC*.

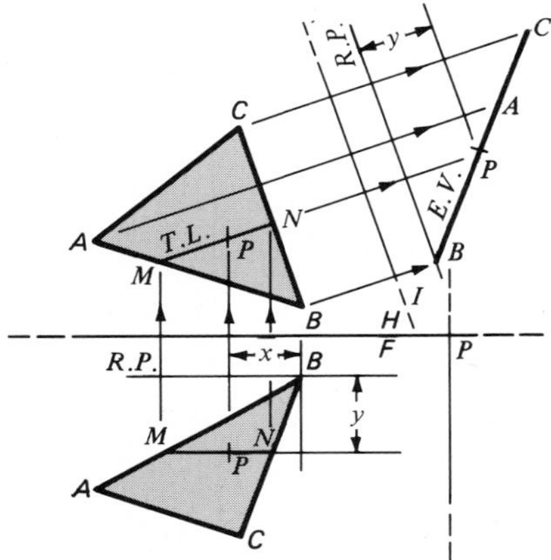

Example 4-11
Relationships: point and plane.

Task: Place point *P* in plane *ABC* so that it is "*y*" distance below and "*x*" distance to the left of point *B*.
Solution:

1. In the front view, show the edge of a horizontal reference plane drawn through *B* and parallel to folding line *H/F*.
2. Construct horizontal line *MN* at "*y*" distance below the horizontal reference plane drawn in step 1.
3. Show the edge of a profile reference plane drawn through *B* and parallel to folding line *F/P*.
4. Place point *P* on line *MN* so that it is "*x*" distance to the left of the reference plane drawn in step 3.
5. Project line *MN* into the top view so the line is in plane *ABC*.
6. Project point *P* into the top view so that *P* is on line *MN*.
7. Point *P* is now on a line in the plane and is "*y*" distance below and "*x*" distance to the left of point *B*.
8. The elevation auxiliary is not necessary for the solution of the problem but is used to check the answer. In this auxiliary, plane *ABC*

appears as an edge view. Point P projects onto the edge view and is "y" distance below point B. ◀

4-4 LINES IN PLANES

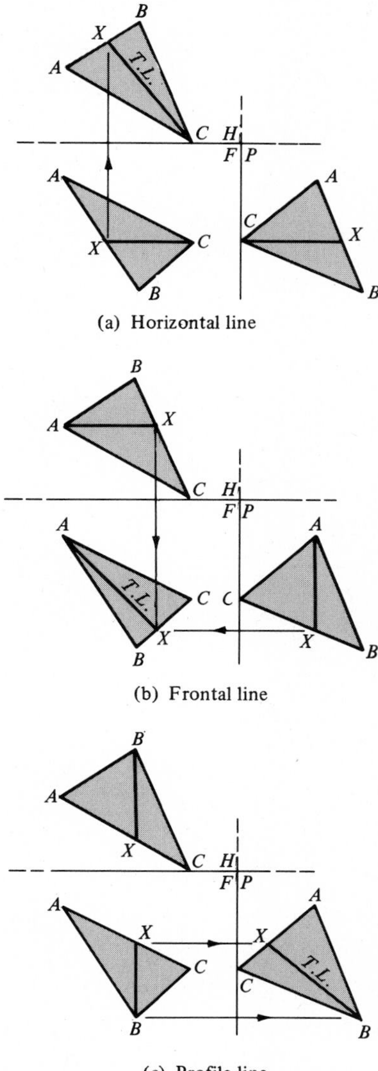

(a) Horizontal line

(b) Frontal line

(c) Profile line

Figure 4-2
Line projections.

One of the many reasons for adding lines in a plane was noted in Sections 3-2 and 4-1 where the true length of a line was found. A line in a plane may be determined by connecting two or more points that are in the plane. However, only one point in the plane is needed if the line is defined as being in the plane and lying through the point in a specified direction.

A straight line may be transferred into any view by projecting two points of the line to their respective positions in the plane as it appears in an adjacent view. Figures 4-2(a), (b), and (c) illustrate the projection of lines in a plane. In each case, the end points of the line were projected into adjacent views.

The following illustrates the addition of a line in a plane where only one point and the bearing of the line is specified.

Example 4-12

Given: Top and front views of triangle *ABC*. Top view of point *P* which is in triangle *ABC*.

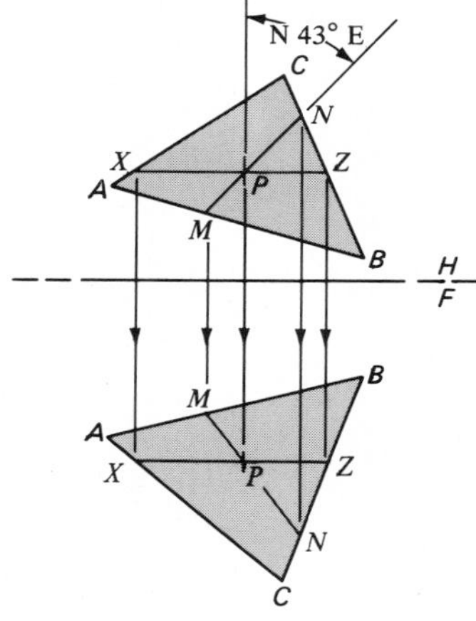

Example 4-12
Line of specified bearing drawn through given point and in given plane.

Task: Through point *P* draw a line that is in plane *ABC* and has a bearing of N43 deg E. Show the top and front views of the point and line, both being in plane *ABC*.

Solution:

1. Draw line *XZ* through point *P* as shown in the top view.
2. Project line *XZ* and point *P* into the front view in the usual manner. Point *P* is now in the plane.
3. In the *top* view (as noted in Section 3-10) draw line *MN* with the required bearing through point *P* so that it intersects the limiting edges of triangle *ABC*.

4. Project line *MN* into the front view where it will pass through point *P* and show the correct *projected* view of the line. ◀

All the principles listed below have been used in the solution of problems studied previously in this text. These fundamental rules are of such value in descriptive geometry that it is well to summarize them as follows:

1. All horizontal lines in a given plane are parallel, (excluding those in a horizontal plane where they will be horizontal but not necessarily parallel).
2. All frontal lines in a given plane are parallel (excluding those in a frontal plane where they cannot be considered parallel without further proof).
3. All profile lines in a given plane will be parallel (excluding those in a profile plane where the lines are not necessarily parallel).
4. Any principal line will appear true length in *one* of the principal views.
5. If a folding line is placed parallel to a line, that line will show true length in the adjacent view.
6. If a folding line is placed perpendicular to a true length line, the line will appear as a point in the adjacent view.
7. When a folding line is placed parallel to the edge view of a plane (which appears as a line) the true size and shape of the plane is shown in the adjacent view.

Geometric modeling of lines and planes in space requires that vector methods be used to quantify spatial relationships. Some of these methods are given in the following section. The mathematical definition of a plane can then be given, based upon vector operations.

4-5 VECTOR DOT PRODUCT

Mathematical representation of points and lines using vector concepts was given in earlier chapters. The same requirement exists for planes if a geometric modeling approach is to be taken. The equation for a plane is also based upon the properties of vectors. One such property, the *dot product* (or inner product) provides the fundamental idea for deriving the equation for a plane. The dot product is now introduced along with a review of the vector cross product. Then equations and supporting software are given which allow planes to be treated in an analytical manner, along with points and lines.

The vector dot product is most easily defined using the rule of matrix multiplication. Represent vector **A** as a single row matrix and vector **B** as a single column matrix.

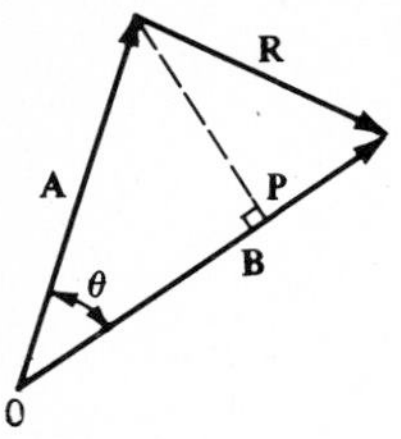

Figure 4-3
Dot product **A** · **B**.

$$\mathbf{A} = [A_x\ A_y\ A_z]; \quad \mathbf{B} = \begin{bmatrix} B_x \\ B_y \\ B_z \end{bmatrix}$$

The vector dot product **A** · **B** is equal to the matrix product given by

$$\mathbf{A} \cdot \mathbf{B} = [A_x\ A_y\ A_z] \begin{bmatrix} B_x \\ B_y \\ B_z \end{bmatrix} = A_xB_x + A_yB_y + A_zB_z \tag{4-1}$$

Notice that the result of the dot product **A** · **B** is a single scalar value.

The dot product has an important graphical significance. Two base vectors **A** and **B** are shown in Figure 4-3, drawn from the origin **O**. Using the law of cosines gives

$$|\mathbf{R}|^2 = |\mathbf{A}|^2 + |\mathbf{B}|^2 - 2|\mathbf{A}||\mathbf{B}| \cos\theta \tag{4-2}$$

where θ is the angle between the vectors as shown, and $\mathbf{R} = \mathbf{B} - \mathbf{A}$. Using the components of $\mathbf{A} = [A_x\ A_y\ A_z]$ and $\mathbf{B} = [B_x\ B_y\ B_z]$ gives

$$(B_x - A_x)^2 + (B_y - A_y)^2 + (B_z - A_z)^2$$
$$= (A_x^2 + A_y^2 + A_z^2) + (B_x^2 + B_y^2 + B_z^2) - 2\,|\mathbf{A}||\mathbf{B}| \cos\theta \tag{4-3}$$

Solving for $\cos\theta$ gives

$$\cos\theta = \frac{(A_xB_x + A_yB_y + A_zB_z)}{|\mathbf{A}||\mathbf{B}|} \tag{4-4}$$

The numerator in Equation (4-4) is the vector dot product **A** · **B**. Thus one can write

$$\mathbf{A} \cdot \mathbf{B} = |\mathbf{A}||\mathbf{B}| \cos\theta \tag{4-5}$$

If **B** has a magnitude of unity, $|\mathbf{B}| = 1.0$, then $\mathbf{A} \cdot \mathbf{B} = |\mathbf{A}| \cos\theta = OP$. Distance OP is the component of **A** projected onto the direction of vector **B**, as shown in Figure 4-3. In general, using Equation (4-5) gives

$$OP = |\mathbf{A}| \cos\theta = \mathbf{A} \cdot \mathbf{B}/|\mathbf{B}| \tag{4-6}$$

where $\mathbf{A} \cdot \mathbf{B}/|\mathbf{B}|$ is the component of **A** projected on the vector **B**. The vector $\mathbf{B}/|\mathbf{B}|$ is a unit vector **n** in the direction of **B**.

Two special cases occur when taking the dot product. If the vector **A** is perpendicular to the vector **B**, then the dot product $\mathbf{A} \cdot \mathbf{B} = 0$, as seen by Equation (4-5). Also, if a vector is dotted with itself, say **B**, then $\mathbf{B} \cdot \mathbf{B} = B_x^2 + B_y^2 + B_z^2 = |\mathbf{B}|^2$.

Both expressions for the vector dot product, the component form and the physical form, should be learned. These are

$$\mathbf{A} \cdot \mathbf{B} = A_xB_x + A_yB_y + A_zB_z = |\mathbf{A}||\mathbf{B}| \cos\theta \tag{4-7}$$

Example 4-13

Given: $\mathbf{A} = [3\ 2\ -1]$ and $\mathbf{B} = [2\ 1\ -2]$.
Task: Calculate the projection of vector **A** onto vector **B**. Calculate the angle θ between the two vectors.
Solution:

$$\mathbf{A} \cdot \mathbf{B} = 3(2) + 2(1) - 1(-2) = 10$$

$$|\mathbf{A}| = \{(3)^2 + (2)^2 + (-1)^2\}^{1/2} = 14^{1/2} = 3.74$$

$$|\mathbf{B}| = \{(2)^2 + (1)^2 + (-2)^2\}^{1/2} = 9^{1/2} = 3.0$$

$$OP = \frac{\mathbf{A} \cdot \mathbf{B}}{|\mathbf{B}|} = \frac{10}{3} = 3.33$$

$$\cos\theta = \frac{\mathbf{A} \cdot \mathbf{B}}{|\mathbf{A}|\,|\mathbf{B}|} = \frac{10}{\{(3.74)(3.0)\}} = 0.891$$

Thus,

$$\theta = 27 \text{ deg}$$ ◀

The operation of the vector dot product using unit vectors is the same as for any other vector pair. Performing matrix multiplication using a single row and a single column matrix leads to:

$$\mathbf{i} \cdot \mathbf{i} = [1\ 0\ 0]\begin{bmatrix} 1 \\ 0 \\ 0 \end{bmatrix} = 1(1) + 0(0) + 0(0) = 1$$

$$\mathbf{i} \cdot \mathbf{j} = [1\ 0\ 0]\begin{bmatrix} 0 \\ 1 \\ 0 \end{bmatrix} = 1(0) + 0(1) + 0(0) = 0$$

The reader should verify that $\mathbf{j} \cdot \mathbf{j} = \mathbf{k} \cdot \mathbf{k} = 1$ and $\mathbf{i} \cdot \mathbf{k} = \mathbf{j} \cdot \mathbf{k} = 0$. Recall that the dot product of *any* two perpendicular vectors is always zero.

The scalar, rectangular components of a vector can be expressed in terms of the dot product and the definition of the direction cosines, Equation (3-13), as follows:

$$R_x = \mathbf{R} \cdot \mathbf{i} = |\mathbf{R}| \cos\theta_x$$

$$R_y = \mathbf{R} \cdot \mathbf{j} = |\mathbf{R}| \cos\theta_y$$

$$R_z = \mathbf{R} \cdot \mathbf{k} = |\mathbf{R}| \cos\theta_z$$

Example 4-14

Given: $\mathbf{S} = [1\ 4\ 2]$ and $\mathbf{T} = [2\ 4\ 1]$.
Task: Calculate the magnitudes $|\mathbf{S} \times \mathbf{T}|$ and $|\mathbf{T} \times \mathbf{S}|$.
Solution:

$$\mathbf{S} \times \mathbf{T} = \det \begin{vmatrix} \mathbf{i} & \mathbf{j} & \mathbf{k} \\ 1 & 4 & 2 \\ 2 & 4 & 1 \end{vmatrix}$$

$$\mathbf{S} \times \mathbf{T} = \mathbf{i}(4 - 8) - \mathbf{j}(1 - 4) + \mathbf{k}(4 - 8) = -4\mathbf{i} + 3\mathbf{j} - 4\mathbf{k}$$

$$\mathbf{T} \times \mathbf{S} = \det \begin{vmatrix} \mathbf{i} & \mathbf{j} & \mathbf{k} \\ 2 & 4 & 1 \\ 1 & 4 & 2 \end{vmatrix}$$

$$\mathbf{T} \times \mathbf{S} = \mathbf{i}(8 - 4) - \mathbf{j}(4 - 1) + \mathbf{k}(8 - 4) = 4\mathbf{i} - 3\mathbf{j} + 4\mathbf{k}$$

$$|\mathbf{S} \times \mathbf{T}| = |\mathbf{T} \times \mathbf{S}| = (16 + 9 + 16)^{1/2} = 6.40$$

Alternatively, making use of the dot product, notice that

$$|\mathbf{S}| = (1 + 16 + 4)^{1/2} = 4.5826$$

$$|\mathbf{T}| = (4 + 16 + 1)^{1/2} = 4.5826$$

$$\mathbf{S} \cdot \mathbf{T} = 2 + 16 + 2 = 20 = |\mathbf{T}||\mathbf{S}| \cos \theta$$

$$\cos \theta = \frac{20}{(4.5826)^2} = 0.9525$$

$$\theta = 17.75 \text{ deg}$$

Then

$$|\mathbf{S} \times \mathbf{T}| = |\mathbf{S}||\mathbf{T}| \sin \theta = (4.5826)(4.5826) \sin (17.75) = 6.40$$ ◀

The subroutine DOTPROD given in Appendix 1 performs the operation of a vector dot product. Scalar components of vector **A** are passed to the subroutine in the array A(1), A(2), A(3), while array B(1), B(2), B(3) contains the scalar components of vector **B**. The dot product is a single scalar value (INNER) which is returned to the main program. The main program VECPRODS given in Appendix 1 is an example of a program which calls both subroutines DOTPROD and VECPROD. These two subroutines are often used tools in geometric modeling.

4-6 VECTOR EQUATION OF A PLANE

Let $\mathbf{Q} = [x\ y\ z]$ represent any point in a plane. This point does not necessarily lie within the rectangular or triangular boundaries formed by the

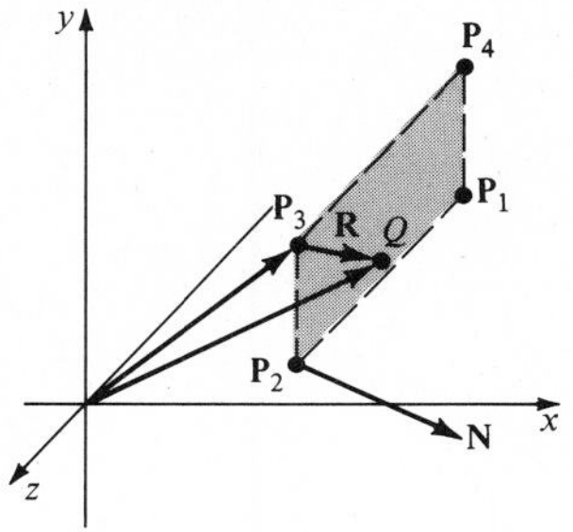

Figure 4-4
Point in a plane.

given points used to define the plane. A plane, like a line, is infinite in extent. The vector difference $\mathbf{Q} - \mathbf{P}_3$ gives a vector $\mathbf{R}$ which lies in the plane as shown in Figure 4-4. Thus, the vector $\mathbf{R}$ will be perpendicular to the normal vector $\mathbf{N}$. Any other position vector such as $\mathbf{P}_1$, $\mathbf{P}_2$, or $\mathbf{P}_4$ can be used in place of $\mathbf{P}_3$ in the above discussion. These vertices are shown in Figure 4-4.

The vector dot product for any two perpendicular vectors is equal to zero. This allows one to write the vector equation for a plane as

$$\mathbf{N} \cdot \mathbf{R} = \mathbf{N} \cdot (\mathbf{Q} - \mathbf{P}_3) = 0 \tag{4-8}$$

Using Equation (4-8) and a cross product for $\mathbf{N}$ gives

$$\{(\mathbf{P}_1 - \mathbf{P}_2) \times (\mathbf{P}_3 - \mathbf{P}_2)\} \cdot (\mathbf{Q} - \mathbf{P}_3) = 0 \tag{4-9}$$

for the vector equation of a plane.

Another way to represent the equation for a plane is to use the matrix representation of a vector dot product. The components of $\mathbf{N}$ are written as a single row matrix and the components of $\mathbf{R} = (\mathbf{Q} - \mathbf{P}_3)$ are written as a single column matrix. Equation (4-8) then becomes

$$[N_1 \ N_2 \ N_3] \begin{bmatrix} (x - x_3) \\ (y - y_3) \\ (z - z_3) \end{bmatrix} = 0 \tag{4-10}$$

Expanding this matrix product gives

$$N_1x + N_2y + N_3z - (N_1x_3 + N_2y_3 + N_3z_3) = 0 \tag{4-11}$$

Since all the terms in the parenthesis in Equation (4-11) will be known for a given plane, let this sum equal a constant k.

$$k = (N_1x_3 + N_2y_3 + N_3z_3) \tag{4-12}$$

Notice that k is equal to the vector dot product

$$\mathbf{N} \cdot \mathbf{P}_3 = [N_1 \ N_2 \ N_3] \begin{bmatrix} x_3 \\ y_3 \\ z_3 \end{bmatrix} = k$$

We now have the scalar equation for a three-dimensional plane given by

$$N_1x + N_2y + N_3z - k = 0 \tag{4-13}$$

The value of k has an important geometrical significance. It can be shown that the ratio $k/|\mathbf{N}|$ is the shortest distance from the origin of the coordinate system to the plane surface. This is discussed further in Section 8-5.

The orientation of a plane can be defined by giving the components of the normal vector to the plane, or by giving the direction cosines for the normal direction. Let the normal vector components be $\mathbf{N} = [N_1 \ N_2 \ N_3]$. The magnitude of the normal vector is

$$|\mathbf{N}| = \{N_1^2 + N_2^2 + N_3^2\}^{1/2} \tag{4-14}$$

The unit normal vector components are

$$n_x = \frac{N_1}{|\mathbf{N}|}; \quad n_y = \frac{N_2}{|\mathbf{N}|}; \quad n_z = \frac{N_3}{|\mathbf{N}|} \tag{4-15}$$

and these three components are equal to the direction cosines for the normal vector.

The subroutine PLANE (Appendix 1) can be used to calculate the three components of the normal vector to a plane and the value of k defined by Equation (4-12). These four values are sometimes called the implicit coefficients for a plane. Especially notice a change in the use of subscripted variables in this subroutine. *Separate* subscripted variables are used to define the components of a vector. The components for the three noncollinear points used to define a plane must now be stored in arrays as follows:

$$\begin{aligned}
\mathbf{P}_1 &= [R(1) \quad S(1) \quad T(1)] \\
\mathbf{P}_2 &= [R(2) \quad S(2) \quad T(2)] \\
\mathbf{P}_3 &= [R(3) \quad S(3) \quad T(3)]
\end{aligned}$$

The subscripts of $R(\,)$, $S(\,)$, and $T(\,)$ now refer to the point number, with all x-components contained in $R(\,)$, all y-components in $S(\,)$, and all z-components in $T(\,)$. This is more efficient when several points (vectors) are to be defined. Only three subscripted variables are needed to contain every component of every vector.

The subroutine PLANE returns the implicit coefficients as variables $N1$, $N2$, $N3$, and K. Another function is also performed, depending upon the value of N supplied by the main program. If $N = 3$ then a fourth vector $[R(4) \ S(4) \ T(4)]$ is defined equal to the first vector $[R(1) \ S(1) \ T(1)]$. This facilitates drawing a three-sided plane segment by moving to point 1 and then drawing to points 2, 3, 4 in order. If $N = 4$, then a fourth point is calculated to create a rectangular facet and a fifth point is set equal to the first. These extra vectors are returned to the main program in the $R(\,)$, $S(\,)$, and $T(\,)$ arrays. The user of subroutine PLANE must specify the value of N to be equal to either 3 or 4. Also notice that PLANE calls VECPROD to calculate the cross product. The input vectors to this subroutine must be in the form $\mathbf{X} = [X(1) \ X(2) \ X(3)]$ and $\mathbf{Y} = [Y(1) \ Y(2) \ Y(3)]$. The components of the returned cross products are $\mathbf{Z} = [Z(1) \ Z(2) \ Z(3)]$.

Vector equations have now been defined to represent both lines and planes. These will be used in subsequent chapters to calculate the inter-

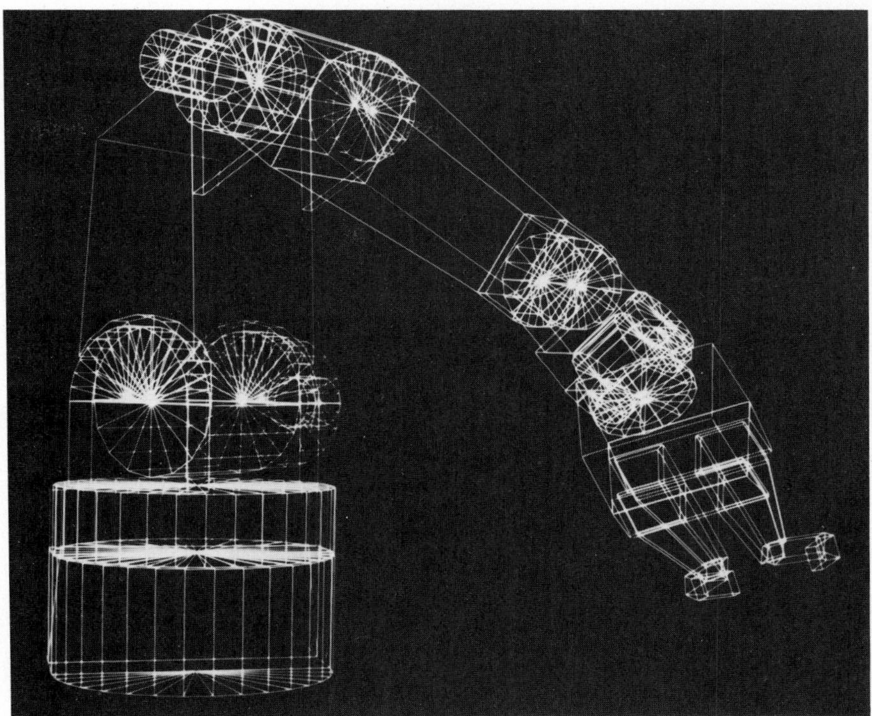

Figure 4-5
Robot arm showing orientation of plane facets. Courtesy of Evans and Sutherland Inc.

section point between a line and plane, as well as other geometric attributes. An example of using subroutine PLANE is given in the following example. Figure 4-5 shows an object with many plane facets.

Example 4-15

Given: A plane is defined by three vectors in space.

$$\mathbf{P}_1 = [\ 5 \quad 7 \quad 6]$$
$$\mathbf{P}_2 = [-7 \quad 4 \quad 6]$$
$$\mathbf{P}_3 = [\ 5 \quad -4 \quad 7]$$

Task: Write a BASIC program which will call subroutine PLANE and print out the components of the normal vector and the coefficient k.
Solution: A sample program is listed below. The solution is

$$N1 = -3 \quad N2 = 12 \quad N3 = 132 \quad k = 861$$

```
100 !*****DEFPLANE*****
110 !
120 ! DEFINE A PLANE WITH 3 NONCOLLINEAR PTS
130 ! USE SUBROUTINE PLANE TO FIND
140 ! COMPONENTS FOR IMPLICIT EQN
150 DIM X(5), Y(5), Z(5)
160 PLOTTER TEK4010
170 OPTION NOLET
180 LIBRARY"PLANE"
190 PRINT"TO DEFINE A TRIANGULAR PLANE"
200 PRINT"ENTER 3 NONCOLLINEAR PTS IN SPACE"
210 FOR I = 1 TO 3
220 PRINT"POINT";I; "X,Y,Z";
230 INPUT X(I), Y(I), Z(I)
240 NEXT I
250 TYPE = 3
260 CALL PLANE (TYPE,X,Y,Z,N1,N2,N3,K)
270 PRINT
280 PRINT"NORMAL COMPONENTS ARE";N1;N2;N3
290 PRINT
300 PRINT"COEFFICIENT K =";K
310 END
```

```
SAMPLE RUN
TO DEFINE A TRIANGULAR PLANE
ENTER 3 NONCOLLINEAR PTS IN SPACE
POINT 1 X,Y,Z? 5,7,6
POINT 2 X,Y,Z? -7,4,6
POINT 3 X,Y,Z? 5, -4,7
NORMAL COMPONENTS ARE 3   12   132
COEFFICIENT K =861
```

4-7 POINT IN A PLANE

It is often necessary to know whether a given point in space lies in a particular plane. Further, does the point lie within a particular plane facet with finite boundaries? Consider the scalar equation for a plane given by Equation (4-13) and define a function $f(x,y,z)$ as follows:

$$f(x,y,z) = N_1x + N_2y + N_3z - k = 0 \tag{4-16}$$

Now consider a point in space represented by the vector $\mathbf{S} = [x\ y\ z]$. If the values of x,y,z are substituted into Equation (4-16) and $f(x,y,z) = 0$, then the vector **S** lies within the plane. Furthermore, if $f(x,y,z) < 0$ the vector **S** lies to one side of the plane, and if $f(x,y,z) > 0$ the vector lies to the other side of the plane. Thus, vectors $\mathbf{S}_1$ and $\mathbf{S}_2$ will lie on the same

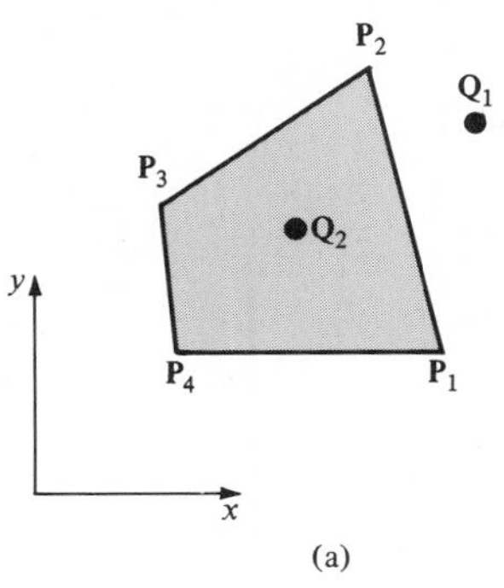

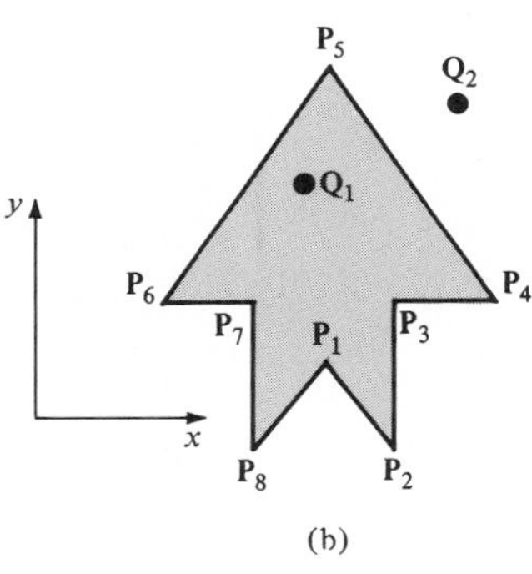

Figure 4-6
Two planar facets.

side of the plane if $f(x_1,y_1,z_1)$ has the same sign as $f(x_2,y_2,z_2)$. If the signs are *opposite* one can conclude that the line joining $\mathbf{S}_1$ and $\mathbf{S}_2$ will always *intersect* the plane since a plane is infinite in extent and the two vectors are on opposite sides of the plane.

If a point lies within a plane a simple test called the *polygon test* can be used to establish whether the point lies within a finite region of interest. Let the finite, planar region of interest be defined by a set of polygon points $\mathbf{P}_i$ which lie in the plane. This planar region can have *any* shape. Two examples are shown in Figure 4-6.

Let $\mathbf{Q} = [x \; y]$ be a general point in the plane containing a polygon defined by n polygon points. In general, the plane of interest will not be parallel to the x-y plane. However, since $\mathbf{Q}$ lies in the plane, the problem can be analyzed by considering the orthographic projection of the facet and the point $\mathbf{Q}$ onto the x-y plane.

Consider a simple triangle and its three polygon vertices in the x-y plane. Two cases are shown in Figure 4-7. In Figure 4-7(a) the point $\mathbf{Q}$ lies outside the triangular region, and in Figure 4-7(b) the point $\mathbf{Q}$ lies within the triangular region. In each case, lines from $\mathbf{Q}$ to each polygon vertex have been constructed. Let $\alpha_1 = \angle P_1QP_3$, $\alpha_2 = \angle P_3QP_2$, and $\alpha_3 = \angle P_2QP_1$. If the sum of $\alpha_1 + \alpha_2 + \alpha_3 = 360$ deg then the point $\mathbf{Q}$ lies within the boundary of the plane, triangular facet. If any angle α_i is 180 deg then $\mathbf{Q}$ lies on a boundary of the facet. If $\mathbf{Q}$ falls outside the triangular facet as in Figure 4-7(a), then the angle sum $\alpha_1 + \alpha_2 + \alpha_3$ is always less than 360 deg. This is also true for an n-sided planar facet.

A polygon facet is called *convex* if a line between any two points within the polygon lies totally within the polygon. Figure 4-6(a) is a convex polygon while Figure 4-6(b) is nonconvex. The polygon test can be used for both polygons if proper signs are used for the α_i angles, positive for counterclockwise and negative for clockwise. Figure 4-8 illustrates the use of the polygon test applied to a nonconvex, planar facet.

To implement an algorithm for the polygon test one needs a subroutine to add the α_i angles. The arctan function ATN gives only principal

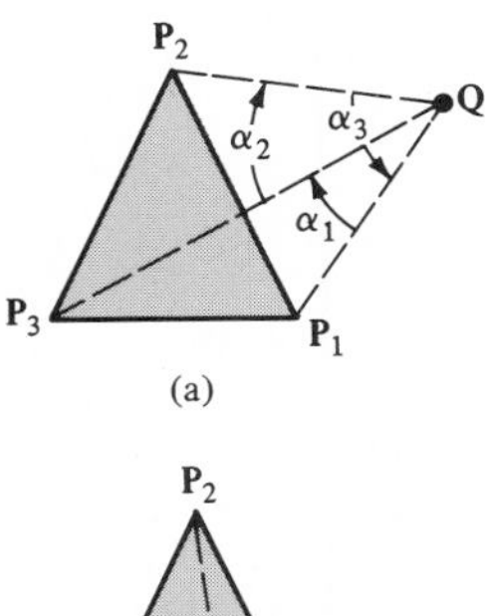

Figure 4-7
Triangular facets.

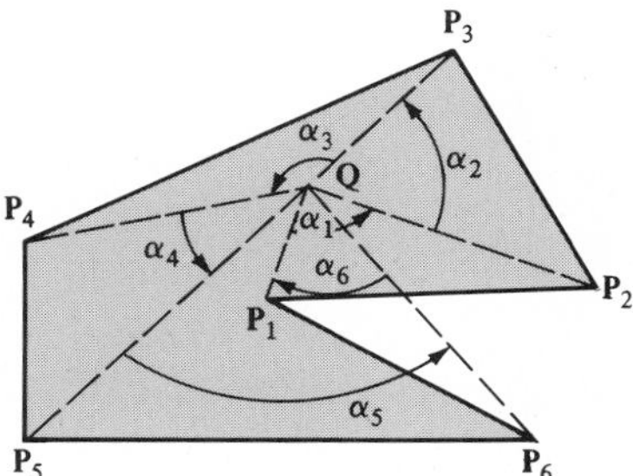

Figure 4-8
Nonconvex facet.

angles between $-\pi/2 \leq \alpha \leq \pi/2$. This is discussed in Appendix 2 where the subroutine ATAN is given. The ANGLE function in BASIC8 can also be used.

4-8 PARAMETRIC EQUATION OF A PLANE

Section 4-6 gave both a vector and scalar equation that can be used to define a plane. A parametric, vector equation can also be used to define points that fall on a given plane surface. The parametric, vector equation for a line was given in Section 3-8.

Let a plane pass through the point **Q** and have a normal vector $\mathbf{N} = \mathbf{A} \times \mathbf{B}$. For a flat surface, a bivariate, parametric (two variable parameters) equation for any point **P** in the plane is then given by: (See Figure 4-9.) At **P**, $s > 0$ and $t > 0$.

$$\mathbf{P} = \mathbf{Q} + s\mathbf{A} + t\mathbf{B} \tag{4-17}$$

The left side of Equation (4-17) is often written $\mathbf{P}(s,t)$ to indicate that **P** is expressed as a bivariate function of two scalar parameters s and t. Notice that vectors **Q** and **P** are fixed vectors, whereas vectors **A**, **B**, **N** are free vectors.

Since a plane is infinite in extent, any pair of values for s,t (that is, $-\infty < s < \infty$ and $-\infty < t < \infty$) will generate a point on the plane surface when used in Equation (4-17). A finite, plane facet can be defined by use of a finite range of parameters such as $0 \leq s \leq 1$ and $0 \leq t \leq 1$. This type of mathematical formulation can be extended to define

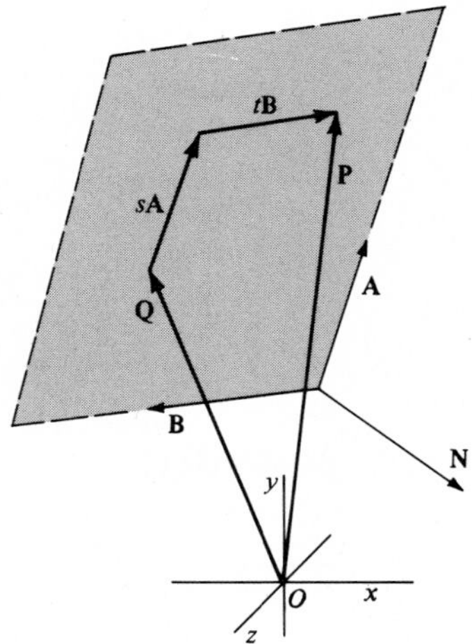

Figure 4-9
Parametric, vector equation for a plane.

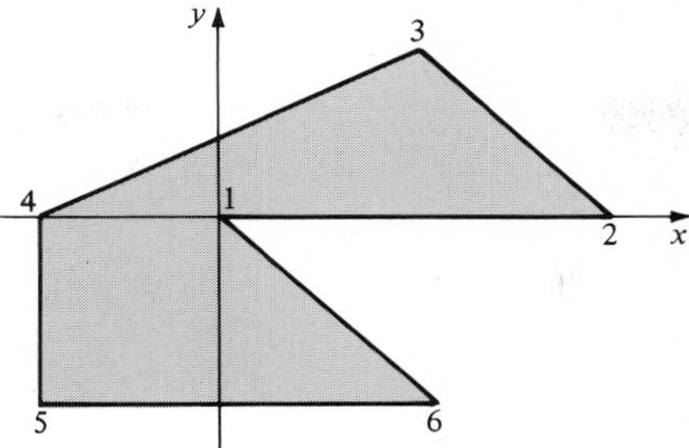

Figure 4-10
Nonconvex polygon facet.

curved surfaces for use in geometric modeling for computer aided design. Many of the listed references at the end of the text contain examples of this type of surface definition.

Example 4-16

Given: A polygon in the *x-y* plane is defined by the six vertices in the following matrix:

$$\begin{bmatrix} 0 & 0 & 0 \\ 2 & 0 & 0 \\ 1 & 1 & 0 \\ -1 & 0 & 0 \\ -1 & -1 & 0 \\ 1 & -1 & 0 \end{bmatrix}$$

The polygon is shown in Figure 4-10.
Task: Check the following points to see if they fall within the defined polygon facet.

$$\mathbf{P}_1 = [1.3 \;\; 0.3]; \quad \mathbf{P}_2 = [0.6 \;\; -0.5]; \quad \mathbf{P}_3 = [1.5 \;\; 0.5]$$

Solution: The reader should verify the values of the variables generated within subroutine POLY as given by the following tables.

Case 1: $\mathbf{Q} = [1.3 \;\; 0.3] = [A(7) \;\; B(7)]$

I	*T(I) at 170*	*I(I) at 200*	*I(I) at 260*	*S + I(I) at 330*
1	192.995	143.807	143.807	143.8
2	336.801	−223.603	136.397	280.2
3	113.199	74.233	74.223	354.4
4	187.431	22.045	22.045	376.5
5	209.476	47.527	47.529	424.0
6	257.005	—	−64.011	360.0 *P*$ = "inside"

Case 2: **Q** = [0.6 −0.5]

I	*T(I) at 170*	*I(I) at 200*	*I(I) at 260*	*S + I(I) at 330*	
1	140.194	−120.542	−120.541	−120.541	
2	19.654	55.415	55.415	−65.126	
3	75.069	87.577	87.577	22.452	
4	162.646	34.708	34.708	57.160	
5	197.354	111.306	111.306	168.465	
6	308.66	—	−168.465	0.0	*P*$ = "outside"

Case 3: **Q** = [1.5 0.5] ◀

I	*T(I) at 170*	*I(I) at 200*	*I(I) at 260*	
1	198.435	116.565	116.565	
2	315.0	−180.0	180.0	*P*$ = "on boundary"
3	135	56.309		
4	191.31	19.654		
5	210.96	40.601		
6	251.57	—		

Figures 4-11 and 4-12 show computer graphics output generated by the program SHOWPOLY. This program checks the status of any two-dimensional point relative to the defined polygon and draws both the polygon boundary and the point location. Subroutine ATAN is listed in Appendix 2.

```
90 !          POLY
100 EXTERNAL
110 SUB POLY (N,A( ),B( ),P$)
120 ! CALLED BY MAIN SHOWPOLY
130 ! CHECK IF POINT A(N + 1), B(N + 1) FALLS WITHIN POLYGON
140 ! PLANE POLYGON IN A( ), B( ) CONTAINS N POINTS
150 OPTION NOLET
160 DIM T(30)
170 DIM J(30)
180 SUM = 0
190 FOR I = 1 TO N
200 CALL ATAN (A(I) − A(N + 1), B(I) − B(N + 1), ANG)
210 T(I) = ANG*180/PI
220 NEXT I
230 FOR I = 1 TO N − 1
240 J(I) = T(I + 1) − T(I)
250 IF ABS(J(I))<180 THEN 270
260 J(I) = J(I) + 360
270 NEXT I
```

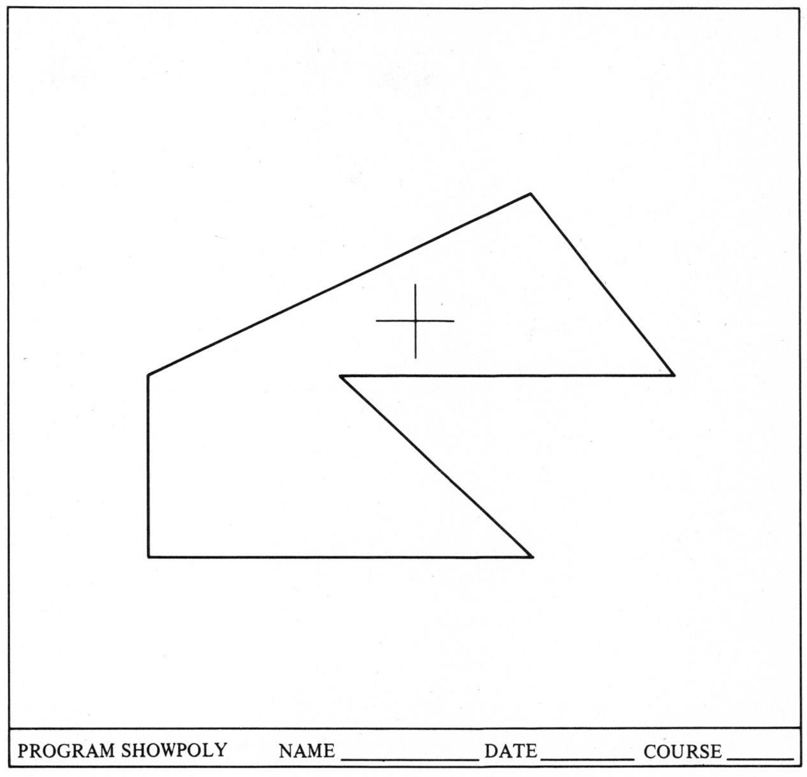

Figure 4-11
Inside point.

```
280 J(N) = T(1) - T(N)
290 FOR I = 1 TO N
300 IF ABS(J(I))<>180 THEN 330
310 P$ = "ON BOUNDARY"
320 GO TO 430
330 NEXT I
340 IF ABS(J(N))<180 THEN 360
350 J(N) = J(N) + 360
360 FOR I = 1 TO N
370 SUM = SUM + J(I)
380 NEXT I
390 IF ABS(SUM - 360)>0.1 THEN 420
400 P$ = "IN"
410 GO TO 430
420 P$ = "OUT"
430 END SUB
```

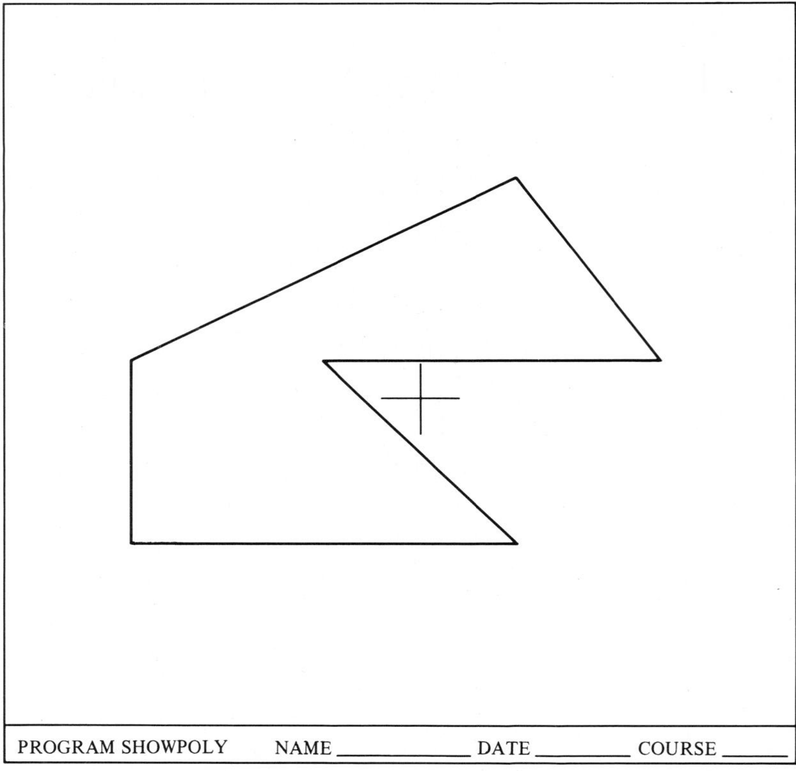

Figure 4-12
Outside point.

```
100 !*****SHOWPOLY*****
110 !
120 ! LIBRARY "ATAN"
130 LIBRARY"POLY"
140 LIBRARY"TITLE"
150 CLEAR
160 OPTION NOLET
170 PLOTTER TEK4010
180 DIM P(2)
190 DIM A(30),B(30),C(30)
200 SET VIEWPORT 30,130,0,100
210 W1 = -2
220 W2 = 2
230 W3 = -2
240 W4 = 2
250 SET WINDOW W1,W2,W3,W4
```

```
260 PRINT"2-D POLYGON VERTICES"
270 READ N
280 FOR I = 1 TO N
290 READ A(I),B(I)
300 PRINT A(I);B(I)
310 NEXT I
320 PRINT
330 PRINT"INPUT CHECK PT X,Y";
340 INPUT P(1),P(2)
350 A(N + 1) = P(1)
360 B(N + 1) = P(2)
370 FOR I = 1 TO N
380 PLOT A(I),B(I);
390 NEXT I
400 PLOT A(1),B(1)
410 PLOT P(1) - 0.05*(W2 - W1),P(2);P(1) + 0.05*(W2 - W1),P(2)
420 PLOT P(1),P(2) - 0.05*(W4 - W3);P(1),P(2) + 0.05*(W4 - W3)
430 DATA 6
440 DATA 0,0,1.75,0,1,1,-1,0,-1,-1,1,-1
450 CALL POLY(N,A,B,P$)
460 PRINT
470 PRINT"STATUS ";P$
480 PRINT
490 PRINT"WINDOW = ";W1;W2;W3;W4
500 TITLE$ = "PROGRAM SHOWPOLY"
510 NAME$ = "NAME____________"
520 DAY$ = "DATE____________"
530 COURSE$ = "COURSE____________"
540 CALL TITLE(W1,W2,W3,W4,TITLE$,NAME$,DAY$,COURSE$)
550 END
```

Example 4-17

Given: A plane region in the shape of a parallelogram, such as generated by subroutine PLANE (using TYPE = 4) and three specified noncollinear points in space. Assume that a point of intersection **Q** between a line and the infinite plane has been found.

Task: For this special case of a parallelogram, devise an alternate method to the polygon test to determine whether the point of intersection lies within the boundary of a parallelogram.

Solution: The following procedure is suggested:

a. Devise a local *u-v* coordinate system (not necessarily orthogonal) along sides 1-2 and 2-3 as shown in Figure 4-13.

b. Calculate the magnitude of vectors $\mathbf{Q}_1 = \mathbf{P}_3 - \mathbf{P}_2$ and $\mathbf{Q}_2 = \mathbf{P}_1 - \mathbf{P}_2$.

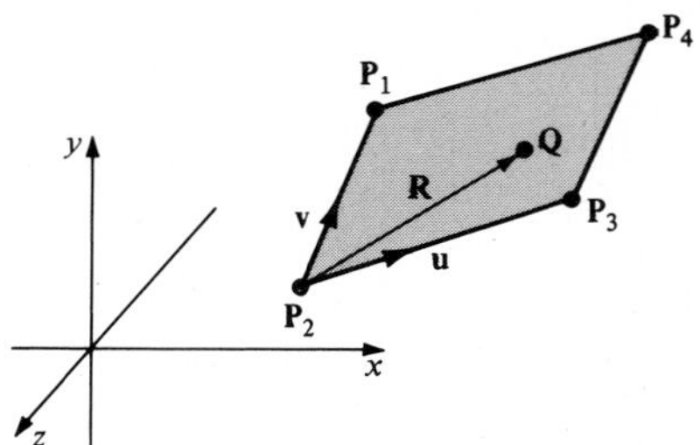

Figure 4-13
Parallelogram test.

c. Calculate the unit vectors **u**, **v** along axes u and v.

d. Calculate the components x,y,z of the direction vector **R** from $\mathbf{P}_2$ to **Q**.

e. Dot vector **R** with unit vectors **u** and **v**. This gives the projection of **R** in the **u** and **v** directions. If $\mathbf{R} \cdot \mathbf{u} < 0$ or $> |\mathbf{Q}_1|$; or $\mathbf{R} \cdot \mathbf{v} < 0$ or $> |\mathbf{Q}_2|$ then point **Q** is *outside* the parallelogram.

The subroutine BOUNDARY given below implements this theory.

```
100 ! *****BOUNDARY*****
110 !
120 EXTERNAL
130 SUB BOUNDARY(R( ),S( ),T( ),Q( ),P$)
140 OPTION NOLET
150 ! CHECK IF INTERSECTION IN Q( )
160 ! FALLS WITHIN BOUNDARY OF A
170 ! PARALLELOGRAM IN R( ),S( ),T( )
180 ! RETURN P$ AS "IN" OR "OUT"
190 DIM U(3),V(3)
200 ! CALC UNIT VECTOR U ON SIDE 2-3
210 X = R(3) - R(2)
220 Y = S(3) - S(2)
230 Z = T(3) - T(2)
240 M = SQR(X*X + Y*Y + Z*Z)
250 MAG1 = M
260 U(1) = X/M
270 U(2) = Y/M
280 U(3) = Z/M
290 ! CALC UNIT VECTOR V ON SIDE 2-1
300 X = R(1) - R(2)
310 Y = S(1) - S(2)
320 Z = T(1) - T(2)
330 M = SQR(X*X + Y*Y + Z*Z)
340 MAG2 = M
```

```
350 V(1) = X/M
360 V(2) = Y/M
370 V(3) = Z/M
380 ! CALCULATE COMPONENTS OF DIRECTIONAL VECTOR P2-Q
390 X = Q(1) − R(2)
400 Y = Q(2) − S(2)
410 Z = Q(3) − T(2)
420 ! USE DOT PRODUCT TO FIND RELATIVE COORDS
430 XO = X*U(1) + Y*U(2) + Z*U(3)
440 YO = X*V(1) + Y*V(2) + Z*V(3)
450 IF XO<O THEN 510
460 IF XO>MAG1 THEN 510
470 IF YO<O THEN 510
480 IF YO>MAG2 THEN 510
490 P$ = "IN"
500 GO TO 520
510 P$ = "OUT"
520 END SUB
```

PROBLEMS: CHAPTER 4

4-1 *Given:* Top, front, and side views of triangle *ABC* and point 1. Vectors **A** = [1 1 3], **B** = [3 6 6], **C** = [6 3 1], and $x_1 = 3$, $y_1 = 3$, $z_1 = 4$.

Task: a. Determine whether point 1 lies in plane *ABC*.

b. Place point 2 in plane *ABC* extended indefinitely. Use short method of solution.

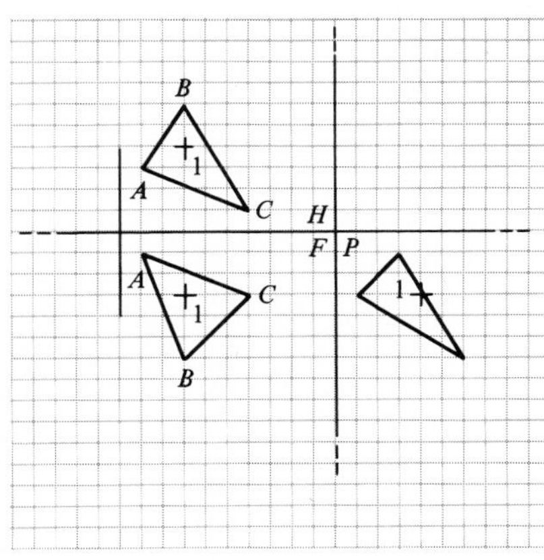

Figure 4-1P

4-2 *Given:* Front and top views of triangle *ABC*.
Task: Find the edge view of the given triangle.

A = [4 10 8] **B** = [16 8 12]

C = [10 4 3]

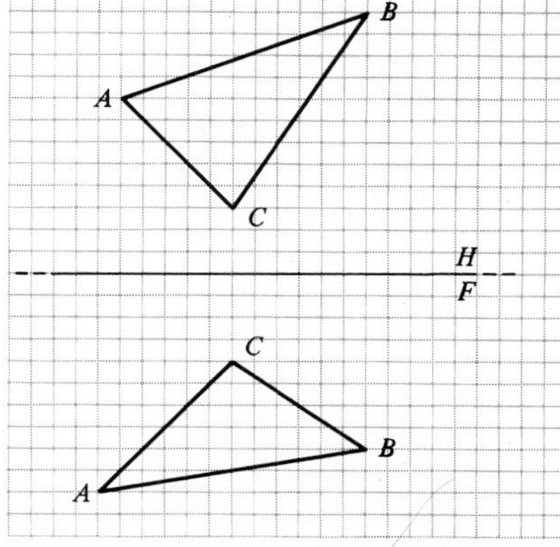

Figure 4-2P

4-3 *Given:* Front and top views of plane *ABC* and point *P*.

A = [7 9 7] **B** = [12.5 6.5 15]

C = [20 12.5 5] **P** = [12 9 10]

Task: Use both descriptive geometry and computation to determine whether point *P* is in, above, or below plane *ABC*.

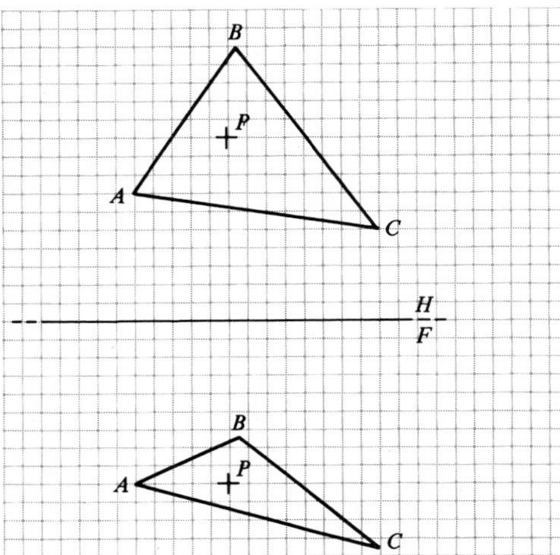

Figure 4-3P

4-4 *Given:* Front and top views of triangle *ABC*.
Task: Place point *P* in the plane and *y* distance below the horizontal plane. Show the point *P* in all views.

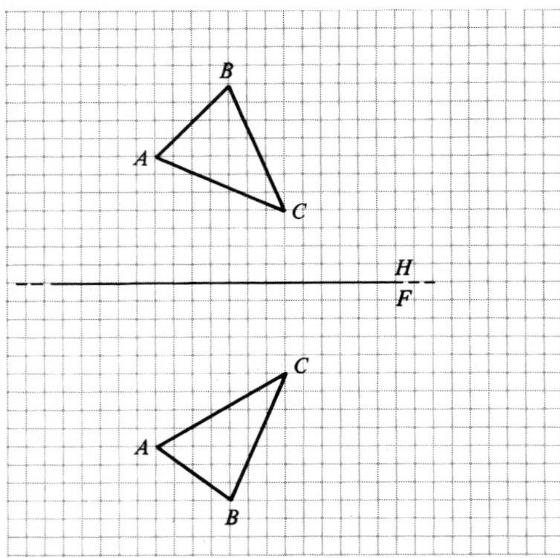

Figure 4-4P

A = [7 9 7] **B** = [12 12 11]

C = [15 5 4]

4-5 *Given:* Front and top views of lines *AB* and *CD*.
Task: Prove whether the given lines form a plane.

A = [5 5 11] **B** = [15 7 9]

C = [9 10 3] **D** = [17 9 7]

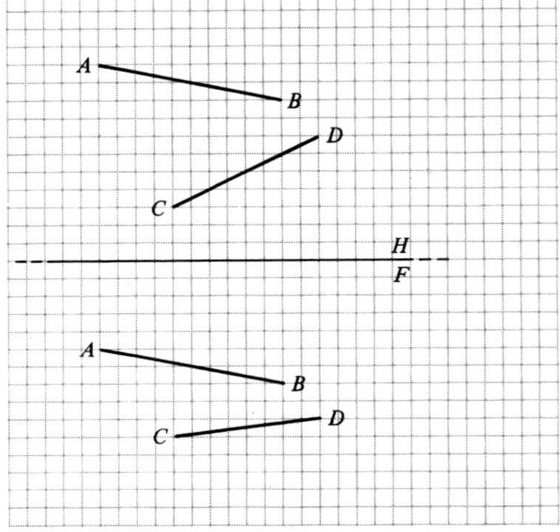

Figure 4-5P

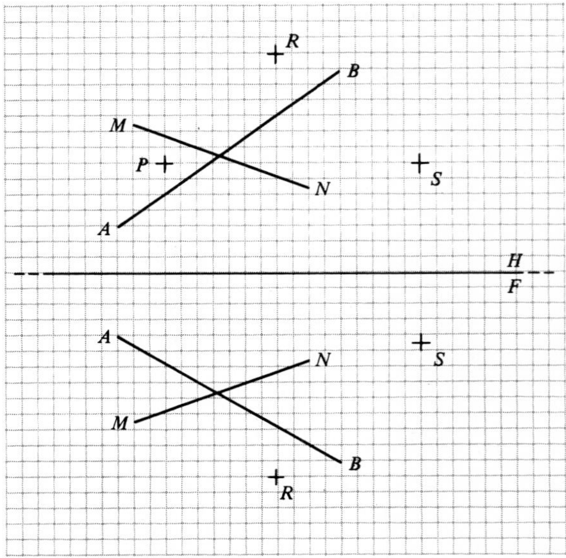

Figure 4-6P

4-6 *Given:* Front and top views of lines MN and CD. Front and top views of points S and R. Top view of point P.
Task: a. Place P in the front view of the plane which is defined by the two intersecting lines MN and AB.
b. Determine whether point R lies in the plane.
c. Determine whether point S lies in the plane.

4-7 *Given:* Front and top views of triangle ABC. Front view of point P which is in plane ABC.
Task: Draw a line in triangle ABC which contains point P and has a bearing of N47 deg E.

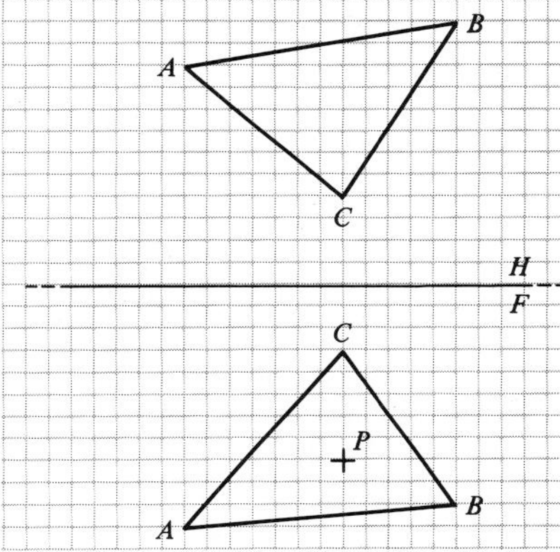

Figure 4-7P

4-8 *Given:* Front and top views of three points A, B, and C.
Task: Prove by graphics that the three given noncollinear points form a plane.

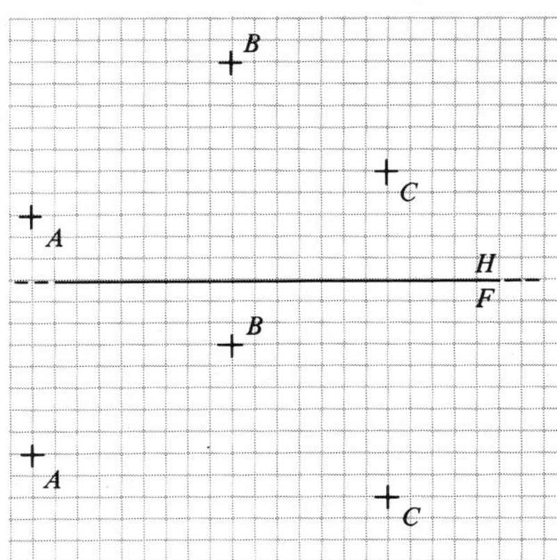

Figure 4-8P

4-9 *Given:* Front and top views of triangle ABC.
Task: As a requirement for future work show the steps needed to place the given plane on edge.

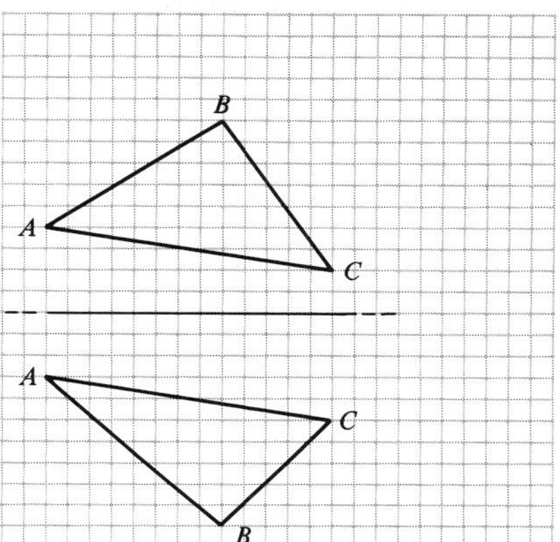

Figure 4-9P

4-10 *Given:* Front and top views of derrick supports (boom not shown) mounted on a barge.
Task: Find the true length and slope of the leg BC and of the anchor cable AC.

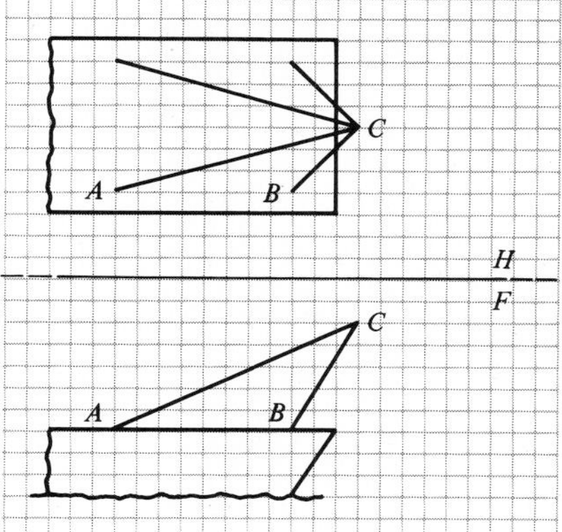

Figure 4-10P

4-11 *Given:* Base vectors $\mathbf{A} = [3\ -1\ 2]$ and $\mathbf{B} = [-2\ 1\ 4]$.
Task: Calculate the following quantities:

a. $\mathbf{A} \cdot \mathbf{B}$
b. $|\mathbf{A}|$ and $|\mathbf{B}|$

c. cos θ, where θ is the angle between the two vectors

d. $\mathbf{A} \times \mathbf{B}$

4-12 *Given:* Vectors $\mathbf{P}_1 = [3\ 2\ 4]$ and $\mathbf{P}_2 = [-2\ 1\ 1]$.
Task: Define the plane which contains the line P_1P_2 and passes through the origin of the coordinate system using the following two ways:

a. Define the normal to the plane at $\mathbf{P}_1$.

b. Find the implicit parameters for the plane.

4-13 *Given:* Vectors $\mathbf{P}_1 = [1\ 0\ 0]$, $\mathbf{P}_2 = [0\ 0\ 0]$, and $\mathbf{P}_3 = [0\ 1\ 0]$.
Task: Write the scalar (implicit) form of the plane equation which is defined by the given three vectors. Then repeat if $\mathbf{P}_3 = [1\ 1\ 0]$.

4-14 *Given:* Vectors $\mathbf{P}_1 = [-1\ -2\ 3]$, $\mathbf{P}_2 = [3\ 2\ 1]$, and $\mathbf{P}_3 = [1\ 4\ -2]$.
Task: a. Calculate the components of the normal vector to the plane containing these three vectors.
b. Calculate the value of k given by Equation (4-12).

4-15 *Given:* A plane in space passes through the following three points: $\mathbf{P}_1 = [1\ 4\ 4]$, $\mathbf{P}_2 = [3\ 3\ 2]$, and $\mathbf{P}_3 = [2\ 1\ 1]$.
Task: Calculate the direction cosines of the normal vector of this plane. Repeat if $\mathbf{P}_3 = [3\ 3\ 3]$.

4-16 *Given:* The three vectors used as input to define the triangular plane in Figure 2-26.
Task: a. Calculate the normal vector components and compare with those given in Figure 2-26.
b. Calculate the area of the triangle and compare with the area given in Figure 2-26.

4-17 *Given:* Vectors $\mathbf{A} = [1\ 0\ 0]$, $\mathbf{B} = [1\ 1\ 0]$, $\mathbf{C} = [1\ 1\ 1]$.
Task: Illustrate that the following general vector relationships are valid for the given vectors.

a. $(\mathbf{A} \times \mathbf{B}) \cdot \mathbf{A} = 0$

b. $(\mathbf{A} \times \mathbf{B}) \cdot \mathbf{B} = 0$

c. $\mathbf{A} \times \mathbf{A} = 0$

d. $\mathbf{A} \times \mathbf{B} = -\mathbf{B} \times \mathbf{A}$

e. $\mathbf{A} \times (\mathbf{B} + \mathbf{C}) = \mathbf{A} \times \mathbf{B} + \mathbf{A} \times \mathbf{C}$

f. $\mathbf{A} \times (\mathbf{B} \times \mathbf{C}) = (\mathbf{A} \cdot \mathbf{C})\mathbf{B} - (\mathbf{A} \cdot \mathbf{B})\mathbf{C}$

g. $|\mathbf{A} \times \mathbf{B}|^2 = |\mathbf{A}|^2\,|\mathbf{B}|^2 - (\mathbf{A} \cdot \mathbf{B})^2$

h.

$$\det \begin{vmatrix} A_x & A_y & A_z \\ B_x & B_y & B_z \\ C_x & C_y & C_z \end{vmatrix} = (\mathbf{A} \times \mathbf{B}) \cdot \mathbf{C}$$

4-18 *Given:* Three noncollinear vectors $\mathbf{P}_1 = [1\ 10\ -1]$, $\mathbf{P}_2 = [1\ 5\ -6]$, $\mathbf{P}_3 = [5\ 15\ -1]$ which define a plane.
Task: Determine the scalar (implicit) equation for the plane. See Equation (4-13).

4-19 *Given:* The plane triangle defined by Problem 2-12.
Task: a. Calculate the normal vector $\mathbf{N}$ to the plane.
b. Calculate the unit normal vector $\mathbf{n}$ to the plane.
c. Show that $\mathbf{N} \cdot \mathbf{A} = \mathbf{N} \cdot \mathbf{B} = \mathbf{N} \cdot \mathbf{C}$
d. Calculate the shortest distance from the origin of the coordinate system to the plane surface.

4-20 *Given:* Program VECPRODS and subroutines DOTPROD and VECPROD in Appendix 1.
Task: Use these algorithms to obtain a computer solution to Problems 4-11(a) and 4-11(d).

4-21 *Given:* Subroutine PLANE in Appendix 1.
Task: Write a program similar to DEFPLANE (Section 4-6) to obtain the implicit coefficients for the plane defined in Problem 4-15.

4-22 *Given:* Plane *ABC* as defined in Problem 4-1 and point 1 where $x_1 = 3$, $y_1 = 3$, $z_1 = 4$.
Task: a. Determine the implicit equation for this plane.
b. Substitute x_1, y_1, z_1 into the implicit equation to see if the equation is satisfied. Does point 1 lie on the plane?

4-23 *Given:* Program SHOWPOLY and subroutines POLY and TITLE in Appendix 1.
Task: Follow Example 4-16 and apply the polygon test to the two-dimensional region and check the status of the following points:

a. $x = -1.5$, $y = -1$

b. $x = -0.5$, $y = -1.5$

c. $x = -0.5$, $y = 1.0$

d. $x = 1.0$, $y = 2.0$

4-24 *Given:* The three points $\mathbf{P}_1$, $\mathbf{P}_2$, $\mathbf{P}_3$ in Problem 4-15.
Task: Choose values for **Q**, **A**, and **B** that can be used to write a bivariate, parametric equation for the plane given by the form of Equation (4-17).

4-25 *Given:* $\mathbf{P}_1 = [1\ 0\ 0]$, $\mathbf{P}_2 = [0\ 0\ 0]$, $\mathbf{P}_3 = [0\ 1\ 0]$.
Task: a. Calculate $\mathbf{P}_4$ needed to define a parallelogram.
b. Write the implicit equation which defines the plane.

4-26 *Given:* The plane defined in Problem 4-2.
Task: Write the vector equation for this plane as given by Equation (4-9).

4-27 *Given:* Triangle *ABC* defined in Problem 4-3.
Task: Calculate the length of the projection of side **B** − **A** onto side **C** − **A**.

4-28 *Given:* Two lines defined in Problem 4-5.
Task: Calculate the projection of line **B** − **A** in the direction of line **D** − **C**. Then use the dot product $\mathbf{R} \cdot \mathbf{R}$ to calculate the length of each line.

4-29 *Given:* A triangle whose sides are defined by $\mathbf{A} = 3\mathbf{i} + 4\mathbf{j} - \mathbf{k}$; $\mathbf{B} = 2\mathbf{i} - \mathbf{j} + 3\mathbf{k}$; $\mathbf{C} = \mathbf{i} + 5\mathbf{j} - 4\mathbf{k}$.
Task: Calculate the length of the projections of side **A** onto **B**, **B** onto **C**, and **C** onto **A**.

4-30[c] *C Given:* Problem 4-10.
Task: Calculate the lengths and slopes of the four derrick supports relative to a chosen scale. Draw the profile view.

4-31 *Given:* A plane defined by $\mathbf{A} = [4\ -1\ 2]$, $\mathbf{B} = [-2\ 3\ 1]$ and $\mathbf{C} = [3\ 1\ -2]$. Two additional points are $\mathbf{P}_1 = [5\ 4\ 4]$ and $\mathbf{P}_2 = [2\ 3\ -2]$.
Task: a. Do points $\mathbf{P}_1$ and $\mathbf{P}_2$ lie on the same or opposite sides of the plane?
b. Does either $\mathbf{P}_1$ or $\mathbf{P}_2$ lie on the same side of the plane as the origin of the coordinate system?
c. Which point is closer to the plane?

4-32[c] *Given:* A polygon plane facet is defined by the following four points in the $z=0$ plane.

Point	*x*	*y*
1	0	1
2	2	0
3	0	4
4	−2	0

Task: Use subroutine POLY to determine whether the following points fall within the facet boundaries.

Point	*x*	*y*
1	0	0.9
2	0.5	0.5
3	0.5	0.8
4	0	1.5

5

Successive Auxiliary Views

When two-dimensional methods of descriptive geometry are used, it is sometimes necessary to draw one or more auxiliary views to expose the geometric attribute of interest. This chapter presents methods for generating successive auxiliaries. Then several examples are given for common problems encountered in descriptive geometry. When a three-dimensional data base exists in a computer, a single orthographic view can usually be selected which shows the geometric attribute of interest. The chapter concludes with several examples of computer generated auxiliary views based upon carefully chosen eye positions in order to show the desired relationships. Sample subroutines are given in Appendix 1 which can be used to produce similar results.

5-1 SECOND AUXILIARY VIEWS

Chapters 3 and 4 were concerned with the methods and procedures used in solving problems by using primary auxiliaries which were projected from the principal planes. Figure 5-1(a) shows an exploded view using a first (or primary) auxiliary plane which, by definition, must be perpendicular to one principal plane and inclined to the other two. This particular figure shows that both the *frontal* plane and the *first* auxiliary plane are perpendicular to the *horizontal* plane. The layout is shown in Figure 5-1(b) as it would appear in the plane of the paper and the solution is given in Figure 5-1(c). It must be remembered that all auxiliary planes are oriented the same as principal planes (i.e., each projection plane must be at right angles with the one to which it is attached.)

The engineer will find that in practical work, not all objects are simple

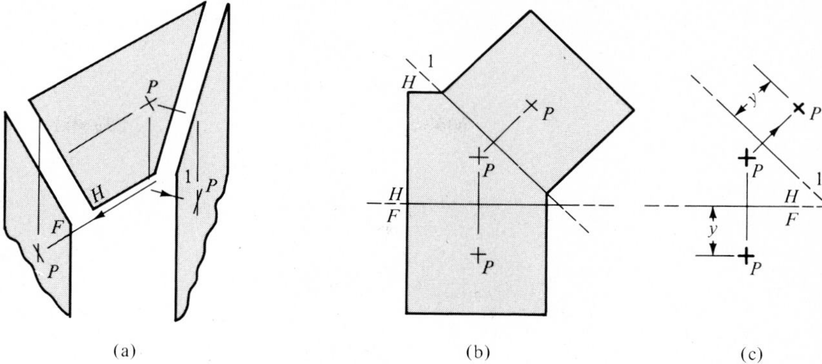

Figure 5-1
First auxiliary.

ones which consist only of principal lines and surfaces. In order to convey the exact details of an object so that replicas or mating parts may be created, it may be necessary to view the object through a plane which is *perpendicular* to a *primary* auxiliary view. The principles that are used in the construction of these views and of successive auxiliary views must always be followed so that each plane in the sequence or "chain" will be perpendicular to the preceding plane. The relationship of planes (perpendicular) must always be observed regardless of the number of views required.

Figure 5-2 is a continuation of Figure 5-1 and shows a second auxiliary plane added to the first auxiliary. Notice that the *horizontal* plane and

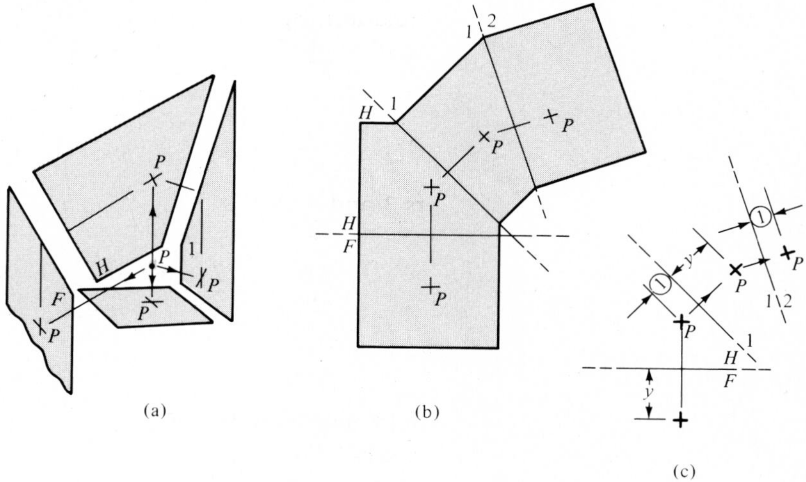

Figure 5-2
Second auxiliary.

the *second* auxiliary plane are *both* perpendicular to the *first* auxiliary plane. Since all planes are at right angles the information needed to produce the first auxiliary in Figure 5-1(c) was obtained from the front view and when the second auxiliary was drawn in Figure 5-2(c) the information from the top view was used.

Figure 5-2(a) shows an exploded view of the second auxiliary with the layout in Figure 5-2(b) and the solution in Figure 5-2(c). If the student has difficulty visualizing the principles involved in this operation it would be wise to make an overlay of Figure 5-2(b) and bend it at right angles on the folding lines so that the relationship of the planes can be seen.

5-2 SUCCEEDING AUXILIARIES

The previous section shows that the use of successive auxiliaries in descriptive geometry is nothing more than the continuation of orthographic projection used in various situations. The sequence of views will be as follows:

a. A primary auxiliary view projected from a principal view.
b. A second, or secondary, view projected from a primary view.
c. A third auxiliary view projected from a secondary view.
d. A continuation of the above sequences.

Future work will necessitate the completion of three, and possibly four, auxiliaries. Example 5-1 shows the steps required to show three principal views and three auxiliary views of a point in space. The example also shows the views in which the various measurements were taken and where they were plotted. Note that the principal planes are labeled *F, H, P,* and the auxiliary planes are numbered in sequence.

Example 5-1

Given: Top and front views of point *P*.
Task: Show the views of point *P* in the three principal views and in three auxiliary views.
Solution:

1. Draw the folding lines and lines of sight for the three principal views and for the first auxiliary view. The folding lines will be labeled *F/P*, *H/F*, and *H/*1, as shown.
2. Draw folding line 1/2 and the required line of sight. Point *P* lies dimension ① from the edge of the first auxiliary plane into the hori-

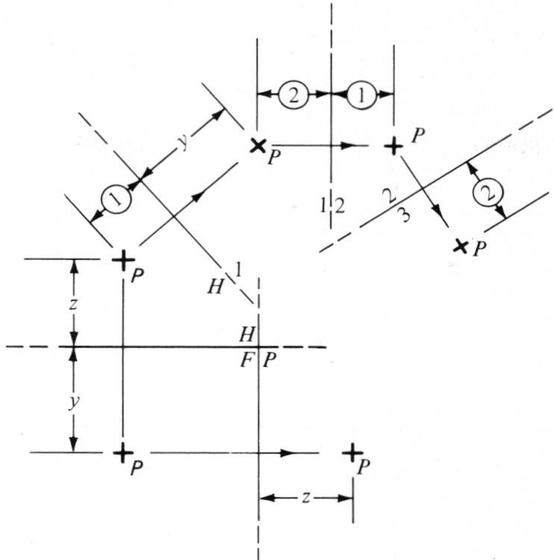

Example 5-1
Successive auxiliaries solution.

zontal plane and will be plotted dimension ① from the edge of the first auxiliary plane into the second auxiliary view.

3. Draw folding line 2/3 and the necessary line of sight. Point *P* lies dimension ② from the edge of the second auxiliary into the first auxiliary view and will lie dimension ② from the edge of the second auxiliary plane into the third auxiliary view. ◀

If more auxiliaries are required, the above procedures should be followed.

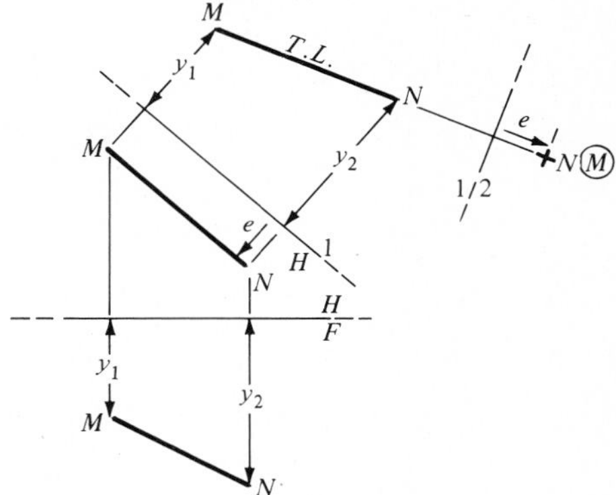

Figure 5-3
Point view of an oblique line.

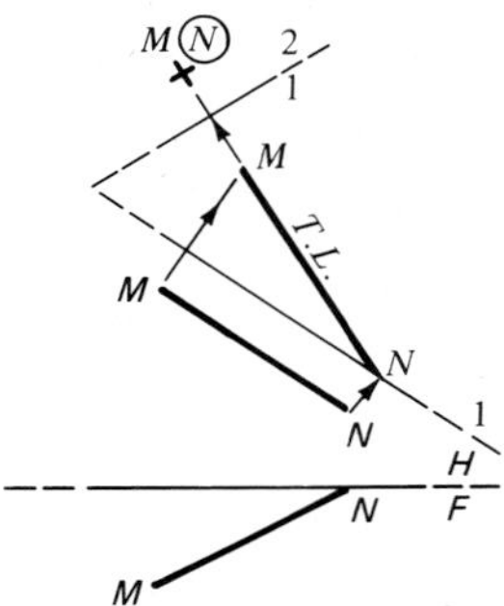

Example 5-2
Point view of an oblique line.

5-3 POINT VIEW OF AN OBLIQUE LINE

Prior material in this text discussed several of the four fundamental concepts of descriptive geometry which were given in Chapter 4. The student should now be familiar with the methods used to graphically determine the true length of all *straight* lines and the point view of principal lines.

The point view of an oblique line can be found by repeating the operations where the lines of sight pass through a plane that is perpendicular to a true length line. Since the lines of sight are then parallel to the line, the line will appear as a point on the projection plane. (See Figure 5-3.)

Example 5-2

Given: Top and front views of oblique line *MN*.
Task: Find the point view of line *MN*.
Solution:

1. In the top view, draw folding line *H*/1 parallel to line *MN*.
2. Project line *MN* into the first auxiliary where it is shown in true length.
3. In the first auxiliary, draw folding line 1/2 perpendicular to true length line *MN*.
4. The point view of line *MN* is shown in the second auxiliary view. ◀

Example 5-3 is given so that the student can follow the graphical solution and describe the procedure.

Example 5-3

Given: Front and right-side views of line *MN*.

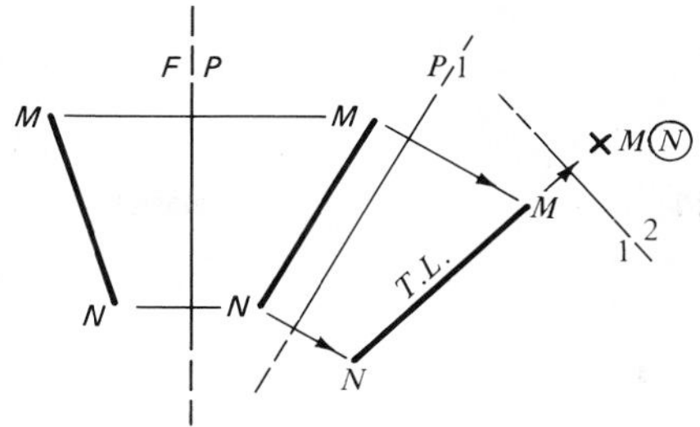

Example 5-3
Point view of a line.

Task: Find the point view of line *MN*.

5-4 NORMAL VIEW OF AN OBLIQUE PLANE

The last *fundamental* concept of descriptive geometry to be discussed is the system of operations required to find the normal view of a plane. This is not only one of the basic operations needed for further work in graphics, but it is probably one of the most widely used sources of information needed in industry. The normal view of an object clearly describes its true size and shape along with any irregularities which might appear. This, along with other design information, is essential to anyone involved in its construction. The engineer will find a better understanding of the conditions involved and problems to be solved when analytical methods are enhanced by graphical analysis.

Past experience in descriptive geometry will enable the student to obtain the edge view of a plane whether it is shown as a principal plane or an oblique plane. If it appears as an edge view, one auxiliary is needed to obtain the normal view. If the edge view is not shown, a principal line may be inserted in the oblique plane and two auxiliaries used.

In all cases the edge view must be either given or obtained so that a plane may be passed parallel to it in order to obtain a normal view of the given plane.

Example 5-4

Given: Top and front views of an oblique triangular plane *ABC*.
Task: Find the true size and shape of triangle *ABC*.
Solution:

1. Draw a horizontal construction line *AX* in the front view of the plane as shown. (See Section 4-1.)

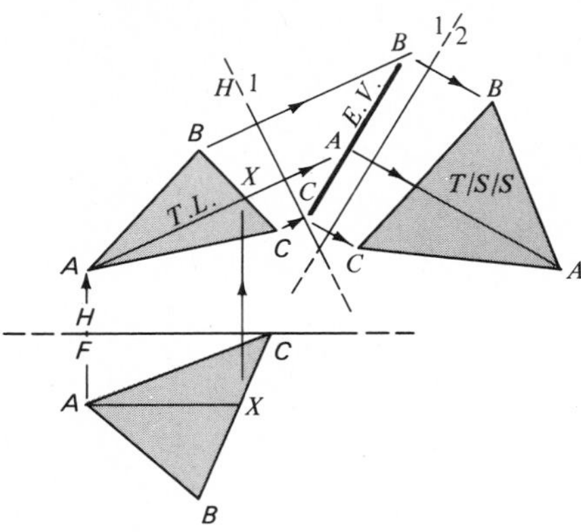

Example 5-4
Normal view of a plane.

2. Project points *A* and *X* into the top view and form a true length line *AX* in the horizontal view of plane *ABC*.
3. Extend lines of sight in the direction of the true length line.
4. Draw a folding line *H*/1 perpendicular to the lines of sight. This is, of course, perpendicular to the true length line.
5. Take dimensions from the *H*/*F* folding line (horizontal plane on edge) to each point in the front view. Using the *H*/1 folding line as a reference plane, plot each point on its respective line of sight in the first auxiliary.
6. This view shows the point view of construction line *AX* and the edge view of plane *ABC*. At this point, disregard the construction line since it was employed only to find direction of projection.
7. Draw folding line 1/2 parallel to the edge view of *ABC*.
8. Construct necessary lines of sight into the second auxiliary.
9. Use folding line *H*/1 as a reference and measure the distance to points in the top view. With the 1/2 folding line as a reference, transfer these distances to the second auxiliary.
10. Connect these points to obtain the normal view of the triangular plane *ABC*. This normal view shows the true size and shape of the plane. ◀

The solution to Example 5-5 is shown so that it may be studied and a step-by-step procedure set down by the student.

Example 5-5

Given: Top and front view of plane *ABCD*.

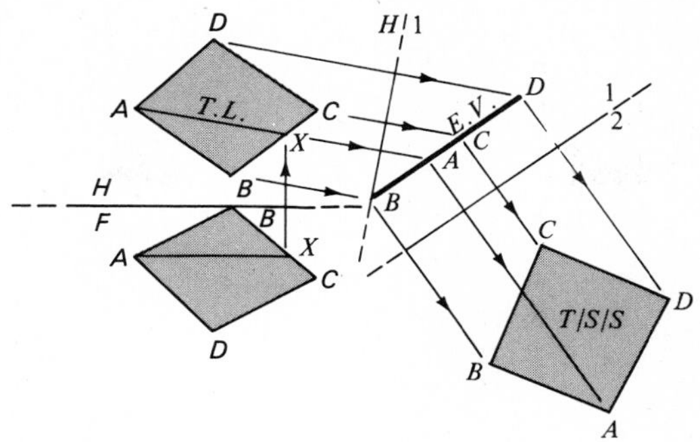

Example 5-5
Normal view of a plane.

Task: Show the true size and shape of plane *ABCD*.

5-5 NORMAL VIEW OF AN IRREGULAR OBLIQUE PLANE

Curves, especially noncircular ones, can be transferred by projecting a sufficient number of points on the curve and drawing a smooth line through them. (See Section 2-6.) The projection of curved shapes is handled the same way in all projection problems and should be treated as a locus of points.

Example 5-6

Given: Front and top views of irregular plane facet *ABC*.
Task: Find the true size and shape of plane facet *ABC*.
Solution:

1. Arbitrarily designate numbered points on the irregular side of plane *ABC*.
2. Insert horizontal line *AX* in the front view and project it into the top view where it becomes true length.
3. Project numbered points to the plane *ABC* in the top view.
4. Project lines of sight from the top view parallel to the true length line *AX* into the first auxiliary.
5. Draw folding line *H*/1 perpendicular to the true length line *AX*.
6. Obtain dimensions in the front view and plot them in the first auxiliary view. Plane *ABC* will appear on edge.
7. Draw folding line 1/2 parallel to the edge view of the plane and draw lines of sight from all points into the second auxiliary.

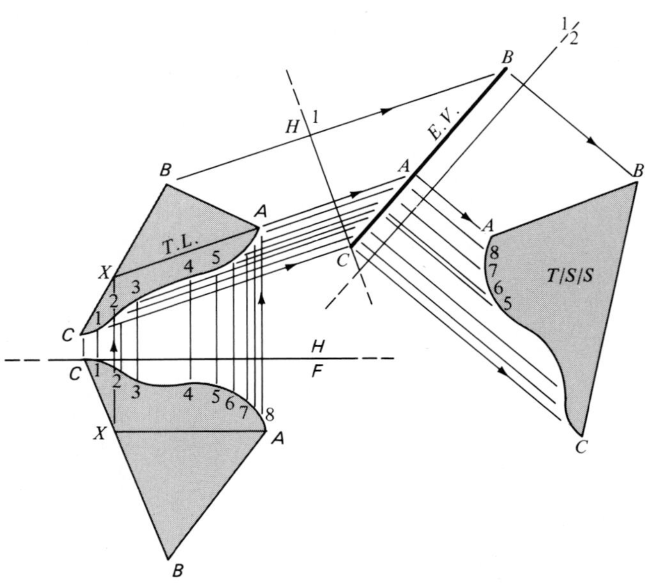

Example 5-6
Normal view of a plane facet.

8. Obtain dimensions from folding line $H/1$ into the top view and plot them from folding line 1/2 into the second auxiliary view.
9. Connect all points to show the normal view of the plane. ◀

Example 5-7 is presented for self-study.

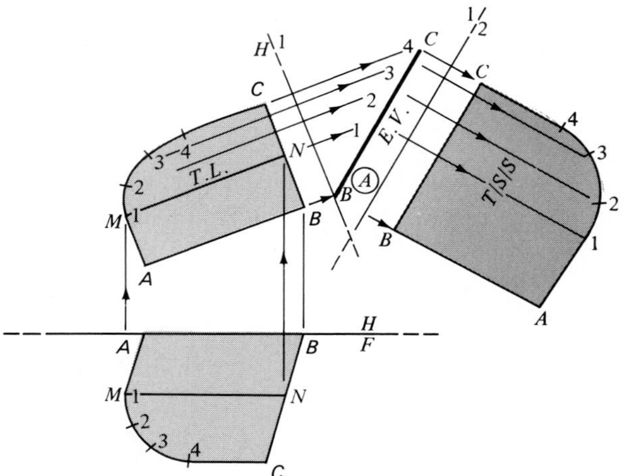

Example 5-7
Normal view of a plane facet.

Example 5-7

Given: Front and top views of a plane facet, *ABC*.
Task: Show the normal view of plane facet *ABC*.

5-6 ORTHOGRAPHIC VIEWS WITH COMPUTER GRAPHICS

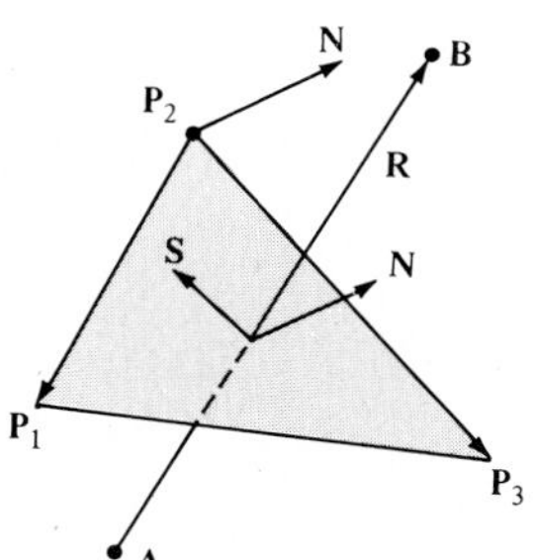

Figure 5-4
Line-plane intersection.

The program ORTHOEYE used in Chapter 2 made use of the matrix theory given in Appendix 2 to generate orthographic views as seen from any eye position specified by the user. Often it is more convenient to calculate the eye position based upon some descriptive geometry requirement.

Consider a line-plane intersection as illustrated in Figure 5-4. The normal to the plane is $\mathbf{N} = (\mathbf{P}_1 - \mathbf{P}_2) \times (\mathbf{P}_3 - \mathbf{P}_2)$. A line of sight along **N** shows the plane in true size, but not the true length of the line between **A** and **B**. The directional vector of this line is $\mathbf{R} = \mathbf{B} - \mathbf{A}$. A line of sight along **R** shows a point view of the line and a distorted view of the plane. A line of sight along any line in the plane, such as the edge $(\mathbf{P}_1 - \mathbf{P}_2)$, shows the plane on edge. However, only one line in the plane will show both the plane on edge and the true length of the line $(\mathbf{B} - \mathbf{A})$. This is the line of sight formed by the vector cross product $\mathbf{S} = \mathbf{N} \times \mathbf{R}$. Since **S** is perpendicular to **N** (it is in the plane) and since **S** is perpendicular to **R**, the true length of $(\mathbf{B} - \mathbf{A})$ is seen along with the edge view of the plane when the line of sight is along **S**.

Example 5-8

Given: Consider a line through $\mathbf{A} = [4\ 2\ 5]$ and $\mathbf{B} = [15\ 13\ 13]$. This line passes through the plane defined by $\mathbf{P}_1 = [3\ 15\ 8]$, $\mathbf{P}_2 = [10\ 2\ 16]$, and $\mathbf{P}_3 = [18\ 9\ 5]$. A vertex file PLATE1 and a line file PLINE1 can be created as shown in Table 5-1. The first two vertices are **A** and **B**. Vertices 3, 4, 5 are the three points which define the plane.
Task: Calculate the line of sight (direction components) which will show the plane on edge and the line in true length.

Table 5-1

(Vertex)	*PLATE1*	*PLINE1*
1	4,2,5	1,2
2	15,13,13	3,4
3	3,15,8	4,5
4	10,2,16	5,3
5	18,9,5	

Solution: Let

$$C_1 = X(5) - X(4) = 18 - 10 = 8$$

$$C_2 = Y(5) - Y(4) = 9 - 2 = 7$$

$$C_3 = Z(5) - Z(4) = 5 - 16 = -11$$

$$D_1 = X(3) - X(4) = 3 - 10 = -7$$

$$D_2 = Y(3) - Y(4) = 15 - 2 = 13$$

$$D_3 = Z(3) - Z(4) = 8 - 16 = -8$$

These are the direction components of the two edges of the plane facet. The normal vector is

$$\mathbf{N} = \mathbf{C} \times \mathbf{D} = \det \begin{vmatrix} \mathbf{i} & \mathbf{j} & \mathbf{k} \\ 8 & 7 & -11 \\ -7 & 13 & -8 \end{vmatrix} = 87\mathbf{i} + 141\mathbf{j} + 153\mathbf{k}$$

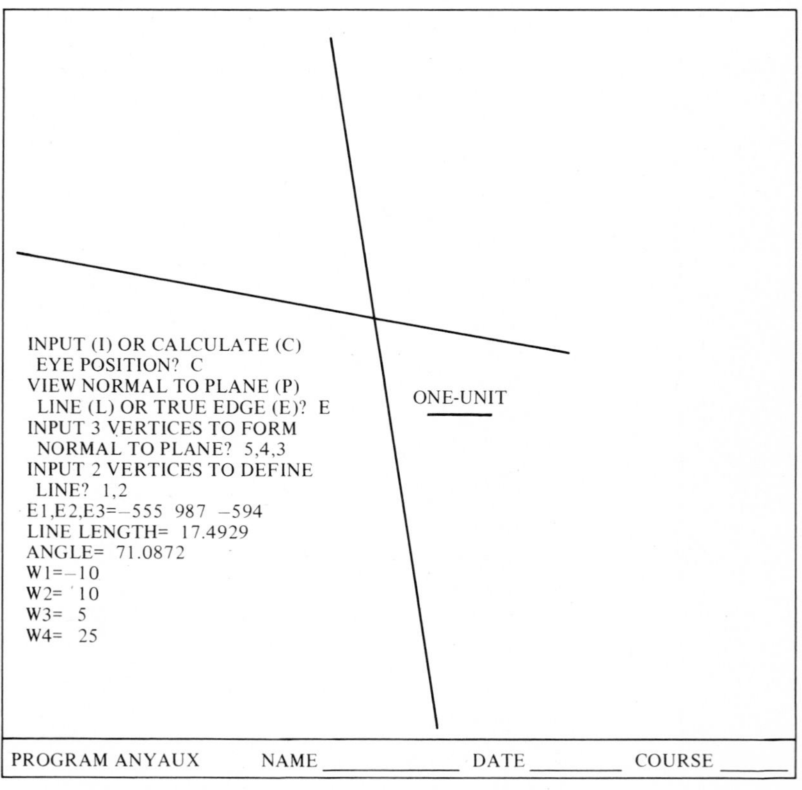

Figure 5-5
Edge view of plane and true length of line.

The direction vector for the line **B** − **A** is **R** = 11**i** + 11**j** + 8**k**
Then

$$\mathbf{S} = \mathbf{N} \times \mathbf{R} = \det \begin{vmatrix} \mathbf{i} & \mathbf{j} & \mathbf{k} \\ 87 & 141 & 153 \\ 11 & 11 & 8 \end{vmatrix} = -555\mathbf{i} + 987\mathbf{j} - 594\mathbf{k}$$

These components give the direction components for the line of sight which will show the required orthographic projection. The program ANYAUX performs these calculations as shown in Figure 5-5. When option *E* is chosen and eye components E1 = −555, E2 = 987, and E3 = −594 are passed to subroutine ORTHOMAT, a 3 × 3 transformation matrix is returned as explained in Appendix 2. ◀

Within the program ANYAUX, the orthographic projection onto the *x*-*y* plane is accomplished by a matrix multiplication of the data matrix and the transformation matrix in lines 870–900. The true length is cal-

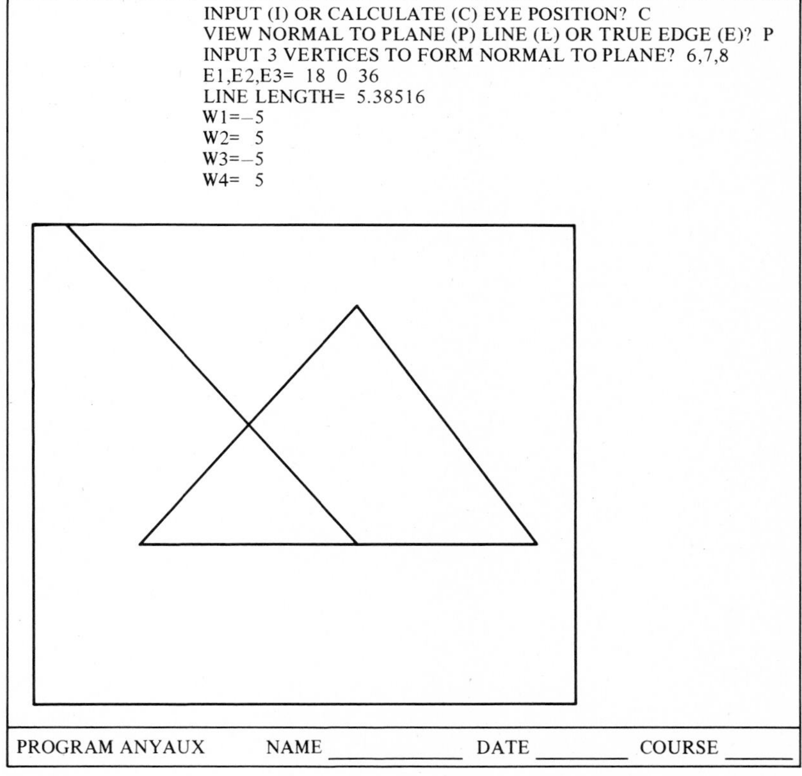

Figure 5-6
True size of rectangle.

culated to be 17.49 and the angle between the line and plane is 71.09 deg as found by use of the ANGLE function in lines 980–1000.

5-7 SPECIAL ORTHOGRAPHIC VIEWS

To further illustrate the capabilities of ANYAUX, consider the vertex file ASSY3PTS and the line file ASSY3LNS used to generate Figure 1-44. These two files are listed below.

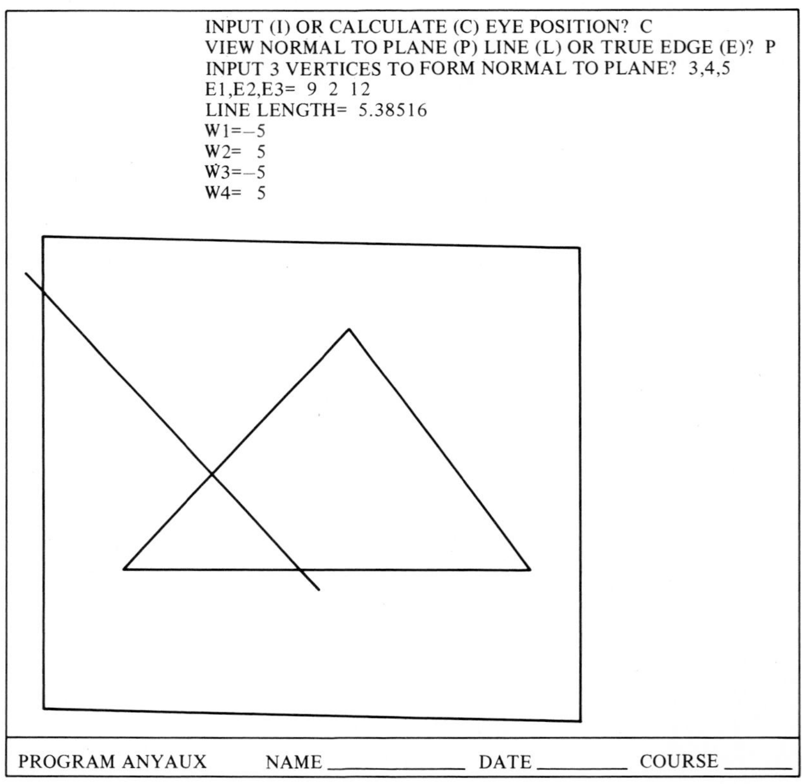

Figure 5-7
True size of a triangle.

ASSY3PTS	ASSY3LNS
−1,3,3	1,2
2,−1,1	3,4
−1,−1,1	4,5

1,2,−1	5,3
3,−1,−2	6,7
−3,−3,0	7,8
−3,3,0	8,9
3,3,−3	9,6
3,−3,−3	

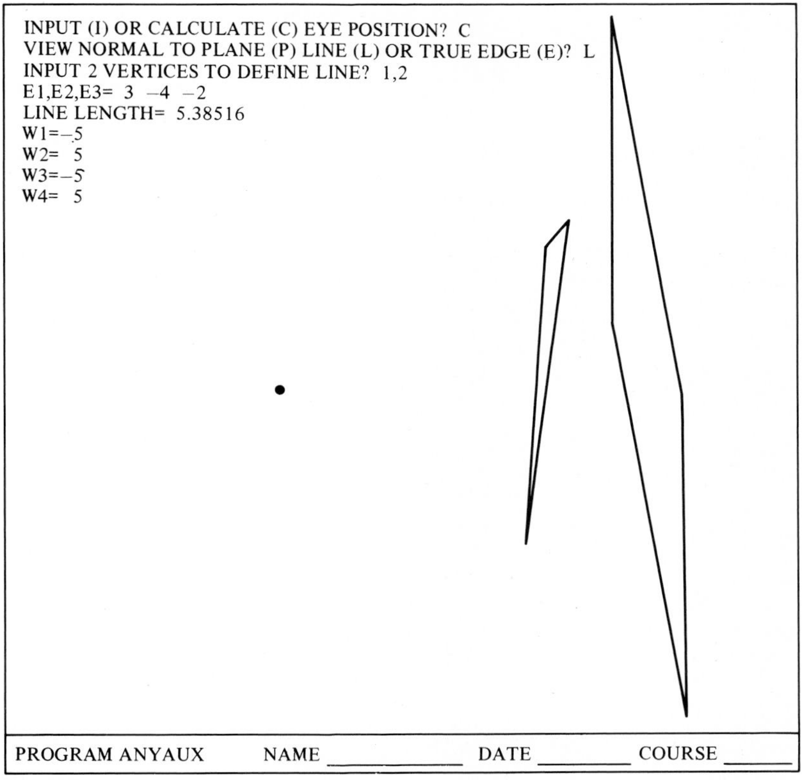

Figure 5-8
Point view of a line.

The lines in ANYAUX which must be changed to display the objects defined in these files are lines 320–350 and lines 380–410.

```
320 W1 = −5
330 W2 = 5
340 W3 = −5
350 W4 = 5
```

```
360 SET WINDOW W1,W2,W3,W4
370 ! CHANGE NEXT 4 LINES
380 VERTICES = 9 !NO OF VERTICES IN FILE #1'
390 LINES = 8 !NO OF LINES IN FILE #2'
400 OPEN #1: name "ASSY3PTS"
410 OPEN #2: name "ASSY3LNS"
```

Recall that the order of the vertices in ASSY3PTS corresponds to the numbers used in ASSY3LNS. Thus, a line is drawn between the first two vertices in ASSY3PTS. Then a triangle is formed using the next three vertices, and finally a square is drawn using the remaining four points.

Figure 5-6 creates a view normal to the plane defined by vertices 6, 7, 8. This is the plane of the rectangle. Thus, the rectangle appears in true

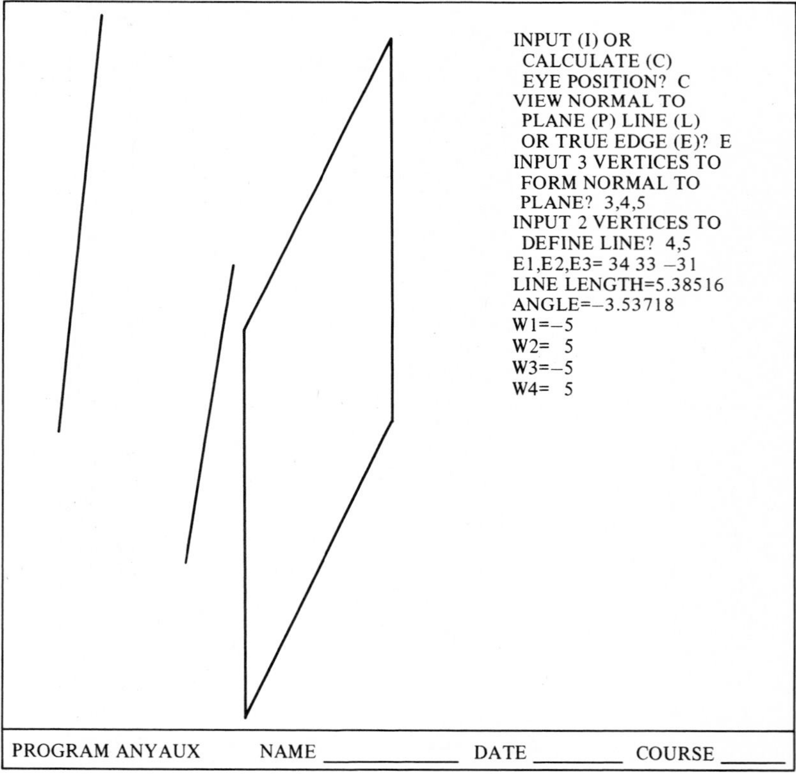

Figure 5-9
Edge view of plane 1.

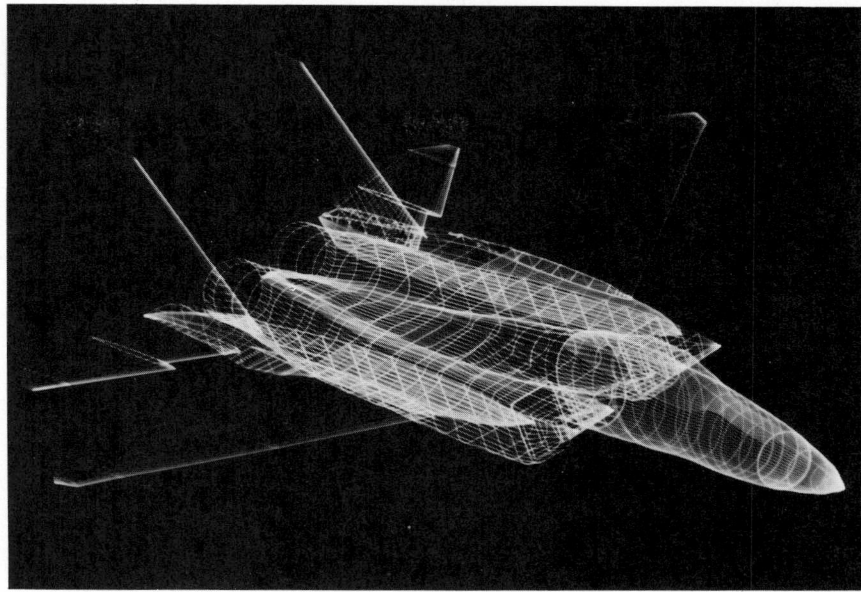

Figure 5-10
Wire frame projection. Courtesy of Evans and Sutherland Inc.

size in Figure 5-6. Notice the values of the WINDOW variables W1, W2, W3, and W4 used in Figure 5-6.

Figure 5-7 shows a similar display where the triangle is shown in true size and shape. By comparing Figures 5-6 and 5-7 it can be seen that the triangle and rectangle are almost parallel, but not exactly.

Figure 5-8 depicts an orthographic projection of a plane perpendicular to the line between vertices 1 and 2. Thus, the line appears as a point. Figure 5-9 shows the plane defined by vertices 3, 4, 5 as an edge and the line between vertices 4, 5 in true length. Many other auxiliaries can be created in a similar manner. The reader should be able to follow the documentation in ANYAUX and understand the mathematical operations used in the solutions.

The use of computer graphics allows one to "walk around" an object and view it from any position. This can be done interactively in a dynamic manner on some computer graphics work stations. Many techniques such as hidden surface removal, shading, ray tracing, surface texturing, animation, and use of colors gives the rendering of solid surfaces exciting possibilities for the user. These topics are discussed in the computer graphics literature. Figure 5-10 shows a wire frame projection produced on a modern engineering work station.

PROBLEMS: CHAPTER 5

5-1 *Given:* Front and right-side views of plane *ABCDEF*.
Task: Draw the normal view of the given plane.

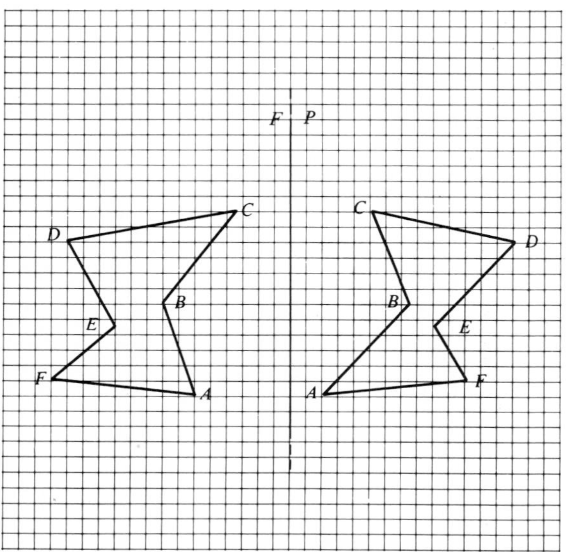

Figure 5-1P

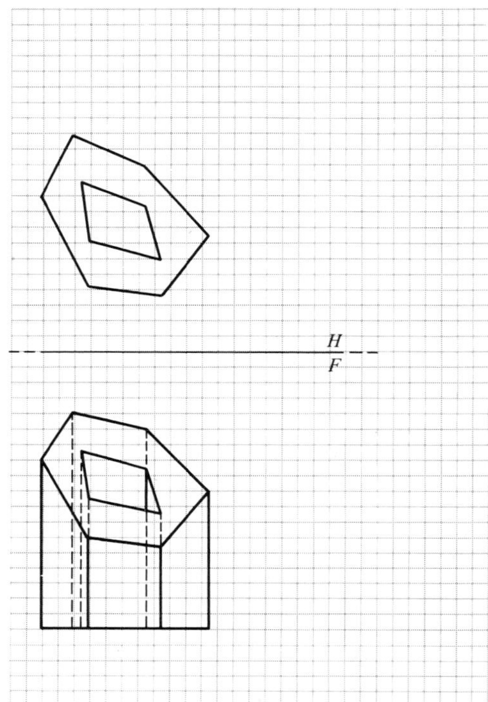

Figure 5-2P

5-2 *Given:* Front and top views of a truncated hollow prism.
Task: Find the normal view of the truncated surface so that a mating part can be made.

5-3 *Given:* Front and right-side views of a mutilated block.
Task: Draw the normal view of surface *ABCDEF*.

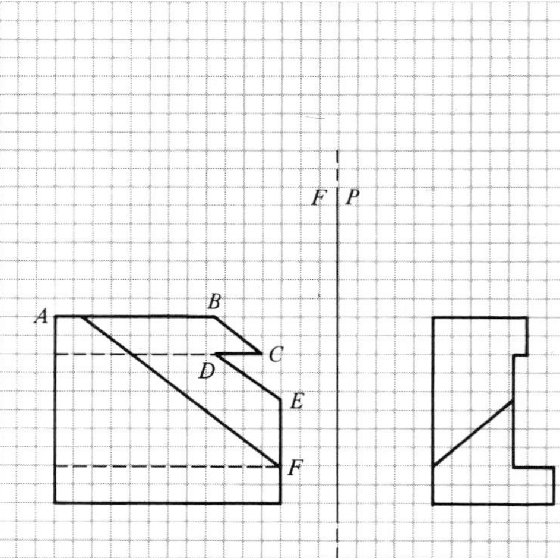

Figure 5-3P

5-4 *Given:* *P* is a point on the circumference of a circle whose axis is *MN*. Front and top views of line *MN* and point *P* are shown.
Task: Draw the top and front views of the circle.

5-5 *Given:* Front and top views of points *A*, *B*, and *C* which define a triangular-shaped ore vein.
Task: Determine the normal view of the ore vein so that the surface area can be determined.

5-6 *Given:* Front and top views of parallel lines *MN* and *AB*.

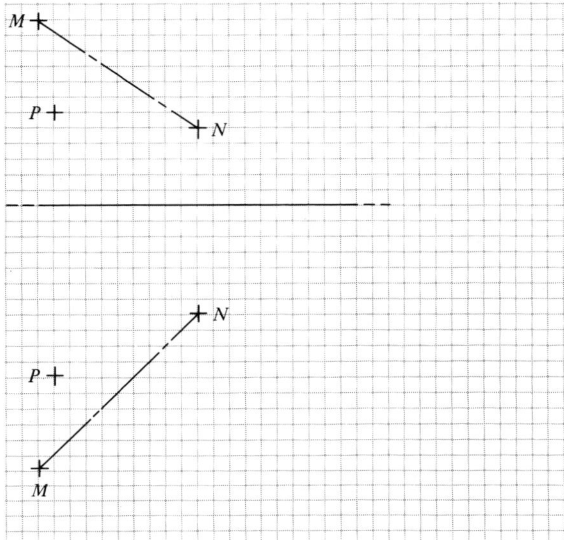

Figure 5-4P

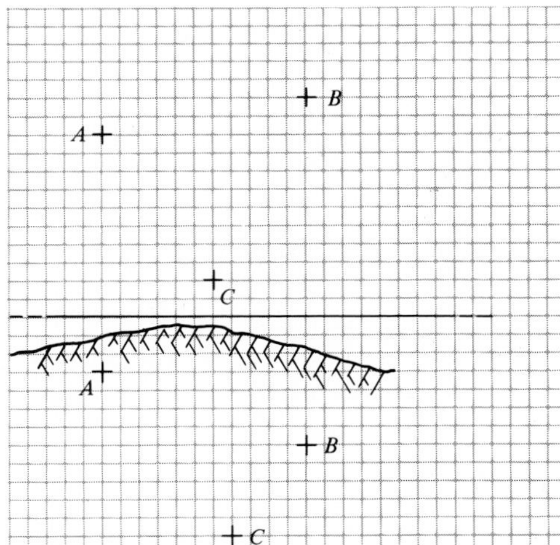

Figure 5-5P

Task: Determine the shortest distance between the two given lines.

5-7 *Given:* Front and top views of a block.
Task: Draw the first and second auxiliaries of the given block in the directions indicated by the folding lines.

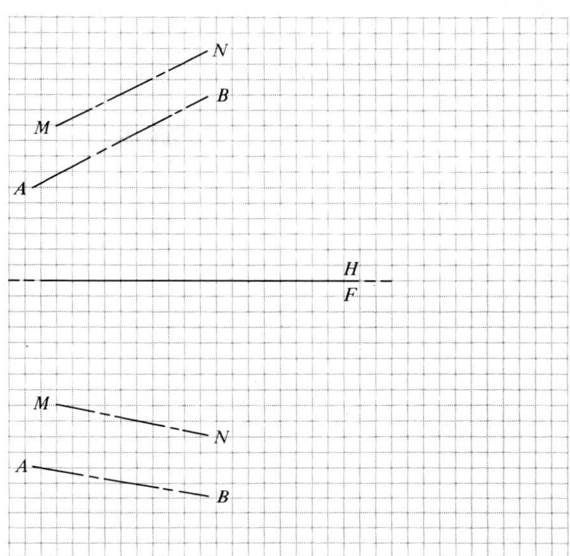

Figure 5-6P

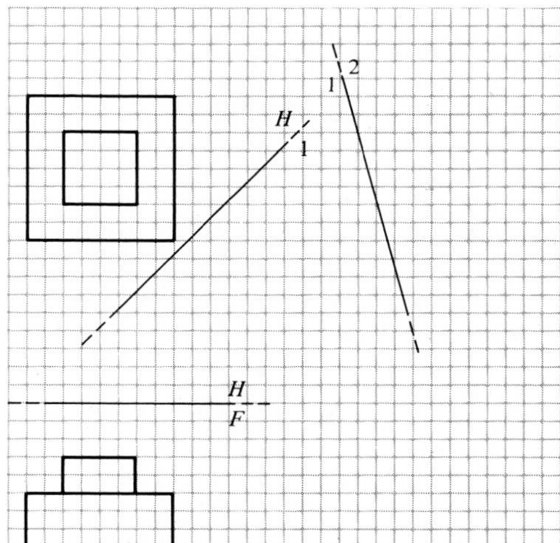

Figure 5-7P

5-8 *Given:* Front and top views of a truncated prism.
Task: Draw the two first auxiliaries of the entire object in the directions indicated by the folding lines.

5-9 *Given:* Front and top views of a pyramid.
Task: Draw three successive auxiliaries as indicated by the three given folding lines.

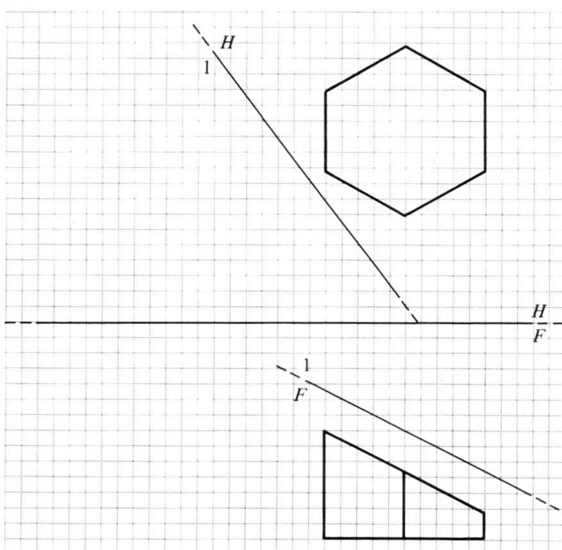

Figure 5-8P

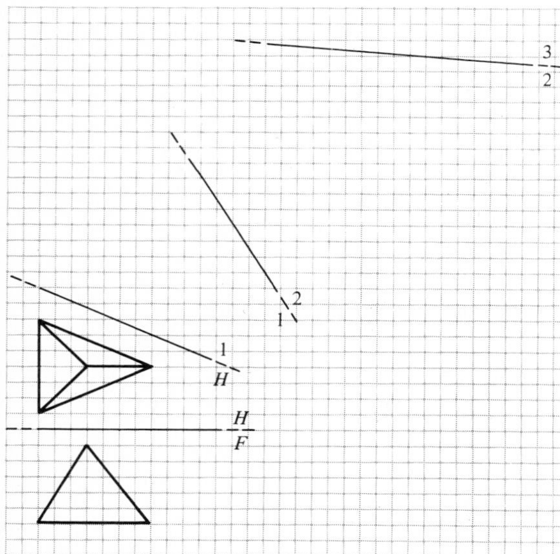

Figure 5-9P

5-10 *Given:* Front and top views of triangle *ABC*.
Task: Determine the angle that plane *ABC* makes with the horizontal plane.

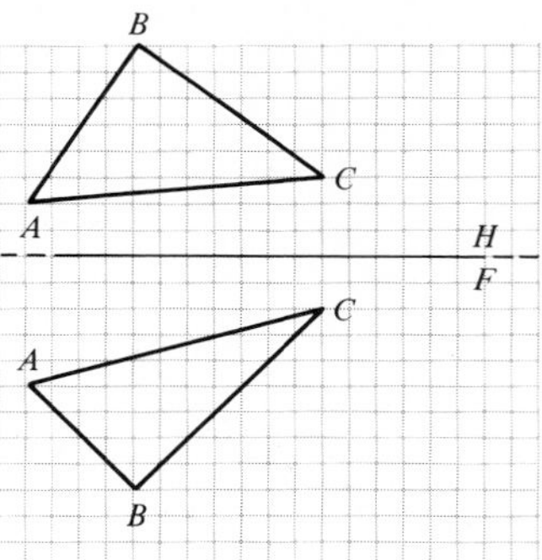

Figure 5-10P

5-11 *Given:* Matrix multiplication as defined in Appendix 2.
Task: Transform the three points represented by the vectors $\mathbf{P}_1 = [1\ 1\ 1]$, $\mathbf{P}_2 = [2\ 1\ 1]$, $\mathbf{P}_3 = [1\ 2\ 2]$ by using the matrix product **[P][T]** as follows:

$$\begin{bmatrix} 1 & 1 & 1 \\ 2 & 1 & 1 \\ 1 & 2 & 2 \end{bmatrix} \begin{bmatrix} 0.886 & 0.5 & 0 \\ -0.5 & 0.866 & 0 \\ 0 & 0 & 1 \end{bmatrix}$$

Plot the three initial and the three transformed points. What effect is caused by the transformation matrix **[T]**?

5-12 *Given:* Matrix multiplication as defined in Appendix 2.
Task: Repeat Problem 5-11 for the following **[T]** matrices:

a.
$$[\mathbf{T}] = \begin{bmatrix} 0.866 & 0 & -0.5 \\ 0 & 1 & 0 \\ 0.5 & 0 & 0.866 \end{bmatrix}$$

b.
$$[\mathbf{T}] = \begin{bmatrix} 1 & 0 & 0 \\ 0 & 0.866 & 0.5 \\ 0 & -0.5 & 0.866 \end{bmatrix}$$

5-13 *Given:* A triangle lying in the *x-y* plane defined by $\mathbf{A} = [5\ 3\ 0]$, $\mathbf{B} = [10\ 3\ 0]$, $\mathbf{C} = [10\ 6\ 0]$.
Task: Specify a 3×3 transformation matrix which will perform the following transformation to the initial object. (See Appendix 2.)

a. Rotate 45 deg counterclockwise about the *z*-axis
b. Rotate 45 deg counterclockwise about the *y*-axis
c. Rotate 45 deg counterclockwise about the *x*-axis.

5-14 *Given:* Matrix multiplication as defined in Appendix 2.
Task: Create a single 3 × 3 matrix which will cause a 30 deg counterclockwise rotation about the x-axis, followed by a 45 deg counterclockwise rotation about the y-axis.

5-15 *Given:* Subroutine EYETOZ in Appendix 1.
Task: What two rotations are combined into the single matrix **R**(,) in subroutine EYETOZ? What alternate matrix combination could be used to produce the same effect?

5-16 *Given:* Subroutine EYETOZ in Appendix 1.
Task: a. If E1 = 1, E2 = 2 and E3 = 3 in line 210, what are the elements in the 3 × 3 **R**(,) matrix which are returned to the main program?
b. Repeat part a if E1 = 2, E2 = −3, and E4 = 5.

5-17 *Given:* A three-dimensional line in space passes between **A** = [−1 4 6] and **B** = [3 −2 5].
Task: What values of the eye position E1, E2, E3 should be defined in subroutine EYETOZ (Appendix 1) to create a 3 × 3 matrix that can be used to make the given line parallel to the z-axis?

5-18[c] *Given:* Solid figure defined in Problem 1-18.[c]
Task: Create an orthographic projection which shows the area of the face 3, 7, 8, 4 in true size. Then create an orthographic projection which shows lines 3, 7 and 4, 8 as points.

5-19[c] *Given:* Program ANYAUX and subroutines VECPROD, AXISROT, and TITLE in Appendix 1.
Task: Use the discussion in Sections 5-6 and 5-7 as a guide and create special orthographic projections of three-dimensional objects of your choice. Commercial software may be used if available.

---

[c]Superscript *c* indicates that a computer solution is required.

6

Parallelism

The subject of parallelism often arises in design. It may be necessary to construct a line parallel to a plane, or a plane parallel to a line. The creation of parallel planes is also a common requirement. This chapter first gives the two-dimensional, descriptive geometry methods for solving these types of problems. The three-dimensional modeling methods are then presented which make use of the vector techniques presented earlier.

6-1 PARALLEL LINES

Parallel lines are those which have the same direction, and are an equal distance apart throughout their entire length. The angle between the two lines is 0 deg. Lines that are parallel in space appear in all views either as parallel lines, points, or collinear lines. Proof of the above is shown in Figure 6-1 where line *BD* was placed in plane *ABCD* in the front and top views. The second auxiliary shows the normal view of the plane and the true length of line *BD*. In this auxiliary, the line *MN* is drawn in the plane and parallel to line *BD*. When projected, the lines *BD* and *MN* appear either parallel or collinear in all views.

Caution must be used in determining whether lines *are* parallel in space even though they *appear* parallel on a drawing or computer terminal screen. *Oblique* lines which appear parallel in two or more views *are* parallel in space and, as stated before, may also be seen as points or collinear lines.

However, principal lines which *appear* parallel must be seen in true length in at least one adjacent view before they can be deemed parallel. In most cases, it is necessary to supplement the drawing by adding an

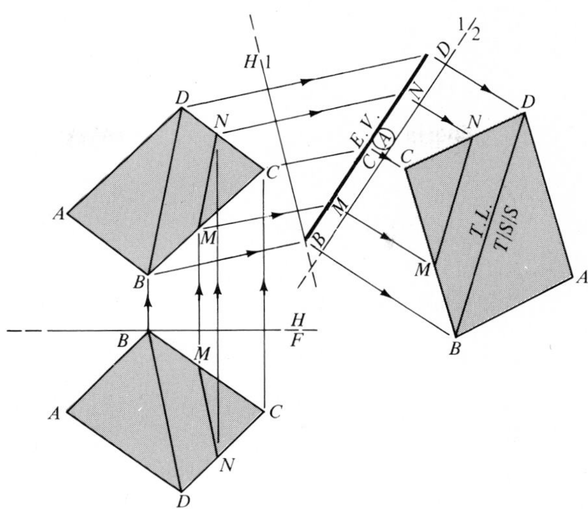

Figure 6-1
Parallel lines.

additional view. Figure 6-2 shows three views of principal lines which appear to be parallel but are shown by the addition of a third view to be nonparallel.

In order to find the distance between parallel lines it is well to proceed as shown in the following example:

Example 6-1

Given: Top and front views of two parallel lines *AB* and *MN*.

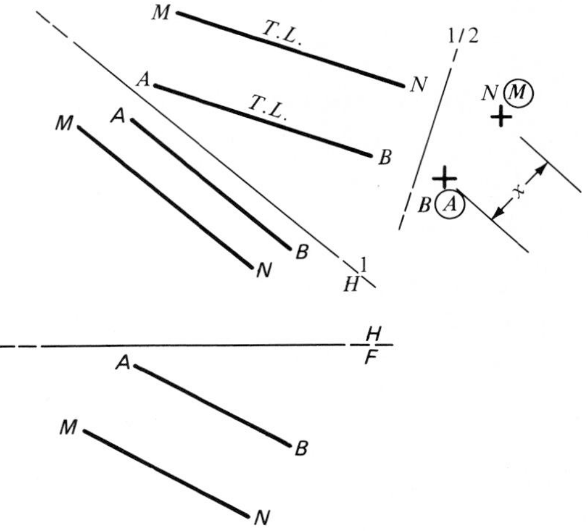

Example 6-1
Distance between parallel lines.

Task: Find the perpendicular distance between the two lines.
Solution:

1. In the first auxiliary, find the true length of the lines *AB* and *MN*.
2. In the second auxiliary, find the point view of the two lines.
3. Measure the required distance.

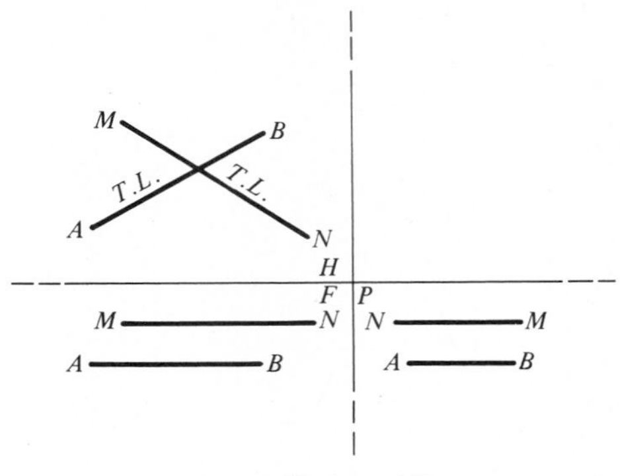

(a) Horizontal lines

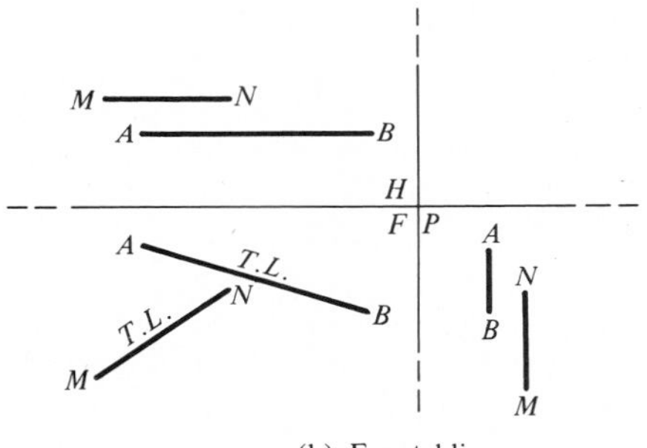

(b) Frontal lines

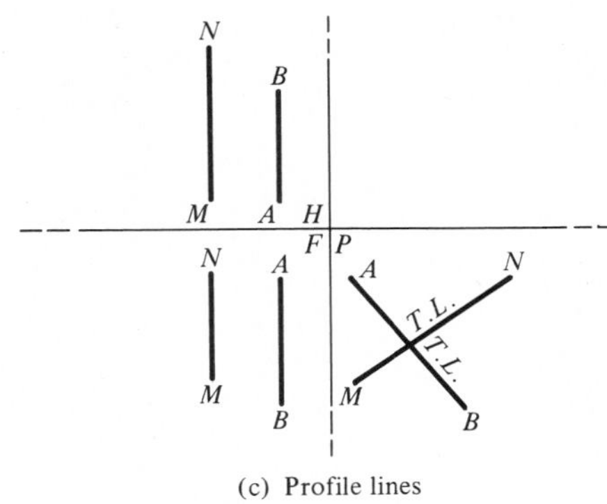

(c) Profile lines

Figure 6-2
Nonparallel lines.

6-2 LINE PARALLEL TO A PLANE

A line is parallel to a plane if it is parallel to any line in that plane.

A. Auxiliary Method

Example 6-2

Given: Top and front views of triangle *ABC* and point "*P*."

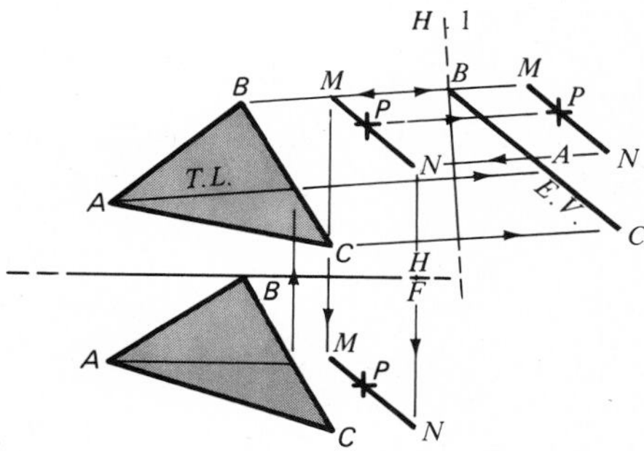

Example 6-2
Line parallel to a plane.

Task: Construct line *MN* through point "*P*" so that it is parallel to plane *ABC*.
Solution:

1. Draw a horizontal line in triangle *ABC* and project it into the top view where it becomes true length.
2. Project in the direction of the true length line and construct a first auxiliary so that point "*P*" and the edge view of triangle *ABC* are shown.
3. Construct a line *MN* through point "*P*" and parallel to the edge view of plane *ABC*. The line is now parallel to one line in the plane.
4. Define line *MN* and project it along with plane *ABC* into the top and front views. Line *MN* can assume any position in the top view. ◀

B. Two-View Method

If a limited amount of space is available, the following approach may be used:

Example 6-3

Given: Top and front views of plane *ABC*.

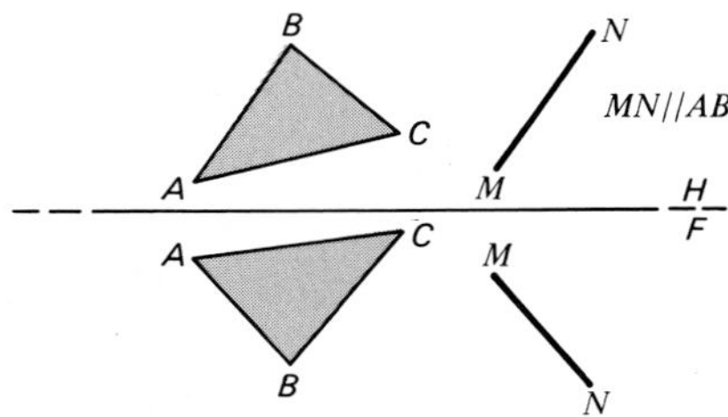

Example 6-3
Line parallel to a plane.

Task: Draw the top and front views of line *MN* so that it is parallel to the given plane.
Solution:

1. Draw line *MN* parallel to line *AB* in the top view.
2. Project both points into the front view and construct line *MN* parallel to line *AB*.

Line *MN* could have been constructed parallel to any line in the plane in order to meet the conditions of the problem. ◀

6-3 PLANE PARALLEL TO A LINE

A plane is parallel to a line if any line in the plane is parallel to the line.

A. Auxiliary Method

Example 6-4

Given: Top and front views of line *MN* and point "*P*."

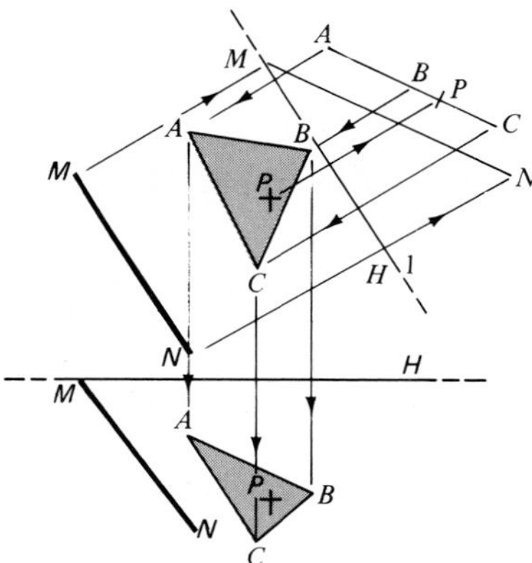

Example 6-4
Plane parallel to a line.

Task: Draw plane *ABC* so that it is parallel to line *MN* and contains point "*P*."
Solution:

1. In the first auxiliary, draw the edge view of plane *ABC* through point "*P*" and parallel to line *MN*. Notice that the point is neither above nor below the plane.
2. Project plane *ABC* into the top view so that points of the plane, although placed arbitrarily, encompass point "*P*."
3. Project plane *ABC* into the front view. ◀

B. Two-View Method

A plane may be constructed parallel to a line using the following method.

Example 6-5

Given: Top and front views of line *MN* and point "*B*."

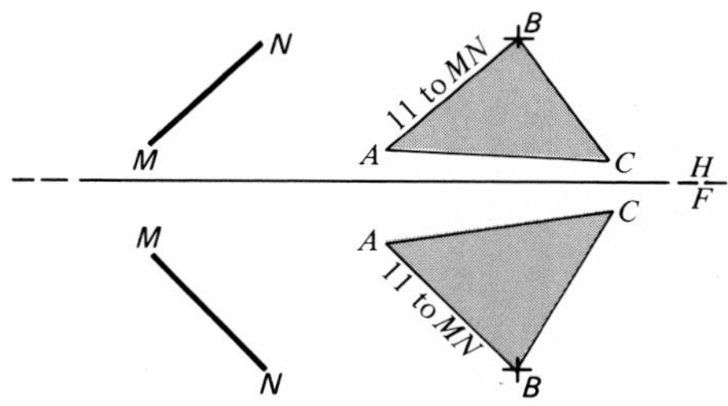

Example 6-5
Plane parallel to a line.

Task: Construct plane *ABC* through point "*B*" and parallel to line *MN*.
Solution:

1. In the top view, draw line *AB* through point *B* and parallel to line *MN*. Since line *AB* was drawn parallel to line *MN*, point "*C*" may be placed arbitrarily.
2. Draw line *AB* through point "*B*" and parallel to line *MN* in the front view. Project point "*C*" into the front view to complete triangle *ABC*. ◀

A plane parallel to two skew lines will contain two lines respectively parallel to the two given lines.

Example 6-6

Given: Top and front views of two skew lines *AB* and *MN*.

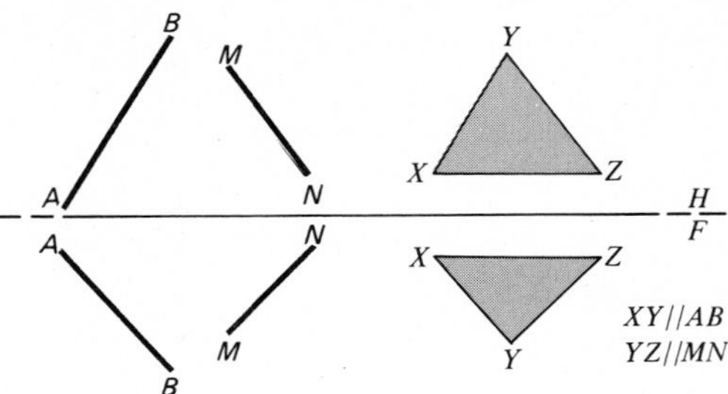

Example 6-6
Plane parallel to two skew lines.

Task: Construct triangle *XYZ* parallel to both line *AB* and line *MN*.
Solution: The foregoing example has been solved graphically using the two-view method which can be studied by the student and the step-by-step procedures noted. ◄

6-4 PARALLEL PLANES

Two planes are parallel when two intersecting lines of one plane are parallel respectively to two intersecting lines of the other plane. (See Figure 6-3.) The angle between the planes can be determined as 0 deg when the planes are seen on edge.

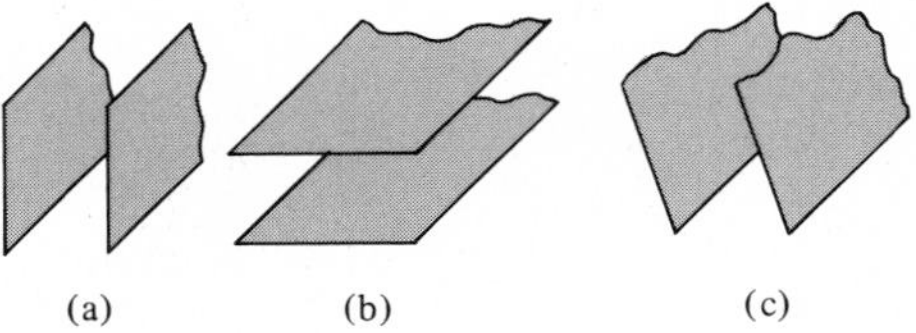

Figure 6-3
Plane parallel to a plane.

A. Auxiliary Method

The following solution is based on the fact that two parallel planes will appear on edge in the same view. It is in this view that the distance between the planes can be measured. Example 6-7, however, deals only with the construction of one plane parallel to another. Distance between planes will be discussed in the following chapter.

Example 6-7

Given: Top and front views of plane *ABCD*.

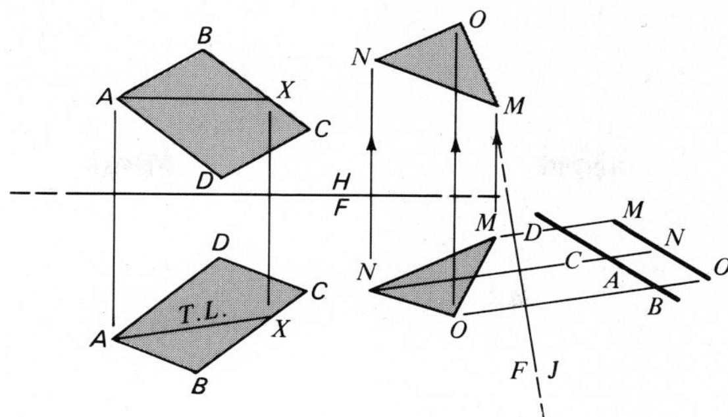

Example 6-7
Plane parallel to a plane.

Task: Construct plane *MNO* parallel to the given plane.
Solution:

1. By using an *F*/1 folding line, show plane *ABCD* on edge in the first auxiliary.
2. Construct the edge view of plane *MNO* parallel to the edge view of plane *ABCD*. Define plane *MNO*.
3. Project plane *MNO* into the front view. The plane may have any triangular shape as long as points *M*, *N*, and *O* lie on their respective lines of sight.
4. Using the distances taken from the edge of the frontal plane in the first auxiliary, project plane *MNO* into the top view. ◀

B. Two-View Method

The solution dealing with parallel planes is practically the same as the one shown in Example 6-6 with the exception of the relationship of intersecting lines required here.

Example 6-8

Given: Top and front views of line *AB*. This is one line of triangle *ABC*.
Task: Construct triangle *XYZ* parallel to triangle *ABC*.
Solution:

1. In the top view, draw line *XY* parallel to given line *AB*. Locate points "*X*" and "*Y*" anywhere on the line.
2. In the front view, draw line *XY* parallel to the given line *AB*.
3. In the top view, draw line *AC* in any convenient direction. In the proposed triangle *XYZ*, draw line *XZ* parallel to line *AC*. Both "*C*" and "*Z*" can be located at any point on their respective lines.
4. Project all points into the front view. Draw parallel lines *AC* and *XZ*.
5. Complete the triangles by drawing lines *BC* and *YZ*. ◀

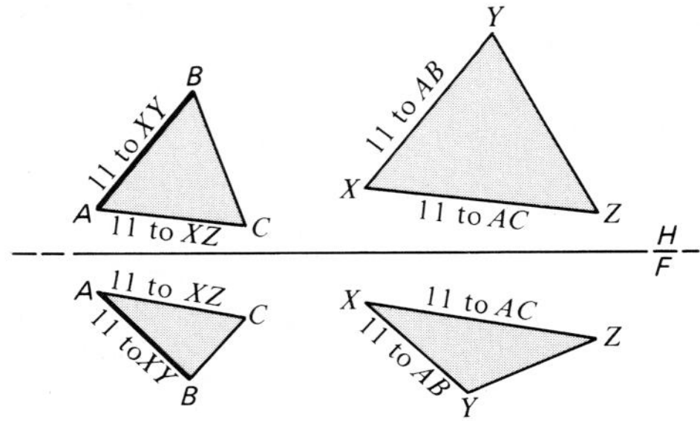

Example 6-8
Plane parallel to a plane.

Many solid objects can be represented by a series of finite, plane facets. These planes may exist only as a series of points in a computer data base. The elements of geometric modeling not only can define lines and planes, but they also check for parallelism, intersection, orientation and other attributes. The same vector operations introduced earlier provide the necessary tools. The remainder of this chapter deals with lines and planes with a required parallelism, using mathematical techniques.

6-5 GEOMETRIC MODELING FOR PARALLEL LINES

One mathematical condition for parallel lines was given in Chapter 3. Recall that if two vectors are parallel, the determinant of the coefficient matrix as given by Equation (3-19) equals zero. Another condition used to check for parallelism between two lines is the vector cross product. Since $\mathbf{C} = \mathbf{A} \times \mathbf{B} = |\mathbf{A}||\mathbf{B}| \sin \theta$, if $\mathbf{C} = 0$ then $\mathbf{A}$ and $\mathbf{B}$ are parallel since $\theta = 0$ is required to meet this condition.

Example 6-9

Given: The nonparametric, implicit equations for two two-dimensional lines are as follows:

$$\text{Line 1:} \quad -2x + 2y + 1 = 0$$

$$\text{Line 2:} \quad 4x - 4y + 6 = 0$$

Task: Show that these lines are parallel.
Solution: For two-dimensional lines, the vector cross product is

$$\mathbf{C} = \mathbf{A} \times \mathbf{B} = \det \begin{vmatrix} \mathbf{i} & \mathbf{j} & \mathbf{k} \\ A_x & A_y & 0 \\ B_x & B_y & 0 \end{vmatrix}$$

$$\mathbf{C} = (A_x B_y - A_y B_x)\mathbf{k}$$

Thus

$$\mathbf{C} = 0 \quad \text{if} \quad A_x B_y = A_y B_x$$

Now, define vectors on the given line with direction components $[A_x \; A_y]$ and $[B_x \; B_y]$. Any two points on each line can be used. Choose $x = 0$ and calculate y; then choose $y = 0$ and calculate x.

Line 1				*Line 2*			
x	y	A_x	A_y	x	y	B_x	B_y
0	$-\frac{1}{2}$	$\frac{1}{2}$	$\frac{1}{2}$	0	$\frac{3}{2}$	$-\frac{3}{2}$	$-\frac{3}{2}$
$\frac{1}{2}$	0			$-\frac{3}{2}$	0		

Therefore, $A_x B_y = A_y B_x = -\frac{3}{4}$ and the lines are parallel. ◀

6-6 GEOMETRIC MODELING FOR PARALLEL PLANES

If two planes are parallel, then their normal vectors will be parallel. If $\mathbf{N}_1$ is the normal to one plane and $\mathbf{N}_2$ is the normal to a second plane, then one can evaluate $\mathbf{C} = \mathbf{N}_1 \times \mathbf{N}_2$. If $\mathbf{C} = 0$, the two planes are parallel.

If the two planes are written in terms of their implicit, nonparametric equations

$$N_{1,1}x + N_{1,2}y + N_{1,3}z - k_1 = 0 \tag{6-1}$$

$$N_{2,1}x + N_{2,2}y + N_{2,3}z - k_2 = 0 \tag{6-2}$$

The planes are parallel only if the direction ratios of the normal vectors are equal. That is,

$$N_{1,1}/N_{2,1} = N_{1,2}/N_{2,2} = N_{1,3}/N_{2,3} \tag{6-3}$$

This result follows from the requirement that $\mathbf{N}_1 \times \mathbf{N}_2 = 0$, where

$$\mathbf{N}_1 \times \mathbf{N}_2 = \det \begin{vmatrix} \mathbf{i} & \mathbf{j} & \mathbf{k} \\ N_{1,1} & N_{1,2} & N_{1,3} \\ N_{2,1} & N_{2,2} & N_{2,3} \end{vmatrix} \tag{6-4}$$

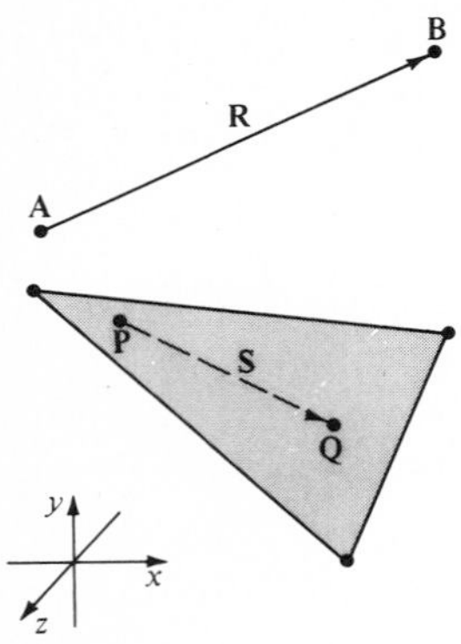

Figure 6-4
Plane parallel to a line.

Example 6-10

Given: Two planes are defined by the equations

$$2x + 8y - 5z - 6 = 0$$

$$6x + 24y - 15z - 10 = 0$$

Task: Are these planes parallel?
Solution: Yes, by inspection: $2/6 = 8/24 = -5/-15 = 1/3$
Note: The same approach could have been used for Example 6-9. ◀

Finally, define a plane through a known point $\mathbf{Q} = [x_1\ y_1\ z_1]$ which is also parallel to a plane given by

$$N_1x + N_2y + N_3z - k_1 = 0 \tag{6-5}$$

Since the normal vector components must be equal for the two planes, let the equation of the desired plane be

$$N_1x + N_2y + N_3z - k_2 = 0 \tag{6-6}$$

The value of k_2 can be found from the scalar dot product.

$$k_2 = \mathbf{Q} \cdot \mathbf{N} = [x_1\ y_1\ z_1] \begin{bmatrix} N_1 \\ N_2 \\ N_3 \end{bmatrix} = N_1x_1 + N_2y_1 + N_3z_1 \tag{6-7}$$

Combining Equations (6-6) and (6-7) leads to the implicit equation for the desired plane.

$$N_1(x - x_1) + N_2(y - y_1) + N_3(z - z_1) = 0 \tag{6-8}$$

6-7 GEOMETRIC MODELING FOR PARALLEL LINE AND PLANE

This section contains another example of mathematically constructing a plane with certain attributes. Consider a line in space which passes through points $\mathbf{A} = [x_1\ y_1\ z_1]$ and $\mathbf{B} = [x_2\ y_2\ z_2]$ as shown in Figure 6-4. Two other points $\mathbf{P} = [x_3\ y_3\ z_3]$ and $\mathbf{Q} = [x_4\ y_4\ z_4]$ are also given. The object is to define a plane which contains the points **P** and **Q** and at the same time is parallel to the given line through **A** and **B**.

Since line $\mathbf{S} = \mathbf{Q} - \mathbf{P}$ lies in the required plane, and since the plane is parallel to the line $\mathbf{R} = \mathbf{B} - \mathbf{A}$, a normal to both lines **S** and **R** can be constructed by the vector cross product.

$$\mathbf{N} = \mathbf{S} \times \mathbf{R} = \det \begin{vmatrix} \mathbf{i} & \mathbf{j} & \mathbf{k} \\ (x_4 - x_3) & (y_4 - y_3) & (z_4 - z_3) \\ (x_2 - x_1) & (y_2 - y_1) & (z_2 - z_1) \end{vmatrix} \tag{6-9}$$

Expanding gives

$$\begin{aligned}\mathbf{N} = &\{(z_2 - z_1)(y_4 - y_3) - (y_2 - y_1)(z_4 - z_3)\}\mathbf{i} \\ &\{(z_4 - z_3)(x_2 - x_1) - (x_4 - x_3)(z_2 - z_1)\}\mathbf{j} \\ &\{(y_2 - y_1)(x_4 - x_3)\} - (x_2 - x_1)(y_4 - y_3)\mathbf{k} = N_1\mathbf{i} + N_2\mathbf{j} + N_3\mathbf{k}\end{aligned} \tag{6-10}$$

Choose point **P** in the plane to form the scalar value of k given by

$$k = \mathbf{N} \cdot \mathbf{P} = N_1x_3 + N_2y_3 + N_3z_3 \tag{6-11}$$

The implicit equation for the required plane is then

$$N_1x + N_2y + N_3z - k = 0 \tag{6-12}$$

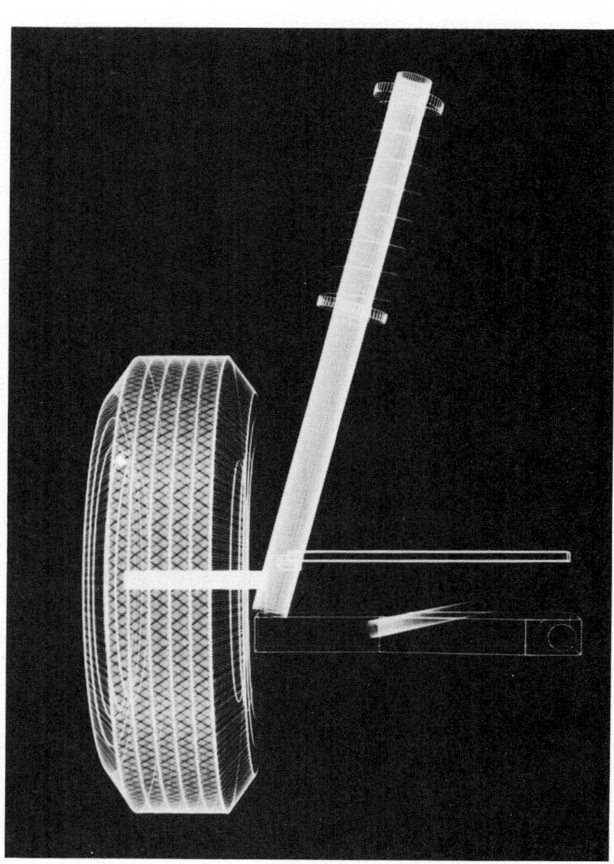

Figure 6-5
A computer graphics display consisting of many parallel lines. Courtesy of Evans and Sutherland Inc.

After obtaining the components for the normal vector **N** in Equation (6-10) a third point **T** on the plane can be found using

$$\mathbf{T} = \mathbf{N} \times \mathbf{S} \tag{6-13}$$

Then points **P, Q,** and **T** are three noncollinear points which can be used to define the plane with the required attributes.

Computer graphics is a powerful tool for designing complex objects and systems. One example is shown in Figure 6-5 where many vectors have been used to display a wheel mechanism. Animation can also be used to simulate the motion of the mechanism when subjected to a variety of external forces. In this manner "experiments" can be made on objects during the preliminary design process which do not actually exist. It is usually also necessary to access this type of data base to obtain solutions to descriptive geometry problems. Vector analysis is a very useful approach for this design requirement.

PROBLEMS: CHAPTER 6

6-1 *Given:* Front and top views of line *MN*. Front view of plane *ABCD* and top view of line *AD*.
Task: Line *MN* is parallel to plane *ABCD*. Complete the top view of the plane.

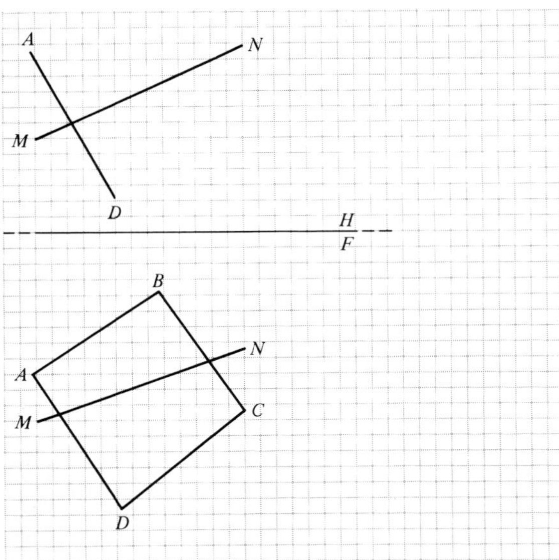

Figure 6-1P

6-2 *Given:* Front and top views of plane *RST*. Front view of point *N* and top view of plane *MNOP*.
Task: Construct plane *MNOP* parallel to plane *RST* and show both planes in the front view.

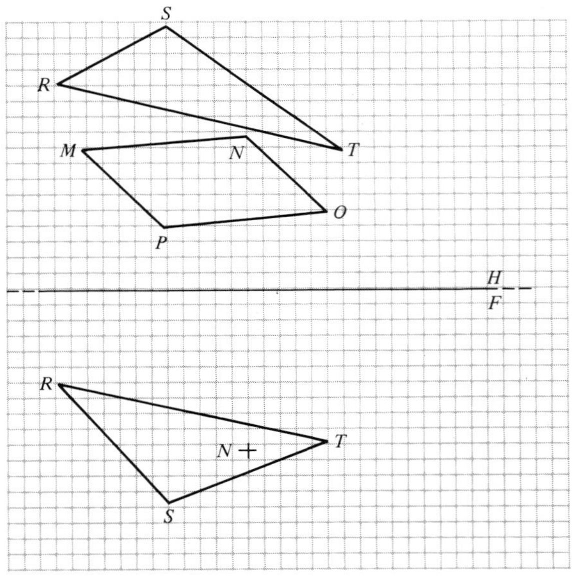

Figure 6-2P

6-3 *Given:* Front and top views of planes *MNO* and *RST*.
Task: Find the perpendicular distance between the two given planes.

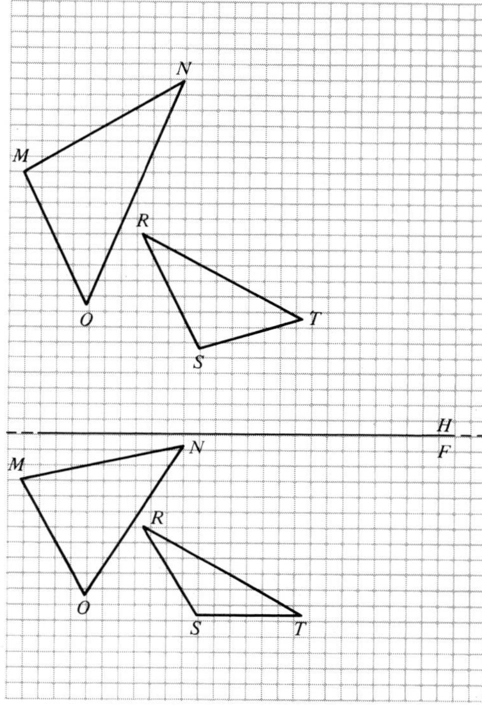

Figure 6-3P

6-4 *Given:* Front and top views of plane *RST*. Top view of plane *ABC* and front view of point *A*.
Task: Plane *ABC* is parallel to plane *RST*. Complete the front view of *ABC*.

6-5 *Given:* Front and top views of plane *ABC*.
Task: Construct a triangular plane *XYZ* parallel to, and four units above, plane *ABC*. Show plane *XYZ* in the front and top views.

6-6 *Given:* Front and top views of plane *ABC* and plane *WXYZ*.
Task: Without using an auxiliary, determine whether the planes are parallel.

6-7 *Given:* Front and top views of plane *ABC* and plane *WXYZ*.
Task: Determine the distance between the two given planes.

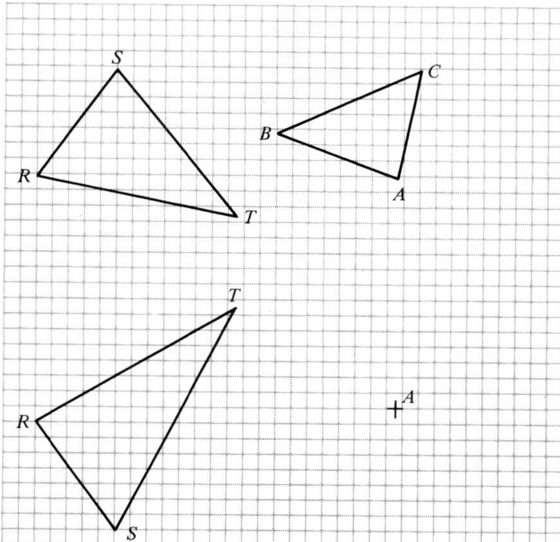

Figure 6-4P

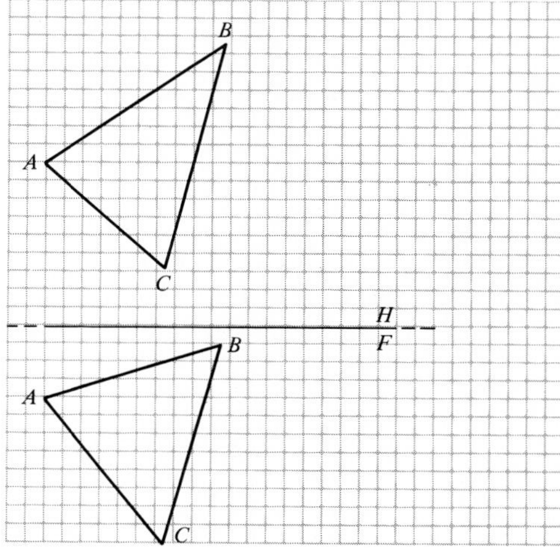

Figure 6-5P

6-8 *Given:* Front and top views of plane *ABC*.
Task: Construct a line *MN* one unit above, and parallel to, plane *ABC*. Show line *MN* in the front and top views.

6-9 *Given:* Front and top views of line *MN* and line *AB*.

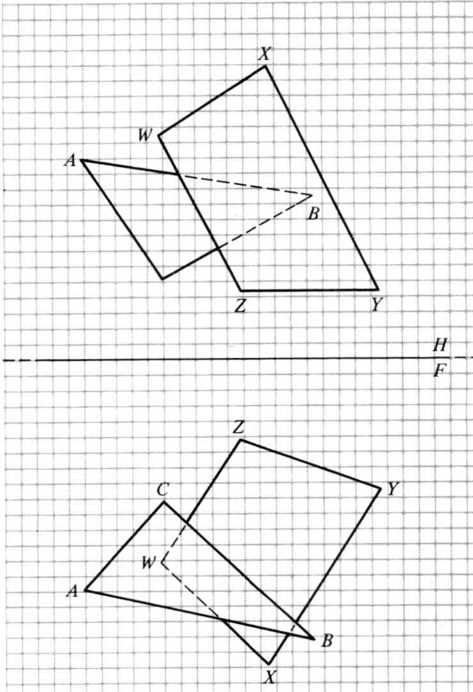

Figure 6-6P

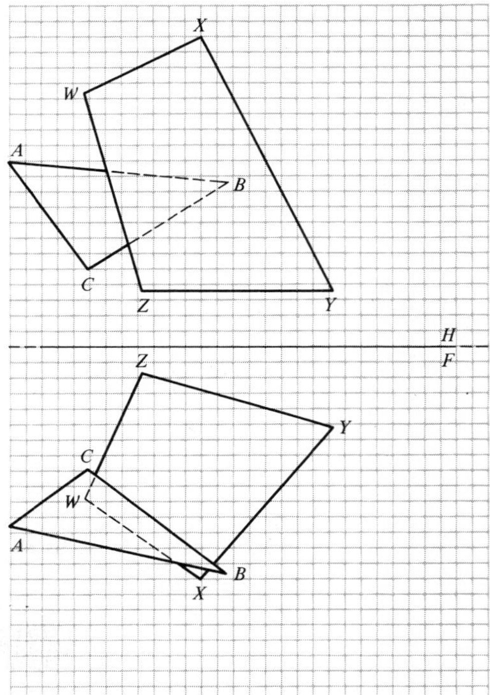

Figure 6-7P

Task: Construct triangle *RST* parallel to both line *MN* and line *AB*.

6-10 *Given:* Front and top views of line *AB* and line *MN*.
Task: Determine whether the two lines are parallel.

6-11 *Given:* Two lines are given in parametric form by

$$\mathbf{P}_1 = \mathbf{A} + t(\mathbf{B} - \mathbf{A})$$

$$\mathbf{P}_2 = \mathbf{C} + s(\mathbf{D} - \mathbf{C})$$

where

$$\mathbf{A} = [3\ 2\ 2],\ \mathbf{B} = [-2\ 1\ -2]$$

$$\mathbf{C} = [-5\ 4\ 3],\ \mathbf{D} = [0\ 3\ -1]$$

Task: Show that the lines are parallel.

6-12 *Given:* Cramer's Rule as given in Appendix 2.
Task: a. Try to use Cramer's Rule to solve for the intersection between the two lines given in Example 6-9. How can you conclude that the lines are parallel?
b. If two general lines are given by

$$a_1x + b_1y + c_1 = 0$$

$$a_2x + b_2y + c_2 = 0$$

when will the lines be parallel?

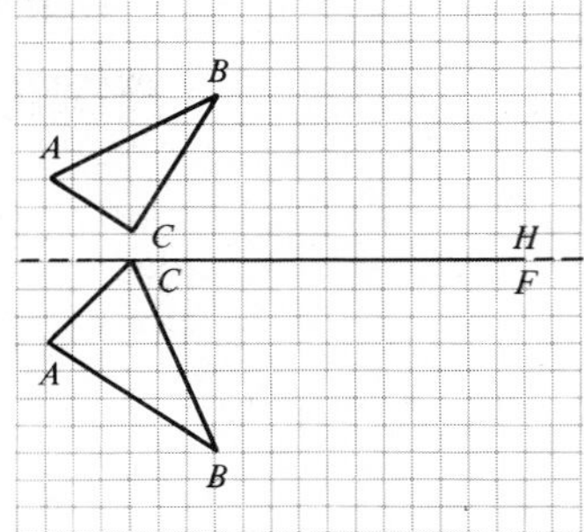

Figure 6-8P

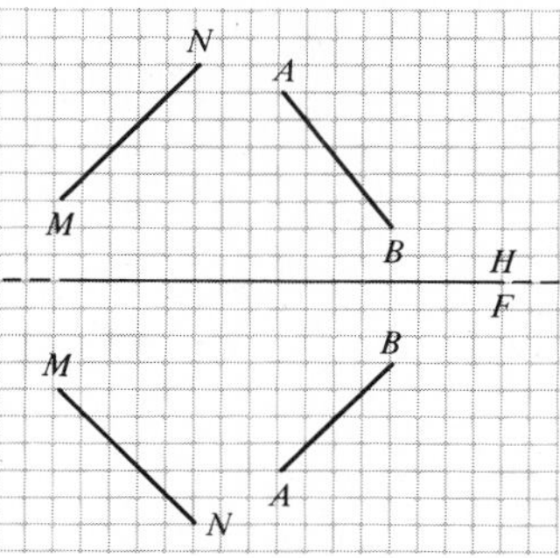

Figure 6-9P

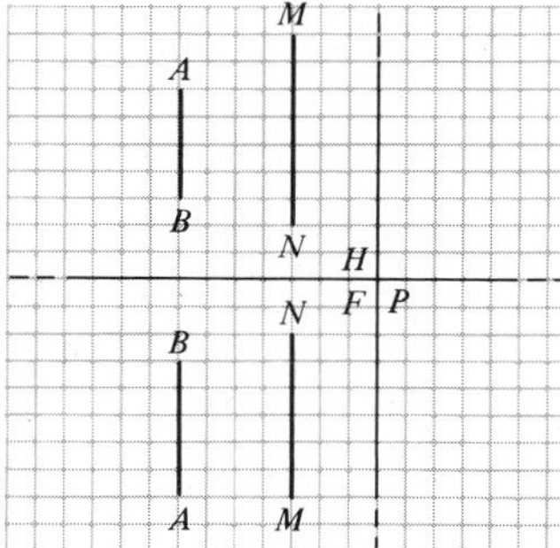

Figure 6-10P

6-13 *Given:* Two planes are given in implicit form by

$$\text{Plane 1: } 3x + 4y + 12z = 18$$

$$\text{Plane 2: } 6x - 8y - z = 9$$

Task: Show that the planes are not parallel.

6-14 *Given:* Plane 1 is defined by the three vectors $\mathbf{A}_1 = [3\ 1\ 2]$, $\mathbf{B}_1 = [1\ -1\ 1]$, $\mathbf{C}_1 = [2\ 2\ 3]$. Plane 2 is defined by the vectors $\mathbf{A}_2 = [-3\ 1\ -2]$, $\mathbf{B}_2 = [2\ -3\ 1]$, $\mathbf{C}_2 = [3\ 2\ 1]$.
Task: Show that the two planes are not parallel.

6-15 *Given:* The object defined in Problem 1-18^c.
Task: a. Show that the plane containing vertices 3, 2, and 9 is parallel to the plane containing vertices 4, 1, 13.

b. Show that the plane containing vertices 10, 11, 15 is not parallel to the plane containing vertices 12, 6, 5.

6-16 *Given:* A plane defined by the implicit equation

$$2x + 8y - 5z - 6 = 0$$

Task: a. Define a plane which is parallel to the given plane and passes through the origin of the coordinate system.

b. Define a plane which is parallel to the given plane and passes through the point $\mathbf{Q} = [1\ 1\ 1]$.

6-17 *Given:* A three-dimensional line passes through $\mathbf{A} = [3\ 2\ 2]$ and $\mathbf{B} = [-2\ 1\ -2]$.
Task: Find an implicit equation for a plane which contains the line $\mathbf{Q} - \mathbf{P}$ and is parallel to the line $\mathbf{B} - \mathbf{A}$, where $\mathbf{P} = [2\ 1\ 5]$ and $\mathbf{Q} = [5\ 6\ 1]$.

6-18 *Given:* Drawings for a wheel assembly similar to Figure 6-5 give frontal and profile views of an axis through *AB* and an attached arm *OC*, with measured angles and distances indicated in Figure 6-18P.
Task: a. Using the given views find the direction cosines of axis *AB* and arm *OC*.

b. Find the angle between the axis and arm. Make use of the direction cosines found in part a and the dot product of two unit vectors along the axis and along the arm.

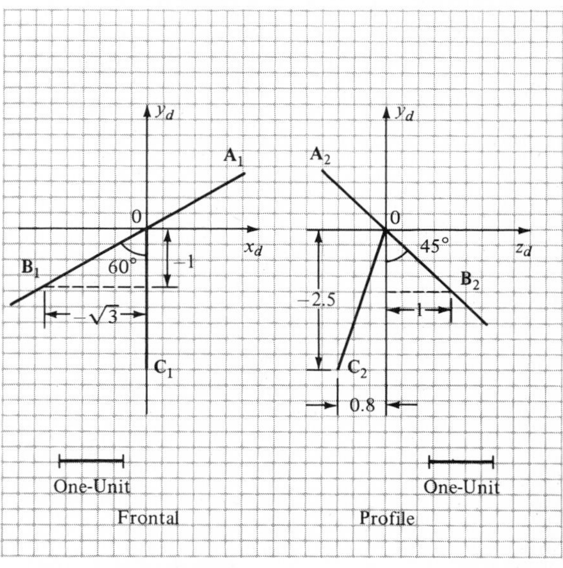

Figure 6-18P

6-19 *Given:* Problem 6-1.
Task: Find the implicit equation for the plane *ABCD.*

6-20 *Given:* Problem 6-2.
Task: Find the implicit equation for plane *RST.*

6-21 *Given:* Problem 6-9.
Task: Find the implicit equation for the plane *RST.*

6-22 *Given:* Problem 6-10.
Task: Use mathematical techniques to find whether the given lines are parallel.

7

Perpendicularity

The condition of perpendicularity is used to find spatial relationships such as the shortest distance from a point to a line, or the shortest distance between two lines. It may also be necessary to construct a line perpendicular to a plane, or a plane perpendicular to a line. Orthogonal planes arise in applications of importance to engineers and architects. Following the pattern established in previous chapters, descriptive geometry methods are first presented, followed by mathematical methods, which are applied to problems of perpendicularity.

7-1 PERPENDICULAR LINES

Before students recognize the relationship between lines found in engineering drawings, they must recognize the basic rules which govern perpendicularity.

All lines in a plane which is 90 deg to a given line are perpendicular to that line (see Figure 7-1(a) and (b)). It can be seen from this figure that the direction of the lines is of prime importance and that the lines can be either intersecting or nonintersecting. The figure also shows that the plane could contain an infinite number of lines which would be perpendicular to the given line. Since perpendicular lines do not show a 90 deg angle between them in all views, it is sometimes difficult to determine an existing right angle relationship between lines. This problem is made more acute by the fact that intersecting lines may *appear* in one or more views to have an angle of 90 deg between them but are not perpendicular.

If the line *MN* in Figure 7-1 were shown in the front view of an orthographic drawing as a true length line, both the planes would be projected

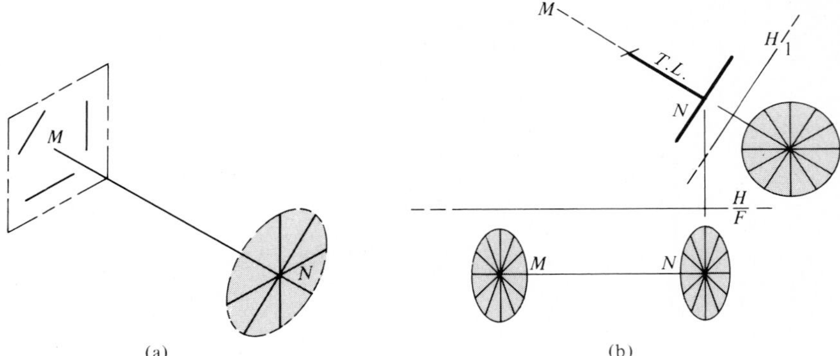

Figure 7-1
Perpendicularity.

as edge views and all lines in those planes would appear perpendicular to line *MN*. When the line *MN* is seen as a point, all lines perpendicular to it will appear in true length. In light of the above, the following corollary can be given: *Perpendicular lines will appear perpendicular in any view which shows either or both of the lines in true length.* It may also be stated that the lines are *not* perpendicular if (1) a view shows two lines at right angles with neither in true length or (2) a view shows one or both lines in true length and an angle between them which is not 90 deg.

Each of the illustrations in Figure 7-2 shows line *PX* drawn through point *"P"* perpendicular to principal line *MN*. The line *MN* shows true length in the adjacent view where the required line may be drawn perpendicular to it.

Figure 7-3 requires a line *PX* to be drawn perpendicular to an oblique line *MN*. Here the line *PX* is drawn as a horizontal line in the front view. When projected to the top view it appears in true length and can be drawn

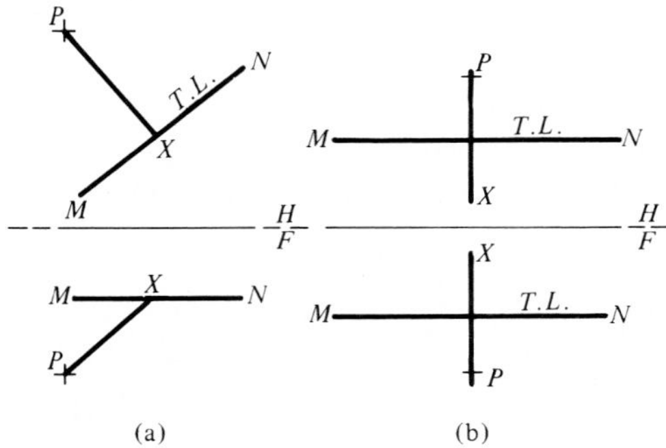

Figure 7-2
Perpendicular lines.

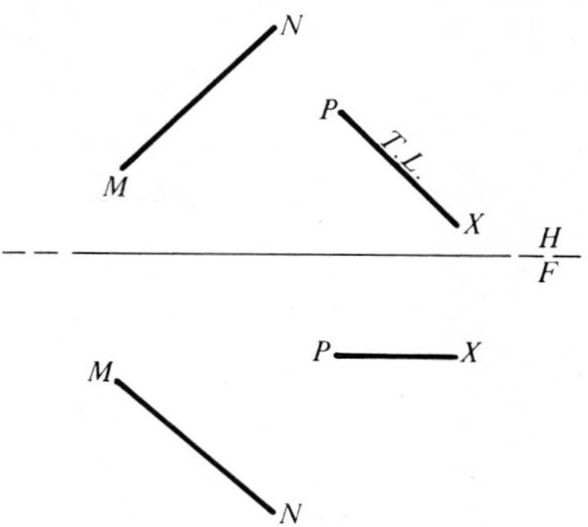

Figure 7-3
Perpendicular lines.

perpendicular to the given oblique line *MN*. This relationship would hold true for any adjacent view in the chain of auxiliaries as long as one line is shown in true length.

The following examples are given to check the visual relationships above.

Example 7-1

Given: Top and front views of lines *AB* and *MN*. These lines *appear* perpendicular in both views.

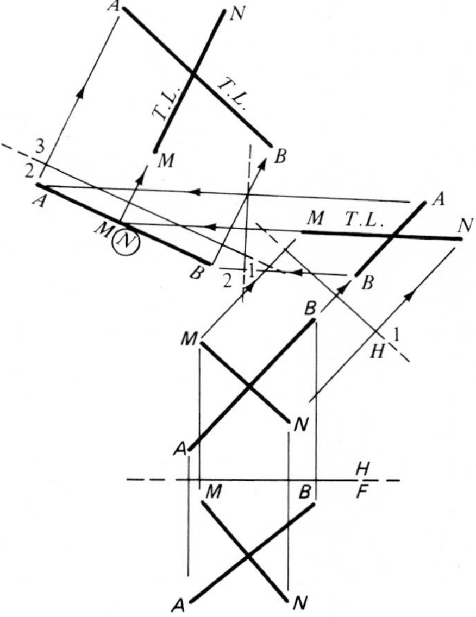

Example 7-1
Perpendicular lines.

Task: Determine whether line *AB* is perpendicular to line *MN*.
Solution:

1. Draw folding line *H*/1 parallel to line *MN* in the top view. The first auxiliary will show line *MN* in true length. It can be seen here that the lines are not perpendicular.
2. The second auxiliary shows the edge view of a plane containing the two lines.
3. The third auxiliary shows both lines in true length and is further proof that the lines are not perpendicular.

Note: The second and third auxiliaries are not needed except for additional proof that line *AB* is not perpendicular to line *MN*. ◀

Example 7-2

Given: Top and front views of line *MN* and point "*P*."

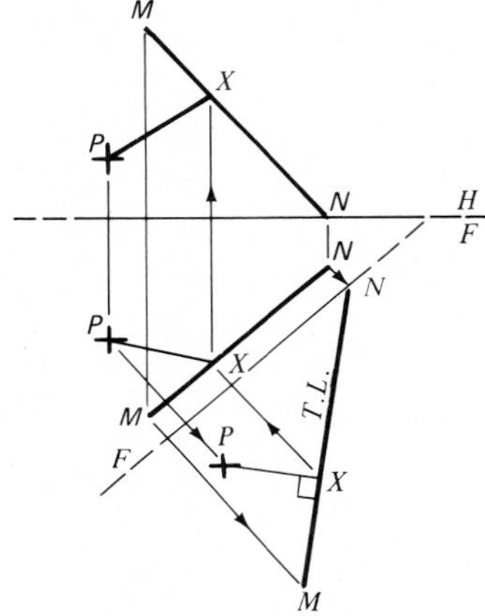

Example 7-2
Perpendicular lines.

Task: Draw line *PX* so that it is perpendicular to line *MN* and intersects it at *X*. Show line *PX* in all views.
Solution: In the first auxiliary, line *MN* appears in true length. Draw a line from "*P*" so that it is perpendicular to line *MN* and intersects it at *X*. Project point *X* into all views and draw line *PX*. ◀

7-2 SHORTEST DISTANCE FROM A POINT TO A LINE

In order to conserve materials and provide clearance, it is often necessary to find the shortest distance from one point to another or from a point to a line in space. In the first case (a point to a point), the true distance can be measured by finding the true length of a line between the two points. A slightly more complicated procedure is involved when finding the shortest distance from a point to a line. Two methods are recommended, both of which are shown in this section. When using these methods it is well to remember the following:

1. The shortest distance between a point and a line can be shown in a view where the line appears as a point.
2. A line appears as a point in any view which shows the line of sight parallel to a true length line.
3. A line must appear in true length in the view adjacent to that in which it appears as a point.
4. Any line may be drawn perpendicular to another if *one* of the lines is shown in true length.

A. Point-View Method

Example 7-3 shows the procedures used in the solution employing the *point-view method.*

Example 7-3

Given: Top and front views of line *AB* and point "*O*."

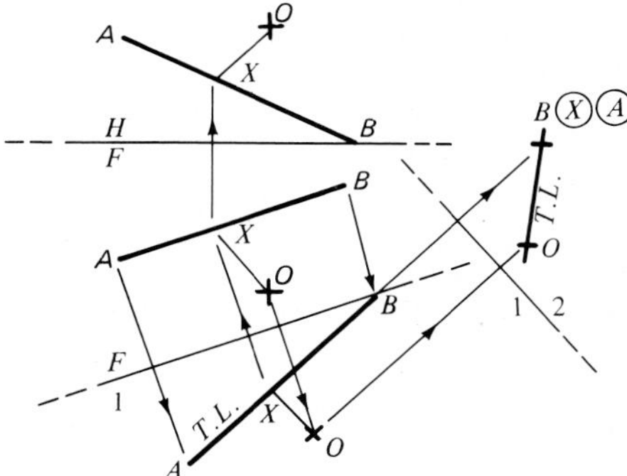

Example 7-3
Point-view method.

Task: Find the shortest distance from point "*O*" to line *AB*. Show the line *OX* in all views.
Solution:

1. Place folding line *F*/1 parallel to *AB* (frontal auxiliary used here).
2. Project line *AB* and point *O* into the first auxiliary. Line *AB* is now in true length. (Lines of sight are perpendicular to folding line *F*/1).

3. Place folding line 1/2 perpendicular to true length line *AB*. Project point *O* and line *AB* into the second auxiliary view. Line *AB* appears as a point in this view. A line *OX* from point *O* to the point view of line *AB* will show the perpendicular distance in question. The *perpendicular* distance from point *O* to line *AB* at *X* was measured in the second auxiliary as a true length line. This means that in the first auxiliary view line, *OX* must be drawn parallel to folding line 1/2 and will be perpendicular to the true length line *AB* in the same auxiliary.
4. Project backward and show the line *OX* in the front and top views.

B. Normal View of Plane

The correct distance from a point to a line will appear in any projection showing the normal view of a plane which contains the point and the line. This method may also be used when a line from a point must intersect another line at a certain point or at a specified angle to the line, as for example, in finding a pipe connection.

Example 7-4

Given: Top and front views of line *AB* and point *O*.

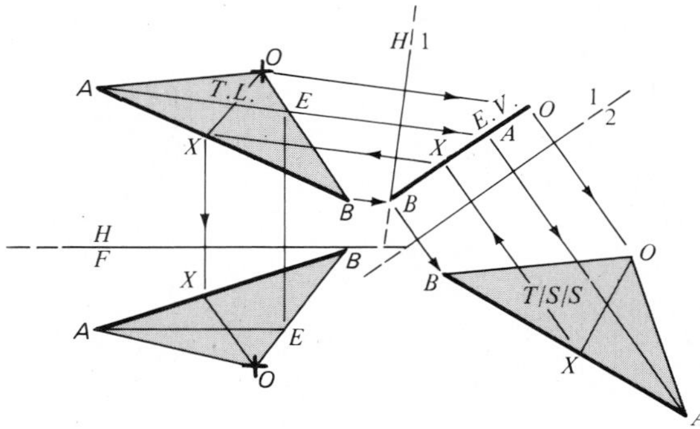

Example 7-4
Normal view of plane method.

Task: Find the shortest distance from point *O* to line *AB*. Show the line *OX* in all views.
Solution:

1. Connect point *O* to points *A* and *B* with construction lines to form triangle *ABO*.
2. Using horizontal construction line *AE* in the front view, obtain true length of *AE* in the top view of plane *ABO*.

3. Place folding line *H*/1 perpendicular to true length line *AE* and project all points into first auxiliary. The constructed plane is now on edge.
4. Place folding line 1/2 parallel to the plane on edge and project the plane into the second auxiliary. The constructed plane is now shown in true size and shape, a normal view.
5. Measure the perpendicular *OX* from point *O* to line *AB* to find the shortest distance as required.
6. Project line *OX* into all views.

7-3 COMMON PERPENDICULAR (SHORTEST DISTANCE) BETWEEN TWO SKEW LINES

Two skew lines can have one, and only one, connector that is perpendicular to both, and it is along this connector that the shortest distance between the lines is measured. There are many engineering situations; civil, industrial, and mechanical, to name a few, where the solutions to this type of problem must be found. These answers can result in huge savings to those who work with materials such as structural steel and pipes, for example. There are two methods of finding the shortest distance between two skew lines, as shown in the following examples.

A. Line Method (Point-View Method)

Example 7-5

Given: Top and front views of two skew lines.

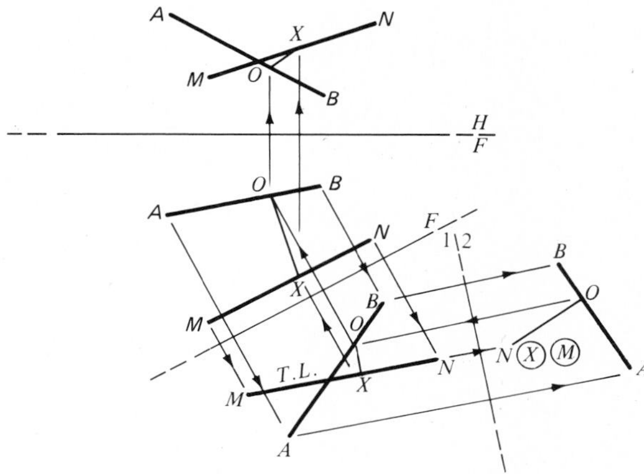

Example 7-5
Common perpendicular: point-view method.

Task: Find the common perpendicular between the given skew lines. Show the common perpendicular in all views.
Solution:

1. Project line *AB* and line *MN* into the first auxiliary. Since folding line *F*/1 is parallel to line *MN*, that line will appear true length when projected.
2. Place folding line 1/2 perpendicular to true length line *MN*. Project both lines into the second auxiliary where *MN* will appear as a point.
3. The common perpendicular (shortest distance) can be drawn from line *MN* to line *AB*. This line *OX*, which is in true length, will intersect line *AB* at *O*, making an angle of 90 deg. The distance can be measured here.
4. Project point *O* back into the first auxiliary. Line *OX* must be shown here at a 90 deg angle to line *MN*. This may be done since line *OX*, which is true length in the second auxiliary, must be parallel to folding line 1/2 in the first auxiliary. This 90 deg angle may be made since line *MN* is in true length.
5. Project line *OX* into all views. ◀

B. Plane Method

Example 7-6

Given: Top and front views of two skew lines *AB* and *MN*.

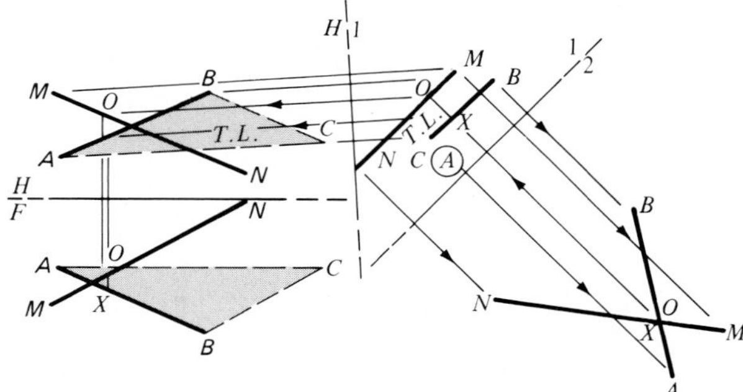

Example 7-6
Common perpendicular: plane method.

Task: Locate the common perpendicular *OX* between the two given skew lines.
Solution:

1. In both views, construct a plane *ABC* that contains given line *AB* and is parallel to given line *MN*. (Reference Section 6-3, "A plane is par-

allel to a line if a line in the plane is parallel to the given line.") When constructing the front view of plane *ABC*, draw the line *AC* horizontal, so that it becomes true length in the top view. Line *BC* in the plane is parallel to the given line *MN* in both views. Therefore, the plane is parallel to the line.

2. Following the usual procedures, show the line and plane parallel in the first auxiliary. (In this view the shortest distance could be measured by drawing a perpendicular between the line and edge view of the plane.)
3. Place folding line 1/2 parallel to line *MN* and *AB* and project both lines into the second auxiliary. The point view of the common perpendicular *OX* will be at the location of the crossed lines. Project this point back into the first auxiliary to show its location and the length of *OX*.
4. Show the location of *OX* by projecting both points into all views. Draw line *OX* in both front and top views. ◀

7-4 SHORTEST HORIZONTAL DISTANCE BETWEEN TWO SKEW LINES

This solution is somewhat the same as that used in Section 7-3-B except that the distance between the lines must be shown and measured parallel to the horizontal plane on edge (*H*/1 folding line). Therefore, the line must be shown as in the given solution.

Example 7-7

Given: Top and front views of two skew lines *AB* and *MN*.

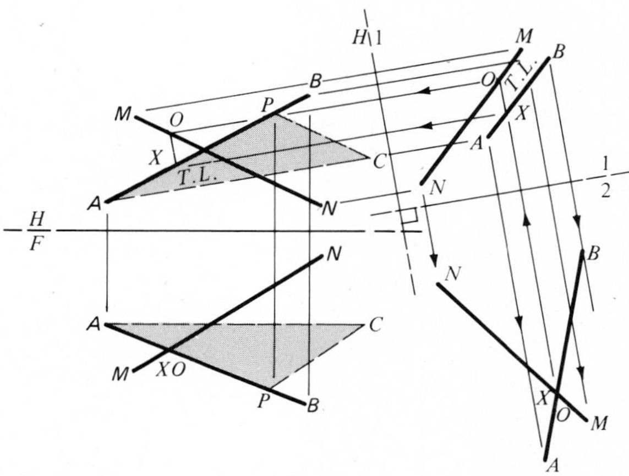

Example 7-7
Shortest horizontal distance between two skew lines.

Task: Find the shortest horizontal distance between the two given skew lines.
Solution:

1. In both top and front views construct plane *ACP* parallel to line *MN* by drawing edge line *PC* parallel to line *MN*. (See Section 6-3.) When constructing the triangle, draw line *AC* parallel to the horizontal plane so that it will project into the top view as a true length line.
2. Project line *MN* and plane *ACP* into the first auxiliary view where the plane will appear on edge and parallel to the line.
3. Project the lines of sight parallel to the horizontal plane (*H*/1) into the second auxiliary view. Draw folding line 1/2 perpendicular to the lines of sight.
4. Project both lines into the second auxiliary. The shortest horizontal connecting line *OX* will appear as a point where line *AB* and line *MN* cross.
5. Project line *OX* into the first auxiliary to show the shortest horizontal line which will connect line *MN* and line *AB*.
6. Project line *OX* into the top and front views. ◀

7-5 SHORTEST LINE OF SPECIFIED GRADE BETWEEN TWO SKEW LINES

The procedure used in the following example combines the method used in prior sections and that shown in Section 3-10.

Example 7-8

Given: Top and front views of two skew lines *AB* and *MN*.
Task: Show the shortest line of 20 per cent grade which connects the two lines.
Solution:

1. Using the procedure in Examples 7-6 and 7-7, construct plane *ABC* parallel to line *MN* in the front and top views.
2. Construct horizontal line *AX* in triangle *ABC* and project into the top view where *AX* will appear in true length.
3. Project plane and line *MN* in the direction of the true length line. The first auxiliary will show the plane on edge and parallel to the line *MN*.
4. Draw a horizontal line (parallel to the *H*/1 folding line) at any convenient location. Using the procedure shown in Section 3-10, lay off 100 units horizontally and 20 units vertically thereby establishing a line having a 20 per cent grade.

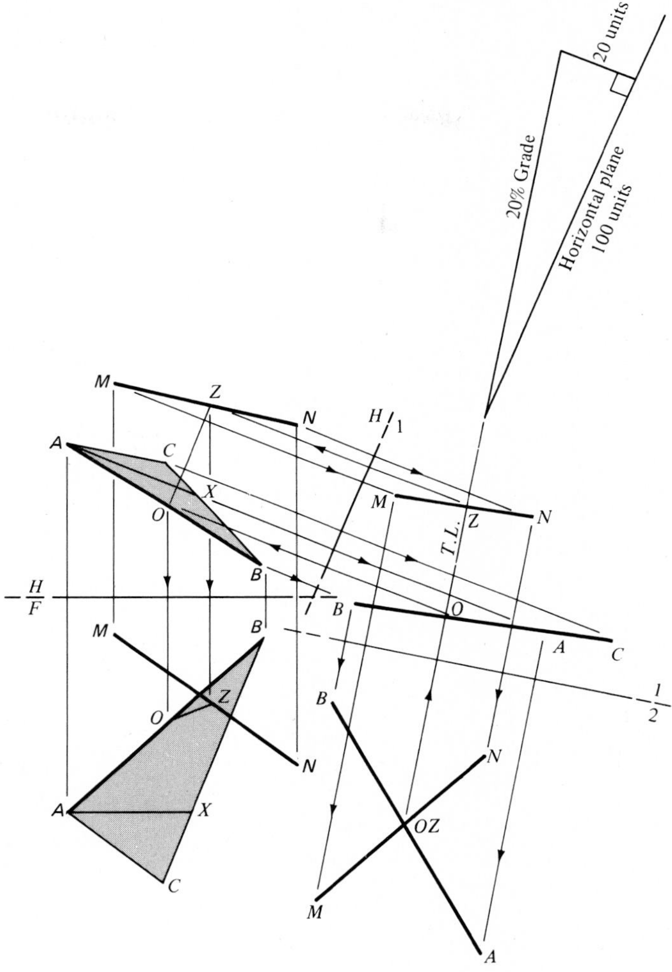

Example 7-8
Shortest connecting line at specified grade.

5. Project in the direction of the grade line into the second auxiliary. The required line *OZ* will appear as a point where line *AB* and line *MN* cross.
6. Project back into the first auxiliary to show the required line.
7. Project line *OZ* into all views. ◀

7-6 LINE PERPENDICULAR TO A PLANE

The three axioms given below form the basis for the procedures used in this section. (See Figure 7-4.)

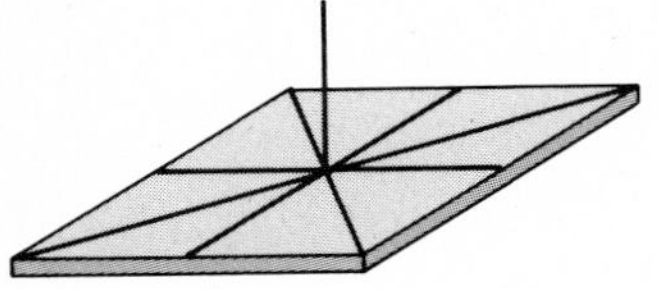

Figure 7-4
Line perpendicular to a plane.

1. If a true length line is perpendicular to the edge view of a plane it is perpendicular to that plane.
2. If a line is perpendicular to a plane it is perpendicular to every line in that plane.
3. A line perpendicular to a plane must be perpendicular to two or more intersecting lines of the plane.

A. Auxiliary Method

Example 7-9

Given: Front and top views of plane *ABC*.

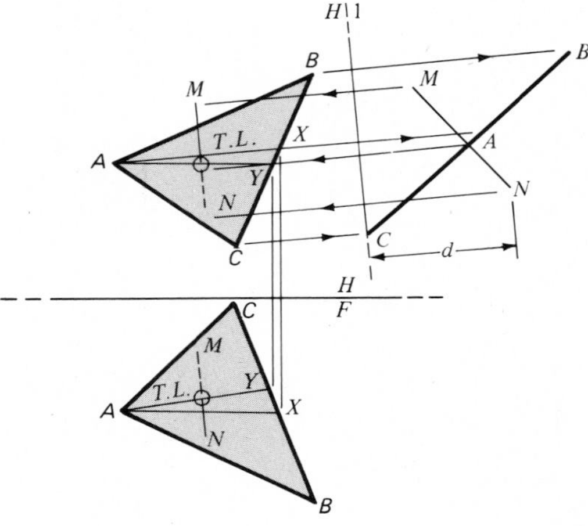

Example 7-9
Auxiliary method: line perpendicular to a plane.

Task: Construct line *MN* perpendicular to plane *ABC*. Show line *MN* in all views.
Solution:

1. Using a horizontal line *AX* in the front view of plane *ABC*, find its true length in the top view. Show plane *ABC* on edge in the first auxiliary. In the first auxiliary draw line *MN* in true length and perpendicular to the edge view of plane *ABC*.
2. Project lines of sight of points *M* and *N* into the top view. Draw line *MN* in the top view so that it is perpendicular to the true length line *AX* (Section 7-6-2). To show line *MN*'s true length in the first auxiliary it must be parallel to folding line *H*/1 in the top view.
3. From the first auxiliary, obtain the distances "*d*" that points *M* and *N* lie below the *H*/1 folding line. Using these measurements, lay off their

respective distances below the *H/F* folding line in the front view. Draw line *MN*. It must be perpendicular to a true length line *AY* constructed in the front view of plane *ABC* (Section 7-6-3). ◀

B. Two-View Method

From Figure 7-4 it can be seen that a line perpendicular to a plane will be at a 90 deg angle to each of two intersecting true length lines in the plane. This is the basis for the following solution.

Example 7-10

Given: Front and top views of plane *ABC*.

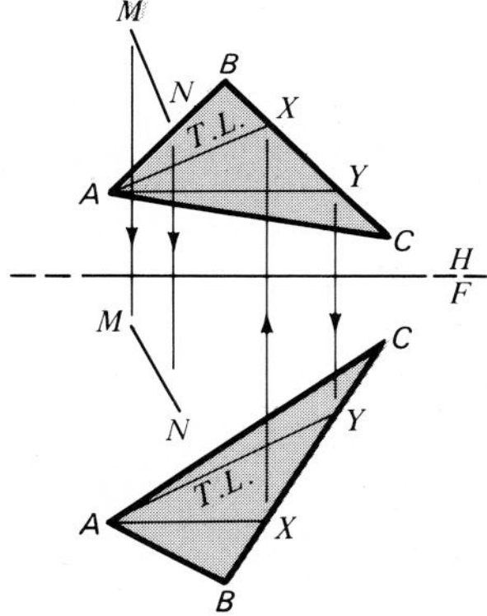

Example 7-10
Two-view method: line perpendicular to a plane.

Task: Using two views only, draw a line *MN* perpendicular to plane *ABC*.
Solution:

1. Construct a horizontal line *AX* in the front view of plane *ABC* and project it into the top view so that it appears true length.
2. Draw line *MN* of any length so that it is perpendicular to true length line *AX*.
3. Construct a frontal line *AY* in the top view of plane *ABC* and project it into the front view so that it appears in true length.
4. Project points *M* and *N* into the front view and draw line *MN* perpendicular to true length line *AY*.
5. Line *AX* and line *AY* intersect. Since *MN* has been drawn perpendicular to each of these intersecting lines it is perpendicular to plane *ABC* (Section 7-6-3). ◀

7-7 PLANE PERPENDICULAR TO A LINE

A. Auxiliary Method

Example 7-11

Given: Top and front views of line *MN*.

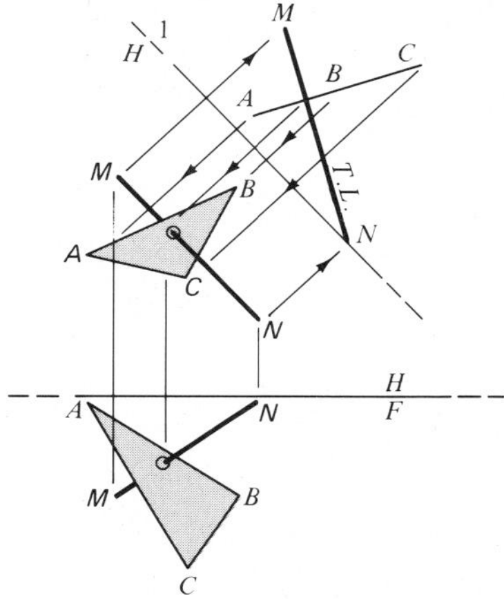

Example 7-11
Plane perpendicular to a line: auxiliary method.

Task: Draw plane *ABC* perpendicular to the given line.
Solution:

1. Draw folding line *H*/1 parallel to line *MN* in the top view.
2. Project line *MN* into the first auxiliary where it will appear in true length.
3. Construct plane *ABC* on edge and perpendicular to true length line *MN*.
4. Lay off points *A*, *B*, and *C* at random on the edge view of *ABC*.
5. Project points *A*, *B*, and *C* into the top view and arbitrarily fix the points on their respective lines of sight to form plane *ABC*.
6. Project points *A*, *B*, and *C* into the front view to complete the required task.
7. Check this solution by constructing true length lines in plane *ABC*. These lines should all be perpendicular to line *MN*.

B. Two-View Method

The converse of Section 7-6-3 would state that a plane perpendicular to a line would contain two intersecting lines in that plane with each line perpendicular to the given line.

Example 7-12

Given: Top and front views of line *MN*.

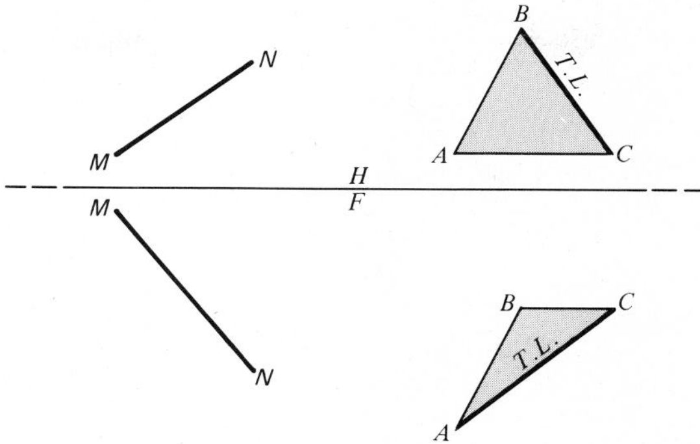

Example 7-12
Plane perpendicular to a line: two-view method.

Task: Using only two views, construct plane *ABC* perpendicular to the given lines.
Solution: The solution is shown here so that the student may fill in the step-by-step procedure.

7-8 PLANE PERPENDICULAR TO A PLANE

If a plane contains a line which is perpendicular to two intersecting lines of another plane, the planes are perpendicular to each other (Figure 7-5 and Section 7-6-3).

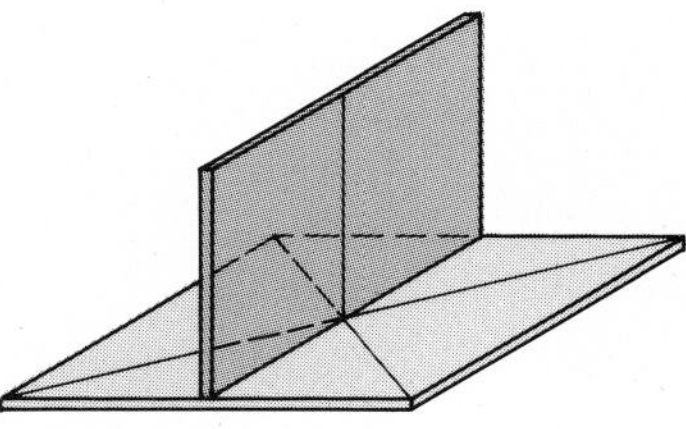

Figure 7-5
Plane perpendicular to a plane.

A. Auxiliary Method

Example 7-13

Given: Top and front views of plane *ABCD*.

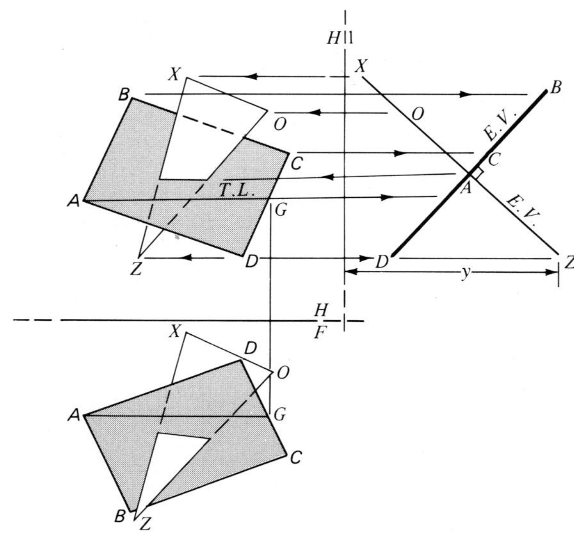

Example 7-13
Plane perpendicular to a plane: auxiliary method.

Task: Using the auxiliary method, draw a plane *XOZ* perpendicular to the given plane. Show both planes, possible intersection, and visibility in all views.
Solution:

1. Using a horizontal line *AG* in the front view of plane *ABCD*, project it into the top view where it will be shown in true length.
2. Place folding line *H*/1 perpendicular to true length line *AG*. Project plane *ABCD* into the first auxiliary where it appears on edge.
3. Draw plane *XOZ* on edge and at a 90 deg angle to the edge view of plane *ABCD*.
4. Place points *XOZ* at random positions on the edge view of the plane and project them into the top view. *X*, *O*, and *Z* may be placed at arbitrary positions in the top view as long as they are on their respective lines of sight.
5. Project the line of intersection into the top view.
6. Using the distances "*y*" in the first auxiliary view, transfer all points into the front view. Show the line of intersection.
7. Show correct visibility in all views.

B. Two-View Method

Example 7-14

Given: Top and front views of plane *ABCD*.

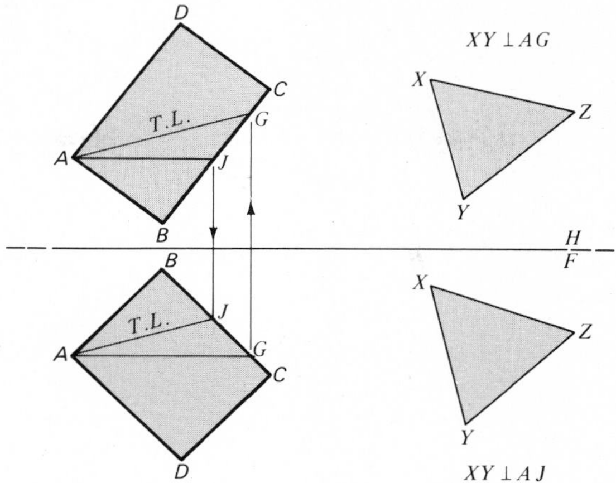

Example 7-14
Plane perpendicular to a plane: two-view method.

Task: Using two views only, draw a plane *XYZ* perpendicular to the given plane. Show both planes in all views.
Solution:

1. Construct horizontal line *AG* in the front view of plane *ABCD*. This line becomes true length in the top view of *ABCD*.
2. Construct frontal line *AJ* in the top view of plane *ABCD*. Line *AJ* becomes true length in the front view of the given plane.
3. Draw line *XY* in the top view so that it is perpendicular to true length line *AG* which is in plane *ABCD*.
4. Project points *X* and *Y* into the front view on their respective lines of sight so that line *XY* is drawn perpendicular to the true length line *AJ* which is in plane *ABCD*.
5. The third point, *Z*, of the plane *XYZ*, can be placed at random in the top view and projected into the front view in the usual manner. The three points *X*, *Y*, and *Z* can now be connected to form the triangular plane *XYZ*. Since line *XY* of the required plane is perpendicular to two intersecting lines of the given plane the planes are perpendicular to each other. ◀

7-9 GEOMETRIC MODELING FOR PERPENDICULARITY

The approach used in this section is to use vector mathematics and computer solutions to help reinforce the basic concepts of descriptive geometry and to apply methods of geometric modeling to meet perpendicularity conditions for lines and planes.

A. Shortest Distance from a Point to a Three-Dimensional Line

Let a three-dimensional line be defined by two points; $\mathbf{A} = [x_1\ y_1\ z_1]$ and $\mathbf{B} = [x_2\ y_2\ z_2]$. A third point is given by $\mathbf{P} = [x_3\ y_3\ z_3]$. The problem is to define the shortest distance from **P** to the line passing through points **A** and **B.**

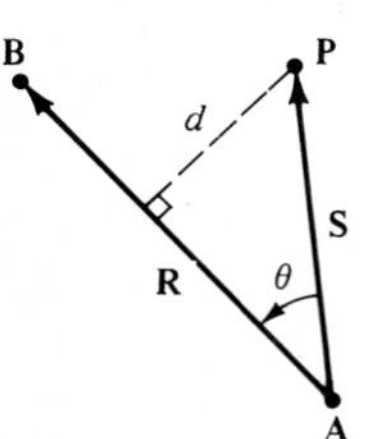

Figure 7-6

The true length can be computed directly from the given vectors using

$$L = |\mathbf{B} - \mathbf{A}| = [(\mathbf{B} - \mathbf{A}) \cdot (\mathbf{B} - \mathbf{A})]^{1/2} \tag{7-1}$$

Vector operations can be used to solve for the shortest distance without use of an orthographic projection. First define $\mathbf{S} = \mathbf{P} - \mathbf{A}$. From Figure 7-6 it can be seen that

$$d = |\mathbf{S}| \sin\theta \tag{7-2}$$

Introducing the relative vector $\mathbf{R} = \mathbf{B} - \mathbf{A}$, and multiplying the numerator and denominator of Equation (7-2) by $|\mathbf{R}|$ gives

$$d = \frac{|\mathbf{S}||\mathbf{R}| \sin\theta}{|\mathbf{R}|} = \frac{|\mathbf{S} \times \mathbf{R}|}{|\mathbf{R}|} \tag{7-3}$$

Notice that this solution depends only upon the vectors **R** and **S** and is independent of the coordinate system origin location.

The distance from **A** to the point of intersection on line $\mathbf{B} - \mathbf{A}$ with the perpendicular line to point **P** is $|\mathbf{S}| \cos\theta$, where $|\mathbf{S}| = [(\mathbf{P} - \mathbf{A}) \cdot (\mathbf{P} - \mathbf{A})]^{1/2}$ and $\cos\theta = \mathbf{R} \cdot \mathbf{S}/|\mathbf{R}||\mathbf{S}|$. Let this distance be denoted by e. The vector, parametric equation for line $\mathbf{B} - \mathbf{A}$ is

$$\mathbf{Q} = \mathbf{A} + t(\mathbf{B} - \mathbf{A}) = \mathbf{A} + t\mathbf{R}$$

If we choose $t = e/|\mathbf{R}|$, where $|\mathbf{R}| = [(\mathbf{B} - \mathbf{A}) \cdot (\mathbf{B} - \mathbf{A})]^{1/2}$, then

$$\mathbf{Q} = \mathbf{A} + \left(\frac{e}{|\mathbf{R}|}\right)\mathbf{R} = \mathbf{A} + e\mathbf{n} \tag{7-4}$$

where **n** is the unit vector along line $\mathbf{B} - \mathbf{A}$. The coordinates of the point of intersection are then given by $\mathbf{Q} = [x_i\ y_i\ z_i]$, where

$$\begin{aligned} x_i &= x_1 + \left(\frac{e}{|\mathbf{R}|}\right)(x_2 - x_1) \\ y_i &= y_1 + \left(\frac{e}{|\mathbf{R}|}\right)(y_2 - y_1) \\ z_i &= z_1 + \left(\frac{e}{|\mathbf{R}|}\right)(z_2 - z_1) \end{aligned} \tag{7-5}$$

This method is implemented in the main program PTLINE.

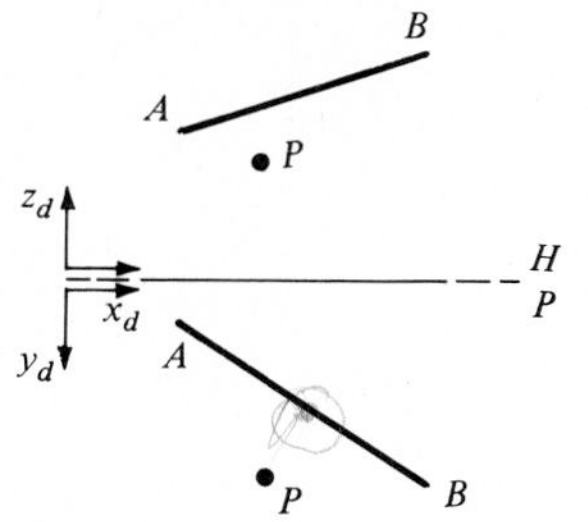

Figure 7-7
Orthographic views of cable and *P*.

Example 7-15

Given: A point in space has coordinates **P** = [4 5 3]. This point can be used to mount a support bracket. Two orthographic views of a three-dimensional cable running between **A** and **B** are shown in Figure 7-7. Task:

a. Find the shortest distance from point **P** to line **R** = **B** − **A.**
b. Find the coordinates of point **Q** on line **B** − **A** which is nearest to point **P.**

Solution: The following table can be constructed using the given orthographic views.

Location	*x*	*y*	*z*
A	2	1	4
B	8	5	6
P	4	5	3

Let **R** = **B** − **A** and **S** = **P** − **A.**
Then,

$$R_x = 8 - 2 = 6; \quad R_y = 5 - 1 = 4; \quad R_z = 6 - 4 = 2$$

$$S_x = 4 - 2 = 2; \quad S_y = 5 - 1 = 4; \quad S_z = 3 - 4 = -1$$

$$\mathbf{R} \cdot \mathbf{S} = 6(2) + 4(4) + 2(-1) = 26$$

$$|\mathbf{R}| = (6^2 + 4^2 + 2^2)^{1/2} = 7.48$$

$$|\mathbf{S}| = (2^2 + 4^2 + 1^2)^{1/2} = 4.58$$

$$\cos\theta = \frac{\mathbf{R} \cdot \mathbf{S}}{|\mathbf{R}||\mathbf{S}|} = \frac{26}{[(7.48)(4.58)]} = 0.759$$

$$\theta = 40.63 \text{ deg}$$

The shortest distance is then

$$d = \frac{|\mathbf{S}||\mathbf{R}| \sin\theta}{|\mathbf{R}|} = \frac{(4.58)(7.48)(0.651)}{7.48} = 2.98 \text{ units}$$

The distance from **A** to the point of intersection is

$$e = |\mathbf{S}| \cos\theta = (4.58)(0.759) = 3.476 \quad \text{(See Figure 7-6.)}$$

Set the parameter value

$$t = \frac{e}{|\mathbf{R}|} = \frac{3.476}{7.48} = 0.465$$

The location of point **Q** on line **B** − **A** is given by **Q** = $[x_i \; y_i \; z_i]$, where

$$x_i = x_1 + t(x_2 - x_1) = 2 + 0.465(8 - 2) = 4.79$$

$$y_i = y_1 + t(y_2 - y_1) = 1 + 0.465(5 - 1) = 2.86$$

$$z_i = z_1 + t(z_2 - z_1) = 4 + 0.465(6 - 4) = 4.93$$

B. Shortest Distance from a Point to a Plane

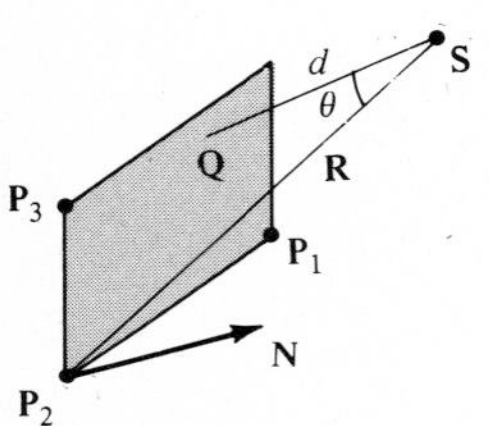

Figure 7-8
Shortest distance point to plane.

Let a plane be defined by three points $\mathbf{P}_1$, $\mathbf{P}_2$, and $\mathbf{P}_3$. A fourth point **S** lies outside the plane. The shortest distance between the point and the plane can be seen when the plane is viewed on edge. The shortest distance is the perpendicular distance from the point to the edge view of the plane.

Define the point **S** relative to a vertex of the plane, say $\mathbf{P}_2$, as shown in Figure 7-8. Let $\mathbf{R} = \mathbf{S} - \mathbf{P}_2$ and observe that $d = |\mathbf{R}| \cos \theta$. Multiplying the right side of this equation by $|\mathbf{N}|/|\mathbf{N}|$ gives

$$d = \frac{|\mathbf{R}|\,|\mathbf{N}| \cos \theta}{|\mathbf{N}|} = \frac{|R \cdot N|}{|N|} \tag{7-6}$$

where

$$\mathbf{N} = (\mathbf{P}_1 - \mathbf{P}_2) \times (\mathbf{P}_3 - \mathbf{P}_2) \text{ and } |\mathbf{N}| = (\mathbf{N} \cdot \mathbf{N})^{1/2}$$

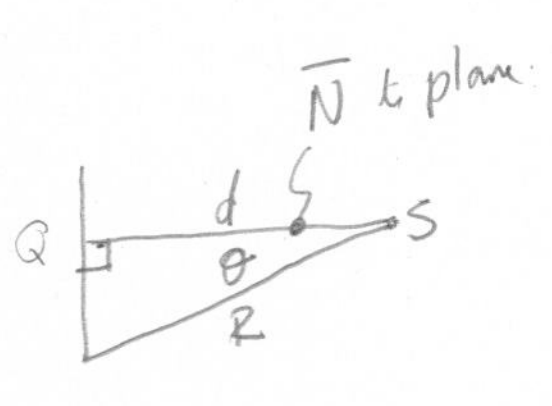

The line of shortest distance from **S** to the plane is parallel to the normal vector **N.** If **S** lies on the positive side of the plane (the same side as vector **N**) then the unit vector **s** from **S** to **Q** points in the opposite direction to the unit normal vector **n.** Otherwise, **s** and **n** point in the same direction. The intersection point **Q** can be found from the equation

$$\mathbf{Q} = \mathbf{S} + d\mathbf{s} \tag{7-7}$$

where **S** can be expressed in relative coordinates with respect to $\mathbf{P}_2$ or in absolute coordinates with respect to the coordinate system origin. Program PTPLANE uses the above method.

If $\mathbf{S} = [x_4 \; y_4 \; z_4]$ then when $F = N_1x_4 + N_2y_4 + N_3z_4 - k = 0$, the point **S** lies on the plane. If $F < 0$ the point lies on one side of the plane, and if $F > 0$ the point is on the other side. Two points on the same side of a plane will always produce the same sign of F when the function is evaluated for each point.

If the given point **S** is on the positive side of the plane, then $\mathbf{S} = -\mathbf{n} = -\mathbf{N}/|\mathbf{N}|$, and the coordinates of point $\mathbf{Q} = [x_i\ y_i\ z_i]$ are

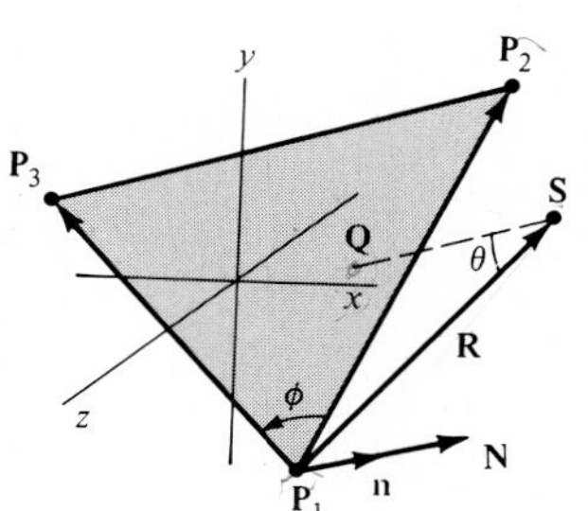

Figure 7-9
Plane for Example 7-16.

$$x_i = S_1 - d\left(\frac{N_1}{|\mathbf{N}|}\right),\ y_i = S_2 - d\left(\frac{N_2}{|\mathbf{N}|}\right),\ z_i = S_3 - d\left(\frac{N_3}{|\mathbf{N}|}\right) \quad (7\text{-}8)$$

Example 7-16

Given: Point $\mathbf{S} = [3\ 1\ 1]$ and a plane defined as in Figure 7-9. $\mathbf{P}_1 = [1\ -2\ 2]$; $\mathbf{P}_2 = [0\ 0\ -2]$; $\mathbf{P}_3 = [-1\ 1\ 1]$.
Task:

a. Calculate the shortest distance from point **S** to the plane.
b. Calculate the intersection point **Q** on the plane.

Solution:

a.

$$\mathbf{N} = \det\begin{vmatrix} \mathbf{i} & \mathbf{j} & \mathbf{k} \\ -1 & 2 & -4 \\ -2 & 3 & -1 \end{vmatrix} = [10 \quad 7 \quad 1]$$

Let $\mathbf{R} = \mathbf{S} - \mathbf{P}_1 = [2\ 3\ -1]$.
Then
$\mathbf{R}\cdot\mathbf{N} = [2\ 3\ -1]\begin{bmatrix} 10 \\ 7 \\ 1 \end{bmatrix} = 40$, and $|\mathbf{N}| = (100 + 49 + 1)^{1/2} = 12.25$

Using Equation (7-6) gives

$$d = \frac{\mathbf{R}\cdot\mathbf{N}}{|\mathbf{N}|} = \frac{40}{12.25} = 3.265$$

b.

$$k = \mathbf{N}\cdot\mathbf{P}_1 = 10(1) + 7(-2) + 1(2) = -2$$

Evaluating F at **S** gives

$$N_1S_1 + N_2S_2 + N_3S_3 - k = 10(3) + 7(1) + 1(1) + 2 = 40 > 0$$

A point on the normal vector is given by $\mathbf{V} = \mathbf{P}_1 + \mathbf{N}$. Thus, $V_1 = 11$, $V_2 = 5$, $V_3 = 3$.
Evaluating F at **V** gives

$$10(11) + 7(5) + 1(3) + 2 > 0$$

Hence, both **S** and **V** are on the same (positive) side of the given plane.

The components of $\mathbf{Q} = [x_i \ y_i \ z_i]$ are then given by Equations (7-8) as

$$x_i = 3 - 3.265\left(\frac{10}{12.25}\right) = 0.333$$

$$y_i = 1 - 3.265\left(\frac{7}{12.25}\right) = -0.866$$

$$z_i = 1 - 3.265\left(\frac{1}{12.25}\right) = 0.733$$

Note: If one had chosen to calculate $\mathbf{N} = (\mathbf{P}_1 - \mathbf{P}_2) \times (\mathbf{P}_3 - \mathbf{P}_2)$, then

$$\mathbf{N} = \det \begin{vmatrix} \mathbf{i} & \mathbf{j} & \mathbf{k} \\ 1 & -2 & 4 \\ -1 & 1 & 3 \end{vmatrix} = [-10 \ -7 \ -1]$$

Then $k = \mathbf{N} \cdot \mathbf{P}_1 = 2$, and evaluating F at $\mathbf{S}$ gives

$$-10(3) - 7(1) - 1(1) - 2 = -40$$

A point on the normal is given by $\mathbf{V} = \mathbf{P}_2 + \mathbf{N}$. Thus, $V_1 = -10$, $V_2 = -7$, $V_3 = -3$.
Evaluating F at $\mathbf{V}$ gives

$$-10(-10) - 7(-7) - 1(-3) - 2 = 150 > 0$$

Thus **S** is on the *negative* side of the plane, **n** and **s** point in the *same* direction, and the point of intersection is found to be

$$x_i = 3 + 3.265\left(\frac{-10}{12.25}\right) = 0.335; \quad y_i = 1 + 3.265\left(\frac{-7}{12.25}\right) = -0.866$$

$$z_i = 1 + 3.265\left(\frac{-1}{12.25}\right) = 0.733$$

This agrees with the previous results.

C. Plane Perpendicular to a Line Passing Through a Point

Sometimes, it may be necessary to construct a plane using conditions other than three known noncollinear points. Once the plane is defined, three points on the plane can be determined to allow additional computations following the techniques used for geometric modeling.

Consider a line defined in the usual sense by two points **A** and **B.** A third point in space $\mathbf{S} = [x_3 \ y_3 \ z_3]$ is known and it does not lie on the given line between **A** and **B.** A plane can be defined with the following two attributes: (See Figure 7-10.)

a. The plane passes through the point **S.**

b. The plane is perpendicular to the given line.

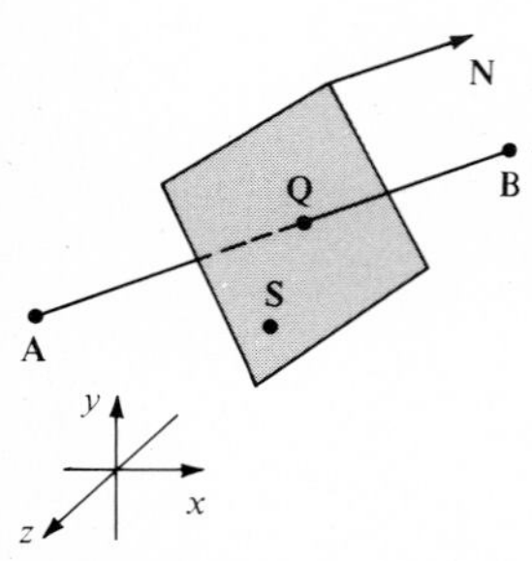

Figure 7-10
Plane perpendicular to a line.

The vector, parametric equation for the line is

$$\mathbf{Q} = \mathbf{A} + t(\mathbf{B} - \mathbf{A}) = \mathbf{A} + t\mathbf{R} \tag{7-9}$$

Since the direction of the line is perpendicular to the required plane, the components of a normal vector **N** can be equated to the components of the directional vector $\mathbf{R} = \mathbf{B} - \mathbf{A}$. If $\mathbf{A} = [x_1\ y_1\ z_1]$ and $\mathbf{B} = [x_2\ y_2\ z_2]$, then $N_1 = x_2 - x_1$; $N_2 = y_2 - y_1$; and $N_3 = z_2 - z_1$. Also, since **S** lies on the plane, the constant k used in the implicit definition of a plane can be calculated using

$$k = \mathbf{N} \cdot \mathbf{S} = N_1x_3 + N_2y_3 + N_3z_3 \tag{7-10}$$

The implicit equation for the required plane is then written as

$$N_1x + N_2y + N_3z - (N_1x_3 + N_2y_3 + N_3z_3) = 0 \tag{7-11}$$

If the line through **A** and **B** is the normal vector to an initial plane, then the newly calculated plane through **S** will be parallel to the initial plane. Thus we have another method for constructing specific planes which are parallel to a reference plane. A test for parallel planes was given in Section 6-6.

D. Test for Perpendicular Planes

There are several ways to mathematically check for the perpendicularity of two planes. If $\mathbf{N_1}$ and $\mathbf{N_2}$ are the two normals, then they will be perpendicular and $\mathbf{N_1} \cdot \mathbf{N_2} = 0$. If the two implicit equations for the planes are known, such as Equations (6-1) and (6-2), then the planes are perpendicular when

$$N_{1,1}N_{2,1} + N_{1,2}N_{2,2} + N_{1,3}N_{2,3} = 0 \tag{7-12}$$

That is, the sum of the product of each pair of coefficients is equal to zero.

PROBLEMS: CHAPTER 7

7-1 *Given:* Top and front views of two skew lines *MN* and *RS*.
Task: Find and locate the shortest line (*OX*) which will connect *MN* and *RS*. Show line *OX* in all views.

7-2 *Given:* Top and front views of two skew lines *MN* and *RS*.
Task: Find and locate the shortest *horizontal* line which will connect lines *MN* and *RS*. Show this line (*OX*) in all views. (Use plane method.)

7-3 *Given:* Top and front views of point *P* and triangle *ABC*.
Task: Construct a 3/4 line from *P*, sloping upward and perpendicular to plane *ABC*.

7-4 *Given:* Top view of line *MN* and line *AB* of plane *ABC*, and front view of line *MN* and line *AC* of plane *ABC*.
Task: Construct plane *ABC* perpendicular to line *MN*. Use only two views.

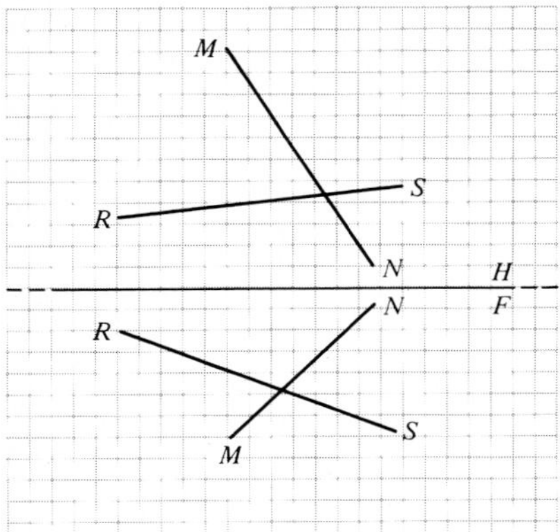

Figure 7-1P

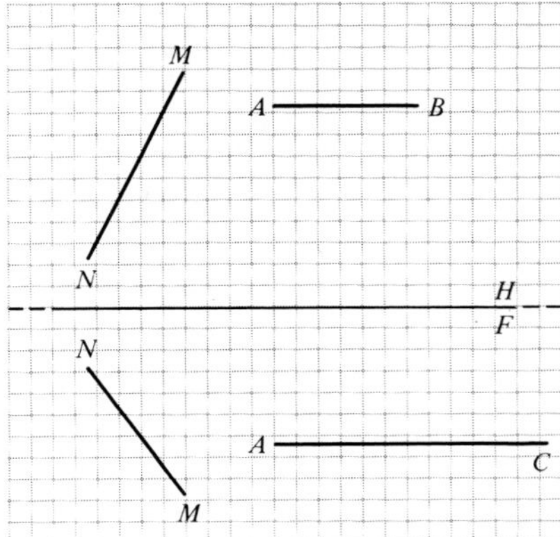

Figure 7-4P

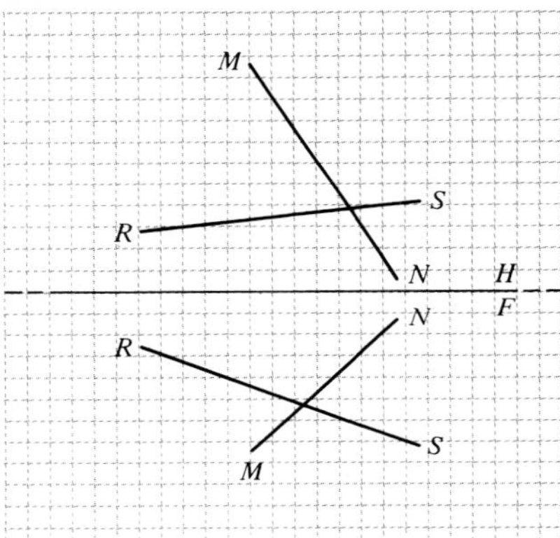

Figure 7-2P

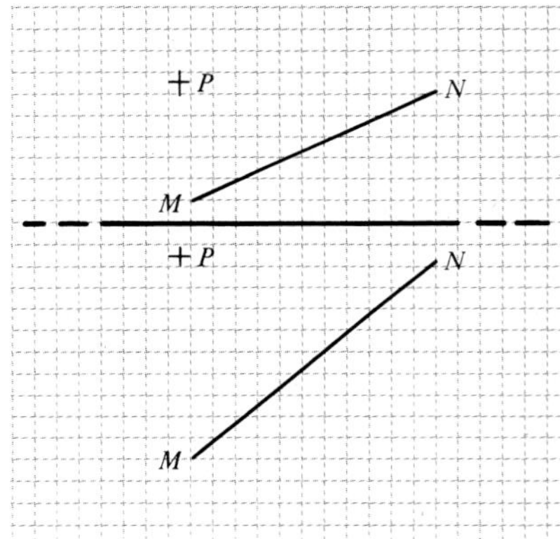

Figure 7-5P

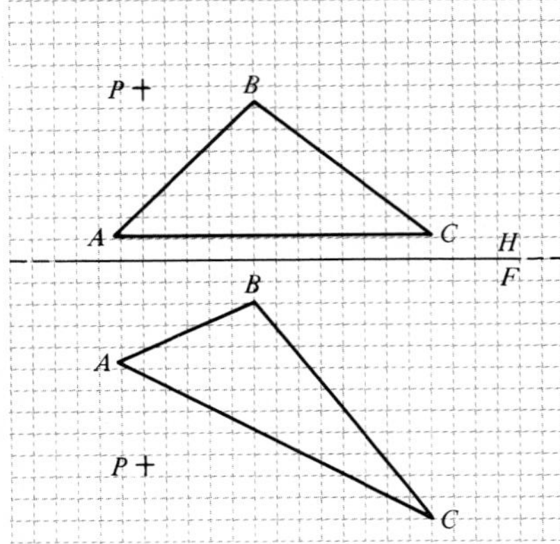

Figure 7-3P

7-5 *Given:* Top and front views of line *MN* and point *P*.
Task: Determine the shortest line which will connect point *P* with line *XY*. Show the projection of this line in all views.

7-6 *Given:* Complete top and partial front views of two high tension wires *MN* and *RS*.
Task: Wire *MN* must pass below *RS* with a clearance of 7 feet. Show both wires in the front view. Scale = ⅛″ = 1′-0″

7-7 *Given:* Top and front views of shaft *MN* and shaft *RS*.
Task: Determine the bearing of the shortest *horizontal* tunnel that can be dug from *MN* to *RS*.

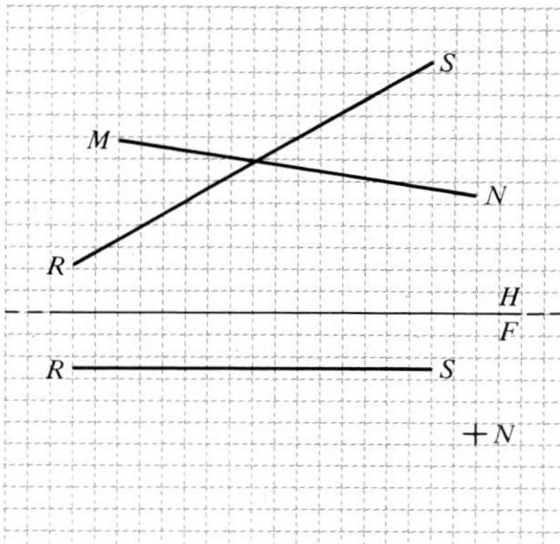

Figure 7-6P

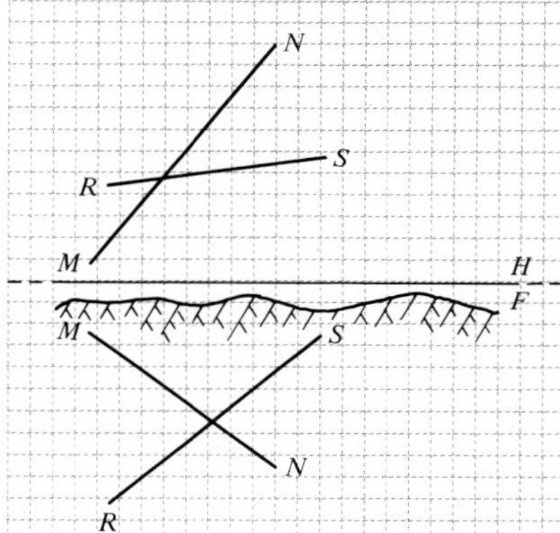

Figure 7-7P

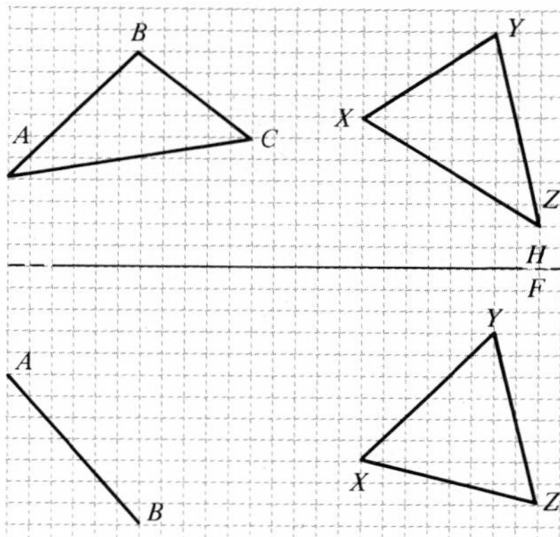

Figure 7-8P

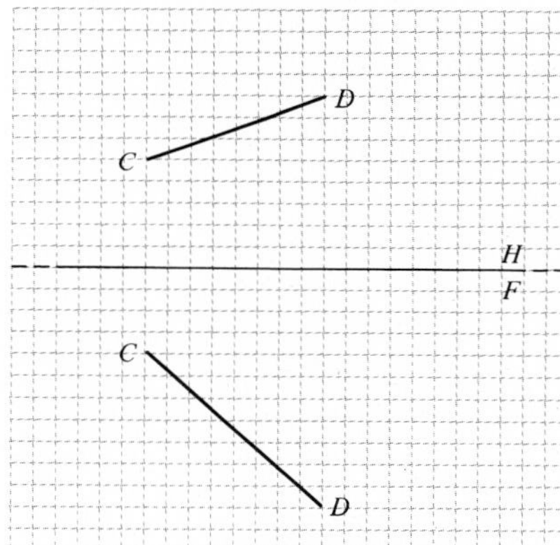

Figure 7-9P

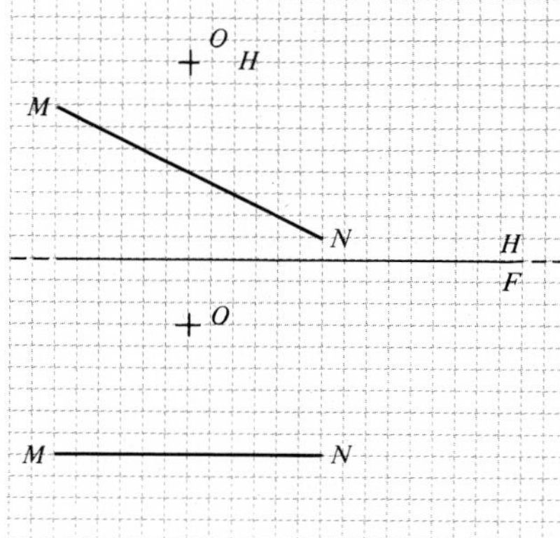

Figure 7-10P

7-8 *Given:* Top view and partial front view of plane *ABC* which is perpendicular to plane *XYZ* shown in top and front view.
Task: Complete the front view of plane *ABC*.

7-9 *Given:* Top and front views of line *CD*.
Task: Construct a plane *ABC* which is perpendicular to given line *CD*. (Use only the two views given.)

7-10 *Given:* Top and front views of line *MN* and point *O*.
Task: Using the point view method find the true length of the shortest line (*OX*) which can be drawn from the point to the line. Show *OX* as an object in all views.

7-11 *Given:* Top and front views of line *MN* and point *O*.

Task: Using the plane method, find the true length of the shortest line (*OX*) which can be drawn from the point to the line. Show *OX* as an object line in all views.

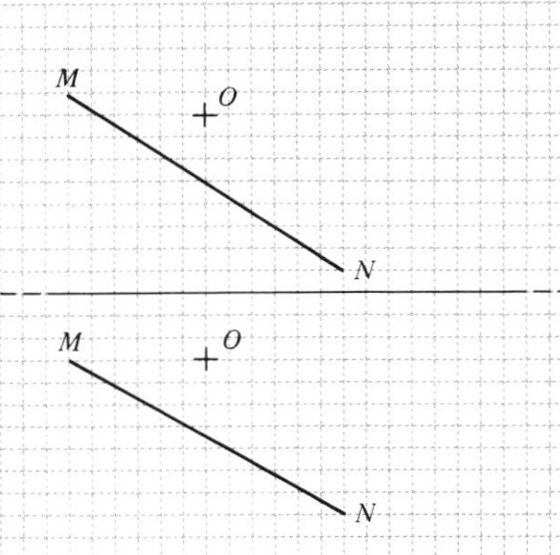

Figure 7-11P

7-12 *Given:* Top and front views of tubing lines *MN* and *XY*.
Task: Find the length of the *shortest* straight segment which could connect the two given lines. Show the connecting segment in all views. Scale: 1″ = 1′-0″

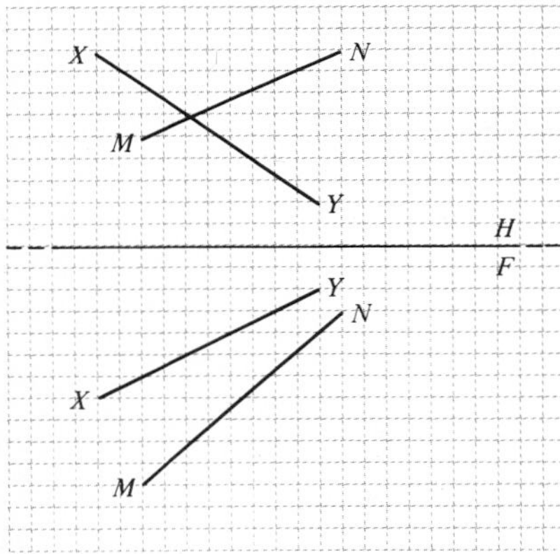

Figure 7-12P

7-13 *Given:* Front view of line *XY* and top view of point *Y*, top and front view of line *MN* which is perpendicular to line *XY*.
Task: Show the projection of line *XY* in the top view. (Use only the given views.)

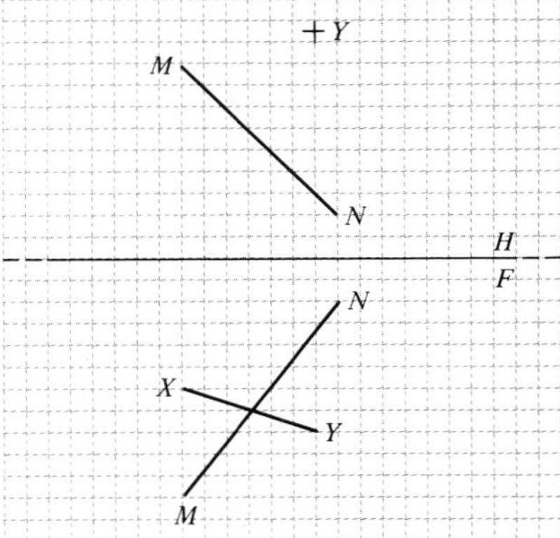

Figure 7-13P

7-14 *Given:* Top and front views of line *MN* and point *Y*. Point *Y* is a point on horizontal line *XY* with point *X* being one inch to the left of point *Y*.
Task: Using only the two given views show the top and front views of line *XY*.

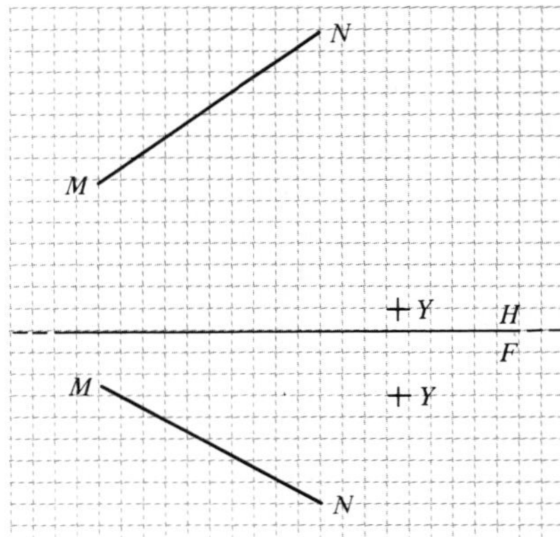

Figure 7-14P

7-15 *Given:* The actual dimensions represented by the reduced scale drawing given in Example 7-1 are as follows:

$$\mathbf{A} = [0\ 29\ 6],\ \mathbf{B} = [29\ 5\ 37],\ \mathbf{M} = [3\ 5\ 31],$$
$$\mathbf{N} = [23\ 29\ 12].$$

Task: Use the vector dot product to show that the lines are not perpendicular.

7-16 *Given:* The actual endpoint coordinates of the line represented in Example 7-4a are **A** = [0 11 22] and **B** = [32 0 7]; the coordinates of point **O** are [11 16 24].
Task: Calculate the shortest distance from point **O** to line **B** − **A**.

7-17 *Given:* The centerline of a pipe passes through the points represented by **A** = [0 0 0] and **B** = [2.6875 2 1.1875]. A point **C** = [0.6875 2.25 1.5] must contact the surface of the pipe.
Task: Calculate the required pipe diameter.

7-18 *Given:* Problem 7-4.
Task: Find the implicit equation for a plane *ABC* which is perpendicular to line *MN*.

7-19 *Given:* Problem 7-5.
Task: Use vector techniques to find the shortest line from point P to line *XY*.

7-20 *Given:* A line is defined between **A** = [8 4 −2] and **B** = [−2 5 6], while a point **S** = [3 3 −3].
Task: Determine the implicit equation for the plane which is perpendicular to the given line and contains the given point **S**.

7-21 *Given:* A plane defined by **A** = [1 1 1], **B** = [1 −1 1], and **C** = [−1 −1 −1].
Task: Define a plane which is parallel to the given plane and which passes through the point **P** = [0 0 4].

7-22[c] *Given:* Example 7-15.
Task: Check the solution using program PTLINE.

7-23[c] *Given:* Example 7-16.
Task: Check the solution using program PTPLANE.

---

[c]Superscript *c* indicates that a computer solution is required.

8

Intersecting Lines and Planes

One of the most common descriptive geometry problems is to find the intersection between a line and a plane. The two descriptive geometry approaches for solving this type of problem are the *auxiliary method* and the *two-view method.* The geometric modeling approach is to equate the vector equations for the line and plane, and solve for the parameter value at the point of intersection. If a surface is represented by a series of lines, then the line of intersection between a plane and the surface can be found by repeated application of the solution technique.

8-1 INTERSECTING LINE AND PLANE-PIERCING POINTS

A line which is neither in a plane nor parallel to it will intersect that plane in one, and only one, point. (See Figure 8-1.) This point is referred to as a *piercing point.* In most cases, the piercing point in question will be within the boundaries of a finite plane and the limits of the line segment. It must be remembered, however, that both a line and a plane are considered to be infinite unless otherwise noted. For this reason, the limits of a plane or a line may be extended if necessary.

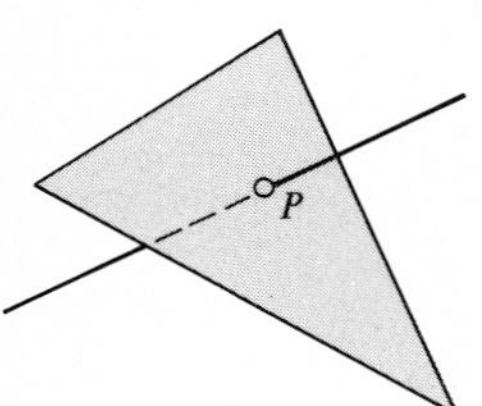

Figure 8-1
Piercing point.

Even though a plane theoretically has no thickness, the following examples consider a plane to be opaque in order that the visibility of the line and the plane can be considered.

A. Auxiliary Method

The auxiliary method used in locating the piercing point of a line and a plane will not be difficult for the student since all the necessary concepts have been discussed in prior chapters of this text. The piercing point may be established in an auxiliary view which shows the line crossing the edge view of the plane. The line does not have to appear in true length.

Example 8-1

Given: Incomplete top and front views of plane *ABC* and line segment *MN*.

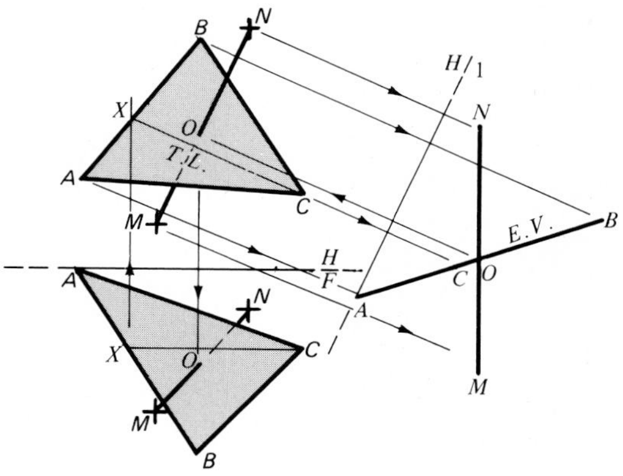

Example 8-1
Piercing point: auxiliary method.

Task: Find the point at which line *MN* pierces plane *ABC*. Show correct visibility in all views.
Solution:

1. Place horizontal line *XC* in plane *ABC* in the front view and find its true length in the top view.
2. Draw folding line *H*/1 perpendicular to the true length line *XC* in the top view.
3. Draw lines of sight parallel to *XC* into the first auxiliary and lay off line *MN* and edge view of triangle *ABC*.
4. Line *MN* will cross the edge view of triangle *ABC* at "*O*." The line of sight determined by point "*O*" is projected into the top view until it meets line *MN* and establishes the point "*O*" at which the line pierces plane *ABC*.
5. Project point "*O*" into the front view.
6. Show correct visibility of line *MN* and plane *ABC*. (See Section 1-11.) ◀

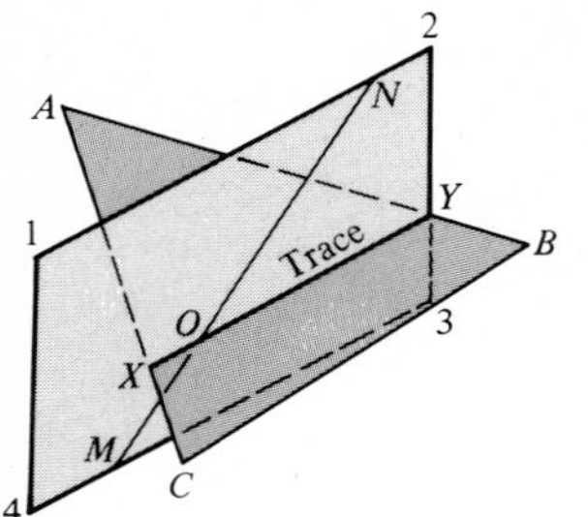

Figure 8-2
Piercing point: cutting-plane method.

B. Two-View Method: Cutting-Plane Method

Although the auxiliary view method is an easy way to determine the piercing point of a line and a plane, it is, because of minimum space and construction requirements, sometimes advantageous to use the cutting-plane method. Figure 8-2 shows a line *MN* intersecting plane *ABC*. A *vertical* cutting-plane 1-2-3-4 is constructed so that it contains the line *MN* and is perpendicular to the *horizontal* projection plane. The point at which the trace (*xy*) of the cutting-plane 1-2-3-4 and the line *MN* intersect is the piercing point. The intersection of one plane with another is called its *trace* on that plane. The step-by-step procedure of the cutting-plane method is shown in the following example.

Example 8-2

Given: Incomplete top and front views of line *MN* and plane *ABC*.

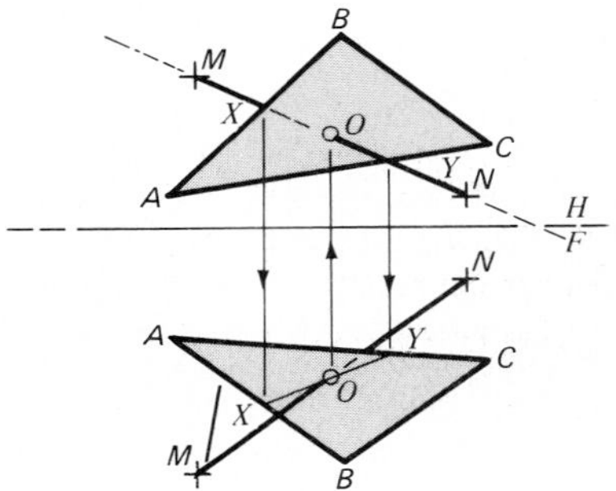

Example 8-2
Piercing point: cutting-plane method.

Task: Using only the given views, find the point at which line *MN* pierces plane *ABC*. Show correct visibility of line and plane in both views.
Solution:

1. Pass a *vertical* cutting-plane so that it contains line *MN* and cuts plane *ABC* at *X* and *Y*. Project *X* and *Y* into the front view of *ABC*. The intersection (*XY*) will show as a trace of the cutting-plane in the front view. (It is suggested that the edge view of the cutting-plane be identified as such. Examples in this text will use the symbol ---————---.)
2. In the front view, the trace (*XY*) and the line *MN* intersect at point *O*, the piercing point.
3. Since "*O*" is a point, it can be projected into the top view.
4. Show visibility in both top and front views. (Reference Section 1-11.)

The same procedure would be followed if the cutting-plane were passed perpendicular to the frontal plane as shown in Example 8-3. This example is inserted here so the reader can use it as a self-testing medium.

Example 8-3

Given: Incomplete top and front views of triangle *ABC* and line segment *MN*.

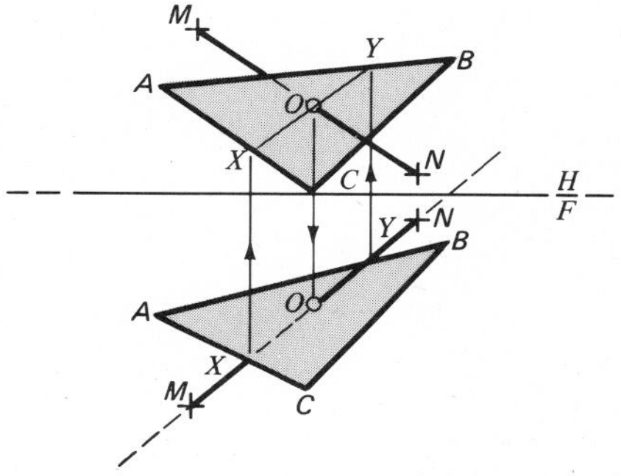

Example 8-3
Piercing point: cutting-plane method.

Task: Find the point at which line *MN* pierces plane *ABC*. Show correct visibility in both views.

8-2 PROJECTION OF A LINE ON A PLANE

The projection of a point, in space, on a plane is obtained by finding the piercing point on the plane of a perpendicular line from the point to the plane.

The projection of a line on a plane can be obtained by finding the projection on the plane of each defining point of the line segment. Figure 8-3 shows that, as in the above definition, the end points of the line seg-

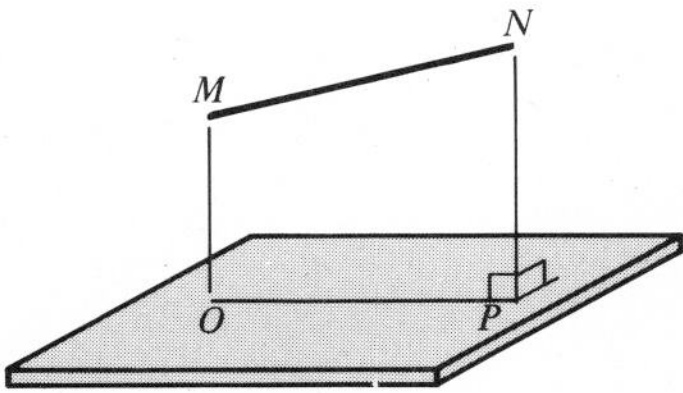

Figure 8-3
Projection of a line on a horizontal plane.

ment have been projected perpendicularly onto the plane and the two piercing points connected to form the projection of the line segment. The following examples illustrate the two methods of determining the projection of a line segment on a given plane.

A. Auxiliary Method

Example 8-4

Given: Top and front views of line *MN* and plane segment *ABCD*.

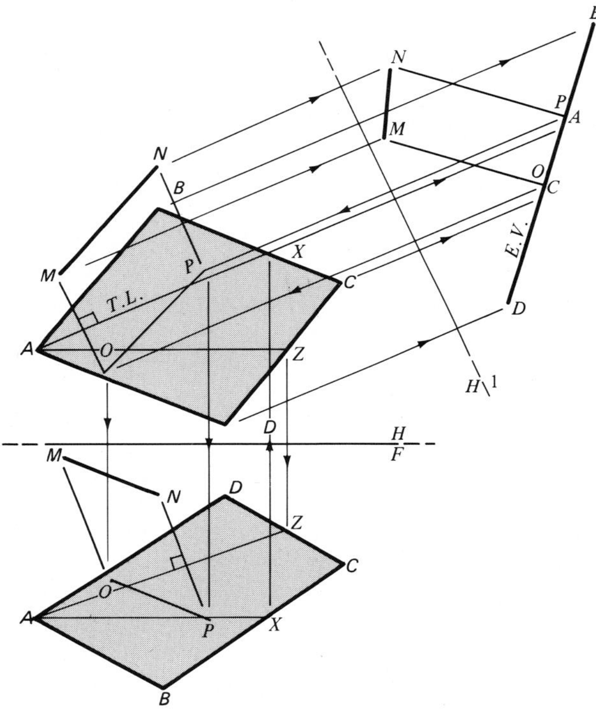

Example 8-4
Projection: line on a plane.

Task: Find the projection (*OP*) of line *MN* on plane segment *ABCD*.
Solution:

1. In a first auxiliary, show line *MN* and the edge view of plane *ABCD*.
2. Construct perpendicular lines from point *M* and *N* to intersect the edge view of the plane at points *O* and *P*.
3. Project the lines of sight of points *O* and *P* into the top view.

4. Since lines *MO* and *NP* are both perpendicular to plane *ABCD* and true length in the first auxiliary they must, in the top view, be drawn parallel to folding line *H*/1. These lines will also be perpendicular to *AX* which is a true length line in the plane. The intersection of the lines of sight of "*O*" and "*P*" from the first auxiliary with these perpendiculars will determine the piercing points "*O*" and "*P*" in the top view.
5. Using the distances below the H/1 folding line in the first auxiliary, project points "*O*" and "*P*" into the front view to obtain the required projection of *OP*. ◀

Check—The construction lines *MO* and *NP* are each perpendicular to two intersecting true length lines in the plane (i.e., *AX* in the top view and *AZ* in the front view). The lines are, therefore, perpendicular to the plane.

B. Two-View Method

Example 8-5

Given: Top and front views of line *MN* and plane segment *ABCD*.

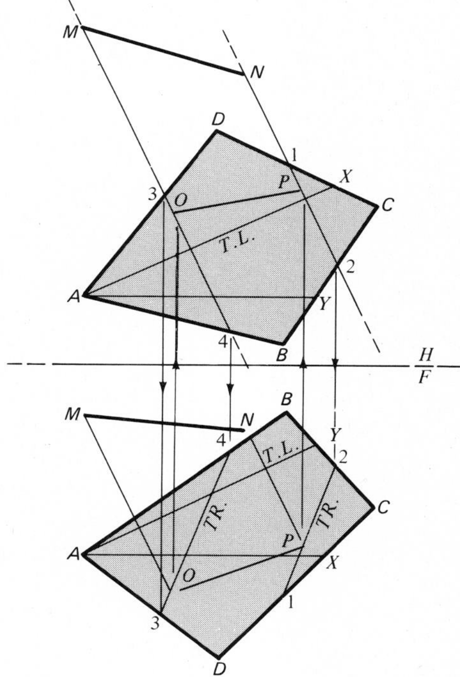

Example 8-5
Projection: line on a plane.

Task: Find the projection (*OP*) of line *MN* on plane segment *ABCD*.
Solution:

1. Using horizontal and frontal lines, draw *AX* true length in the top view and *AY* true length in the front view.
2. In each view, draw lines from *M* and *N* perpendicular to their respective true length line. *AX* and *AY* intersect and are in the plane, therefore the lines from *M* and *N* are perpendicular to the plane.
3. Pass two vertical cutting-planes each containing a perpendicular from line *MN* and determine the pierce points "*O*" and "*P*" in the front view. Line *OP* is the required projection.
4. Project *OP* into the top view. ◀

8-3 LINE OF INTERSECTION: INTERSECTING PLANE SEGMENTS GIVEN

Any two planes in space must be parallel or intersecting. Although most of the problems in this text which deal with intersecting planes will be concerned with defined planar facets, it must be remembered that planes are infinite unless limited by specific boundaries. Two intersecting planes will show their intersection as a straight line which connects two or more points, each of which is common to both planes. Since the two points involved can be determined by the piercing point of the lines of one plane with another plane, the problem can be solved by using the procedures shown in Section 8-1.

A. Auxiliary Method

In Section 8-1-A, the view which showed the edge of the plane also indicated the point at which the line pierced the plane. This method can also be used here to show the intersection of two plane segments since the boundaries of the planes can be considered as lines. The points at which these lines pierce the plane will be revealed in a view which shows one of the planes on edge.

Example 8-6

Given: Incomplete top and front views of two intersecting plane segments *ABC* and *XYZ*.
Task: Using the auxiliary method, find the intersection of the two given triangles. Show complete visibility in all views.

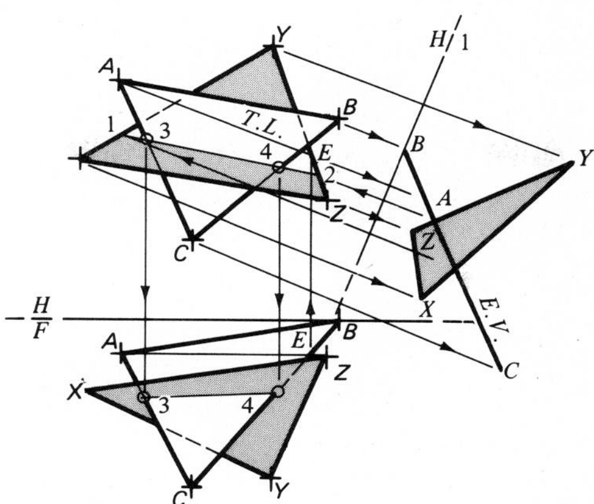

Example 8-6
Intersection of planes: auxiliary method.

Solution:

1. Construct the first auxiliary so that triangle *ABC* shows as an edge view (E.V.) This auxiliary indicates that a cutting-plane containing the edge view would cut plane *XYZ* at points 1 and 2.
2. Project points 1 and 2 into the top view. A line (trace) 1-2 intersects lines *AC* and *BC* at the piercing points 3 and 4.
3. Draw line of intersection, 3-4. Project it into the front view, and determine visibility in both views as illustrated.

B. Two-View Method

It may be advisable, if the situation warrants, to use the two-view method of finding the intersection of two plane segments. The line of intersection is found by discerning where the lines of one plane pierce the other. As stated before, the boundaries of the planes are generally considered as the lines in question. Since this method is identical to that used in Section 8-1-B, the student should be well-versed in its use. It is sometimes difficult to determine which lines will pierce the opposite plane, since the selection of lines is arbitrary. This means that in some cases the selection of a line will not produce a pierce point. If this is the case, select other lines which will provide the required pierce points and the subsequent intersection.

Example 8-7

Given: Incomplete top and front views of two plane segments *ABC* and *XYZ*.

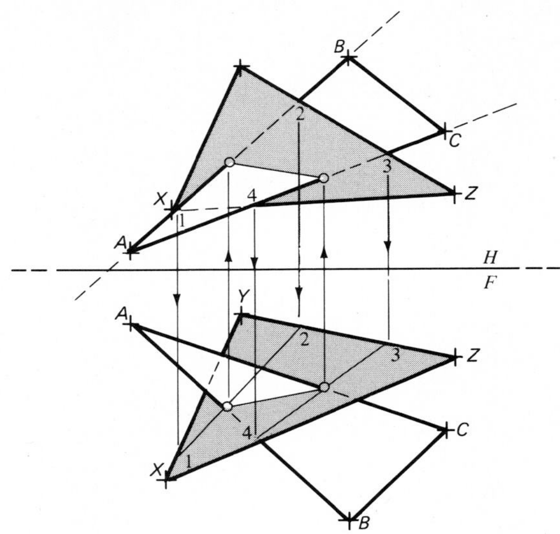

Example 8-7
Intersection of planes: cutting-plane (two-view) method.

Task: Using the two-view method, find the intersection of the two given triangles. Show complete visibility in both views.
Solution:

1. In order to find the two piercing points required for the line of intersection, apply the cutting-plane method to line *AB* and to line *AC* in the top view.
2. Project points and find ends of the trace at points 1, 2 and 3, 4. Find the pierce points.
3. The line between the pierce points of *AB* and *AC* is the required line of intersection.
4. Complete the visibility in both views. ◀

8-4 LINE OF INTERSECTION: NONINTERSECTING PLANE SEGMENTS GIVEN

Since planes are infinite it is necessary, especially in civil engineering, to find the intersection, in space, of two planes which are defined on paper by two nonintersecting segments. Where this situation occurs, the cutting-plane method is used as shown in Figure 8-4. One point on the line of intersection of two planes will be shown by the intersection of the traces of the cutting-plane on the two given triangles.

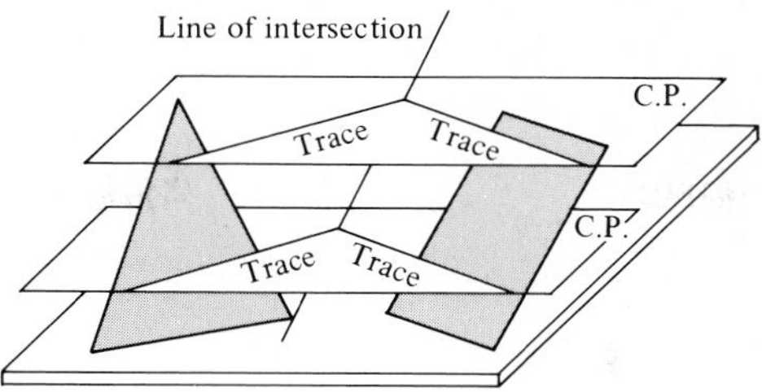

Figure 8-4
Intersection of infinite planes: nonintersecting segments shown.

Example 8-8

Given: Top and front views of two nonintersecting plane segments *ABC* and *MNO*.

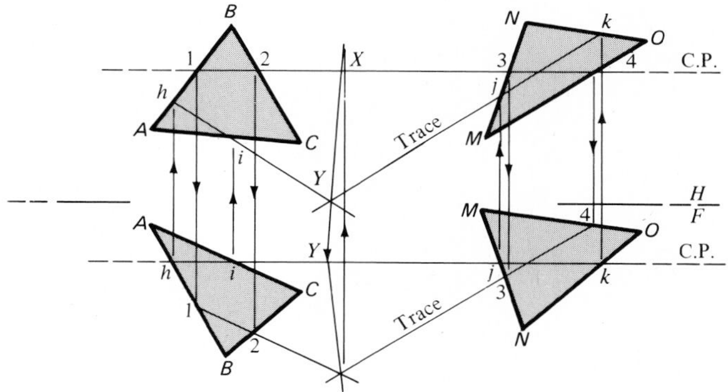

Example 8-8
Intersection of planes: nonintersecting segments given.

Task: Find the intersection, in space, of the two planes denoted by the given plane segments.
Solution:

1. Draw at random a vertical cutting-plane which passes through both the plane segments in the top view at 1, 2, 3, and 4.
2. Project these points (cutting-plane and given planes) to show the traces of the cutting-plane in the front view.
3. The intersection of the traces of the cutting-plane gives one point (*X*) of the required intersection.
4. Project point *X* to the cutting-plane shown in the top view.
5. Draw another random cutting-plane in either view cutting the plane segments at *h*, *i*, *j*, and *k*. Project these points into the top view to form traces of the horizontal cutting-plane.

6. The intersection of the traces will give the second point (Y) of the line of intersection.
7. Project Y to the edge of the horizontal cutting-plane in the front view.
8. Draw a line through X and Y in both views to obtain the line of intersection of the two given planes. ◀

Descriptive geometry offers several methods of solving a given type of problem. The same is true for geometric modeling. This is necessary because of the various ways that geometric objects can be represented.

The topic of intersection is now presented from a geometric modeling point of view.

8-5 MODELING THE INTERSECTION OF A LINE AND PLANE

The vector, parametric equation for a three-dimensional line was given by Equation (3-3) as: $\mathbf{Q} = \mathbf{A} + t(\mathbf{B} - \mathbf{A}) = \mathbf{A} + t\mathbf{R}$. The vector equation for a plane was given by Equation (4-8) as $\mathbf{N} \cdot (\mathbf{Q} - \mathbf{P}_1) = 0$. To find the point which is common to both the line and plane combine these two equations to obtain

$$\mathbf{N} \cdot (\mathbf{A} + t(\mathbf{B} - \mathbf{A}) - \mathbf{P}_1) = 0 \tag{8-1}$$

Solving for the scalar parameter t gives

$$t = \frac{(\mathbf{N} \cdot \mathbf{P}_1 - \mathbf{N} \cdot \mathbf{A})}{(\mathbf{N} \cdot (\mathbf{B} - \mathbf{A}))} \tag{8-2}$$

Using $\mathbf{N} \cdot \mathbf{P}_1 = k$ and $(\mathbf{B} - \mathbf{A}) = \mathbf{R}$ gives

$$t = \frac{(k - \mathbf{N} \cdot \mathbf{A})}{(\mathbf{N} \cdot \mathbf{R})} \tag{8-3}$$

Using this value of t in the equation for the line gives the point of intersection $\mathbf{Q}_i$ as

$$\mathbf{Q}_i = \mathbf{A} + \left(\frac{(k - \mathbf{N} \cdot \mathbf{A})}{(\mathbf{N} \cdot \mathbf{R})}\right)\mathbf{R} \tag{8-4}$$

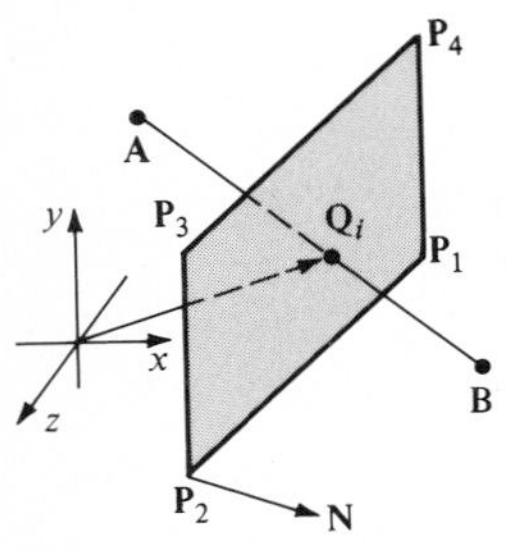

Figure 8-5
Intersection between line and plane.

The coordinates of $\mathbf{Q}_i = [x_i\ y_i\ z_i]$ give the components of the point of intersection. An intersection point is shown in Figure 8-5. In general, the point of $\mathbf{Q}_i$ may lie outside the boundary formed by vertices $\mathbf{P}_1$, $\mathbf{P}_2$, $\mathbf{P}_3$, $\mathbf{P}_4$; or $\mathbf{Q}_i$ may lie on an extension to the line segment between **A** and **B.**

If a plane is given in implicit mathematical form by

$$ax + by + cz + d = 0 \tag{8-5}$$

then use $N_1 = a$, $N_2 = b$, $N_3 = c$, and $k = -d$ in Equation (8-4) when calculating the coordinates of the piercing point $\mathbf{Q}_i$.

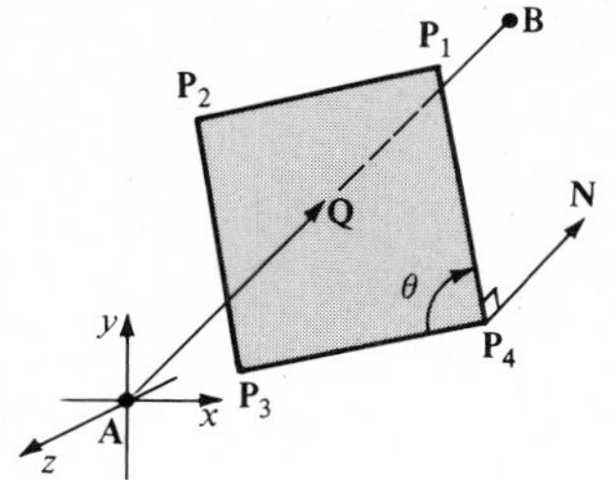

Figure 8-6
Line through origin normal to a plane.

Previously, we used $k/|\mathbf{N}|$ to find the shortest distance from the origin of a coordinate system to a plane. To show why this is true, define a line segment between $\mathbf{B} - \mathbf{A}$, with base vector $\mathbf{A}$ at the origin of the coordinate system. Let this line be *perpendicular* to the plane defined by points $\mathbf{P}_1$, $\mathbf{P}_2$, $\mathbf{P}_3$ as shown in Figure 8-6. Note carefully the numbering of the defining vertices for the plane, and the direction of the vector normal $\mathbf{N}$, given by the cross product

$$\mathbf{N} = (\mathbf{P}_3 - \mathbf{P}_4) \times (\mathbf{P}_1 - \mathbf{P}_4) \tag{8-6}$$

The angle θ is clockwise and $\mathbf{N}$ points away from an observer at a large positive z-location.

The parametric, vector equation for this normal line can be written in terms of a parameter t as

$$\mathbf{Q} = \mathbf{A} + t(\mathbf{B} - \mathbf{A}) = t\mathbf{B} \tag{8-7}$$

since $\mathbf{A} = [0\ 0\ 0]$. Because the directional vector $(\mathbf{B} - \mathbf{A}) = \mathbf{B}$ is in the same direction as the normal vector to the plane, Equation (8-7) can also be written in terms of another scalar parameter s as

$$\mathbf{Q} = s\mathbf{N} \tag{8-8}$$

Although the direction of the vector $\mathbf{B}$ is the same as the vector $\mathbf{N}$ in Figure 8-6, the magnitude of these two vectors is not the same. The second scalar s was defined such that $t|\mathbf{B}| = s|\mathbf{N}|$. Since $\mathbf{B}$ was chosen perpendicular to the plane, the shortest distance from the origin to the plane is measured along the vector $\mathbf{B}$. This distance can be related to the value of k.

Returning to the vector equation for a plane,

$$\mathbf{N} \cdot (\mathbf{Q} - \mathbf{P}_1) = 0$$

Using Equation (8-8) for $\mathbf{Q}$ gives

$$\mathbf{N} \cdot (s\mathbf{N} - \mathbf{P}_1) = 0$$

Alternatively,

$$s\mathbf{N} \cdot \mathbf{N} - \mathbf{N} \cdot \mathbf{P}_1 = 0$$

Since $k = \mathbf{N} \cdot \mathbf{P}_1$ and $\mathbf{N} \cdot \mathbf{N} = |\mathbf{N}|^2$, solving for the parameter s gives

$$s = \frac{k}{|\mathbf{N}|^2} \tag{8-9}$$

Using Equation (8-8), the point of intersection $\mathbf{Q}$ between the plane and the line normal to the plane can now be expressed as

$$\mathbf{Q} = s\mathbf{N} = \left\{\frac{k}{|\mathbf{N}|^2}\right\}\mathbf{N} = \left\{\frac{k}{|\mathbf{N}|}\right\}\mathbf{n} \tag{8-10}$$

where **n** is the unit normal vector $\mathbf{N}/|\mathbf{N}|$. Now $k/|\mathbf{N}|$ is the magnitude and **n** gives the direction of the position vector **Q**. Thus, $|\mathbf{Q}| = k/|\mathbf{N}|$ is the shortest distance from the origin to the plane.

Example 8-9

Given: $\mathbf{P}_1 = [5\ 3\ 3]$, $\mathbf{P}_2 = [4\ 2\ -3]$, $\mathbf{P}_3 = [1\ 3\ -2]$.
Task: Calculate the shortest distance from the origin to the plane defined by the three given points.
Solution:

$$\mathbf{N} = (\mathbf{P}_1 - \mathbf{P}_2) \times (\mathbf{P}_3 - \mathbf{P}_2) = \det \begin{vmatrix} \mathbf{i} & \mathbf{j} & \mathbf{k} \\ 1 & 1 & 6 \\ -3 & 1 & 1 \end{vmatrix}$$

$$\mathbf{N} = [-5 \quad -19 \quad 4]$$

$$k = \mathbf{N} \cdot \mathbf{P}_1 = [-5 \quad -19 \quad 4] \begin{bmatrix} 5 \\ 3 \\ 3 \end{bmatrix} = -70$$

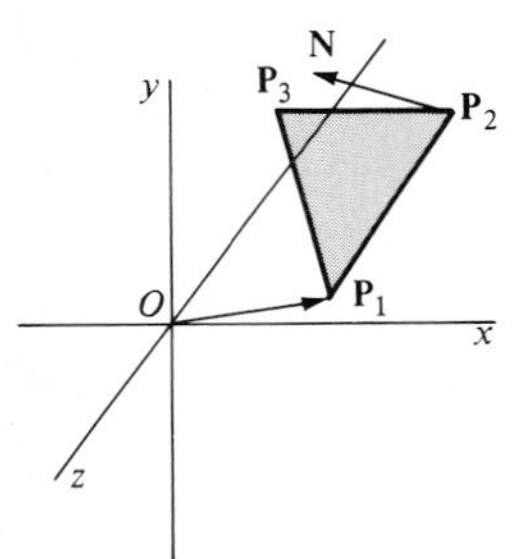

Figure 8-7
Plane defined by three points.

The negative value of the dot product means that the vector $\mathbf{P}_1$ points in the opposite direction to vector **N**. This can be seen in Figure 8-7. The magnitude of the normal vector is

$$|\mathbf{N}| = \{5^2 + 19^2 + 4^2\}^{1/2} = \{25 + 361 + 16\}^{1/2} = 20.05$$

The shortest distance from the plane to the origin is then

$$d = \frac{k}{|\mathbf{N}|} = \frac{-70}{20.05} = -3.49$$

Note: If the origin is on the opposite side of the plane, a positive answer is obtained. Also, note that if $\mathbf{P}_2$ had been used in the plane equation rather than $\mathbf{P}_1$ then

$$k = \mathbf{N} \cdot \mathbf{P}_2 = [-5 \quad -19 \quad 4][4 \quad 2 \quad -3]^T = -70$$

If $\mathbf{P}_3$ had been used, then $k = \mathbf{N} \cdot \mathbf{P}_3 = [-5 \quad -57 \quad -8] = -70$. Identical answers are obtained using *any* point on the plane to calculate k. ◀

Subroutine INTLP in Appendix 1 can be used to compute the intersection point between a line and a plane. This program in turn calls subroutines PLANE and BLINE. A sample program MAIN1 shows how to calculate the piercing point when the plane is defined by the following three points:

$$\mathbf{P}_1 = [\mathrm{R}(1) \quad \mathrm{S}(1) \quad \mathrm{T}(1)] = [1 \quad 4 \quad 2.625]$$

$$\mathbf{P}_2 = [\mathrm{R}(2) \quad \mathrm{S}(2) \quad \mathrm{T}(2)] = [2.375 \quad 1.75 \quad 4]$$

$$\mathbf{P}_3 = [\mathrm{R}(3) \quad \mathrm{S}(3) \quad \mathrm{T}(3)] = [4.5 \quad 2.875 \quad 1.875]$$

and the three-dimensional line in MAIN1 is defined by the following two points:

$$\mathbf{A} = [X(1) \quad Y(1) \quad Z(1)] = [0.625 \quad 5.5 \quad 4.625]$$

$$\mathbf{B} = [X(2) \quad Y(2) \quad Z(2)] = [4.25 \quad 1.875 \quad 1.0]$$

The solution gives the point of intersection as

$$\mathbf{Q}_i = [P(1) \quad P(2) \quad P(3)] = [2.88615 \quad 3.23885 \quad 2.36385]$$

and the length of the line from **A** to $\mathbf{Q}_i$ as LENGTH = 3.91642. The angle between the line and plane is returned as −34.1129 deg. Angle calculations are discussed in the next chapter. Note that if the vector dot product between the normal vector to the plane and the directional vector for the line is zero, then no intersection exists.

Example 8-10

Given: The plane defined by $\mathbf{P}_1 = [1.2 \quad 1.2 \quad 1.2]$, $\mathbf{P}_2 = [-1.2 \quad 1.2 \quad 1.2]$, and $\mathbf{P}_3 = [-1.2 \quad -1.2 \quad 0.8]$.
Task: Calculate by hand the first piercing point given for the line between $\mathbf{A} = [1 \quad 1 \quad -1.8]$ and $\mathbf{B} = [-1 \quad -1 \quad 1.8]$.
Solution:

$$\mathbf{R} = [-2 \quad -2 \quad 3.6]$$

$$\mathbf{N} = (\mathbf{P}_2 - \mathbf{P}_1) \times (\mathbf{P}_3 - \mathbf{P}_1) = \det \begin{vmatrix} \mathbf{i} & \mathbf{j} & \mathbf{k} \\ -2.4 & 0 & 0 \\ -2.4 & -2.4 & -0.4 \end{vmatrix}$$

$$\mathbf{N} = [0 \quad -0.96 \quad 5.76]$$

$$k = \mathbf{N} \cdot \mathbf{P}_1 = [0 \quad -0.96 \quad 5.76] \begin{bmatrix} 1.2 \\ 1.2 \\ 1.2 \end{bmatrix} = 5.76$$

$$\mathbf{N} \cdot \mathbf{A} = [0 \quad -0.96 \quad 5.76] \begin{bmatrix} 1 \\ 1 \\ -1.8 \end{bmatrix} = -11.328$$

$$\mathbf{N} \cdot \mathbf{R} = [0 \quad -0.96 \quad 5.76] \begin{bmatrix} -2 \\ -2 \\ 3.6 \end{bmatrix} = 22.656$$

Equation (8-4) gives

$$\mathbf{Q} = \mathbf{A} + \left\{\frac{(k - \mathbf{N} \cdot \mathbf{A})}{(\mathbf{N} \cdot \mathbf{R})}\right\} \mathbf{R} = [1 \quad 1 \quad -1.8] + (0.754)[-2 \quad -2 \quad 3.6]$$

$$\mathbf{Q} = [-0.508 \quad -0.508 \quad 0.915]$$

This agrees with a computer solution using program PIERCE in Appendix 1. ◀

8-6 COMPUTATIONAL DESCRIPTIVE GEOMETRY

Computational descriptive geometry may sound like a contradiction in terms. Descriptive geometry uses graphical solution techniques, and computational geometry uses numerical techniques. The term is used to define an approach based upon both the projection techniques of descriptive geometry and a computational solution using the resulting planar points and lines. The use of parametric equations allows calculated points on these two-dimensional lines to be extrapolated to the corresponding three-dimensional points in space.

Usually advanced geometric modeling is best done in three-dimensional space using vector techniques. However, if simple subroutines for obtaining solutions to descriptive geometry problems are needed, the computational descriptive geometry approach allows solutions to be obtained with a minimum use of vector operations. The reader should now be able to understand the approach to problems taken in this manner. This section outlines four such solutions to typical descriptive geometry problems.

A. Pierce Point Between a Line and Plane

To find the pierce point between a line and plane graphically by means of the edge view method, an auxiliary view is taken to show the plane on edge. Then the intersection between the line and plane edge can be projected back to the horizontal and frontal views.

When computational methods are used to find the pierce point between a line and plane, the line of sight can be chosen to see the plane on edge by looking parallel to *any* edge of the plane. The resulting orthographic projection gives two two-dimensional lines, one of which is the plane edge.

The intersection between two two-dimensional lines can be found using subroutine INTER2D. The parameter values t and s at the point of intersection are the key items of interest. Since parametric equations are axis-independent, the parameter value obtained for the intersection between the line and the edge view of the plane will be the same for the intersection of the line with *any* view of the plane. If the given line segment lies between **A** = $[x_4\ y_4\ z_4]$ and **B** = $[x_5\ y_5\ z_5]$, and $s = s_1$ is the calculated parameter value for intersection with the edge view of the plane, then the three-dimensional point of intersection between the line and plane is given by the following parametric equations:

$$x = x_4 + s_1(x_5 - x_4)$$
$$y = y_4 + s_1(y_5 - y_4) \tag{8-11}$$
$$z = z_4 + s_1(z_5 - z_4)$$

This procedure is implemented in program CDGPLANE (Appendix 1). A solution to Example 8-10 obtained from this program is given.

```
PROGRAM CDGPLANE

INPUT DATA
  1.2   1.2    1.2
 -1.2   1.2    1.2
 -1.2  -1.2     .8
  1     1     -1.8
 -1    -1      1.8

E1,E2,E3=-2.4  0  0
T1 = .711864
S1 = .754237
USING STD COORDINATES:
PIERCE POINT AT
-.508475 -.508475 .915254
```

The vertices are defined relative to the x,y,z coordinate system shown in Figure 1-46. When using a descriptive geometry solution, the x_d, y_d, z_d coordinates are used, based upon a chosen HEIGHT and DEPTH of the ortho-box as required by Equation (1-2). If a common need exists for this type of calculation, the reader may want to rewrite CDGPLANE as a subroutine which will accept the definition of any line and plane, and then return the piercing point.

B. Shortest Distance Between Skew Lines

The solution to this problem was discussed in Section 7-3. In the plane method, the first auxiliary view is taken to show the edge view of a specially defined plane, and the true length of the shortest distance between the edge view and the other given line. A second auxiliary which shows the plane in true size and the given line in true length is needed to find the actual points of intersection.

In a computational descriptive geometry approach, the eye coordinates can be chosen to directly produce the view which shows the true size of the plane and the true length of the line, once the required plane is carefully defined. In program CDGSKEW the endpoints of two skew lines 1-2 and 3-4 are given in DATA statements. Vertex 5 is calculated such that line 4-5 is parallel to the *F*/*H* folding line in the frontal plane, and line 3-5 is parallel to line 1-2 in both the frontal and horizontal planes. The trace equation for line 3-5 in the *H*-plane is

$$z_5 = M_2(x_5 - x_3) + z_3 \tag{8-12}$$

where

$$M_2 = \frac{(z_2 - z_1)}{(x_2 - x_1)} \tag{8-13}$$

The calculation of vertex 5 is documented in the listing of CDGSKEW. The vertices 3-4-5 form the plane needed for obtaining the solution.

The eye direction is given by the direction of the normal vector $\mathbf{N} = (\mathbf{P}_3 - \mathbf{P}_4) \times (\mathbf{P}_5 - \mathbf{P}_4)$. Once the five vertices are projected orthographically along $\mathbf{E} = [N_1\ N_2\ N_3]$, both two-dimensional lines appear in true length. The subroutine INTER2D returns the parameter values t_1 and s_1, at the point of intersection between the two-dimensional lines. (See line 870.) Then these parameter values are used to calculate the points along the two given three-dimensional lines where intersection occurs with the line of shortest distance. The actual distance is found using these points of intersection. Details are given in the program. Notice that the actual point of intersection between the two two-dimensional lines returned by the subroutine INTER2D is not used. Only the two parameter values at this intersection point are used.

A solution obtained from CDGSKEW is given below. It is suggested that the reader check this solution by using the plane method. Remember to convert the endpoint coordinates of the two given lines to x_d, y_d, z_d coordinates. It is suggested that one let HEIGHT = DEPTH = 3 in Equation (1-2).

```
PROGRAM CDGSKEW

SHORTEST DISTANCE BETWEEN TWO SKEW LINES

ENDPOINTS FOR LINE 1 ARE
 1  1.5  2.125 , 2.6875  2.75  2.5
ENDPOINTS FOR LINE 2 ARE
 1.125  2  1.125 , 0  2.875  2.4375

E1,E2,E3= .91875 -1.8457  2.01797

T1= 8.31844e-2
S1= .263627
```

USING STD COORDINATES:
INTERSECTION ON LINE1 = 1.14037 1.60398 2.15619
INTERSECTION ON LINE2 = .82842 2.23067 1.47101
SHORTEST DISTANCE BETWEEN SKEW LINES = .979559

C. Shortest Horizontal Distance Between Two Skew Lines

The approach to this problem is very similar to the previous one as can be seen by studying the listing of program CDGHORIZ. The only difference is that the line of sight is now taken along a direction given by $\mathbf{N} = \mathbf{j} \times (\mathbf{P}_5 - \mathbf{P}_4)$ where $\mathbf{j} = [0\ 1\ 0]$ is the unit normal vector for a horizontal plane.

A solution obtained from CDGHORIZ is given below for the two skew lines indicated. The reader should produce the corresponding descriptive geometry solution.

PROGRAM CDGHORIZ

ENDPOINTS FOR LINE 1 ARE
1.5625 1 1.4375 ; 1.375 2.375 2.75
ENDPOINTS FOR LINE 2 ARE
.125 1.125 .125 ; 1.25 2.1875 .8125

E1,E2,E3 = .326705 0 1.26989
T1 = .691364
S1 = .777059
USING STD COORDINATES:
INTERSECTION ON LINE1 = 1.43287 1.95063 2.34492
INTERSECTION ON LINE2 = .999191 1.95063 .659228
SHORTEST HORIZONTAL DISTANCE BETWEEN SKEW LINES = 1.74058

D. Shortest Line of a Given Slope Between Two Skew Lines

Consider the skew lines between vertices 1-2 and 3-4 respectively. It is required to find the line of shortest distance between the two lines which makes a given angle (say 10 deg) with the horizontal, slanting upward from line 3-4 to line 1-2. The approach is identical to the previous problem except that the line of sight is rotated about the x-axis by 10 deg so that the required shortest length appears as a point in the orthographic projection. The matrix transformation required for a positive (counterclockwise) rotation about the x-axis is given by Equation (A-12) in Appendix 2. The rotation is performed in lines 760–790 in program CDGSLOPE.

A typical solution obtained from CDGSLOPE is given below. The corresponding descriptive geometry solution for this problem is given on page 119 of Reference 16. (See References.) Notice that the calculated eye coordinates are $\mathbf{E} = [-0.5575 \quad -0.359886 \quad 2.04101]$. This allows the connector, which slants upward, to be seen as a point.

```
PROGRAM CDGSLOPE

SHORTEST LINE OF GIVEN SLOPE (THETA) BETWEEN TWO SKEW LINES
SLOPE=10

INPUT DATA IS
 2.4375   1.25   1.25
 1.4375   2.8125   1.75
 .5   .875   1.375
 1.8125   2.0625   2.3125

E1,E2,E3=-.5575 -.359886  2.04101
T1= .537956
S1= .91731
USING STD COORDINATES:
INTERSECTION ON LINE1= 1.89954   2.09056   1.51898
INTERSECTION ON LINE2= 1.70397   1.96431   2.23498
SHORTEST REQUIRED DISTANCE BETWEEN SKEW LINES= .752891
```

8-7 PIERCING POINT USING PARAMETRIC EQUATION FOR PLANE

The bivariate parametric equation for a plane introduced in Section 4-8 can be extended to represent a curved surface in advanced modeling problems. It is useful to gain experience with such a surface representation. For this reason, consider again the intersection of a line and a plane.

In Figure 8-8 let $\mathbf{F} = \mathbf{P}_2 - \mathbf{P}_1$ and $\mathbf{E} = \mathbf{P}_3 - \mathbf{P}_1$. A general point on the plane is given by

$$\mathbf{P} = \mathbf{P}_1 + s\mathbf{E} + t\mathbf{F}; \quad 0 \leq s \leq 1; \quad 0 \leq t \leq 1 \tag{8-14}$$

A three-dimensional line is defined by a parametric vector equation

$$\mathbf{Q} = \mathbf{A} + u(\mathbf{B} - \mathbf{A}); \quad 0 \leq u \leq 1 \tag{8-15}$$

At the point where the line intersects the plane, $\mathbf{P} = \mathbf{Q}$, or

$$\mathbf{P}_1 + s\mathbf{E} + t\mathbf{F} = \mathbf{A} + u(\mathbf{B} - \mathbf{A}) = \mathbf{A} + u\mathbf{D} \tag{8-16}$$

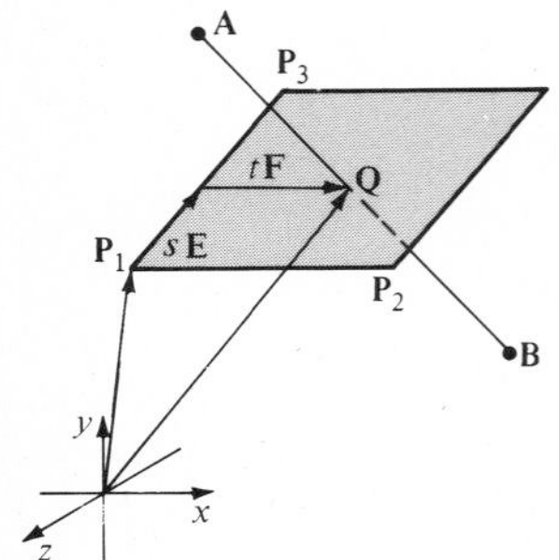

Figure 8-8
Parametric representation of a plane.

The scalar components of Equation (8-16) give three equations in three unknown parameters (s,t,u). Rather than solving three simultaneous equations, a vector technique can be used to isolate each variable in turn. To isolate u, dot Equation (8-16) with the cross product $(\mathbf{E} \times \mathbf{F})$. This gives

$$(\mathbf{E} \times \mathbf{F}) \cdot (\mathbf{P}_1 + s\mathbf{E} + t\mathbf{F}) = (\mathbf{E} \times \mathbf{F}) \cdot (\mathbf{A} + u\mathbf{D})$$

Using the fact that $\mathbf{E} \times \mathbf{F}$ is perpendicular to both $\mathbf{E}$ and $\mathbf{F}$ eliminates two terms and gives

$$(\mathbf{E} \times \mathbf{F}) \cdot \mathbf{P}_1 = (\mathbf{E} \times \mathbf{F}) \cdot \mathbf{A} + u(\mathbf{E} \times \mathbf{F}) \cdot \mathbf{D}$$

Solving for u gives

$$u = \frac{(\mathbf{E} \times \mathbf{F}) \cdot \mathbf{P}_1 - (\mathbf{E} \times \mathbf{F}) \cdot \mathbf{A}}{(\mathbf{E} \times \mathbf{F}) \cdot \mathbf{D}} \tag{8-17}$$

Once u is found, then Equation (8-15) gives the point of intersection. Equation (8-17) requires the evaluation of three scalar triple products.

One can isolate s by dotting Equation (8-16) with $(\mathbf{F} \times \mathbf{D})$, and t by dotting Equation (8-16) with $(\mathbf{E} \times \mathbf{D})$. However, this is not necessary to find the piercing point as explained above. The vector technique to isolate a single variable is a useful "trick" in geometric modeling.

The combination of the three vectors which appear in Equation (8-17) is called the *scalar triple product* because the result is a single scalar quantity resulting from an operation on three vectors. The rules of vector algebra show that $(\mathbf{A} \times \mathbf{B}) \cdot \mathbf{C} = (\mathbf{A} \cdot (\mathbf{B} \times \mathbf{C})$. Also, $(\mathbf{A} \times \mathbf{B}) \cdot \mathbf{C} = 0$ only if $\mathbf{A}$, $\mathbf{B}$, and $\mathbf{C}$ are coplanar. However, when a vector is repeated, such as $\mathbf{A} \cdot (\mathbf{A} \times \mathbf{B})$ or $\mathbf{B} \cdot (\mathbf{A} \times \mathbf{B})$, then the scalar triple product is always equal to zero. The value of a scalar triple product such as $(\mathbf{A} \times \mathbf{B}) \cdot \mathbf{C}$ can be easily evaluated by expanding a determinant. The expression is:

$$(\mathbf{A} \times \mathbf{B}) \cdot \mathbf{C} = \det \begin{vmatrix} A_x & A_y & A_z \\ B_x & B_y & B_z \\ C_x & C_y & C_z \end{vmatrix} \tag{8-18}$$

$$= A_x(B_yC_z - B_zC_y) - A_y(B_xC_z - B_zC_x) + A_z(B_xC_y - B_yC_x)$$

Example 8-11

Given: The problem in Example 8-10

Task: Find the piercing point using Equation (8-17) to calculate the parameter u.

Solution:

$$\mathbf{E} = \mathbf{P}_2 - \mathbf{P}_1 = [-2.4 \quad 0 \quad 0]$$

$$\mathbf{F} = \mathbf{P}_3 - \mathbf{P}_1 = [-2.4 \quad -2.4 \quad -0.4]$$

$$\mathbf{D} = [-2 \quad -2 \quad 3.6]$$

$$(\mathbf{E} \times \mathbf{F}) \cdot \mathbf{P}_1 = \det \begin{vmatrix} -2.4 & 0 & 0 \\ -2.4 & -2.4 & -0.4 \\ 1.2 & 1.2 & 1.2 \end{vmatrix} = -2.4(-2.88 + 0.48) = 5.76$$

$$(\mathbf{E} \times \mathbf{F})\cdot\mathbf{A} = \det \begin{vmatrix} -2.4 & 0 & 0 \\ -2.4 & -2.4 & -0.4 \\ 1 & 1 & -1.8 \end{vmatrix} = -2.4(4.32 + 0.4) = -11.328$$

$$(\mathbf{E} \times \mathbf{F})\cdot\mathbf{D} = \det \begin{vmatrix} -2.4 & 0 & 0 \\ -2.4 & -2.4 & -0.4 \\ -2 & -2 & 3.6 \end{vmatrix} = -2.4(-8.64 - 0.8) = 22.656$$

Equation (8-17) gives $u = (5.76 + 11.328)/22.656 = 0.754$. Equation (8-15) then gives the point of intersection as $\mathbf{Q} = [1 \ \ 1 \ \ -1.8] + 0.754[-2 \ \ -2 \ \ 3.6] = [-0.508 \ \ -0.508 \ \ 0.915]$. This agrees with the solution found in Example 8-10. ◀

8-8 MODELING THE PROJECTION OF A LINE[†]

Consider the orthographic projection of a line onto an arbitrary plane, with normal **N** defined by P_1, P_2, and P_3. Let the line through **A** normal to the plane have a piercing point **A***. Likewise, **B*** is the piercing point for the normal line through **B**.

The distances d_1 and d_2 shown in Figure 8-9 are the minimum distances from the respective points to the plane. If $\mathbf{R}_1 = \mathbf{A} - \mathbf{P}_1$ then $d_1 = \mathbf{R}_1\cdot\mathbf{N}/|\mathbf{N}|$ by Equation (7-8). Similarly, if $\mathbf{R}_2 = \mathbf{B} - \mathbf{P}_1$ then $d_2 = \mathbf{R}_2\cdot\mathbf{N}/|\mathbf{N}|$.

One method of finding **A*** and **B*** is to use the method of Section 7-9B. Equation (7-8) can be written

$$\mathbf{A}^* = \mathbf{A} - d_1\mathbf{n}$$

and

$$\mathbf{B}^* = \mathbf{B} - d_2\mathbf{n} \tag{8-19}$$

where **n** is the unit vector $\mathbf{N}/|\mathbf{N}|$.

If $\mathbf{A} = [x_1 \ y_1 \ z_1]$ and $\mathbf{B} = [x_2 \ y_2 \ z_2]$, then the projected endpoints on the plane are $\mathbf{A}^* = [x_3 \ y_3 \ z_3]$ and $\mathbf{B}^* = [x_4 \ y_4 \ z_4]$ where

$$\begin{aligned} x_3 &= x_1 - d_1\left(\frac{N_1}{|\mathbf{N}|}\right) \\ y_3 &= y_1 - d_1\left(\frac{N_2}{|\mathbf{N}|}\right) \\ z_3 &= z_1 - d_1\left(\frac{N_3}{|\mathbf{N}|}\right) \end{aligned} \tag{8-20}$$

†The last two sections in Chapter 8 may be omitted without loss in continuity of the textual presentation.

and

$$x_4 = x_2 - d_2(\frac{N_1}{|\mathbf{N}|})$$

$$y_4 = y_2 - d_2(\frac{N_2}{|\mathbf{N}|}) \qquad (8\text{-}21)$$

$$y_4 = z_2 - d_2(\frac{N_3}{|\mathbf{N}|})$$

The projected line, of course, is the line in the plane between **A*** and **B***.

Example 8-12

Given: An infinite plane is given by the implicit equation

$$3x + 4y + 2z - 6 = 0$$

and a three-dimensional line segment in space lies between **A** = [4 6 2] and **B** = [6 3 4].
Task: Find the orthographic projection of the three-dimensional line onto the given plane by finding the projected points **A*** and **B***, and then writing the parametric vector equation for the line between **A*** and **B***. Calculate the length of the projected line segment.
Solution:

$$N_1 = 3; \quad N_2 = 4; \quad N_3 = 2; \quad |\mathbf{N}| = (9 + 16 + 4)^{1/2} = 5.39$$

Let $\mathbf{R}_1 = \mathbf{A}_1 - \mathbf{P}_1$ and let $\mathbf{P}_1$ = point on x-axis ($y = z = 0$). Then

$$\mathbf{P}_1 = [2\ 0\ 0]; \quad \mathbf{R}_1 = [2\ 6\ 2]$$

Also, let $\mathbf{R}_2 = \mathbf{B}_1 - \mathbf{P}_1 = [4\ 3\ 4]$.

$$d_1 = \mathbf{R}_1 \cdot \mathbf{N}/|\mathbf{N}| = (2(3) + 6(4) + 2(2))/5.39 = 6.31$$

$$d_2 = \mathbf{R}_2 \cdot \mathbf{N}/|\mathbf{N}| = (4(3) + 3(4) + 4(2))/5.39 = 5.94$$

$$n_1 = \frac{3}{5.39} = 0.557; \quad n_2 = \frac{4}{5.39} = 0.742; \quad n_3 = \frac{2}{5.39} = 0.371$$

$$\mathbf{A}^* = \mathbf{A} - d_1\mathbf{n} = [4\ 6\ 2] - 6.31\ [0.557\quad 0.742\quad 0.371]$$

$$\mathbf{A}^* = [0.485\quad 1.318\quad -0.341]$$

$$\mathbf{B}^* = \mathbf{B} - d_2\mathbf{n} = [6\ 3\ 4] - 5.94\ [0.557\quad 0.742\quad 0.371]$$

$$\mathbf{B}^* = [2.691\quad -1.407\quad 1.796]$$

The equation for the projected line is

$$\mathbf{Q} = \mathbf{A}^* + t(\mathbf{B}^* - \mathbf{A}^*) \quad 0 \le t \le 1$$

The length of the projected line segment is

$$d = \{(2.691 - 0.485)^2 + (1.407 + 1.318)^2 + (1.796 + 0.341)^2\}^{1/2} = 4.106$$

Note: The magnitude of the given line is

$$d_i = (2^2 + 3^2 + 2^2)^{1/2} = 4.123$$

Thus, the given line is almost parallel to the plane since the projected length is almost the same as the original length. ◀

8-9 SHORTEST DISTANCE BETWEEN TWO SKEW LINES

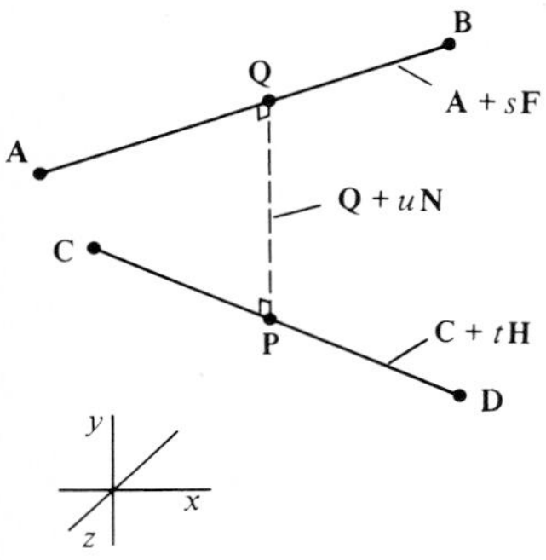

Figure 8-9
Shortest distance between two skew lines.

Two approaches to the problem of calculating the shortest distance between two nonintersecting three-dimensional lines were discussed in Sections 7-4 and 8-6-B. Another approach is to use the power of vector algebra to calculate the solution. Consider two skew lines $\mathbf{B} - \mathbf{A}$ and $\mathbf{D} - \mathbf{C}$ shown in Figure 8-9. The line of shortest distance connects two unknown points **P** and **Q** such that it is perpendicular to both lines.

Let the directional vectors for the two given skew lines be **F** and **H**, respectively. The directional vector **N** for line $\mathbf{Q} - \mathbf{P}$ must be normal to the plane that contains the vector $\mathbf{H} = \mathbf{D} - \mathbf{C}$ and is parallel to the vector $\mathbf{F} = \mathbf{B} - \mathbf{A}$. For this plane whose normal is **N**, the shortest distance between the plane and any point on vector **F** is the same as the shortest distance between the two skew lines. This distance is given by Equation (7-6). Using the notation in Figure 8-9 the shortest distance is (using **A** as any point on vector **F**)

$$d = |(\mathbf{A} - \mathbf{C})\cdot\mathbf{N}|/|\mathbf{N}| \tag{8-22}$$

where $\mathbf{N} = \mathbf{F} \times \mathbf{H}$, and **C** is a vertex on the plane containing vector **H**.

The numerator in Equation (8-22) is the scalar triple product $\mathbf{E}\cdot\mathbf{F} \times \mathbf{H} = \mathbf{E} \times \mathbf{F}\cdot\mathbf{H}$, where $\mathbf{E} = \mathbf{A} - \mathbf{C}$. Using Equation (8-18) this is evaluated by

$$\mathbf{E} \times \mathbf{F}\cdot\mathbf{H} = \det \begin{vmatrix} (A_x - C_x) & (A_y - C_y) & (A_z - C_z) \\ (B_x - A_x) & (B_y - A_y) & (B_z - A_z) \\ (D_x - C_x) & (D_y - C_y) & (D_z - C_z) \end{vmatrix}$$

and the denominator $|\mathbf{F} \times \mathbf{H}|$ is evaluated by the absolute value of

$$\mathbf{F} \times \mathbf{H} = \det \begin{vmatrix} \mathbf{i} & \mathbf{j} & \mathbf{k} \\ F_x & F_y & F_z \\ H_x & H_y & H_z \end{vmatrix}$$

If $\mathbf{n}_1$ and $\mathbf{n}_2$ are the unit vectors along the directional vectors **F** and **H**, respectively, then Equation (8-22) can be written

$$\mathrm{d} = |(\mathbf{A} - \mathbf{C})\cdot(\mathbf{n}_1 \times \mathbf{n}_2)|/|\mathbf{n}_1 \times \mathbf{n}_2| \tag{8-23}$$

When the given skew lines are parallel, then the equation for the shortest distance from any point on one of the lines to the other can be used to find the shortest distance between the two lines. This is given by Equation (7-3).

Equation (8-22) or (8-23) only gives the magnitude of the shortest distance between the two skew lines. If the actual coordinates of intersection represented by points **P** and **Q** are required, then further computations are necessary. This is best illustrated by the following examples.

Example 8-13

Given: Two skew lines in space between **A**, **B**, and **C**, **D**, respectively.

$$\mathbf{A} = [100 \quad 60 \quad 60]; \quad \mathbf{B} = [40 \quad 80 \quad 30]$$

$$\mathbf{C} = [80 \quad 50 \quad 40]; \quad \mathbf{D} = [-10 \quad 20 \quad 50]$$

Task: Find the shortest distance between the given lines.
Solution:

$$\text{Let } \mathbf{F} = \mathbf{B} - \mathbf{A} = [-60 \quad 20 \quad -30]$$

$$\mathbf{H} = \mathbf{D} - \mathbf{C} = [-90 \quad -30 \quad 10]$$

$$\mathbf{F} \times \mathbf{H} = \det \begin{vmatrix} \mathbf{i} & \mathbf{j} & \mathbf{k} \\ -60 & 20 & -30 \\ -90 & -30 & 10 \end{vmatrix} = -700\mathbf{i} + 3300\mathbf{j} + 3600\mathbf{k}$$

$$|\mathbf{F} \times \mathbf{H}| = \{7^2 + 33^2 + 36^2\}^{1/2}\,(\mathrm{x}10^2) = 4934$$

The scalar triple product required for Equation (8-22) is given by

$$(\mathbf{A} - \mathbf{C}) \cdot (\mathbf{F} \times \mathbf{H}) = \det \begin{vmatrix} 20 & 10 & 20 \\ -60 & 20 & -30 \\ -90 & -30 & 10 \end{vmatrix} = 91{,}000$$

The shortest distance as given by Equation (8-22) is then

$$d = \frac{91{,}000}{4{,}934} = 18.4$$

Example 8-14

Given: The centerlines of two pipes are defined by two endpoints on each centerline. For pipe AB,

$$\mathbf{A} = [0.375 \quad 3.375 \quad 0] \text{ and } \mathbf{B} = [3.625 \quad 0 \quad 3.25]$$

For pipe CD,

$$\mathbf{C} = [0 \quad 1.375 \quad 3.375] \text{ and } \mathbf{D} = [3.625 \quad 3.5 \quad 1.625]$$

Task: (a) Calculate the point of intersection on each centerline with the line of shortest distance between the pipes. (b) Calculate the magnitude of the shortest distance between pipe centerlines.

Solution: Direction vector $\mathbf{F} = \mathbf{B} - \mathbf{A} = [3.25 \quad -3.375 \quad 3.25]$
Direction vector $\mathbf{H} = \mathbf{D} - \mathbf{C} = [3.625 \quad 2.125 \quad -1.75]$

$$\mathbf{N} = \mathbf{F} \times \mathbf{H} = \det \begin{vmatrix} \mathbf{i} & \mathbf{j} & \mathbf{k} \\ 3.25 & -3.375 & 3.25 \\ 3.625 & 2.125 & -1.75 \end{vmatrix} = [-1.0 \quad 17.47 \quad 19.14]$$

$$|\mathbf{N}| = \{(-1)^2 + (17.47)^2 + (19.14)^2\}^{1/2} = 25.93$$

The unit vector along the line of shortest distance is then

$$\mathbf{n} = [-0.0386 \quad 0.6737 \quad 0.738]$$

Point $\mathbf{Q}$ along line AB is given by $\mathbf{Q} = \mathbf{A} + s\mathbf{F}$. The three components of this equation give

$$Q_x = 0.375 + s(3.25) \tag{a}$$

$$Q_y = 3.375 - s(3.375) \tag{b}$$

$$Q_z = s(3.25) \tag{c}$$

The intersection point $\mathbf{P}$ can be defined as the intersection between two lines whose direction vectors are $\mathbf{N}$ and $\mathbf{H}$. We write $\mathbf{Q} + u\mathbf{n} = \mathbf{C} + t\mathbf{H}$. The three components of this equation can be rearranged to give the following three equations:

$$Q_x = \mathrm{u}(0.0386) + t(3.625) \tag{d}$$

$$Q_y = -\mathrm{u}(0.6737) + 1.375 + t(2.125) \tag{e}$$

$$Q_z = -\mathrm{u}(0.738) + 3.375 - t(1.75) \tag{f}$$

The above six equations can be combined as follows:

(a) and (d) $\quad 3.25s - 0.0386u - 3.625t = -0.375$

(b) and (e) $\quad 3.375s - 0.6737u + 2.125t = 2$

(c) and (f) $\quad 3.25s + 0.738u + 1.75t = 3.375$

The set of three linear algebraic equations in three unknowns can be solved by using Cramer's Rule. In matrix form we write

$$\begin{bmatrix} 3.25 & -0.0386 & -3.625 \\ 3.375 & -0.6737 & 2.125 \\ 3.25 & 0.738 & 1.75 \end{bmatrix} \begin{bmatrix} s \\ u \\ t \end{bmatrix} = \begin{bmatrix} -0.375 \\ 2.0 \\ 3.375 \end{bmatrix}$$

The determinant of the square matrix is

$$\det \begin{vmatrix} 3.25 & -.0386 & -3.625 \\ 3.375 & -.6737 & 2.125 \\ 3.25 & 0.738 & 1.75 \end{vmatrix} = -25.93 = D$$

To solve for s using Cramer's Rule

$$Ds = \det \begin{vmatrix} -0.375 & -.0386 & -3.625 \\ 2.0 & -0.6737 & 2.125 \\ 3.375 & 0.738 & 1.75 \end{vmatrix} = -12.7$$

Thus, $s = -12.7/(-25.93) = 0.4899$.
To solve for t using Cramer's Rule

$$Dt = \det \begin{vmatrix} 3.25 & -0.0386 & -0.375 \\ 3.375 & -0.6737 & 2.0 \\ 3.25 & 0.7380 & 3.375 \end{vmatrix} = -13.753$$

Thus, $t = -13.753/(-25.93) = 0.5303$.
To solve for u using Cramer's Rule

$$Du = \det \begin{vmatrix} 3.25 & -.375 & -3.625 \\ 3.375 & 2.0 & 2.125 \\ 3.25 & 3.375 & 1.75 \end{vmatrix} = -30.089$$

Thus, $u = -30.089/(-25.93) = 1.1583$.

To calculate the point of intersection **Q** on pipe *AB* use the vector equation $\mathbf{Q} = \mathbf{A} + s\mathbf{F}$. The components give

$$Q_x = 0.375 + .4899(3.25) = 1.967$$

$$Q_y = 3.375 + .4899(-3.375) = 1.722$$

$$Q_z = .4899(3.25) = 1.592$$

Thus, $\mathbf{Q} = [1.967 \quad 1.722 \quad 1.592]$

To calculate the point of intersection **P** on pipe *CD* use the vector equation $\mathbf{P} = \mathbf{C} + t\mathbf{H}$. The components give

$$P_x = .5303(3.625) = 1.922$$

$$P_y = 1.375 + .5303(2.125) = 2.502$$

$$P_z = 3.375 + .5303(-1.75) = 2.447$$

Thus, $\mathbf{P} = [1.922 \quad 2.502 \quad 2.447]$.

The shortest distance between pipe centerlines can be found by calculating the magnitude of $u\mathbf{n}$. This vector is

$$1.1583[-0.0386 \quad 0.6737 \quad 0.738] = [-0.04471 \quad 0.78035 \quad 0.85483]$$

The magnitude is then

$$d = \{(-.04471)^2 + (.78035)^2 + (.85483)^2\}^{1/2} = 1.1583 \text{ (same as } u)$$

Equation (8-22) can be used to check this value. Using the notation in Figure 8-9 the following values are obtained:

$$\mathbf{F} \times \mathbf{H} = \det \begin{vmatrix} \mathbf{i} & \mathbf{j} & \mathbf{k} \\ 3.25 & -3.375 & 3.25 \\ 3.625 & 2.125 & -1.75 \end{vmatrix} = [-1.0 \quad 17.4688 \quad 19.14]$$

$$|\mathbf{F} \times \mathbf{H}| = \{1 + 305.2 + 366.3\}^{1/2} = 25.93$$

The scalar triple product is

$$\mathbf{E} \times \mathbf{F} \cdot \mathbf{H} = \det \begin{vmatrix} 0.375 & 2.0 & -3.375 \\ 3.25 & -3.375 & 3.25 \\ 3.625 & 2.125 & -1.75 \end{vmatrix} = -30.03$$

Equation (8-22) then gives $d = 30.03/25.93 = 1.1583$ ◀

PROBLEMS: CHAPTER 8

8-1 *Given:* Top and front views of house showing proposed location of antenna installation.
Task: Using only the given views determine

a. The location of the support for the antenna and for the moorings used by cables *A*, *B*, and *C*.
b. Using a scale 1″ = 30′−0″ find the length of cable required for the job.

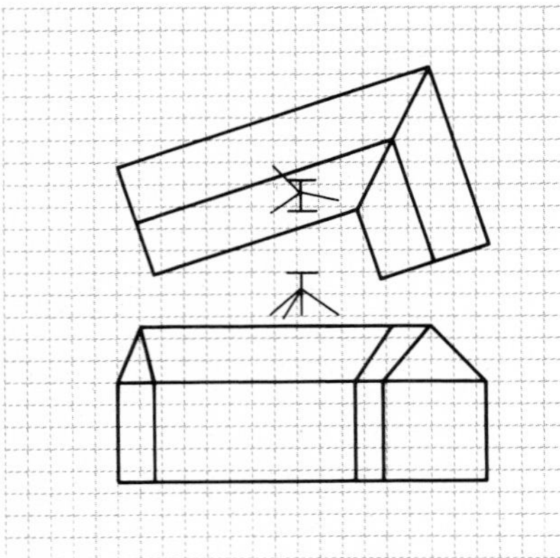

Figure 8-1P

8-2 *Given:* Top and front view of *ABCDEF* (a portion of a bulkhead) strengthened by reinforcing rod *MN*.
Task: Determine the points at which rod *MN* pierces plane *ABEF* and plane *BCDE*. Show correct visibility in both views.

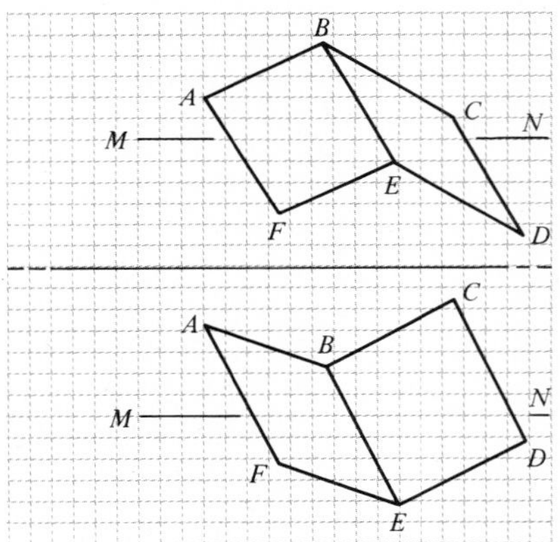

Figure 8-2P

8-3 *Given:* Top and front views of hopper *A*, building *B*, and a segment of conveyor belt *EF*.
Task: Complete the views showing the points at which the belt enters the hopper and the building.

8-4 *Given:* Top and front views of line *MN* and plane *ABCD*.
Task: Show the top and front views of the point at which line *MN* pierces plane *ABCD*. Show complete visibility.

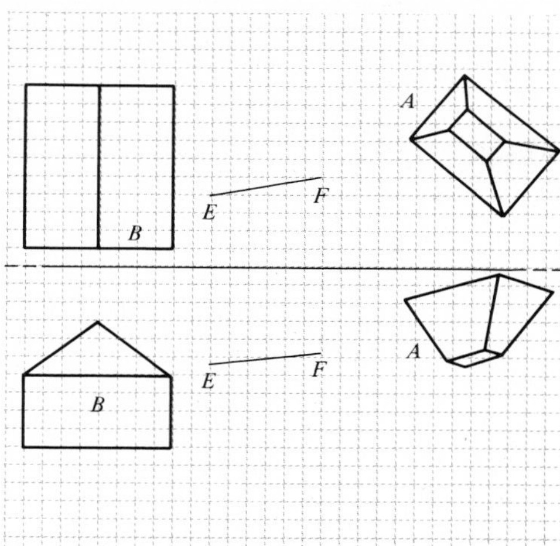

Figure 8-3P

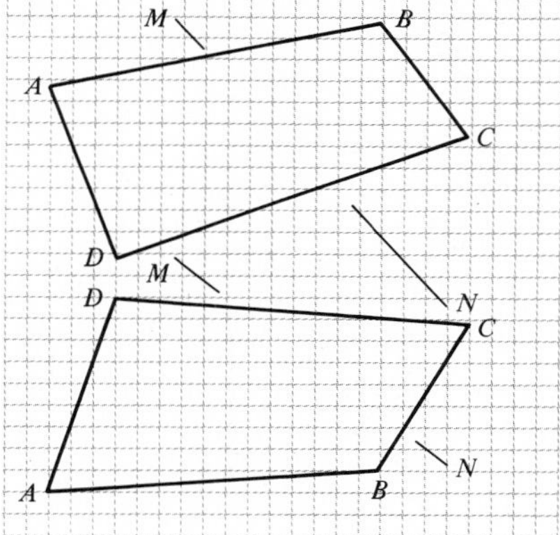

Figure 8-4P

8-5 *Given:* Top and front views of triangle *ABC* and line *MN.*
Task: Find point P_H and P_F which shows the location at which line *MN* pierces triangle *ABC* in the top and front views. Show correct visibility of the line.

8-6 *Given:* Front and profile views of *ABCD* and line *MN.*
Task: Find the point where line *MN* pierces figure *ABCD* in the front and profile views. Show correct visibility of line *MN.*

8-7 *Given:* Top and front views of incomplete figures *ABCD* and *XYZ.*

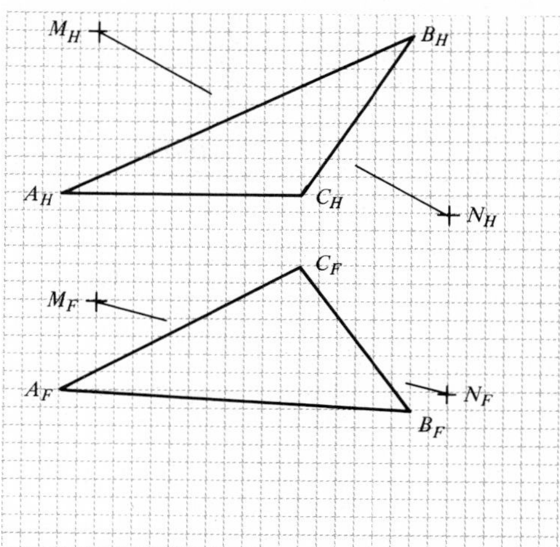

Figure 8-5P

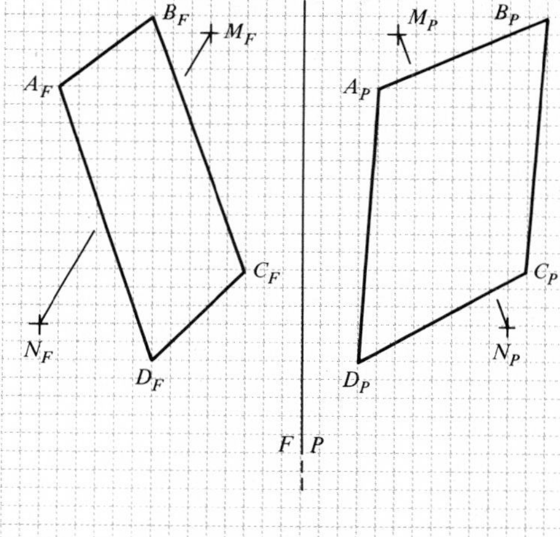

Figure 8-6P

Task: Find the intersection of the two figures. Show complete visibility.

8-8 *Given:* Front and top views at a triangular ore vein defined by points *ABC.*
Task: A shaft is to be drilled from point *M* in the direction of *MN.* Show (in both views) the location at which the shaft strikes the ore vein.

8-9 *Given:* Top and front views of a tripod resting on the side of a hill.

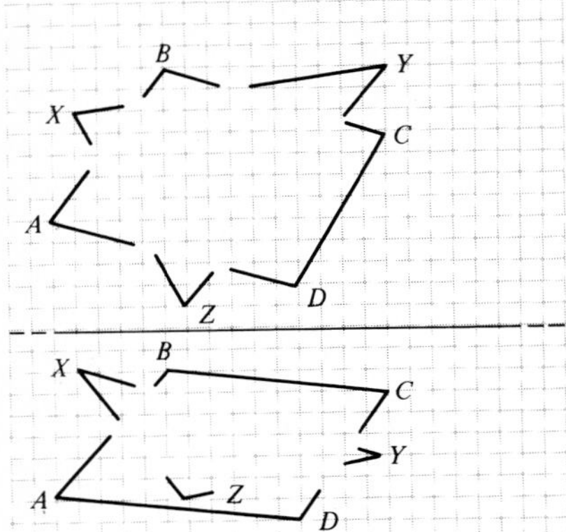

Figure 8-7P

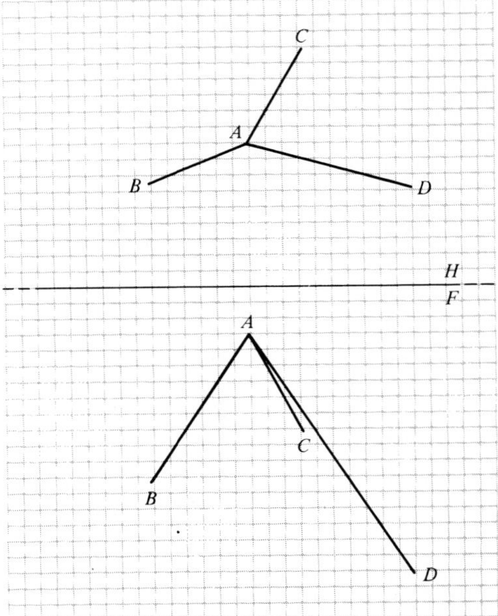

Figure 8-9P

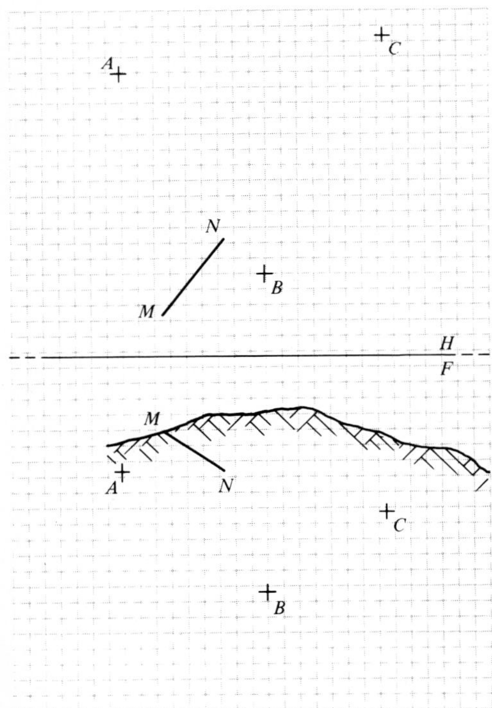

Figure 8-8P

Task: Show the location of the intersection of a plumb line with the ground if it were suspended from point *A*.

8-10 *Given:* Front and top views of the path of a plane as shown by line *AB*. Top view of the contour map of a hill toward which the plane is flying.
Task: If the plane continues on its given course at what altitude will it crash into the side of the hill?

8-11 *Given:* A plane defined by $\mathbf{P}_1 = [0\ 1\ 1]$, $\mathbf{P}_2 = [1\ 2\ 3]$, and $\mathbf{P}_3 = [-2\ 3\ -1]$ and a line which passes through the points $\mathbf{A} = [1\ 0\ 1]$ and $\mathbf{B} = [0\ 0\ -1]$.
Task: a. Calculate the direction cosines of the unit vector normal to the plane.
b. Calculate the point of intersection between the line and the plane.

8-12 *Given:* A line between [1 3 6] and [−1 1 2] intersects with a plane defined by $\mathbf{V}\cdot\mathbf{X} = 5$, where $\mathbf{V} = [1\ 1\ 1]$ and $\mathbf{X} = [x\ y\ z]$.
Task: Calculate the point of intersection between the line and plane.

8-13 *Given:* Three points on a plane are given by $[1\ 4\ 2\tfrac{5}{8}]$, $[2\tfrac{3}{8}\ 1\tfrac{3}{4}\ 4]$, and $[4\tfrac{1}{2}\ 2\tfrac{7}{8}\ 1\tfrac{7}{8}]$. A line is defined by the two points $\mathbf{A} = [\tfrac{5}{8}\ 5\tfrac{1}{2}\ 4\tfrac{5}{8}]$ and $\mathbf{B} = [4\tfrac{1}{4}\ 1\tfrac{7}{8}\ 1]$.

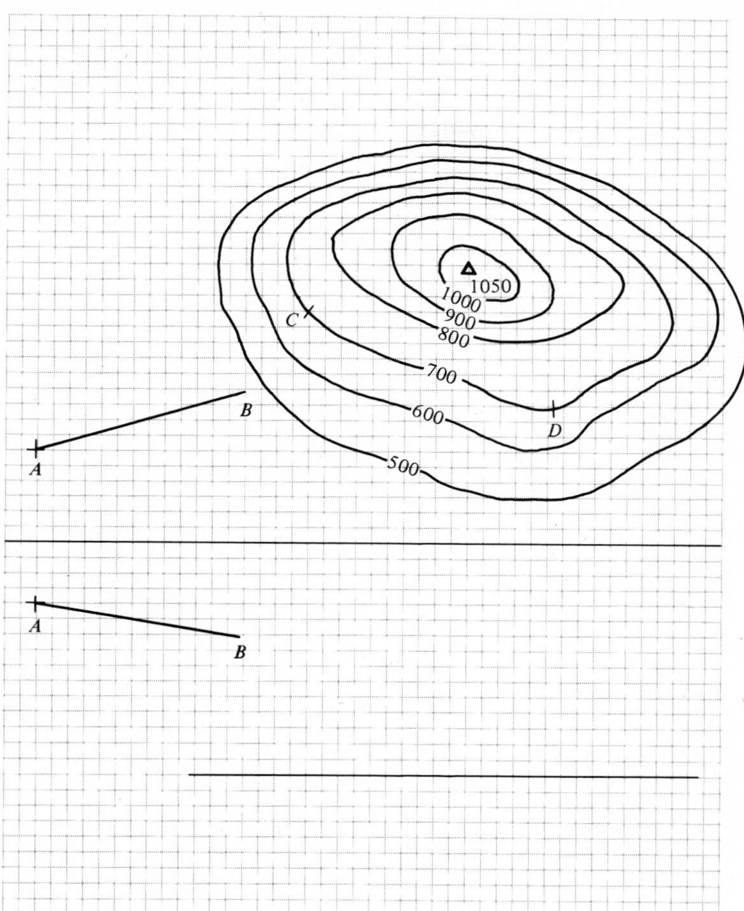

Figure 8-10P

Task: a. Calculate the point of intersection between the line and the plane.
b. Calculate the distance between **A** and the point of intersection.

8-14 *Given:* A control cable passes through **A** = [0 1 2] and **B** = [4½ 0 1], and also through three plane facets bounded by structural members in an airframe. The three planes are defined by three corners each.
Task: Calculate the points in space where the control cable intersects each plane.

8-15 *Given:* A plane is perpendicular to a line between **A** = [−1 2 3] and **B** = [5 3 1] and contains the point **S** = [2 4 3].

Planes for Problem 8-14

Plane 1	*Plane 2*	*Plane 3*
[0 ½ 2½]	[1 0 ½]	[1 0 ½]
[1 0 ½]	[2 1½ 2]	[3½ 2½ 0]
[2 1½ 2]	[4 0 3]	[4 0 3]

Task: a. Find the implicit equation for the plane.
b. Calculate the point of intersection between the line and the plane.
c. Calculate the point of intersection between the plane and a line through **C** = [−1 −1 −1] and **D** = [1 1 1].

8-16 *Given:* Two points $\mathbf{P} = [-3\ -4\ -5]$ and $\mathbf{Q} = [5\ 4\ 3]$ lie in a plane which is parallel to the z-axis.
Task: a. Find the implicit equation for the plane.
b. Calculate the distance from the plane to the origin.

8.17 *Given:* The plane and line defined in Problem 8-11.
Task: Calculate the piercing point using Equation (8-17) to first find the parameter value for u.

8-18 *Given:* The plane defined in Example 8-12 and a three-dimensional line segment between $\mathbf{A} = [-4\ -6\ -2]$ and $\mathbf{B} = [6\ 3\ 4]$.
Task: Find the orthographic projection length of the given line onto the plane.

8-19 *Given:* A plane defined by $\mathbf{A} = [1\ 0\ 0\,]$, $\mathbf{B} = [0\ 1\ 0]$ and $\mathbf{C} = [0\ 0\ 1]$; and a line segment between the two points $\mathbf{M} = [5\ 1\ 1]$ and $\mathbf{N} = [4\ 3\ -1]$.
Task: Find the orthographic projection of the given line onto the given plane.

8-20[c] *Given:* Program CDGPLANE in Appendix 1.
Task: Write a similar program and calculate the piercing point for the line and plane given in Problem 8-11.

8-21[c] *Given:* Program CDGSKEW in Appendix 1.
Task: Write a similar program and solve Problem 7-1.

8-22[c] *Given:* Program CDGHORIZ in Appendix 1.
Task: Write a similar program and solve Problem 7-2.

8-23[c] *Given:* Program CDGSLOPE in Appendix 1.
Task: Write a similar program and calculate the shortest distance between the two skew lines defined in Problem 7-15 for each of the following line slopes:
a. 10 degrees
b. 20 degrees
c. 30 degrees

8-24 *Given:* Lines *MN* and *RS* defined in Problem 7-1.
Task: Calculate the length of the common perpendicular between the two lines without the aid of a computer.

8-25 *Given:* Tubing centerlines *MN* and *XY* defined in Problem 7-12.
Task: *a.* Calculate the point of intersection of each centerline wih the line of shortest distance between the tubes.
b. Calculate the magnitude of the shortest distance between tube centerlines.

8-26 *Given:* A line passes through $\mathbf{A} = [1\ -2\ -1]$ and $\mathbf{B} = [2\ 3\ 1]$.
Task: Calculate the intersection of the line with the x-y plane.

8-27[c] *Given:* The endpoints of two skew lines and a fifth point in space.
Task: Use the concepts of computational descriptive geometry and write a computer program which will accept the five given points and calculate the shortest connector between the two lines which also passes through the fifth point. Note. A descriptive geometry solution for this type of problem can be found on page 106 of Reference 16.

8-28[c] A storage bin consists of three surfaces in the shape of an inverted pyramid. The bin passes through a plane roof defined by the following three corner points.
$\mathbf{A} = [1\ 1/8\quad 1\ 3/8\quad 3\ 3/4]$; $\mathbf{B} = [4\ 1/8\quad 1\ 3/8\quad 2]$; $\mathbf{C} = [3\ 0\ 0]$. Each edge of the bin consists of a straight line passing through two known points.
Edge 1: $\mathbf{P}_1 = [1\quad 2.375\quad 0.75]$; $\mathbf{P}_2 = [2.625\quad -0.5\quad 3.125]$
Edge 2: $\mathbf{P}_3 = [1.625\quad 2.375\quad 2.75]$; $\mathbf{P}_4 = [2.625\quad -0.5\quad 3.125]$
Edge 3: $\mathbf{P}_5 = [3\ 2.375\ 1.25]$; $\mathbf{P}_6 = [2.625\ -0.5\ 3.125]$
The apex of the inverted pyramid is at $\mathbf{P}_2 = \mathbf{P}_4 = \mathbf{P}_6$.
a. Calculate the intersection between the roof and each edge of the bin
b. Calculate the length from the roof to the base point of each edge.
c. Calculate the area of the roof plane which must be removed to insert the bin.
d. Generate three principal views using computer techniques.

8-29[c] *Given:* The horizontal and frontal views of four roof planes are available from two views such as shown in Figure 8-1P, along with four lines which give antenna and guy-wire locations.
R(1) = 0, S(1) = 1.75, T(1) = 1.75, R(2) = 0.375, S(2) = 3, T(2) = 0.875, R(3) = 4.5, S(3) = 3, T(3) = 2.375, R(4) = 5, S(4) = 1.75, T(4) = 3.5 define roof plane I.
R(1) = 3.875, S(1) = 1.75, T(1) = 1.25, R(2) = 4.5, S(2) =

[c]Superscript *c* indicates that a computer solution is required.

3, T(2) = 2.375, R(3) = 5.75, S(3) = 3,T(3) = 0.25, R(4) = 4.375, S(4) = 1.75, T(4) = 0 define roof plane II.
R(1) = 0.625, S(1) = 1.75, T(1) = 0, R(2) = 0.375, S(2) = 3, T(2) = 0.875, R(3) = 4.5, S(3) = 3,T(3) = 2.375, R(4) = 3.875, S(4) = 1.75, T(4) = 1.25 define roof plane III.
R(1) = 0.625, S(1) = 1.75, T(1) = 0, R(2) = 0.375, S(2) = 3, T(2) = 0.875, R(3) = 4.5, S(3) = 3,T(3) = 2.375, R(4) = 3.875, S(4) = 1.75, T(4) = 1.25 define roof plane IV.
Guy-wire PA: X(1) = 3.125, Y(1) = 3.875, Z(1) = 1.625, X(2) = 2.625, Y(2) = 3.125, Z(2) = 1.75
Guy-wire PB: X(1) = 3.125, Y(1) = 3.875, Z(1) = 1.625, X(2) = 4, Y(2) = 3.25, Z(2) = 1.25
Guy-wire PC: X(1) = 3.125, Y(1) = 3.875, Z(1) = 1.625, X(2) = 2.375, Y(2) = 3.125, Z(2) = 1.25
Antenna PP: X(1) = 3.125, Y(1) = 3.875, Z(1) = 1.625, X(2) = 3.125, Y (2) = 3, Z(2) = 1.625
Task: Use vector methods to calculate the piercing points for the antenna and guy-wires with the roof planes, and calculate the total length of guy-wire required. Generate a wire frame drawing for the three principal views of the assembly. Let WIDTH = 3.5 for the ortho-box.

8-30[c] *Given:* A plane surface on the side of a hill is defined by vectors **B**, **C**, and **D**; where **B** = [0 1.1875 0.0625], **C** = [2 1.75 1.875] and **D** = [3.3125 0 2.0].
Task: Write a computer program which will calculate the length of a plumb line from any point **A** to the plane, as well as the intersection between the line and plane. First use **A** = [1.25 3.125 0.625]. Let WIDTH =

8-31[c] *Given:* By use of seismology it has been determined that a vein of ore lies in the plane *ABC*. An existing shaft lies between points **M** and **N**. Vector components are **A** = [0.25 1.5 0.5] , **B** = [2 0 3.0625], **C** = [3.5 1 0] , **M** = [0.75 2.0625 3.625], and **N** = [1.5 1.5 2.625]
Task: *a.* Calculate the length of the straight line extension of the shaft *MN* needed to reach the plane of the vein. (The vein in Spain lies mainly in the plane!) *b.* Investigate other possible new shaft starting points at various inclinations to see if a shorter path is possible from the surface to the vein. A vertical shaft is not practical.

9

Angles

If the angle between two lines is to be measured graphically, then a view must be taken normal to the plane that contains the two lines. Several descriptive geometry methods are given to solve this problem. The angle between a line and plane, and the angle between two planes are considered. Vector techniques can be used to calculate these angles directly, without the need for any graphical presentation. The problem of two intersecting three-dimensional lines is presented in this chapter from a geometric modeling point of view.

9-1 ANGLE BETWEEN TWO INTERSECTING LINES

Two intersecting lines form a plane. (See Section 4-2-A.) *Any normal view of the plane will show the lines in true length and the true size of the angle between them.* Since the principles used in finding the information required in the following example have been discussed in Section 7-2-B, the explanation will be brief and will enable the student to review the problem solving techniques involved.

Example 9-1

Given: Top and front views of angle *MNO*.
Task: Find the number of degrees in the true angle *MNO*.
Solution:

1. Consider *MNO* as a triangle. Find the edge view of the triangle in the first auxiliary.

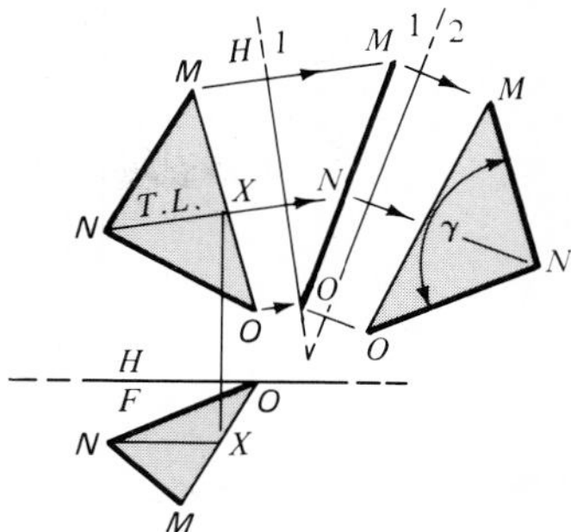

Example 9-1
Angle between two intersecting line segments.

2. Show the normal view of triangle *MNO* in the second auxiliary.
3. Measure the true size of angle *MNO*.

9-2 ANGLE BETWEEN TWO NONINTERSECTING SKEW LINES

Two lines that do not intersect have an angular relationship between them that can be determined when the angle is measured in a view that shows both lines in true length. A method different from that shown in Example 9-1 must be used since the lines will not lie in the same plane. If both lines are shown as oblique lines it will require three auxiliaries to solve the problem as shown in the following example.

Example 9-2

Given: Top and front views of line *AB* and line *MN.*
Task: Find the angle between line *AB* and line *MN.*
Solution:

1. Draw the *H*/1 folding line parallel to the top view of line *AB.* Project both lines into the first auxiliary where line *AB* is shown in true length.
2. Construct folding line 1/2 perpendicular to the true length line *AB.* Project both lines into the second auxiliary where line *AB* appears as a point.
3. Draw folding line 2/3 parallel to the second auxiliary view of line *MN.* Project both lines into the third auxiliary where line *AB* and line *MN* are shown in true length.
4. Measure the angle θ between the true length lines *AB* and *MN.*

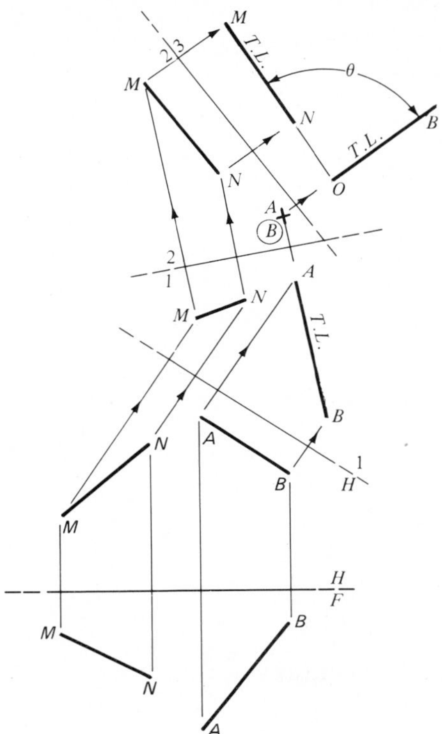

Example 9-2
Angle between two intersecting line segments.

9-3 ANGLE BETWEEN AN OBLIQUE LINE AND PRINCIPAL PLANE

The angle between a line and a plane is shown in the view where the line appears in true length and the plane appears on edge.

An *inclined* line is defined in Section 1-8-B as one which is parallel to a principal plane and forms an angle with the other two. The angle the inclined line makes with the principal plane can be measured in one of the principal views.

The principles involving the angle between an *oblique* line and a principal plane were given in Section 3-10-B where the slope of a line was measured using the true length of a line and its angle with the *horizontal* plane. This section will use the same methods of solution but will be concerned with the frontal and profile planes.

The student should be familiar with operations used in solving Examples 9-3 and 9-4 and be able to determine the step-by-step procedures.

A. Auxiliary Method

Example 9-3

Given: Front and side views of line *MN*.

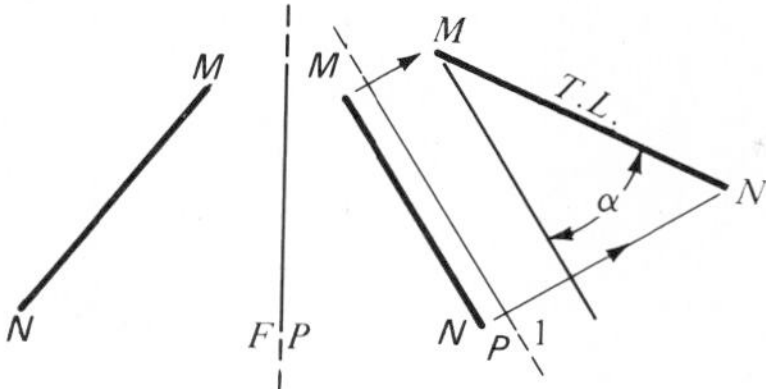

Example 9-3
Angle between line and principal plane: auxiliary method.

Task: Find the angle between line *MN* and the profile plane.

B. Revolution of Line

Example 9-4

Given: Front and top views of line *MN*.

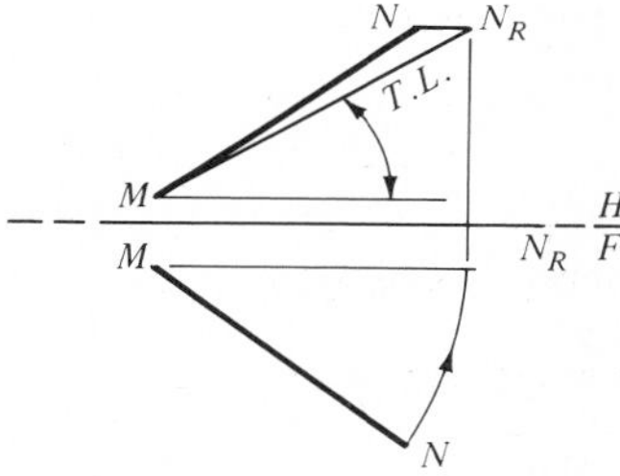

Example 9-4
Angle between line and principal plane: revolution method.

Task: Find the angle between line *MN* and the frontal plane.

9-4 ANGLE BETWEEN AN OBLIQUE PLANE AND PRINCIPAL PLANE

The angle between two planes is shown, and can be measured directly, when both planes appear on edge in the same view. Since any auxiliary projected from the top view will show the horizontal plane on edge, a problem calling for the angle between an oblique plane and a horizontal plane may be solved as shown below.

Example 9-5

Given: Top and front views of oblique triangle *ABC*.

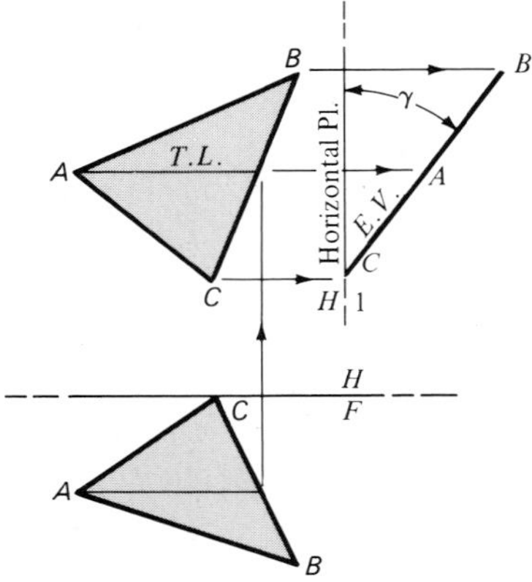

Example 9-5
Angle between oblique plane and principal plane.

Task: Find the angle between plane *ABC* and the horizontal plane.
Solution:

1. Place a horizontal line in the front view of triangle *ABC* and find its true length in the top view.
2. Show the edge view of triangle *ABC* in the first auxiliary.

In this view, the horizontal plane will also show on edge. The required angle can be measured. ◀

Using similar procedures, the angle of an oblique plane with any principal plane can be found: (1) angle with the frontal plane (project from the front view), and (2) angle with the profile plane (project from the side view).

9-5 ANGLE BETWEEN A LINE AND AN OBLIQUE PLANE

As stated before in this text, the angle between a line and a plane can be measured when *one* view shows *both* the line in true length and the plane on edge.

A. Line Method

Example 9-6

Given: Top and front views of plane *ABC* and line *MN*.

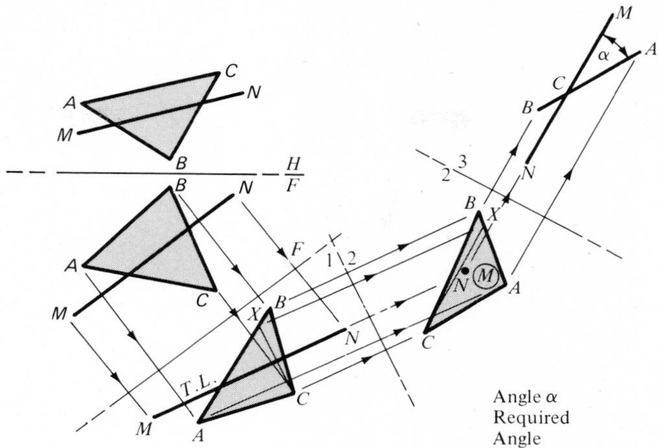

Example 9-6
Angle between line and oblique plane: line method.

Task: Find the angle between line *MN* and plane *ABC*.
Solution:

1. Draw folding line *F*/1 parallel to line *MN* in the front view. Project plane and line into the first auxiliary where line *MN* appears in true length.
2. Place folding line 1/2 perpendicular to true length line *MN* and project the line into the second auxiliary where line *MN* appears as a point. In the first auxiliary, construct line *CX* parallel to folding line 1/2. This line in plane *ABC* will appear true length in the second auxiliary.
3. Place folding line 2/3 perpendicular to the true length line *CX* and project both line and plane into the third auxiliary. The line *MN* will be in true length and the plane *ABC* will appear on edge.
4. Measure the required angle in the third auxiliary. ◀

B. Plane Method

Example 9-7

Given: Top and front views of plane *ABC* and line *MN*.
Task: Find the angle between line *MN* and plane *ABC*.
Solution:

1. Using horizontal line *BX* in the front view of *ABC* find its true length in the top view.
2. Project in the direction of the true length line and find, in the first auxiliary, the projection of line *MN* and the edge view of the plane *ABC*.

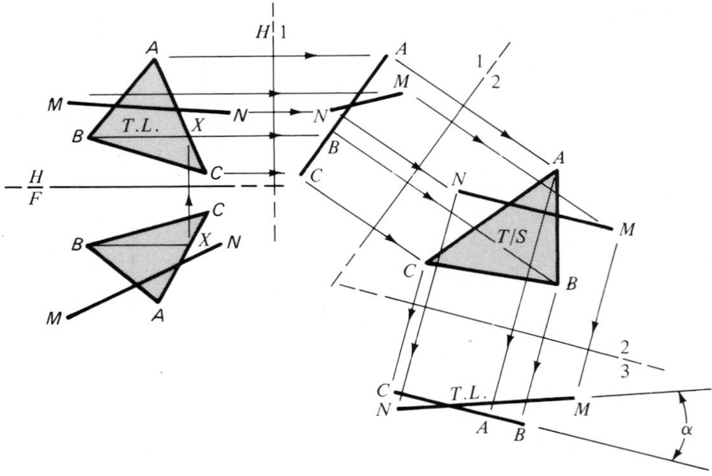

Example 9-7
Angle between line and oblique plane: plane method.

3. Place folding line 1/2 parallel to the edge view of *ABC* and project both line and plane into the second auxiliary. Plane *ABC* is shown in true size and shape.
4. Draw folding line 2/3 parallel to line *MN* and project all points into the third auxiliary. Line *MN* will appear in true length and plane *ABC* will be shown as an edge view.
5. Measure the angle between the line and plane. ◀

C. Complementary-Angle Method

Figure 9-1 shows a line *MN* which has been drawn so that it makes an angle α with plane *ABCD.* An angle of 90 deg is formed at point "*O*" when a line is drawn from point *N* perpendicular to plane *ABCD.* The projection of line *MN* on the plane completes a right triangle in which angle α and

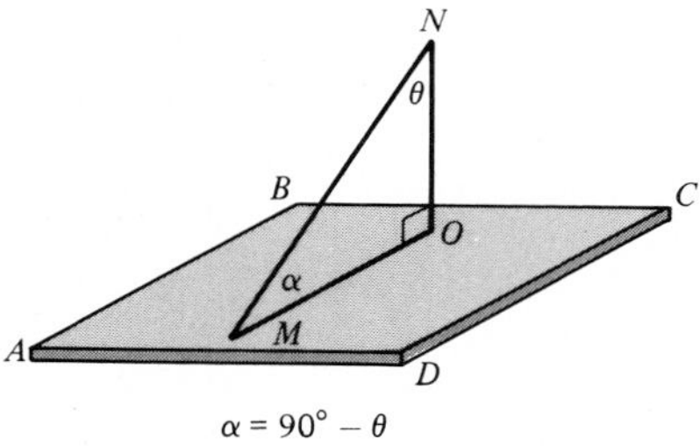

Figure 9-1
Angle between line and plane: complementary angle method.

angle θ are complementary. In the solution of the problem, the length of line *NO* is immaterial as long as it is drawn perpendicular to the plane and establishes the critical angle θ. Line *MO* in Figure 9-1 represents the projection of line *MN* on plane *ABCD* and is shown only to complete the right triangle. Since angle α and angle θ are complementary, angle α can be found by subtracting angle θ from 90 deg as shown in the following example:

Example 9-8

Given: Front and top views of line *MN* and plane *ABCD*.

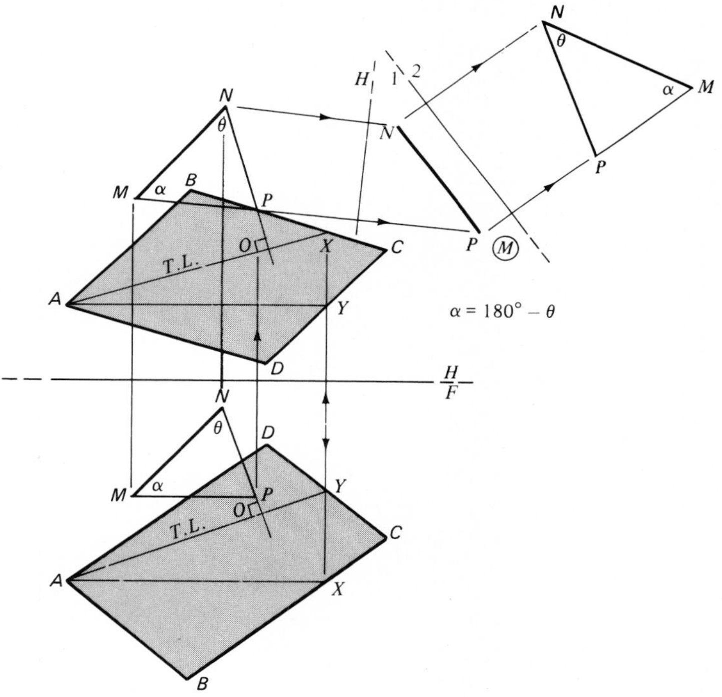

Example 9-8
Angle between line and plane: complementary angle method.

Task: Determine the angle between line *MN* and plane *ABCD*.
Solution:

1. Construct horizontal line *AX* and frontal line *AY* in the front and top views. Their projections will show true length in their respective adjacent views.
2. In the front view, draw a line from point "*N*" perpendicular to true length line *AY*. In the top view, draw a line from point "*N*" perpendicular to the true length line *AX*. The line from point "*N*" has now been drawn perpendicular to the plane and forms angle θ at "N."

3. Since the angle has been formed, point "P" can be placed anywhere on the line perpendicular to "N." Construct *MP* as a horizontal line so it shows as a true length line in the top view.
4. Project in the direction of the true length line *MP* and show the triangle *MNP* on edge in the first auxiliary.
5. Project triangle *MNP* into the second auxiliary where angle θ appears in true size. Measure angle θ and subtract it from 90 deg. The result will determine the number of degrees in the angle formed by line *MN* and the plane *ABCD*. ◀

9-6 DIHEDRAL ANGLE: ANGLE BETWEEN TWO PLANES

Any angle formed by two intersecting planes is called a *dihedral* angle and is measured on a plane that is perpendicular to both planes at the same time, as shown in Figures 9-2(a) and 9-2(b). Since the line of intersection is a straight line which is common to both planes, a view showing this line as a point will show the planes on edge. Two angles are formed, one acute and one obtuse, when two planes intersect. Although both are dihedral angles, it is common practice, unless otherwise specified, to measure the smaller or acute angle.

A. Line of Intersection Shown: Edge-View Method

Example 9-9

Given: Top and front views of two intersecting triangles *ABC* and *ACD*.

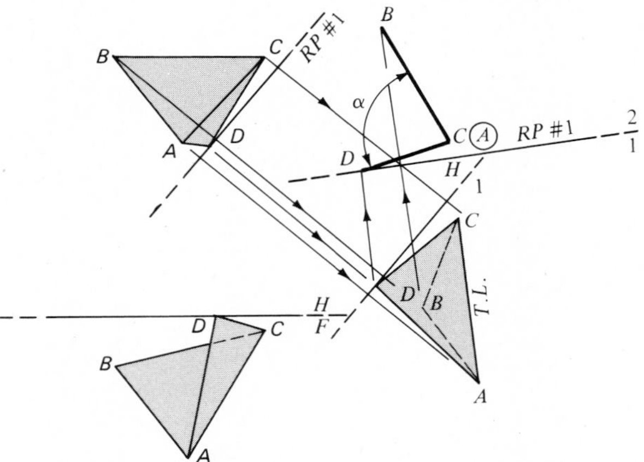

Example 9-9
Dihedral angle.

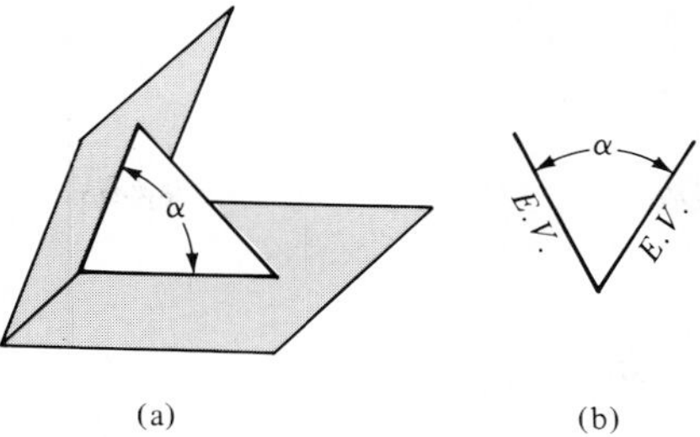

Figure 9-2
Dihedral angle (angle between planes).

Task: Find the dihedral angle between the two given triangles.
Solution:

1. By inspection, we see that *AC* is the line of intersection and is common to both planes.
2. Draw folding line *H*/1 and RP#1 parallel to the line of intersection and project all points into the first auxiliary. This view shows both planes with *AC* in true length.
3. Draw folding line 1/2 perpendicular to *AC* and project all points into the second auxiliary. In this view, line *AC* appears as a point and both planes are shown on edge.
4. Measure the given angle α. ◀

B. Line of Intersection Not Shown: Edge-View Method

Several methods are used to find the angle of intersection between two planes that do not intersect. One of these procedures, an auxiliary method, is given in the following example.

Example 9-10

Given: Top and front views of two nonintersecting planes *ABC* and *DEF*.
Task: Find the dihedral angle between the two given planes.
Solution:

1. Using the true length of a principal line in the plane, project all points into the first auxiliary where triangle *ABC* appears on edge.
2. Draw folding line 1/2 parallel to the edge view of plane *ABC* so the plane will appear as a normal view in the second auxiliary. In the first auxiliary, draw line *EM* parallel to folding line 1/2 so that in the second auxiliary line *EM* will appear true length in plane *DEF*.
3. Since plane *ABC* is in true size and shape, an auxiliary projected from it in any direction will show it on edge. Folding line 2/3 is drawn

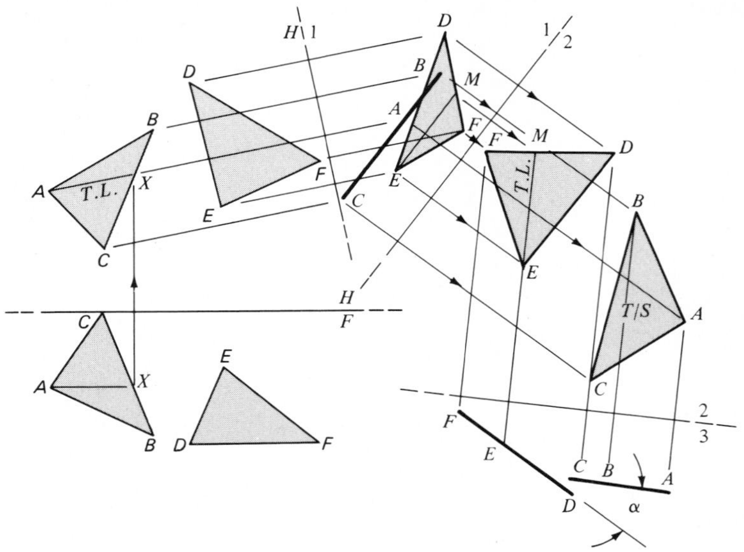

Example 9-10
Angle between two nonintersecting plane segments: edge view method.

perpendicular to the true length line *EM*. The third auxiliary view will show the edge views of both planes so that the dihedral angle can be measured. ◀

C. Line of Intersection Not Shown: Supplementary-Angle Method

Figure 9-3 illustrates the approach used in the following example to determine the angle between two planes. Intersecting lines *BA* and *BC* are each drawn perpendicular to their respective planes forming plane *ABCD* with 90 deg angles at "*A*" and "C." The number of degrees in angle α = 180 deg minus its supplemental angle θ.

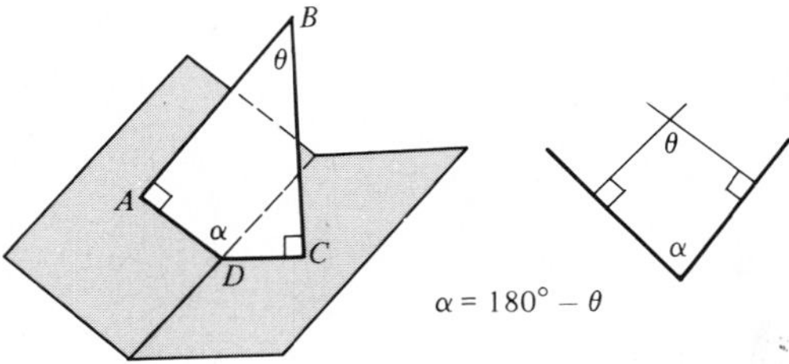

Figure 9-3
Dihedral angle: supplementary angle method.

Example 9-11

Given: Front and top views of two nonintersecting plane segments *ABC* and *XYZ*.

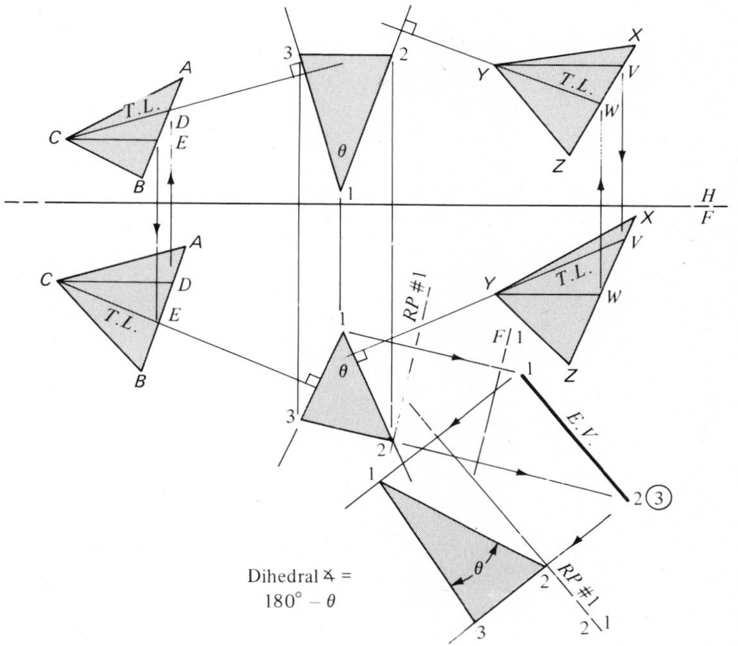

Example 9-11
Angle between planes (supplementary angle): perpendicular line method.

Task: Determine the angle between the two given plane segments.
Solution:

1. Draw horizontal lines *CD* and *YW* in the front views of planes *ABC* and *XYZ*. Project these lines into the top view of the given planes so that the lines appear in true length.
2. Using frontal lines in the top view, repeat the procedure and show the true length of lines *CE* and *YV* in the front view of planes *ABC* and *XYZ*.
3. Follow the procedures of Section 7-6-B and construct lines 1-3 and 1-2 perpendicular in the front view to true length lines *CE* and *YV*, and in the top view to true length lines *CD* and *YW*. Lines 1-3 and 1-2 are now perpendicular to their respective planes.
4. Form a triangular plane by joining the point of intersection of the perpendiculars (1) with two arbitrary points, 2 and 3. Project points 1, 2, and 3 into the front and top views.

5. Find the true size and shape of plane 1,2,3 and measure angle 2, 1, 3(θ).
6. Subtract this angle from 180 deg in order to determine the angle between the two given planes (see Figure 9-3).

A procedure in which the method of revolution is used to find the size of a dihedral angle will be discussed in the following chapter. First, mathematical techniques for calculating angles between lines and planes needed to support geometric modeling are presented.

9-7 ANALYTICAL CALCULATION OF ANGLES

As personal computers become faster and more powerful, vendors are developing three-dimensional drafting packages which can be a great help in creating a three-dimensional data base for defining objects. This, in turn, will greatly increase the need for geometric modeling, both to create the data base and obtain needed information from the data base. Extreme care must be taken when accurate angle measurements are to be obtained using traditional descriptive geometry methods. Analytical methods offer a more accurate technique if the defining points for lines and planes are accurately known. Vectors again provide the foundation for angle calculations for objects stored in a three-dimensional data base.

A. Intersecting Three-Dimensional Lines

Solving for the intersection point between two three-dimensional lines was briefly discussed in Section 3-16. Let the two parametric vector equations be given by

$$\mathbf{P} = \mathbf{A}_1 + t(\mathbf{A}_2 - \mathbf{A}_1), \quad 0 \leq t \leq 1$$
$$\mathbf{Q} = \mathbf{B}_1 + s(\mathbf{B}_2 - \mathbf{B}_1), \quad 0 \leq s \leq 1 \tag{9-1}$$

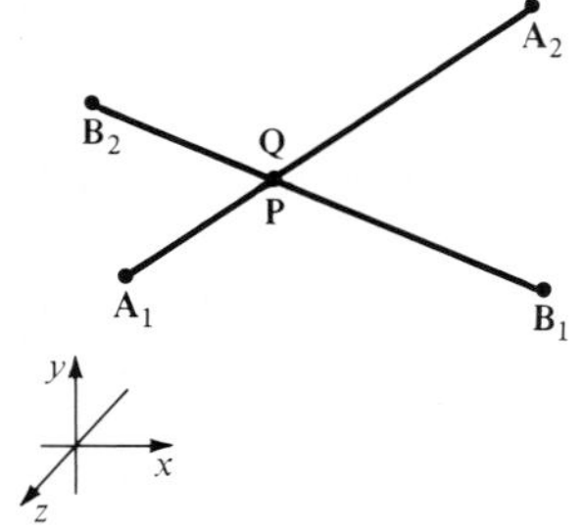

Figure 9-4
Intersecting three-dimensional lines.

Define the directional vectors $\mathbf{D} = \mathbf{A}_2 - \mathbf{A}_1$ and $\mathbf{R} = \mathbf{B}_2 - \mathbf{B}_1$. If the lines actually intersect as shown in Figure 9-4, then the equality

$$\mathbf{A}_1 + t\mathbf{D} = \mathbf{B}_1 + s\mathbf{R} \tag{9-2}$$

must be satisfied.

The three scalar equations represented by Equation (9-2) result in three equations in two unknown parameters t and s. This condition is called *overconstrained.* Any two equations are used to solve for t and s. Cramer's Rule was used in Section 3-16. The two lines intersect only when these calculated values of t and s also satisfy the third equation. (See Program 3DLINE in Appendix 1.)

A more direct approach for solving for t and s is possible by isolating each parameter separately. To isolate t, dot both sides of Equation (9-2) with $\mathbf{B} \times \mathbf{R}$. This gives

$$(\mathbf{B}_1 \times \mathbf{R})\cdot(\mathbf{A}_1 + t\mathbf{D}) = (\mathbf{B}_1 \times \mathbf{R})\cdot(\mathbf{B}_1 + s\mathbf{R}) \tag{9-3}$$

Now, by definition, $\mathbf{B}_1 \times \mathbf{R}$ is perpendicular to the plane containing $\mathbf{B}_1$ and $\mathbf{R}$. Thus, $(\mathbf{B}_1 \times \mathbf{R})\cdot\mathbf{B}_1 = 0$ and $(\mathbf{B}_1 \times \mathbf{R})\cdot s\mathbf{R} = 0$, giving zero for the right side of Equation (9-3). Solving for t gives

$$t = \frac{-(\mathbf{B}_1 \times \mathbf{R})\cdot\mathbf{A}_1}{(\mathbf{B}_1 \times \mathbf{R})\cdot\mathbf{D}} \tag{9-4}$$

Now, to isolate s, dot both sides of Equation (9-3) with $(\mathbf{A}_1 \times \mathbf{D})$.

$$(\mathbf{A}_1 \times \mathbf{D})\cdot(\mathbf{A}_1 + t\mathbf{D}) = (\mathbf{A}_1 \times \mathbf{D})\cdot(\mathbf{B}_1 + s\mathbf{R}) \tag{9-5}$$

The left side of this equation is equal to zero. Solving for s gives

$$s = \frac{-(\mathbf{A}_1 \times \mathbf{D})\cdot\mathbf{B}_1}{(\mathbf{A}_1 \times \mathbf{D})\cdot\mathbf{R}} \tag{9-6}$$

The solution for s and t requires the evaluation of two scalar triple products in both Equations (9-4) and (9-6). See Equation (8-18). This approach avoids the use of Cramer's Rule. The resulting values for t and s must satisfy *all* three scalar components of Equation (9-2) for the two lines to *actually* intersect. These are

$$\begin{aligned} A_{1,x} + t(A_{2,x} - A_{1,x}) &= B_{1,x} + s(B_{2,x} - B_{1,x}) \\ A_{1,y} + t(A_{2,y} - A_{1,y}) &= B_{1,y} + s(B_{2,y} - B_{1,y}) \\ A_{1,z} + t(A_{2,z} - A_{1,z}) &= B_{1,z} + s(B_{2,z} - B_{1,z}) \end{aligned} \tag{9-7}$$

The angle between two intersecting lines can be found by use of the vector dot product. If two three-dimensional lines intersect, then a single plane exists which contains both lines. The angle between the two lines is the angle at the point of intersection measured in the common plane containing both lines.

Define a three-dimensional line between vectors $\mathbf{A}_1$ and $\mathbf{A}_2$, where $\mathbf{A}_1 = [x_1\ y_1\ z_1]$ and $\mathbf{A}_2 = [x_2\ y_2\ z_2]$. Let the three scalar components of the directional vector be $A_x = (x_2 - x_1)$, $A_y = (y_2 - y_1)$, and $A_z = (z_2 - z_1)$. In a similar manner, a second three-dimensional line lies between $\mathbf{B}_1$ and $\mathbf{B}_2$, where $\mathbf{B}_1 = [x_3\ y_3\ z_3]$ and $\mathbf{B}_2 = [x_4\ y_4\ z_4]$. The directional vector of this second line has scalar components $B_x = (x_4 - x_3)$, $B_y = (y_4 - y_3)$, and $B_z = (z_4 - z_3)$. These components are shown in Figure 9-5. When these components are used in Equation (4-4) to calculate $\cos\theta$, the value of θ will be the angle between the two intersecting lines as measured in the plane containing both lines. For example,

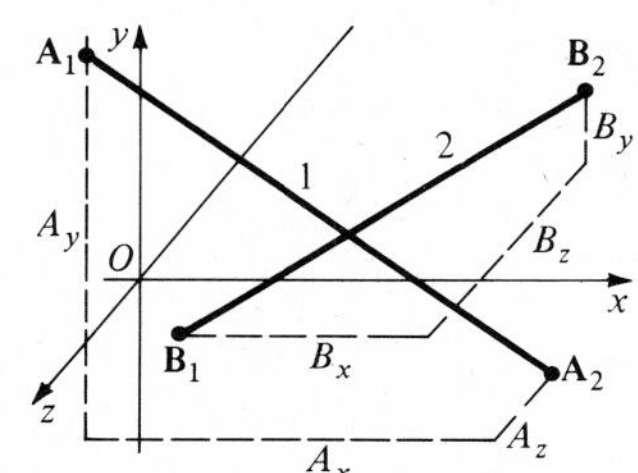

Figure 9-5
Scalar components of directional vectors.

$$\cos\theta = \frac{A_xB_x + A_yB_y + A_zB_z}{|\mathbf{A}_2 - \mathbf{A}_1|\ |\mathbf{B}_2 - \mathbf{B}_1|} \tag{9-8}$$

Equation (9-8) can also be formulated in terms of the direction cosines for each line. Let $\mathbf{R}_1 = \mathbf{A}_2 - \mathbf{A}_1$, and $\mathbf{R}_2 = \mathbf{B}_2 - \mathbf{B}_1$, and write

$$\cos\theta = \frac{(A_x)\,(B_x)}{|\mathbf{R}_1|\;|\mathbf{R}_2|} + \frac{(A_y)\,(B_y)}{|\mathbf{R}_1|\;|\mathbf{R}_2|} + \frac{(A_z)\,(B_z)}{|\mathbf{R}_1|\;|\mathbf{R}_2|} \tag{9-9}$$

Using the definitions of the direction cosines in Equation (9-9) gives

$$\cos\theta = (\cos\theta_x)_A(\cos\theta_x)_B + (\cos\theta_y)_A(\cos\theta_y)_B + (\cos\theta_z)_A(\cos\theta_z)_B \tag{9-10}$$

When two lines intersect at right angles, $\cos\theta = 0$. This condition can be used in Equation (9-10) to see what changes in direction cosines are necessary to create a perpendicular intersection of two lines in space.

We define an *apparent* angle of intersection as the angle between two lines in space which do not intersect. The angle between two skew lines is defined as the angle measured in the plane which is normal to the common perpendicular line between the two skew lines. Refer to Sections 7-4, 7-9-C, and 8-9 for discussions of the line of shortest distance between two skew lines.

If three-dimensional lines are translated in space, without rotation, the direction cosines do not change. Hence, Equations (9-8) and (9-9) can be used to calculate the cosine of the angle between two lines in space, whether they actually intersect or not.

Example 9-12

Given: The frontal and horizontal views of two pipe centerlines are obtained from a blueprint and plotted in Figure 9-6. It can be seen that the pipes do not intersect. The measured coordinates of the endpoints of each pipe are as follows:

$$\mathbf{A}_1 = [5 \quad 2 \quad 4], \quad \mathbf{A}_2 = [15 \quad 4 \quad 0]$$

$$\mathbf{B}_1 = [7 \quad 7 \quad 8], \quad \mathbf{B}_2 = [13 \quad 1 \quad 2]$$

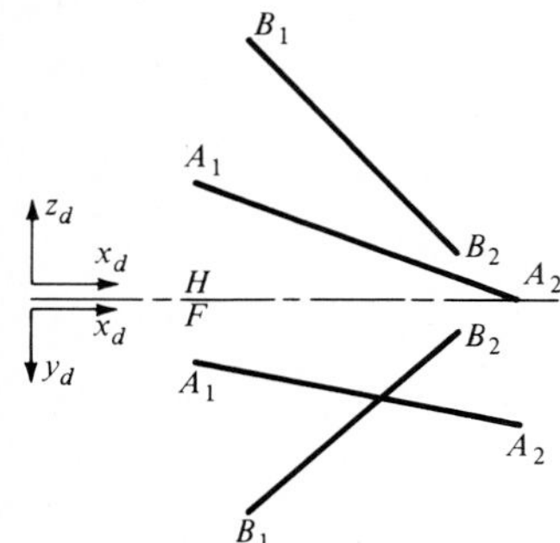

Figure 9-6
Pipeline centerlines.

Task: Find the apparent angle of intersection.
Solution:

$$|\mathbf{A}_2 - \mathbf{A}_1| = (10^2 + 2^2 + 4^2)^{1/2} = 10.95$$

$$|\mathbf{B}_2 - \mathbf{B}_1| = (6^2 + 6^2 + 6^2)^{1/2} = 10.39$$

Using Equation (9-9) gives

$$\cos\theta = \left(\frac{10}{10.95}\right)\left(\frac{6}{10.39}\right) + \left(\frac{2}{10.95}\right)\left(\frac{-6}{10.39}\right) + \left(\frac{-4}{10.95}\right)\left(\frac{-6}{10.39}\right)$$

$$= 0.527 - 0.105 + 0.211 = 0.633$$

Thus, $\theta = 50.7$ deg.

B. Angle Between Planes

Determination of the angle between planes is a special case of the angle between lines as discussed in the previous section. The angle between two planes is the same as the angle between their normal vectors. Once the normal vectors are found, the problem is one of obtaining the angle between these two normal lines.

If a plane is given in implicit form

$$N_1x + N_2y + N_3z - k = 0$$

then the intercept form, Equation (6-25), is

$$\frac{x}{a_1} + \frac{y}{b_1} + \frac{z}{c_1} = 1$$

where $a_1 = k/N_1$, $b_1 = k/N_2$, $c_1 = k/N_3$. The perpendicular distance from the origin to the plane is $p = k/|\mathbf{N}|$. Denote the direction cosines by $l_1 = p/a_1$, $m_1 = p/b_1$, $n_1 = p/c_1$. A second plane may have direction cosines l_2, m_2, n_2. Equation (9-10) then gives

$$\cos\theta = l_1l_2 + m_1m_2 + n_1n_2 \tag{9-11}$$

where θ is the angle between the two planes.

C. Angle Between Line and Plane

The angle of intersection θ between a line and a plane can also be found by use of a dot product. Referring to the angles shown in Figure 9-7

$$\cos\theta = \frac{\mathbf{R}\cdot\mathbf{n}}{|\mathbf{R}|} = \sin\theta \tag{9-12}$$

Here, θ is the acute angle between the line and the plane at the point of intersection, $\mathbf{R}$ is the directional vector along $\mathbf{B} - \mathbf{A}$, and $\mathbf{n}$ is the unit normal vector to the plane. The dot product $\mathbf{R}\cdot\mathbf{n}$ is the projection of line $\mathbf{R} = \mathbf{B} - \mathbf{A}$ in the normal $\mathbf{N}$ direction.

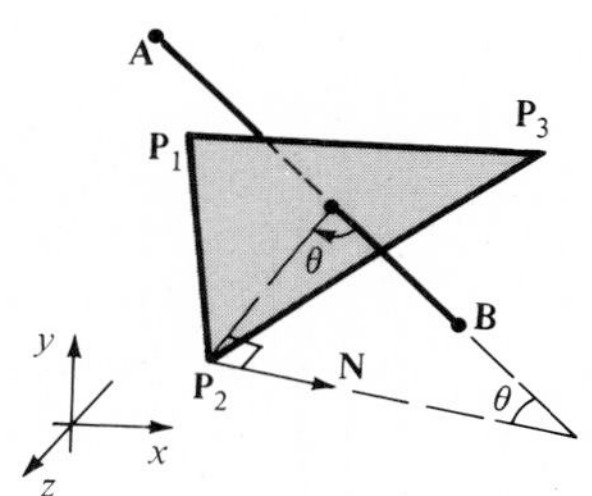

Figure 9-7
Angle of intersection between line and plane.

PROBLEMS: CHAPTER 9

9-1 *Given:* Top and front views of line *MN* and plane *ABCD*.

Task: Determine the true angle between the line *MN* and plane *ABCD*.

9-2 *Given:* Top and front views of line *MN* and triangle *ABC*.
Task: Find the true length of line *MN* and the angle it makes with plane *ABC*.

9-3 *Given:* Top and front views of a sheet metal truncated pyramid.

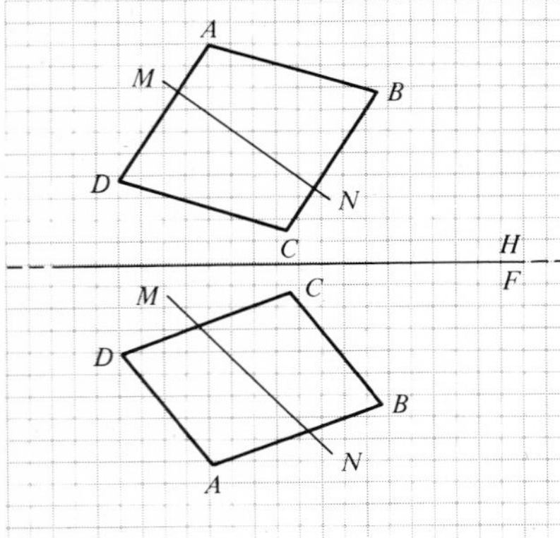

Figure 9-1P

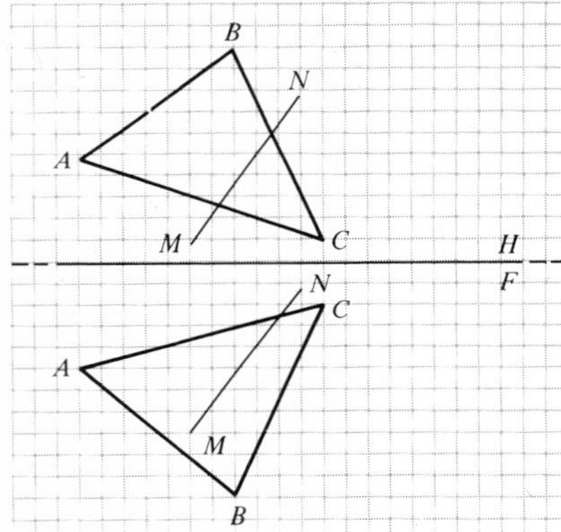

Figure 9-2P

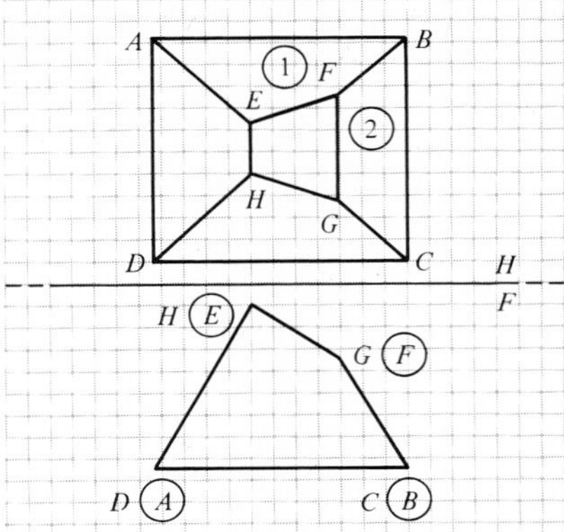

Figure 9-3P

Task: Determine the dihedral angle between surfaces 1 and 2 so that braces can be formed.

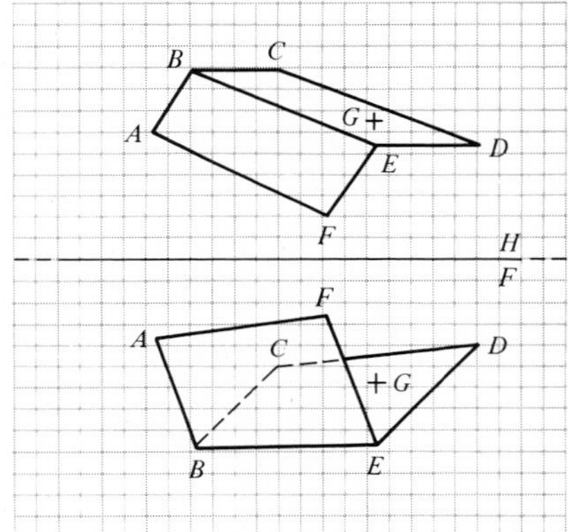

Figure 9-4P

9-4 *Given:* Top and front views of two connected bulkheads, *ABEF* and *BCDE.*

Task: a. Determine the dihedral angle between bulkheads.

b. Show, in both views, the location of the shortest horizontal rod which can be installed from "G" to the opposite bulkhead.

9-5 *Given:* Top and front views of two intersecting planes *ABD* and *BCD.*

Task: Determine the dihedral angle between the two given planes.

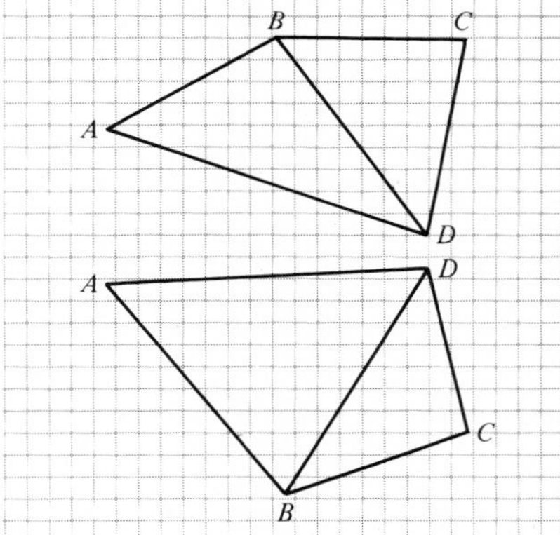

Figure 9-5P

9-6 *Given:* Front and top views of line *AB* and point *C*, the vertex of the required angle.
Task: Draw an isosceles triangle whose base *RT* lies on line *AB* and whose base angles are 70 deg. Show the projection of the triangle in all views.

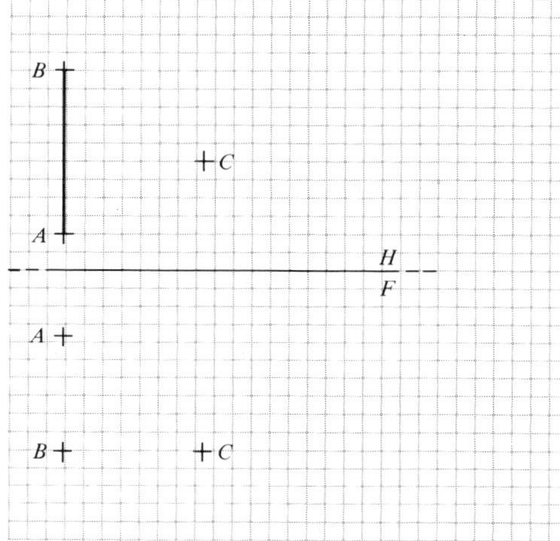

Figure 9-6P

9-7 *Given:* Front and top views of line *MN* and point *P*.
Task: Construct a line from point *P* to intersect and make an angle of 60 deg with line *MN*. Show the constructed line in all views.

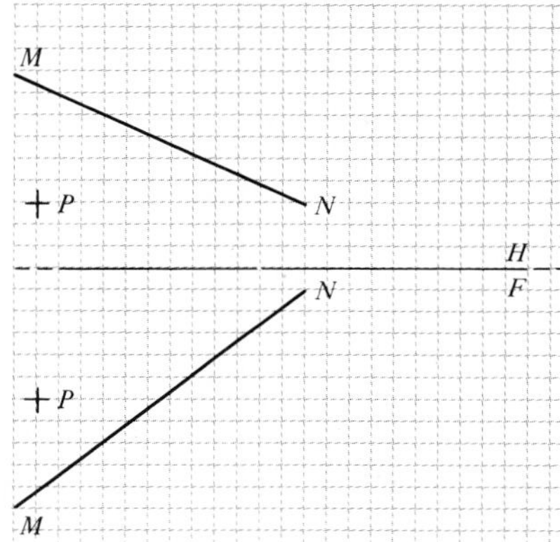

Figure 9-7P

9-8 *Given:* Front and top views of pipe *DEF*.
Task: Determine the angle *DEF*.

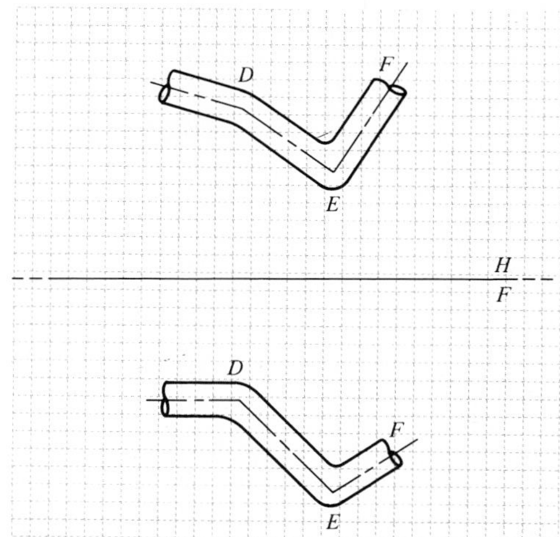

Figure 9-8P

9-9 *Given:* Front and top view of steel members which form angle *ABC*.
Task: Determine the true angle formed at *B* so a gusset plate can be cut.

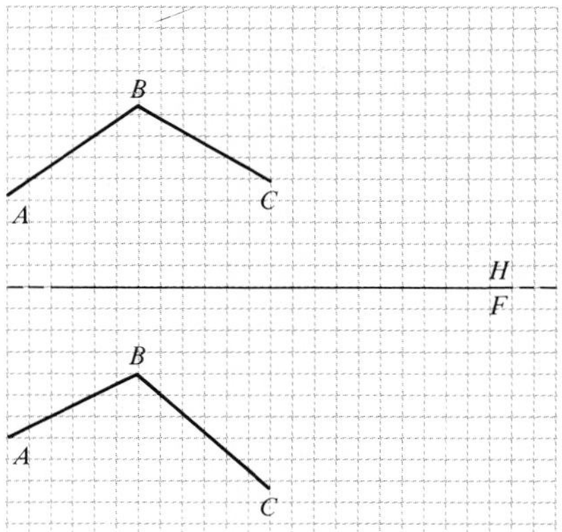

Figure 9-9P

9-10 *Given:* Front and top views of two nonintersecting plane segments *ABC* and *RST*.
Task: Find the angle between the two planes.

9-11 *Given:* Any two two-dimensional vectors **A** and **B**.
Task: Illustrate that the sign of the dot product $\mathbf{A} \cdot \mathbf{B}$ is always the same as the sign of cos θ, where θ is the angle between the two vectors.

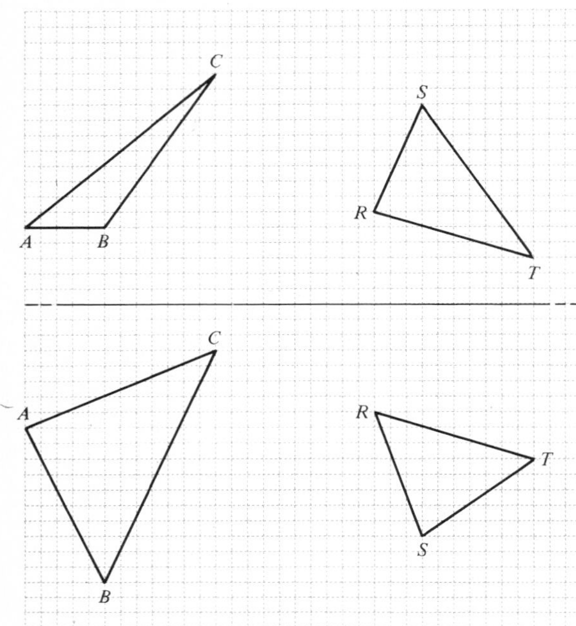

Figure 9-10P

9-12 *Given:* The x-y plane is divided into two halves by the two-dimensional line $\mathbf{P} = \mathbf{A} + (\mathbf{B} - \mathbf{A})t$. The left half is defined as the region to the left when standing at the base vector $\mathbf{A}$ looking along the directional vector $\mathbf{D} = \mathbf{B} - \mathbf{A}$. Let $\mathbf{A} = [2\ 5]$ and $\mathbf{D} = [1\ -3]$.
Task: Use the property of the vector cross product and devise a method to check whether a point, such as $\mathbf{P} = [5\ 8]$, is in the left half or right half of the plane.

9-13 *Given:* The intersection of the line segments between $\mathbf{A}_1 = [1\ 0\ 0]$ and $\mathbf{A}_2 = [-1\ 2\ 0]$ and between $\mathbf{B}_1 = [0\ 0\ 1]$ and $\mathbf{B}_2 = [0\ 2\ -1]$ can be easily determined graphically.
Task: Use Equations (9-4), (9-5), and (9-7) to calculate the point of intersection.

9-14 *Given:* Two three-dimensional line segments in space between $\mathbf{A}_1 = [-1\ 1\ 2]$ and $\mathbf{A}_2 = [5\ 4\ -1]$ and between $\mathbf{B}_1 = [4\ 0\ 3]$ and $\mathbf{B}_2 = [-2\ 4\ -1]$. It is known that these two lines intersect.
Task: Use Equations (9-4), (9-6), and (9-7) to calculate the point of intersection.

9-15 *Given:* Two planes are defined by two sets of three noncollinear points as follows:

Plane 1: $[-1\ 1\ -1]$, $[0\ -1\ -2]$, $[4\ 0\ 2]$

Plane 2: $[-1\ 1\ 1]$, $[4\ -3\ -1]$, $[6\ -1\ 2]$

Task: a. Calculate the direction cosines of the line of intersection between the two planes.
b. Calculate the angle between the planes.

9-16 *Given:* The implicit equation for a plane is given by

$$x - 2y - 2z + 3 = 0$$

Task: a. Calculate the coordinates of the point of intersection between this plane and an infinite line passing through the two points $\mathbf{A} = [-2\ 3\ 0]$ and $\mathbf{B} = [-3\ 4\ -2]$.
b. Calculate the angle of intersection between the line and the plane.
c. Calculate the shortest distance from the origin of the coordinate system to the plane.

9-17 *Given:* Two three-dimensional line segments given in Problem 9-14.
Task: Calculate the angle between the two lines.

9-18 *Given:* Two planes in implicit form are given by

Plane 1: $3x + 4y - 2z + 1 = 0$

Plane 2: $4x - 2y + 3z - 2 = 0$

Task: a. Calculate the angle between the two planes.
b. Repeat if plane 2 is given by

$$5x + y + 2z + 3 = 0$$

9-19 *Given:* The three implicit equations for three planes in Problem 9-18.
Task: Use Cramer's Rule (Appendix 2) to find the point of intersection between these three planes.

9-20 *Given:* A sketch of an anchor block is shown in Figure 9-20P. It must be designed such that the dihedral angle between surfaces 1 and 2 is 120 deg. Design requirements specify that corners 1, 2, 4, and 5 are given by: (L.H.C.) $\mathbf{C}_1 = [0\ 0\ 0]$, $\mathbf{C}_2 = [3.875\ 0\ 0]$, $\mathbf{C}_3 = [4.5\ 1.25\ 2]$, and $\mathbf{C}_4 = [1.125\ 1.25\ 3.125]$.
Task: If plane 2 is horizontal, determine the coordinates of corners 3 and 6 to meet the design requirements.

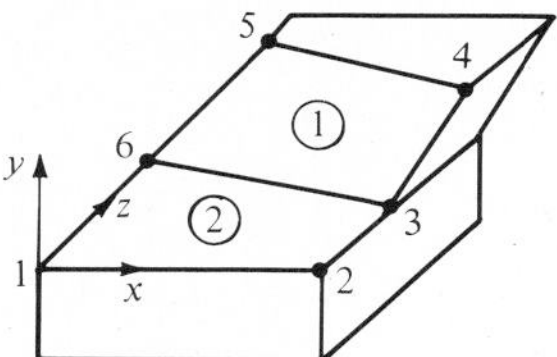

Figure 9-20P
Anchor block.

9-21 *Given:* An isometric view of a storage bin is shown in Figure 9-21P. The four corners of plane 1 are known to be given by [4 2.5 0], [4 2.5 2.5], [2.25 0 1.625], and [2.25 0 0.875]. The four corners of plane 2 are given by [4 2.5 2.5], [2.25 0 1.625], [0.5 0.875 1.9375], and [0 2.5 2.5].
Task: Calculate the angle between plane 1 and plane 2.

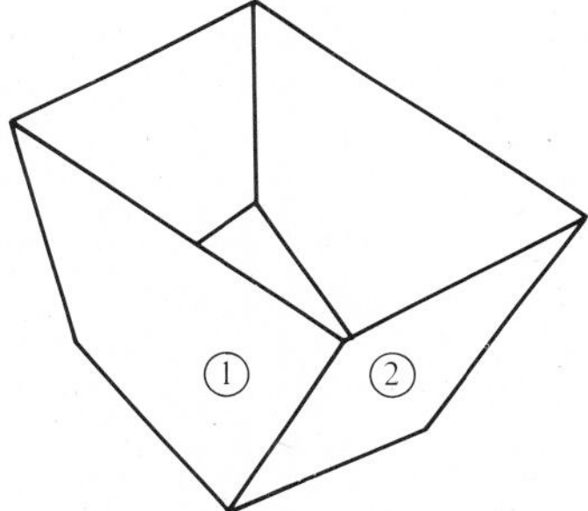

Figure 9-21P
Hopper.

9-22[c] *Given:* The horizontal and frontal views of a fireplace hood are given in Figure 9-22P. Angle braces are required to support the intersection of planes 1 and 2. The following corner coordinates are available: $\mathbf{C}_1$ = [0 0 3], $\mathbf{C}_2$ = [2.3125 0 3], $\mathbf{C}_3$ = [1.375 0.8125 2], $\mathbf{C}_4$ = [0.75 0.8125 2], $\mathbf{C}_5$ = [1.375 0.8125 1.375], and $\mathbf{C}_6$ = [2.3125 0 0.625].
Task: a. Calculate the angle between planes 1 and 2.
b. Generate the principal views and an orthographic view as seen from **E** = [1 3 9].
c. Study angle sensitivity to changes in *y*-values for corners 1, 2, and 6.

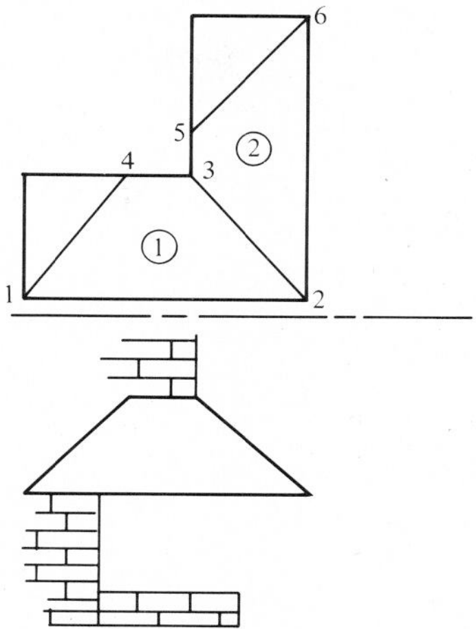

Figure 9-22P
Fireplace hood.

9-23[c] *Given:* Two endpoints of a line and a third point in space.
Task: Use computational descriptive geometry and write a computer program which will accept the three given points and determine the connector between the point and line at a specified angle.
Note: A descriptive geometry solution to this type of problem is given on page 108 of Reference 16. (See References.)

9-24[c] *Given:* Two endpoints of a line and a third point in space.
Task: Use computational descriptive geometry and write a computer program which will accept the three given points and determine the steepest connection between the given point and line.
Note: A descriptive geometry solution to this type of problem is given on page 105 of Reference 16.

9-25 *Given:* A triangle whose sides are defined by the vectors $\mathbf{A} = 3\mathbf{i} + 4\mathbf{j} - \mathbf{k}$; $\mathbf{B} = 2\mathbf{i} - \mathbf{j} + 3\mathbf{k}$; $\mathbf{C} = \mathbf{i} + 5\mathbf{j} - 4\mathbf{k}$.
Task: Calculate the three included angles of the triangle.

---

[c]Superscript c indicates that a computer solution is required.

10

Revolution

In the previous constructions of auxiliary views, the object usually remained stationary and the observer moved about in space. Revolution, however, deals with situations in which the object itself is moved by revolving it about an axis which is perpendicular to a plane. Although Section 3-9 gave a general explanation of the procedure used to find the true length of a line and its angle with a principal plane, further study is necessary in order to understand the theory involved.

When obtaining descriptive geometry solutions, time may be saved and space may be conserved by revolving points, lines, plane surfaces, or solids around specified axes which must be definitely located. Caution must be used, however, since the revolution of any object changes its position relative to that of other objects which remain unchanged in the problem. Surfaces of revolution may also be created as shown later in this chapter.

The procedures given in the following sections will show that the method of revolution may be used in descriptive geometry wherever the axis is shown true length in one view and as a point in an adjacent view.

10-1 REVOLUTION OF A POINT

Although other paths of revolution may be possible, the discussion in this text will assume that any point which revolves about a straight line axis will create a circular path about that axis. This path will have the axis as its center and will lie in a plane which is perpendicular to it. Figure 10-

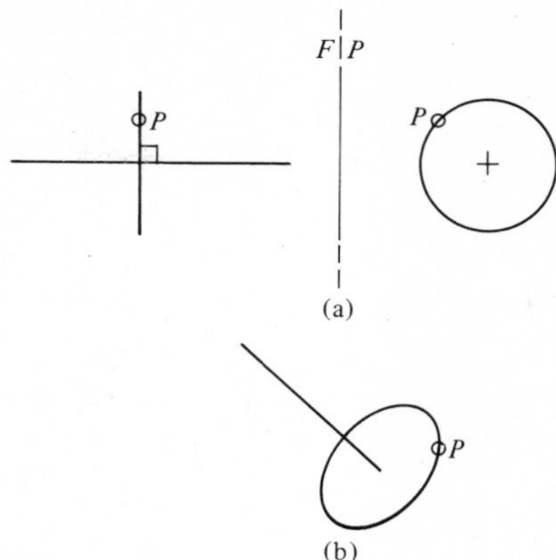

Figure 10-1
Revolution of a point.

1(a) and (b) demonstrate three ways that the path of the point may appear:

1. as a circle or an arc with the axis as its center when the axis appears as a point
2. as the edge view of a plane which is seen as a line that is perpendicular to the true length view of the axis, and
3. as an ellipse when the view of the axis is foreshortened.

The following example is given to enhance understanding of the above concept.

Example 10-1

Given: Front and top views of line *MN* and point *P*.
Task: View the line axis from *M* to *N* and revolve point *P* clockwise to a new position (*P*′) 120 deg from its present location. Show *P*′ in all views.
Solution:

1. Draw folding line *H*/1 parallel to line *MN* in the top view and project both line and point *P* into the first auxiliary where line *MN* appears in true length.
2. Draw folding line 1/2 perpendicular to true length line *MN* and project both the line and point *P* into the second auxiliary where line *MN* appears as a point.
3. In the second auxiliary view, using the point view of the line from *M* to *N* as a center, scribe the circular path of point *P* about that center. On the circle measure 120 deg clockwise and locate the new point *P*′. Project its line of sight into the first auxiliary.

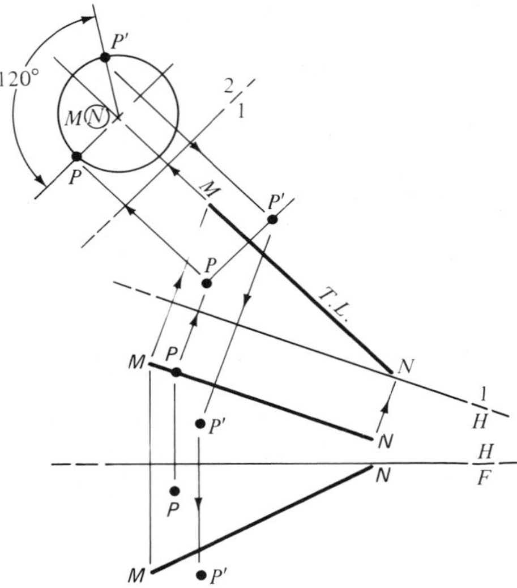

Example 10-1
Revolution of a point.

4. Locate P' in the first auxiliary by moving P perpendicular to the line-axis and parallel to folding line 1/2 until it meets the line of sight of P'. ◀

10-2 REVOLUTION OF A LINE

A. Line Intersecting Axis of Revolution

Section 3-9 discussed the procedures used to find the true length of a line by considering it to be a lateral element of a cone whose axis was perpendicular to one of the principal planes. The theory involved is repeated below.

An oblique line will not project onto any principal plane as a true length line because it is not parallel to any one of them. Figure 10-2 shows an oblique line *MN* viewed through the frontal picture plane. This line can be seen in true length by:

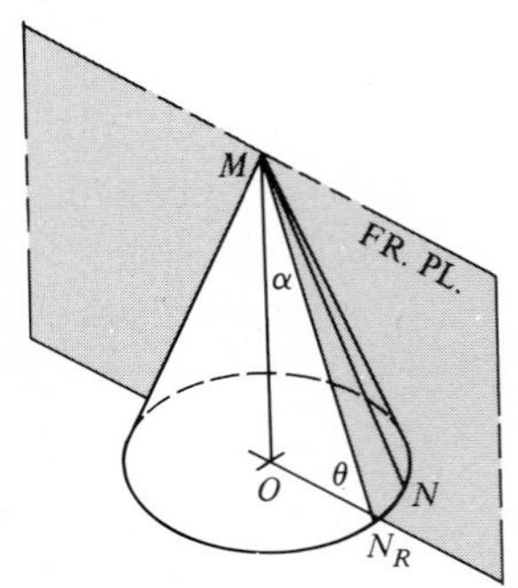

Figure 10-2
Revolution of a line.

1. Passing an axis *MO* through point *M* and perpendicular to the horizontal plane.
2. Letting line *MN* be a lateral element of a cone whose axis *MO* is parallel to the frontal plane and whose base is parallel to the horizontal plane. The radius of the base will be the same as the projection of the line on the horizontal plane.

3. Revolving the cone so that all points on line *MN* pass through the same number of degrees until the line is in the frontal plane and will appear true length in the front view.

Since all points on the line lie in planes which are perpendicular to the axis of rotation, they retain their original distance from the horizontal plane. Since they also retain their original distance from the axis of rotation, the angle θ between the line and the horizontal plane can be measured. Note of caution: Points on line *MN* have not retained their original distance from the profile plane. Therefore, angle α *cannot* be measured.

Examples 3-11, 3-12, and 3-13 show how the true length of a line may be found by considering it to be an element of a cone with its axis perpendicular to one of the three principal planes. The following example shows that any two adjacent views may be used in this type of solution as long as the axis of the cone is perpendicular to one of the planes. If a line is revolved about an axis as stated above, the angle between the line and that plane will not be changed.

Example 10-2

Given: Top, front, and frontal auxiliary views of line *MN*.

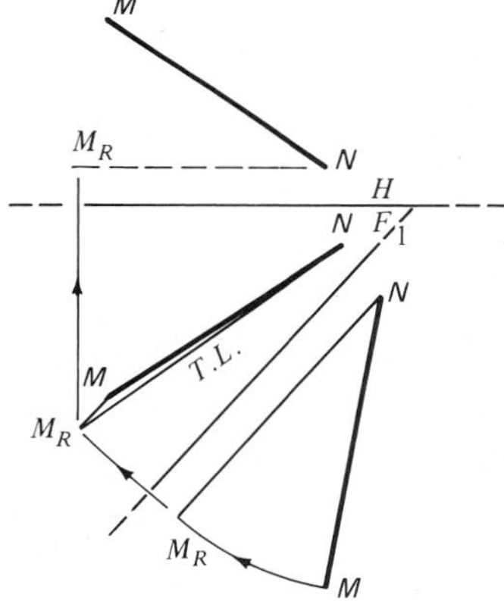

Example 10-2
True length of a line by revolution.

Task: Find the true length of line *MN*. Show the rotated line in all views.
Solution:

1. In the first auxiliary, pass an axis through point *N* and perpendicular to the first auxiliary plane.
2. Using the point of the axis as a center, and the projection of line *MN* on the first auxiliary plane as a radius, locate point M_R so that line M_RN lies parallel to the frontal plane (folding line *F*/1).

3. In the frontal plane, move point M parallel to the folding line $F/1$ until it intersects the line of sight of M_R. M_RN is the required true length.
4. The projection of line M_RN into the top view shows that the revolution of the line has relocated it in space. ◀

B. Line Not Intersecting Axis of Revolution

The axis of revolution may be located anywhere in space and is not required to intersect the line which is to be revolved. So that the given line itself will not be altered while being relocated, all points on the line must be revolved through the same angle.

Example 10-3 shows that a given line may be resolved about an axis and its true length determined by selecting two points on the line and moving them through the same angular displacement.

Example 10-3

Given: Front and frontal auxiliary views of line *MN*.

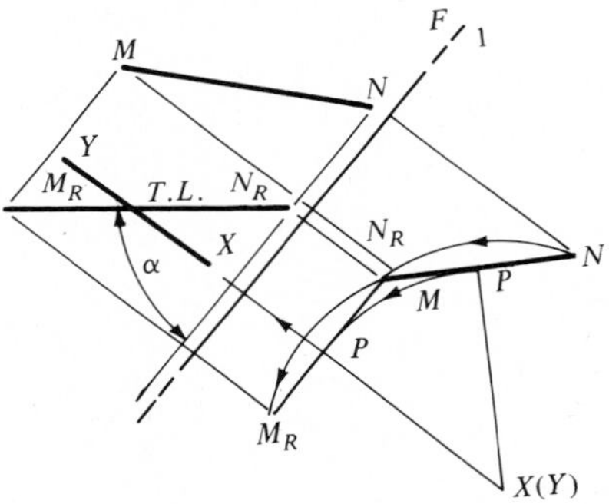

Example 10-3
Angle line makes with a principal plane by revolution.

Task: Find the true length of line *MN* and the angle it makes with the first auxiliary plane.
Solution:

1. Arbitrarily select the point view of axis *XY* in the first auxiliary view and draw the true length axis line into the front view so that both it and its line of sight are perpendicular to the folding line $F/1$.
2. Draw a line $X(Y)P$ from the point view of axis $X(Y)$ so that it is perpendicular to the given line *MN* and intersects it at *P*.
3. Using $X(Y)P$ as a radius, scribe an arc from *P* to intersect the line of sight which was drawn from the point view of the axis into the front view. The revolved radius $X(Y)P$ will now coincide with the line of sight of the axis. Since the infinite line defined by points *M* and *N* is

perpendicular to line $X(Y)P$, the revolved line will lie parallel to the folding line $F/1$.

4. Locate point M_R by revolving point M about axis $X(Y)$ until it intersects the revolved line. Locate point N_R by revolving point N about axis $X(Y)$ until it intersects the revolved line. Note: The studies given in Section 3-7 could be used here since the length of the line MN and the ratio of the points on it will not change because it was revolved.
5. Lines of sight are projected from points M_R and N_R into the front view.
6. Points M and N were revolved about axis $X(Y)$ on planes perpendicular to it in the first auxiliary view. The points M and N are therefore moved perpendicular to the axis in the front view until they intersect their respective lines of sight at M_R and N_R.
7. M_R and N_R, when connected, show the true length of line MN. The angle between the true length line and the first auxiliary plane on edge can now be measured. ◀

10-3 EDGE VIEW OF A PLANE BY REVOLUTION

By revolving all the points of a plane through the same angle the plane itself can be shown on edge without the use of an auxiliary view.

Example 10-4

Given: Top and front views of triangle *ABC*.

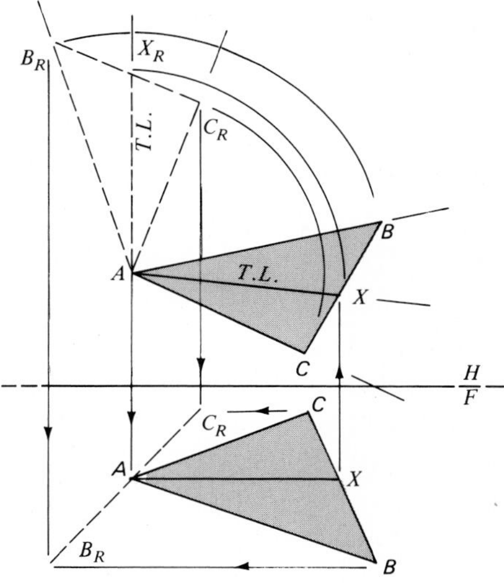

Example 10-4
Edge view of a plane by revolution.

Task: Show the edge view of triangle *ABC* without using an auxiliary view.
Solution:

1. In the front view, draw horizontal line *AX* in triangle *ABC*. Project *AX* into the top view where it becomes a true length line in triangle *ABC*.
2. Assume an axis through point *A* and perpendicular to the horizontal plane. This axis will be shown in true length in the front view and as a point in the top view.
3. In the top view, revolve point *X* about the axis until line *AX* is perpendicular to the frontal plane and locates point X_R. When the true length line *AX* is seen as a point in the front view, the triangle *ABC* will appear on edge.
4. In the top view, revolve all points of the triangle about the axis and through the same angle to obtain points B_R and C_R.
5. Project lines of sight from the revolved points into the front view.
6. In the top view, the points were revolved about the axis on planes perpendicular to the axis. Therefore, move the points in the front view on a line perpendicular to the axis until they intersect their respective lines of sight from the top view. Point *A* and the relocated points B_R and C_R show the edge view of plane *ABC* as required. Recall that in Chapter 2 the program PLANEDGE was used to create an edge view of a plane without use of an auxiliary view. ◀

10-4 NORMAL VIEW OF A PLANE BY REVOLUTION

A. Using One Auxiliary

The normal view of a plane can be shown by first placing the plane on edge in the usual way. It may then be revolved about an axis which shows as a point in that view. The adjacent view will show the axis in true length and the plane in true size as shown in Example 10-5.

Example 10-5

Given: Top and front views of plane *ABC*.
Task: Using only one auxiliary, find the normal view of plane *ABC*.
Solution:

1. Use a horizontal line *AX* in plane *ABC* and by the usual methods show the edge view of the plane in the first auxiliary.
2. Consider line *AX* as an axis which is perpendicular to the first auxiliary plane. This axis shows as a point in the first auxiliary and as a true length line in the top view.

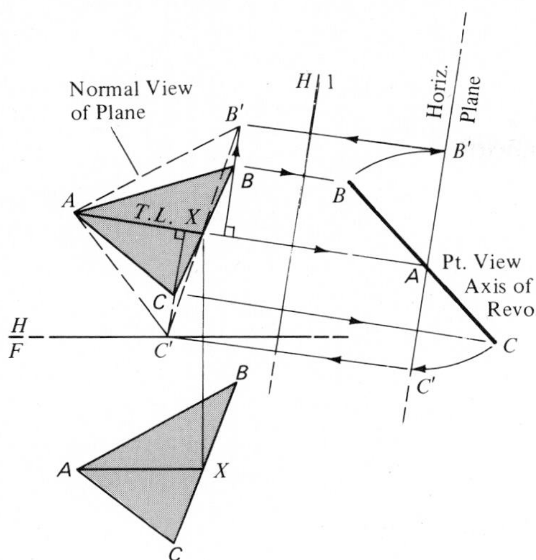

Example 10-5
Plane: normal view by revolution.

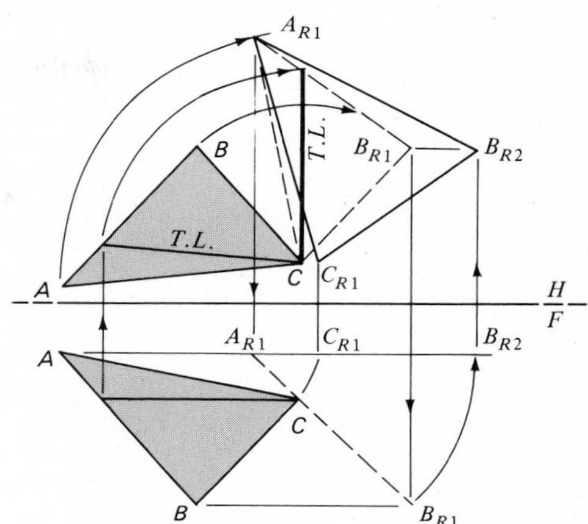

Example 10-6
Normal view of plane by revolution.

3. In the first auxiliary, revolve both point *B* and point *C* about the axis *AX* until points *B′* and *C′*, along with *AX*, lie in the horizontal plane.
4. Project lines of sight from the revolved points into the top view.
5. Move the points of the triangle parallel to folding line *H*/1 and perpendicular to axis *AX* until they intersect their respective lines of sight. Point *A* and the translated points *B′* and *C′* determine the normal view of triangle *ABC*. ◀

B. Two-View Method—No Auxiliaries

The true size of a plane can be found by first putting the plane on edge by revolution as in Section 10-3, and then finding the true normal view as in Section 10-4-A. Since this is a combination of methods studied previously, the solution to Example 10-6 is presented so that the step-by-step procedure can be set down by the student.

Example 10-6

Given: Top and front views of oblique triangle *ABC*.
Task: Using only revolution (no auxiliaries), show the normal view of the triangle as determined by the revolved points A_{R1}, B_{R2}, and C_{R1}. ◀

10-5 REVOLUTION OF A SOLID

When a solid is revolved, all points on the solid must revolve through the same angle and about the same axis. The axis may be placed at any convenient place on the object as long as it (the axis) shows as a point in one view and true length in an adjacent view.

Figure 10-3 shows a rectangular solid revolved about a vertical axis which is perpendicular to the horizontal plane. The axis shows as a point in the top view and true length in the front view. The top view changes *position* while the other view changes *shape* because of regular projection methods.

Figure 10-4 shows the same solid as used before with a horizontal axis of revolution which is perpendicular to the frontal plane. Here the *position* of the solid is changed in the front view while its shape stays the same. The projection of points changes the shape of the image in other views.

In Figure 10-5, the axis of revolution is horizontal and perpendicular to the profile plane. When the points are rotated about the axis, the shape stays the same in the side view since the surface remains parallel to the profile plane. The image is changed in the other views.

The following example shows the method used to revolve a solid when the axis is not perpendicular to any principal plane.

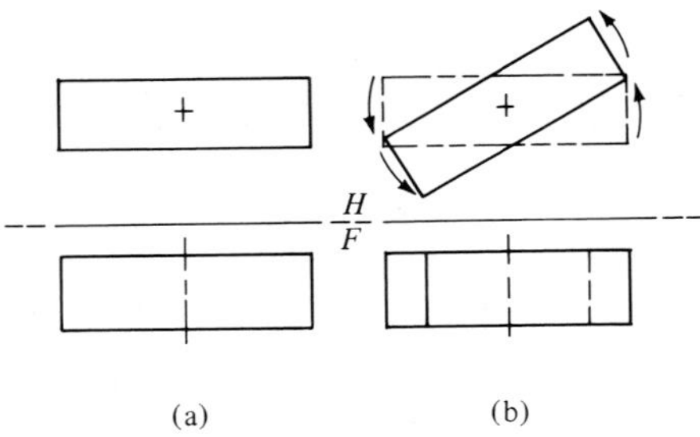

Figure 10-3
Revolution: top view.

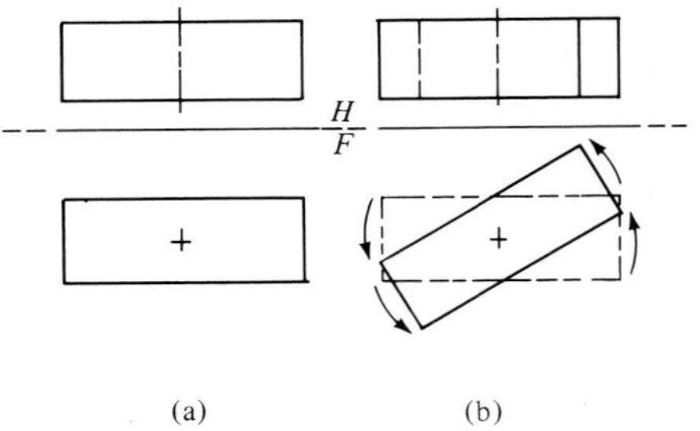

Figure 10-4
Revolution: front view.

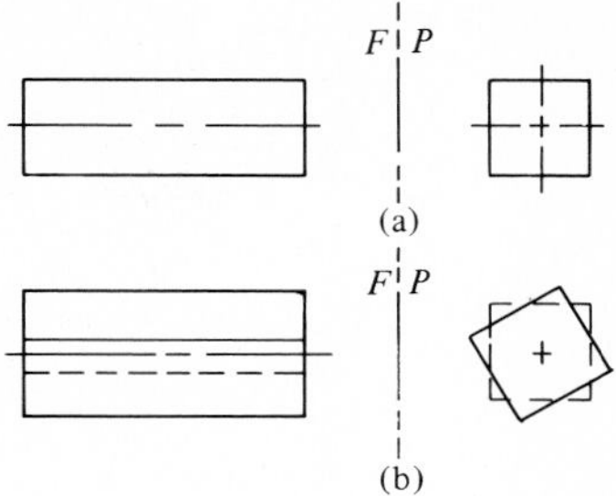

Figure 10-5
Revolution: side view.

Example 10-7

Given: Top and front views of a mutilated block.

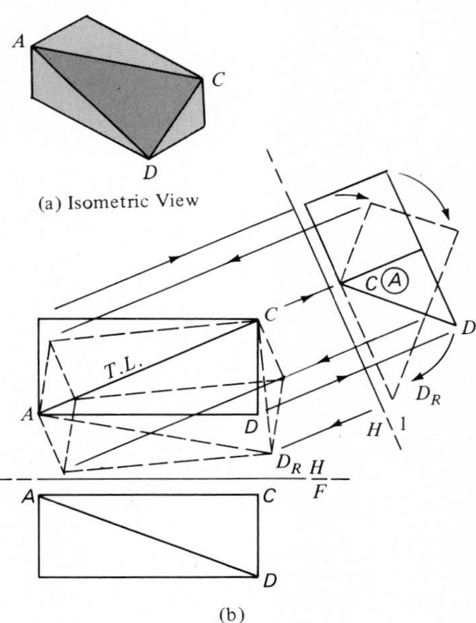

Example 10-7
Revolution of a solid.

Task: To finish the surface, revolve the block so that surface *ACD* becomes parallel to the horizontal plane. Show the revolved object in the top view. Solution:

1. Line *AC* is true length in the top view and will show as a point in the first auxiliary. Plane *ACD* will show as an edge view.
2. Use line *AC* as an axis and revolve point *D* so that it, along with points *A* and *C*, lies in the horizontal plane.

3. Revolve all other points about *AC* and through the same displacement as point *D*.
4. Project all points into the top view and draw the entire object in that view. ◀

10-6 ANGLE BETWEEN OBLIQUE LINE AND OBLIQUE PLANE BY REVOLUTION

To find the angle between an oblique line and an oblique plane it is best to use a combination of the auxiliary method and revolution method. The plane is placed on edge by the use of an auxiliary and the line is revolved about an axis which is perpendicular to the plane.

Example 10-8

Given: Top and front views of oblique plane *ABCD* and oblique line *MN.*

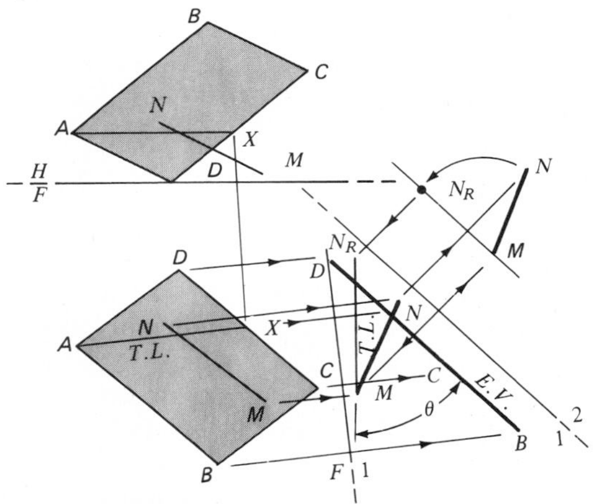

Example 10-8
Angle between line and plane by revolution.

Task: Determine the angle θ between the line *MN* and plane *ABCD.*
Solution:

1. Construct a frontal line *AX* in the top view of plane *ABCD* and by the usual procedures show line *MN* and the edge view of plane *ABCD* in the first auxiliary view.
2. Draw folding line 1/2 parallel to the edge view of plane *ABCD* and project line *MN* into the second auxiliary. Assume an axis through

point M so that the axis appears as a point in the second auxiliary. Rotate point N about the axis until it lies in a plane parallel to folding line 1/2 and shows true length in the first auxiliary.

3. Project the new line MN_R into the first auxiliary where the required angle can be measured. A vector method for finding the angle between a line and any plane was given in Section 9-7. ◀

10-7 DIHEDRAL ANGLE BY REVOLUTION

As stated in Section 9-6, a dihedral angle can be measured where the two given planes are seen on edge in the same view and their line of intersection appears as a point. When the angle is found by revolution, the cutting plane is passed so that it is perpendicular to the line of intersection of the given planes. The dihedral angle is projected onto the cutting plane which is revolved into the picture plane where the angle will be shown in true size.

Example 10-9

Given: Front and top views of two intersecting triangles ABC and ABD.
Task: Using only one auxiliary find the dihedral angle between the two triangles.
Solution:

1. Using the regular projection, show the first auxiliary view of the given triangles where the line of intersection AB appears in true length.
2. In the first auxiliary, pass a cutting plane CP so that it will appear on edge and will be perpendicular to both the line of intersection AB and the first auxiliary plane. It can be drawn anywhere along the intersection line AB. Choose a wide region of each plane to separate the intersection points. Plane ABD is cut along line ZX and plane ABC is cut along ZC. These two intersecting lines form a plane whose included angle is the dihedral angle between the two given planes.
3. Projection into the top view shows the plane MN cutting line AB at Z and line AD at X.
4. Pass an axis through C so that it appears as a point in the first auxiliary. Revolve points X and Z about this axis until C, X_R, and Z_R are in a line parallel to the folding line $H/1$. Then project the revolved points into the top view to show the dihedral angle in true size.
5. In the top view, the method of revolution requires that one project points X and Z parallel to the folding line $H/1$ (perpendicular to line axis through C) until they intersect lines of sight from X_R and Z_R in the auxiliary view. The resulting points C, Z_R, and X_R give the true dihedral angle as shown. ◀

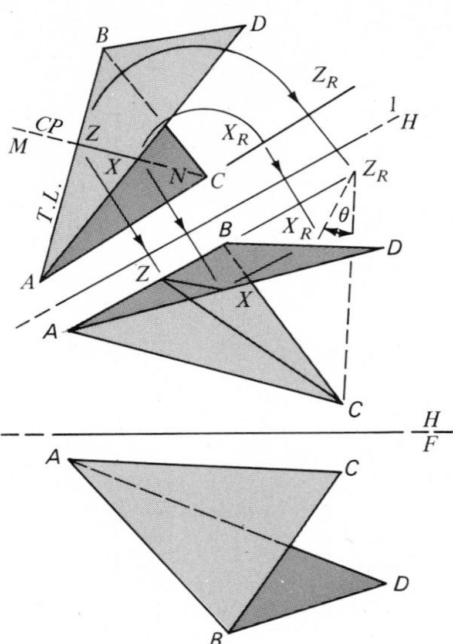

Example 10-9
Dihedral angle by revolution.

10-8 LINE MAKING SPECIFIED ANGLES WITH TWO PERPENDICULAR PLANES

A conical surface which has its axis perpendicular to its base will be the locus of all lateral elements which pass through its vertex and have the same base angle. If a line *PZ* has a specified angle with each of two perpendicular planes, that line will be the intersection of two cones each made up of lateral elements having the required base angles. (See Figure 10-6.)

The following example illustrates the procedure used in constructing a line at a specified angle with each of two perpendicular planes.

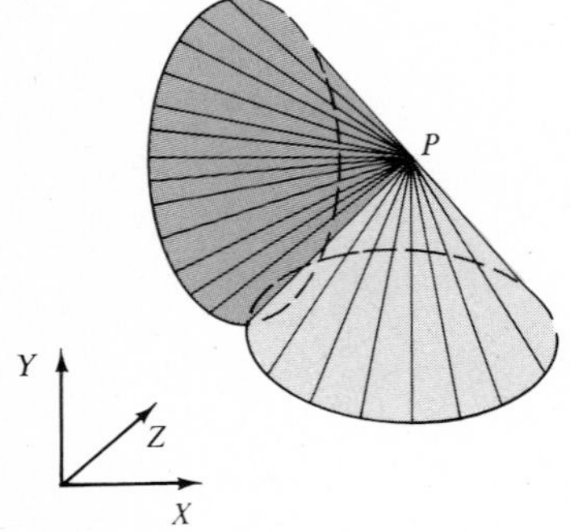

Figure 10-6
Intersection of two cones.

Example 10-10

Given: Front and top views of point *P*.
Task: From point *P* establish a line which has an angle of 50 deg with a horizontal plane and an angle of 25 deg with a frontal plane.
Solution:

1. In the front view, establish line *PM* of length *X* so that it is an element of a cone having an axis perpendicular to the horizontal plane and a

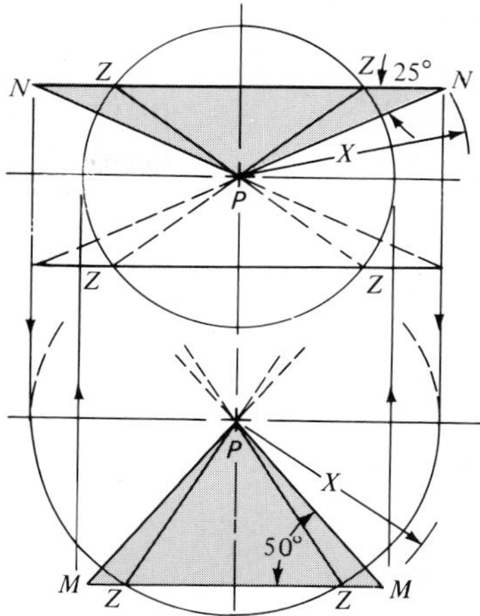

Example 10-10
Line making specified angle with two planes.

base angle of 50 deg. This cone is the locus of all lines through point *P* having an angle of 50 deg with a horizontal plane.

2. Draw the top view of the cone showing the axis as a point at *P*.
3. In the top view, establish line *PN* of length *X* so that it is the element of a cone having an axis perpendicular to a frontal plane and a base angle of 25 deg. This cone is the locus of all lines through point *P* having an angle of 25 deg with a frontal plane.
4. Draw the frontal view of the cone showing the axis as point *P*.
5. Since the elements are the same length, the bases of the cones will intersect at points labeled *Z*. Each line *PZ* is common to both cones and fulfills the requirements of the problem.
6. By using reverse cones, a total of eight solutions could be constructed.

Note: Visibility was ignored in solution 10-10 so that the lines represented by *PZ* could be shown clearly. ◀

When the specifications of a problem as stated above consider two perpendicular planes, the sum of the given angles must be 90 deg or less. If the base angles of the cones total more than 90 deg there would be no intersection; hence no solution. (See Figure 10-7.) An analytical approach to this problem makes use of direction cosines.

Revolution has another important use in geometric modeling. By revolving a line or curve around an axis in space, three-dimensional surfaces can be defined analytically. This leads into the topic of solid modeling, which is an important tool for computer-aided design.

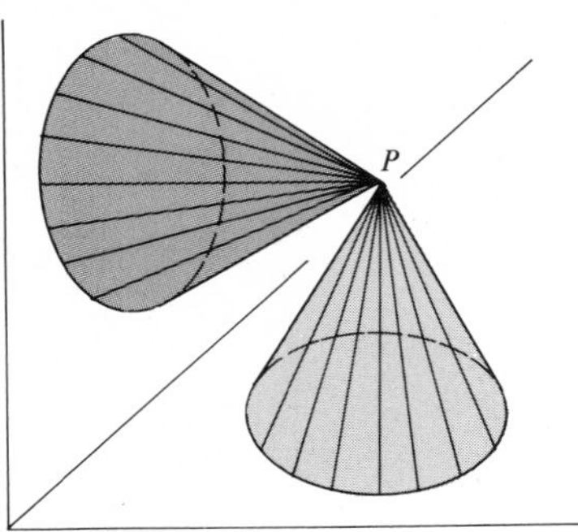

Figure 10-7
Cones with base angles totaling more than 90° (no intersection).

10-9 MODELING SURFACES OF REVOLUTION

Surfaces of revolution are often found in practical design applications because of their structural rigidity and ease of manufacture. Computer-aided design systems for solid modeling usually provide "primitive" solids of revolution which can be combined with other solids to create compound structures. Cones, cylinders, spheres, ellipsoids, and paraboloids are common types of surfaces of revolution.

A. Surfaces with Line Profile

Define a line in the x-y plane in the usual manner.

$$\begin{aligned} x(t) &= x_1 + t(x_2 - x_1) = x_1 + R_x t \\ y(t) &= y_1 + t(y_2 - y_1) = y_1 + R_y t \end{aligned} \qquad 0 \le t \le 1 \qquad (10\text{-}1)$$

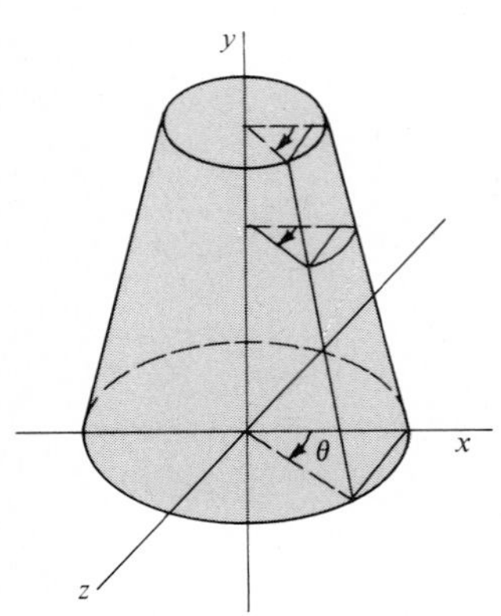

Figure 10-8
Frustum of a cone.

A surface of revolution can be created by rotating the line about the y-axis in a clockwise manner for $0 \le \theta \le 360$ deg. Figure 10-8 shows the frustum of a cone created by a two-dimensional line segment in the x-y plane.

Points on the surface can be expressed as a bivariate, parametric equation given by

$$\mathbf{P}(t, \theta) = [x \;\; y \;\; z] = [x(t)\cos\theta \;\; y(t) \;\; x(t)\sin\theta] \qquad (10\text{-}2)$$

Combining Equations (10-1) and (10-2) gives

$$\mathbf{P}(t, \theta) = (x_1 + R_x t)\cos\theta\mathbf{i} + (y_1 + R_y t)\mathbf{j} + (x_1 + R_x t)\sin\theta\mathbf{k} \qquad (10\text{-}3)$$

Equation (10-3) defines points on the surface of the frustum of a cone, or cylinder, whose axis is the y-axis. Figure 10-9 shows such a cone with $\mathbf{A} = [2\ 1]$ and $\mathbf{B} = [0.5\ 8]$ created by program BODYC (Appendix 1), which calls subroutine AROUND. The x,y coordinates of the N points on the line are passed to AROUND in A( ), B( ), along with the number of horizontal divisions (HORIZ) and the 3×3 projection matrix [R]. The surface equation is used in AROUND to generate the points on the surface. These points are then transformed to create the orthographic projection, and the points are plotted in the x,y plane.

A series of straight lines in the x,y plane can be used as a profile to create a variety of surfaces of revolution. Figures 10-10 and 10-11 show two examples.

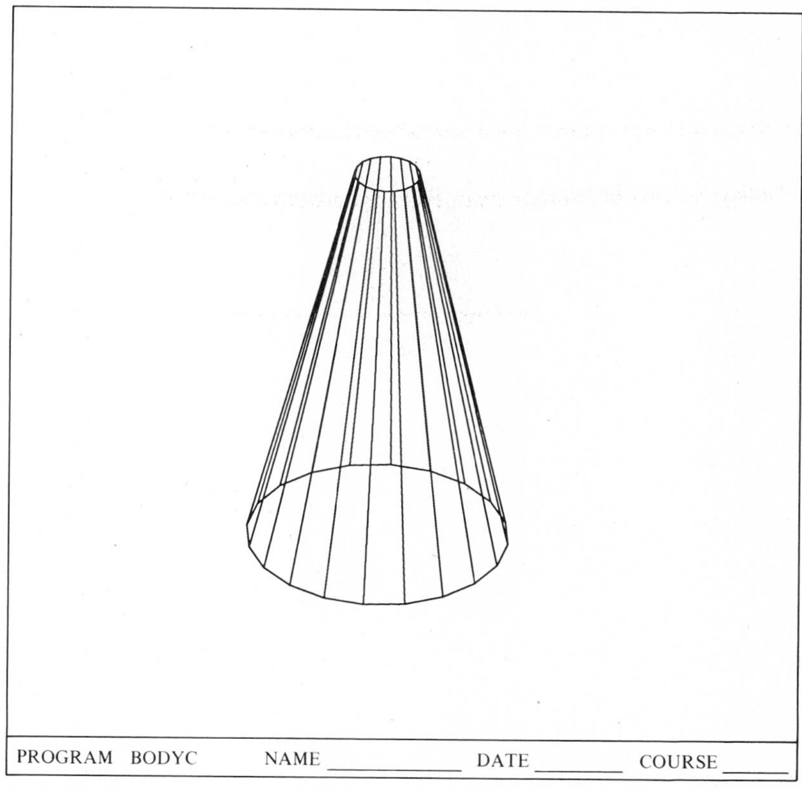

Figure 10-9
Conical surface of revolution.

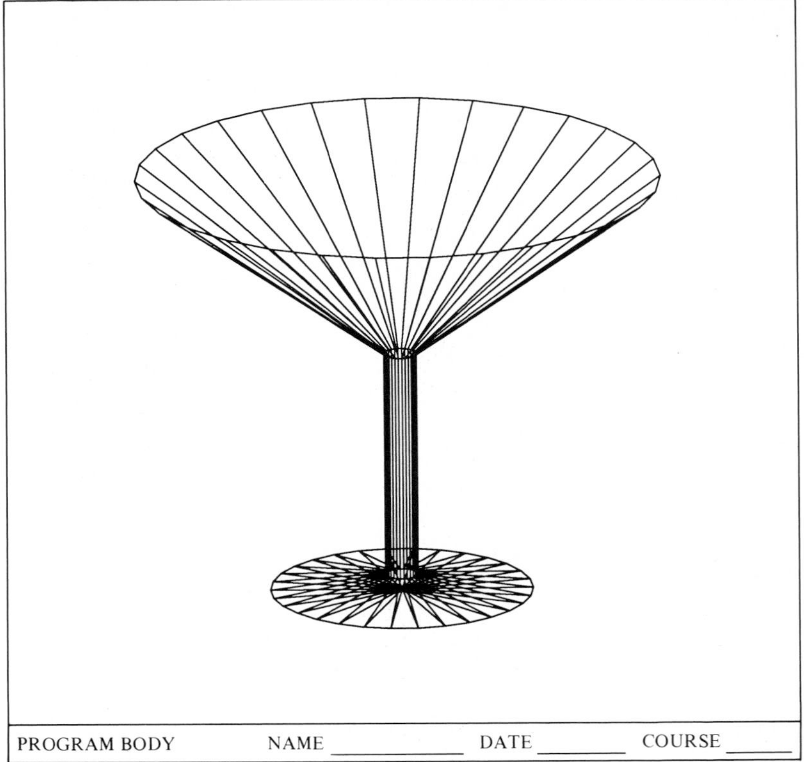

Figure 10-10
Surface of revolution.

B. Surfaces with Curve Profile

Curves can be represented by parametric equations. The subjects of computational geometry and advanced geometric modeling contain many types of curves which have been found useful for shape definition. The list of references gives sources of such work. Simple conic curves can be used to define quadric surfaces of revolution. For example, a parabola can be represented in parametric form by

$$x = 2a\phi \qquad (10\text{-}4)$$
$$y = a\phi^2$$

Figure 10-12 shows a paraboloid of revolution for $a = 1.0$ and $0 \leq \phi \leq \pi$. Only a few modifications to BODYC are needed to create this figure.

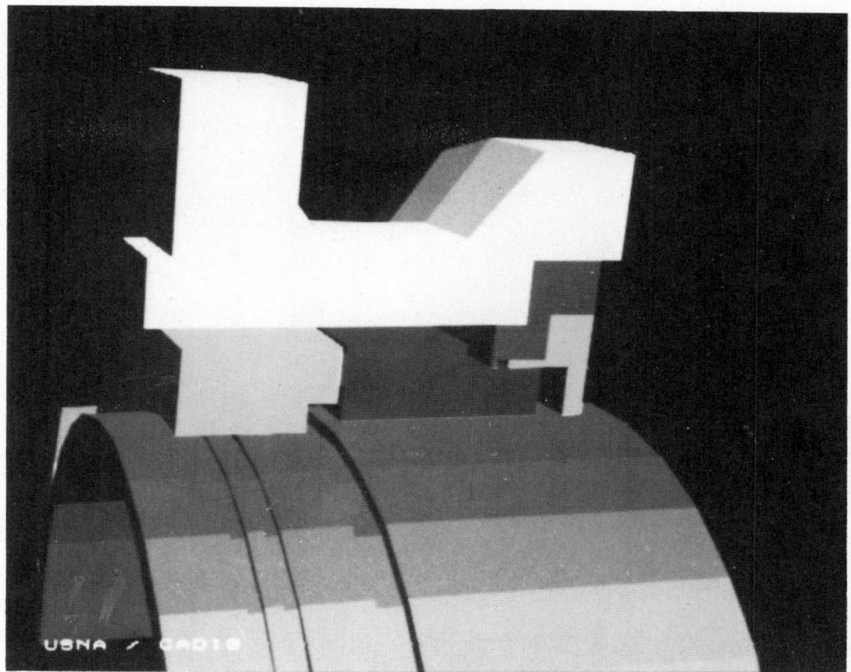

Figure 10-11
Shaft sealing design. Courtesy of CADIG Group. U.S. Naval Academy. Database prepared by Charles Amos.

The surface equation for this paraboloid is

$$\mathbf{P}(\phi,\theta) = 2a\phi \cos\theta\mathbf{i} + a\phi^2\mathbf{j} + 2a\phi \sin\theta\mathbf{k};\quad 0 \leq \phi \leq \pi;\quad 0 \leq \theta \leq 2\pi \qquad (10\text{-}5)$$

A sphere can be created by revolving a semicircle about the y-axis. In spherical coordinates, ϕ is normally measured from the vertical as shown in Figure 10-13. The parametric equations for the circle are then

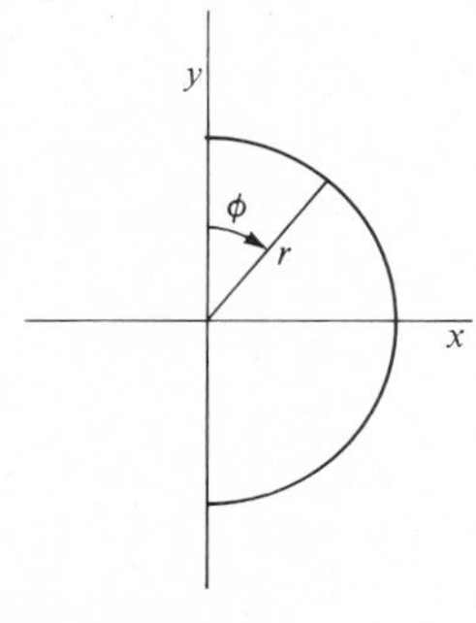

Figure 10-13
Semicircle in x,y plane.

$$\begin{aligned} x &= r\sin\phi \\ y &= r\cos\phi \end{aligned} \quad 0 \leq \phi \leq \pi \qquad (10\text{-}6)$$

The surface equation for the sphere $0 \leq \phi \leq 2\,\pi$ is

$$\mathbf{P}(\phi,\theta) = r\sin\phi\cos\theta\mathbf{i} + r\cos\phi\mathbf{j} + r\sin\phi\sin\theta\mathbf{k} \qquad (10\text{-}7)$$

Figure 10-14 shows a sphere defined in this manner.

An ellipse can be defined by using the following parametric curves to define the profile in the x-y plane.

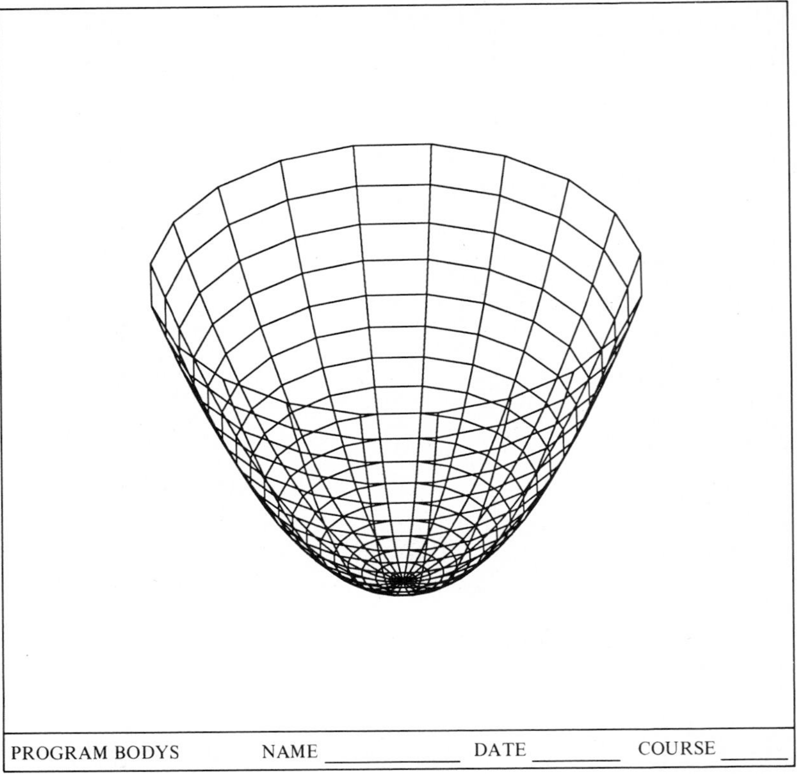

Figure 10-12
Paraboloid of revolution.

$$\begin{aligned} x &= a \sin \phi \\ y &= b \cos \phi \end{aligned} \qquad 0 \leq \phi \leq \pi \tag{10-8}$$

Here a is the semimajor axis and b is the semiminor axis.

10-10 RECAPITULATION

The method of revolution was used in Chapter 2 to align the line of sight OE with the z-axis to produce an orthographic projection as seen from any point in space. Recall that matrix multiplication was used to create a rotation about the y- and x-axes as shown in Figure 2-19. The theory to support these rotations was given in Appendix 2. The computer program ORTHOEYE implements these transformations and it was used often to create a computer graphics display.

The representation of points, lines, and planes have now been dis-

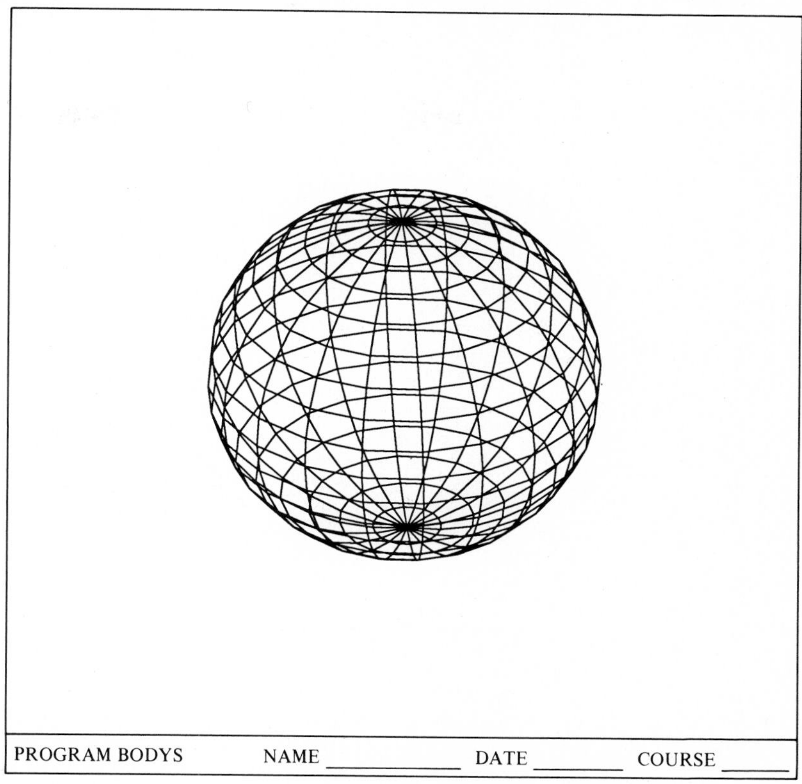

Figure 10-14
Sphere of revolution.

cussed from both a graphical and mathematical point of view. Geometric solutions have been obtained using both manual and computational methods. The reader should now be prepared to proceed to advanced topics of surface intersections and solid modeling.

PROBLEMS: CHAPTER 10

10-1 *Given:* Front and top views of line *AB* and point *X*. *X* is a point on the circumference of a circle whose axis is *AB*.
Task: View the axis from *B* toward *A* and rotate point *X* counterclockwise 80 deg on the circumference of the circle. Show the new location of point *X* in both the top and front view.

10-2 *Given:* Front and top views of a pole *P* supported by three guy wires.
Task: By revolution, determine the length of each guy wire and the angle it makes with the ground.

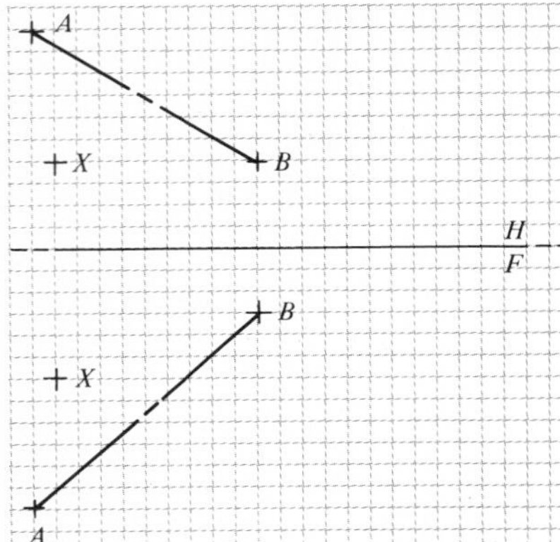

Figure 10-1P

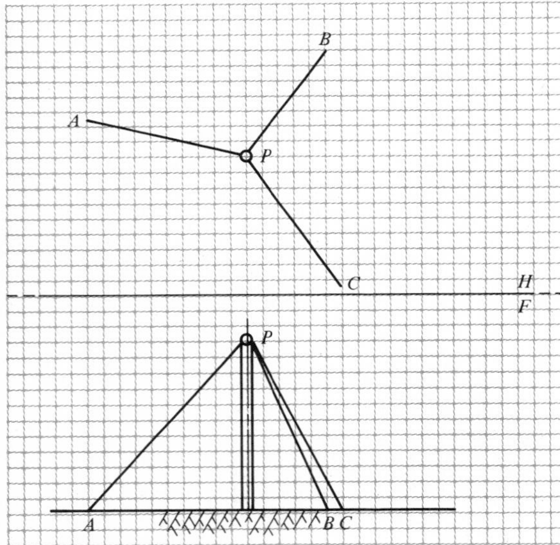

Figure 10-2P

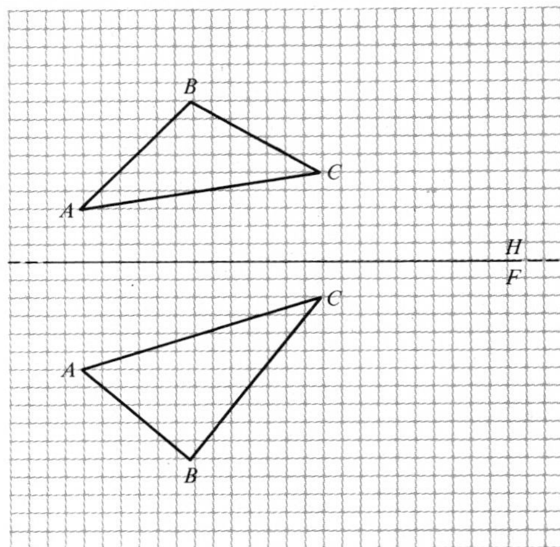

Figure 10-3P

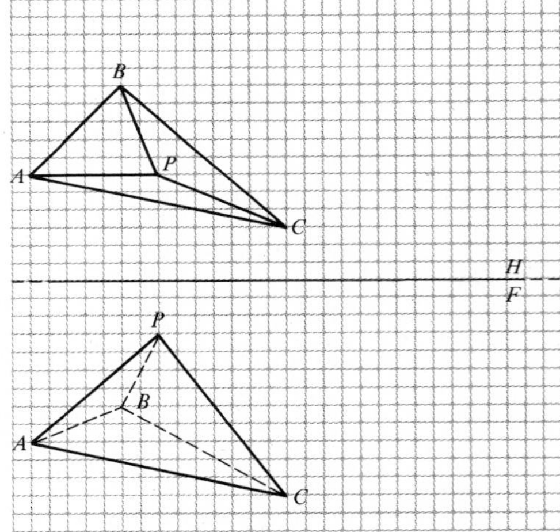

Figure 10-4P

10-3 *Given:* Front and top views of triangle *ABC*. *Task:* Without the use of an auxiliary view, show the edge view of the triangle.

10-4 *Given:* Front and top views of pyramid *ABCP*. *Task:* Using one auxiliary, show the normal view of plane *PBC*.

10-5 *Given:* Front and top views of triangle *ABC*. *Task:* Using revolution only, show the normal view of the given triangle.

10-6 *Given:* Front, top, and right-side views of a mutilated block. *Task:* In order that it may be machined, show the block in a rotated position with face *A* parallel to the horizontal plane.

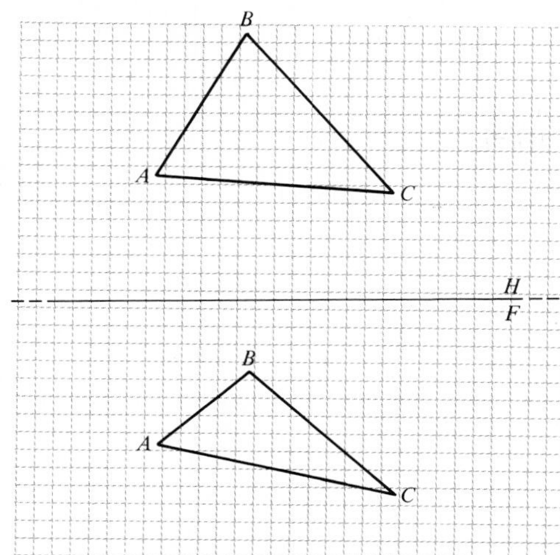

Figure 10-5P

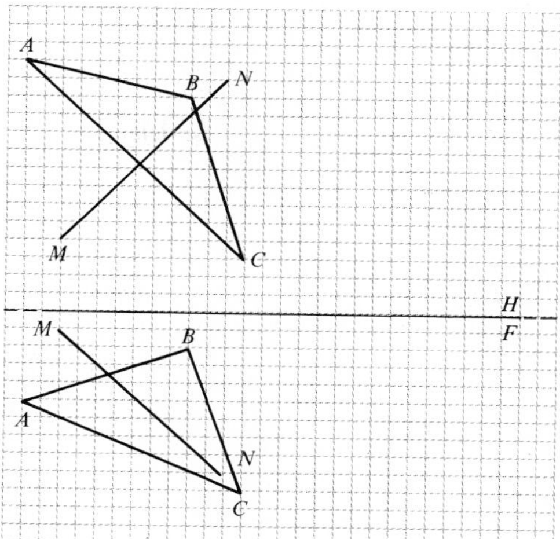

Figure 10-7P

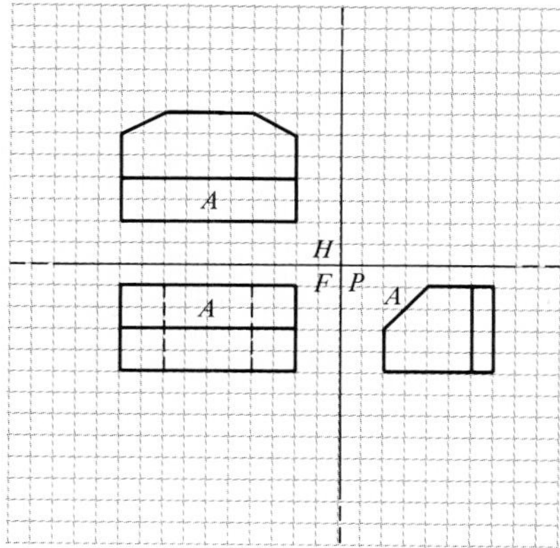

Figure 10-6P

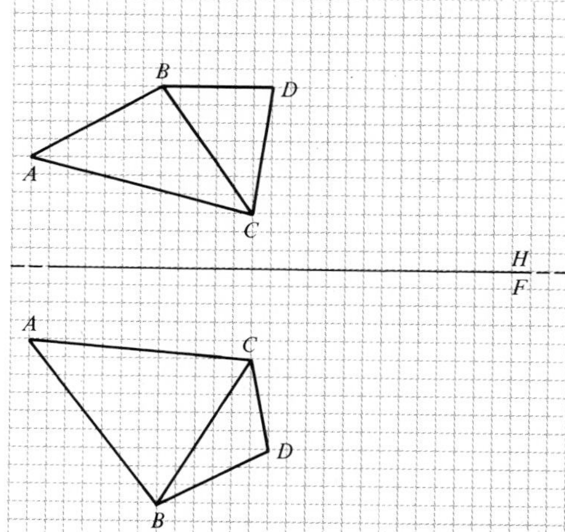

Figure 10-8P

10-7 *Given:* Front and top views of triangle *ABC* and line *MN*.
Task: Using revolution, determine the angle between the line and the plane. (Use one auxiliary only).

10-8 *Given:* Front and top views of two intersecting triangles *ABC* and *BCD*.
Task: Using only one auxiliary, find by revolution, the angle between the two planes.

10-9 *Given:* Front and top views of point *P*.
Task: Through point *P* construct a line which has an angle of 47 deg with a horizontal plane and an angle of 27 deg with a frontal plane.

10-10 *Given:* Front and top views of a sheet metal hopper.
Task: Use revolution to determine the dihedral angle between surface *A* and surface *B*.

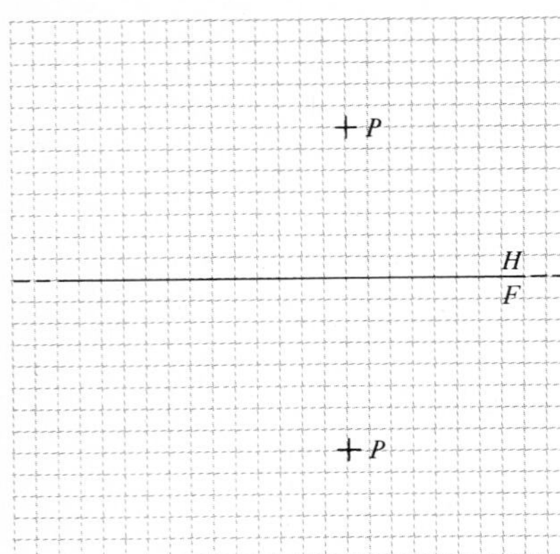

Figure 10-9P

10-11[c] *Given:* Figure 10-9.
Task: a. Generate and display a cone frustum whose axis lies along the x-axis.
b. Generate and display a cone frustum whose axis lies along the z-axis.

10-12[c] *Given:* Program BODYC and subroutine AROUND in Appendix 1, and Figures 10-10 and 10-11.
Task: Design and display a surface of revolution.

10-13[c] *Given:* The parametric equations for a hyperbola are given by $x = a \sec \phi$; $y = a \tan \phi$. Let $a = 1.0$ and $0 \leq \phi \leq 80$ deg.
Task: Create and display a hyperboloid of revolution by rotating a 2 two-dimensional hyperbola about the y-axis.

[c]Superscript *c* indicates that a computer solution is required.

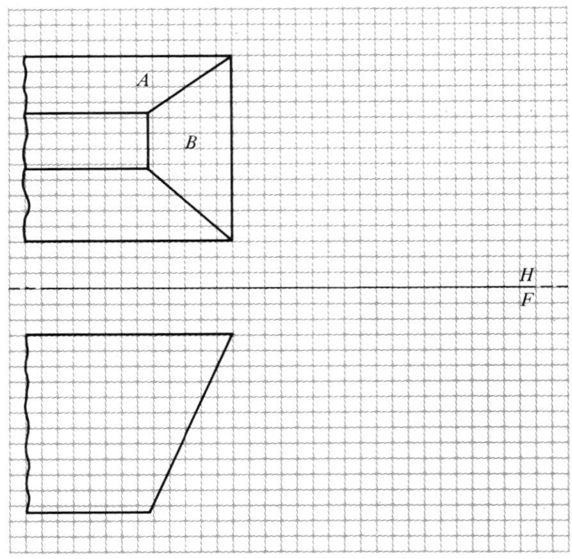

Figure 10-10P

10-14[c] *Given:* Figures 10-12 and 10-13.
Task: a. Write a main program which calls subroutines AXISROT, AROUND, EYETOZ, and TITLE (Appendix 1) and produces an output such as Figure 10-12.
b. Do the same to produce Figure 10-13.

10-15[c] *Given:* Program BODYS.
Task: a. Generate a torus.
b. Write the vector equation for the surface of the torus.

11

Intersections—Line and Solid

The last five chapters in this text are referred to as advanced geometric analysis because the descriptive geometry techniques are complex and the geometric modeling techniques require a sophisticated understanding of mathematics. This is due to the fact that surfaces are considered, rather than lines and single planes. When solids are represented by a series of planes, then the descriptive geometry problem of intersecting lines and solids is a direct extension of the material presented in the previous chapters. Auxiliary and two-view methods are used. Examples of geometric modeling and computer graphics are given to illustrate advanced techniques of vector analysis and computational geometry.

11-1 INTERSECTION—LINE AND PRISM

The study of piercing points in Section 8-1 gave procedures for finding the point where a line pierces a plane. Solutions to the following problems show that the same principles are used here since the prisms are made up of intersecting planes. This section, however, will treat the prisms as solid objects, and will not show the lines inside the geometric figures. The following examples illustrate the techniques involved.

A. Auxiliary Method

Example 11-1

Given: Incomplete front and top views of prism *ABCDEF* and endpoints of intersecting line *MN*.

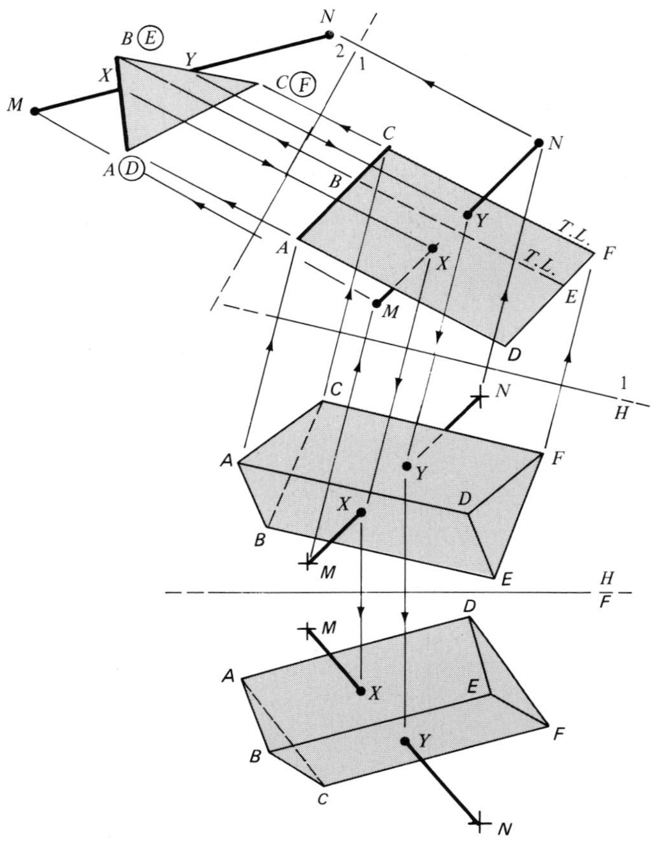

Example 11-1
Intersection of line and prism: auxiliary method.

Task: Use the auxiliary method to find the piercing points for the entrance and exit of line *MN* as it passes through prism *ABCDEF*. Show visibility of the line *MN* in all views.
Solution:

1. Draw folding line *H*/1 parallel to prism edges *AD* shown in the top view. Project line *MN* and the prism into the first auxiliary where the prism edges are shown in true length.
2. Construct folding line 1/2 perpendicular to the true length lines. Project line *MN* and the prism into the second auxiliary where the edges show as points and the lateral faces of the prism show on edge.
3. Establish the piercing points (*X* and *Y*) of line *MN* on the lateral faces of the prism.

4. Project the piercing points into all other views.
5. Show correct visibility in all views. ◀

B. Two-View Method

Example 11-2

Given: Incomplete front and top views of prism *ABCDEF* and intersecting line *MN*.
Task: Use only the given views to find the piercing points for the entrance and exit of line *MN* as it passes through prism *ABCDEF*. Show piercing points and correct visibility in all views.
Solution:

1. In the front view, work only with line MN and plane *ABED*. Pass a cutting plane through line *MN* and perpendicular to the frontal plane so that it is seen on edge in the front view. The cutting plane, in the top view, will cut line *AD* at 1 and line *BE* at 2. Line *MN* intersects trace 1-2 at point *X* which is the piercing point of line *MN* and plane *ABED*.

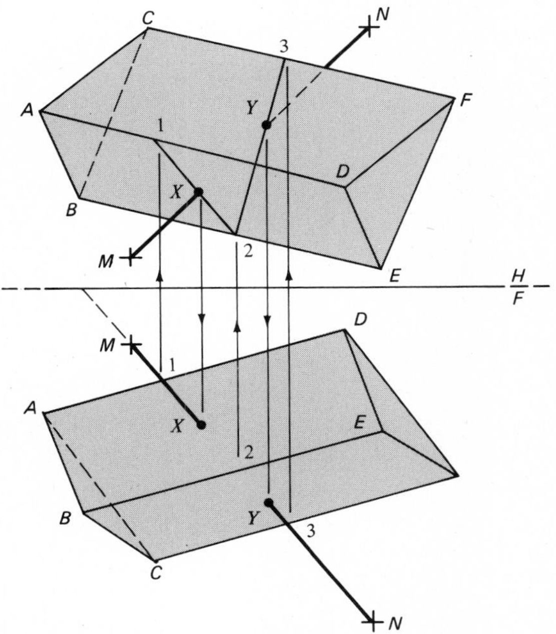

Example 11-2
Intersection of line and prism: two-view method.

2. Work with plane *BCFE* and repeat the operation above. The cutting plane will cut line *BE* at 2 and line *CF* at 3. Trace 2-3 is intersected by line *MN* at piercing point *Y*.
3. In the top view, the two points *M* and *X* are visible. Therefore, the line will be visible from point *M* to point *X*. The line from *X* to *Y* is inside the prism and is not shown. Point *Y* is invisible; therefore, line *MN* will be invisible from point *Y* until it emerges from under the prism. It is then visible to its end at point *N*. (See Section 1-11.)
4. Project points *X* and *Y* into the front view. Check the visibility and draw the required line. Since both points are visible, the entire line is visible except for the portion inside the prism. ◄

11-2 INTERSECTION—LINE AND CYLINDER

A cylinder is generated by moving a generatrix from one position to another to form an infinite number of planes along a curved directrix with each plane being parallel to the original position of the line. The cylinder, then, is a prism with an infinite number of sides. One can *approximate* a cylinder using a finite number of sides. This being true, the piercing points of a line and a cylinder can be found by a procedure similar to that used in Section 11-1-A.

A. Auxiliary Method

Example 11-3

Given: Incomplete front and top views of right cylinder and endpoints of intersecting line MN.
Task: Show the piercing points where line *MN* intersects the given right cylinder.
Solution:

1. Draw folding line *H*/1 parallel to the centerline of the given cylinder. Project the cylinder and line *MN* into the first auxiliary where the centerline and lateral elements appear in true length.
2. Draw folding line 1/2 perpendicular to the true length lines of the first auxiliary and project the cylinder and line *MN* into the second auxiliary. The right section of the cylinder will be shown along with line *MN*. The piercing points *X* and *Z* are seen to lie on lateral elements 1 and 2 respectively.
3. Project lateral elements 1 and 2 into the first auxiliary and determine their intersections with the line *MN*. *X* and *Z* are the required piercing points and should be projected into all views.
4. Show correct visibility in all views. ◄

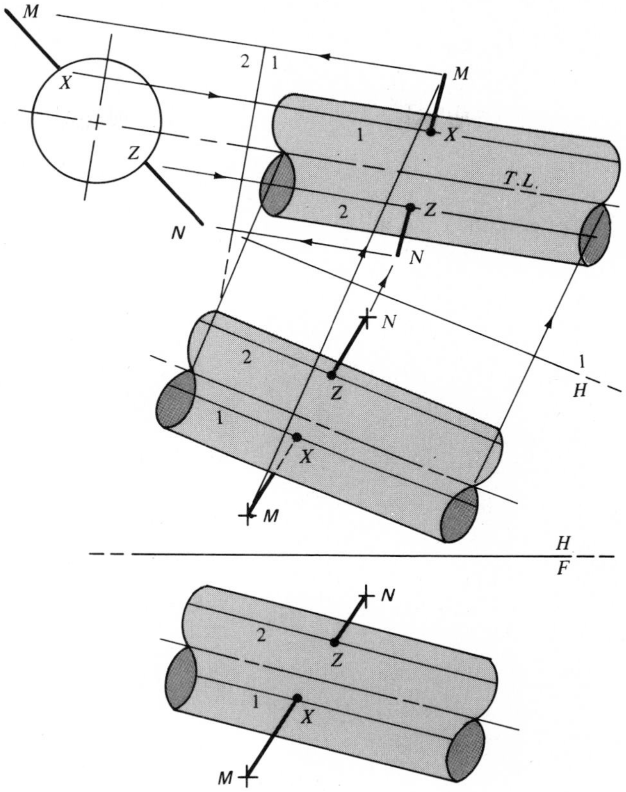

Example 11-3
Intersection of line and cylinder: auxiliary method.

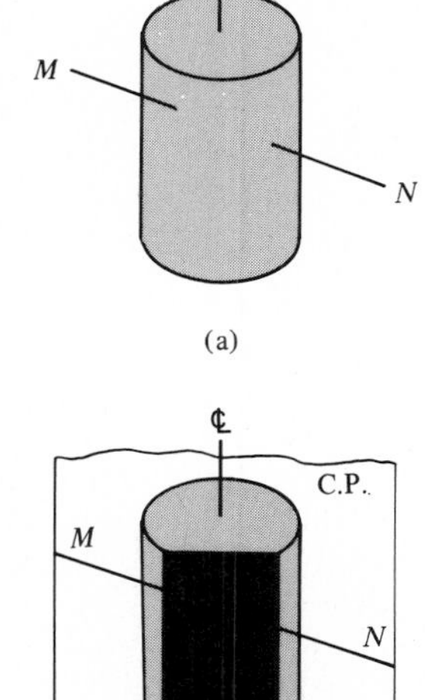

Figure 11-1
Intersection of line and cylinder.

B. Two-View (Cutting-Plane) Method

Figure 11-1 illustrates the theory used to find the intersections of a line and a cylinder. The cutting plane which contains line *MN* and is parallel to the axis of the cylinder will cut straight line elements on the surface of the cylinder from points *X* and *Y* which are located on the base. These straight lines (drawn parallel to the axis of the cylinder from points *X* and *Y* to intersect line *MN*) will show the piercing points of the line with the cylinder. Again, the cylinder is treated as a solid object.

Example 11-4

Given: Front and top views of line *MN* and a cylinder having axis *AB*.
Task: Find the piercing points and show complete visibility of the intersection of line *MN* and the given cylinder.

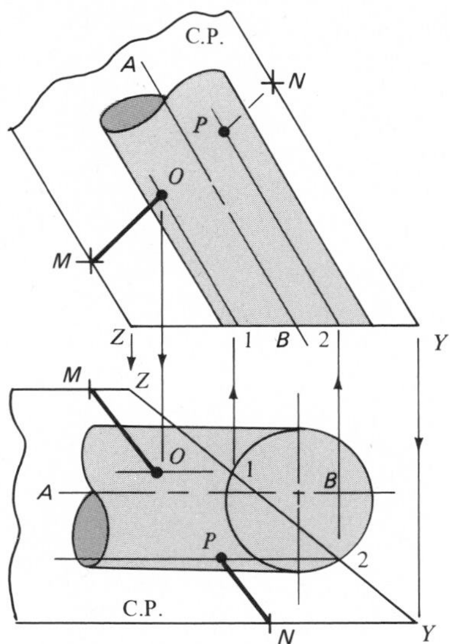

Example 11-4
Intersection of line and cylinder: two-view method.

Solution:

1. Through points *M* and *N* in the top view, draw a cutting plane parallel to the axis of the cylinder by drawing line *NY* and line *MZ* parallel to axis *AB*. (If one line of the cutting plane is parallel to the axis, then the plane is parallel to the axis. See Section 6-3.)
2. In the top view, the cutting plane edges *NY* and *MZ* intersect the base plane of the cylinder at *Y* and at *Z*.
3. Project all points and lines to the front view.
4. Draw trace of cutting plane from point *Y* to point *Z* in the front view. The trace intersects the cylinder at points 1 and 2. Project the points to the cylinder in the top view.
5. From points 1 and 2 in both views, construct straight line elements parallel to axis *AB* until they intersect line *MN* at piercing points *O* and *P*.
6. Line *MN* intersects the cylinder at piercing points *O* and *P*. Show its visibility in all views. ◀

11-3 INTERSECTION—LINE AND PYRAMID

The piercing points found here are basically the same as those shown in Section 11-1-B where the intersections were found when the sides of a prism were not shown on edge. Since the geometric shape of a pyramid makes the auxiliary view method both lengthy and impractical, this section will deal only with the more frequently used two-view method.

Example 11-5

Given: Incomplete front and top views of pyramid *ABCD* and its intersecting line *MN*.
Task: Show the points of intersection and complete the visibility of line *MN* as it intersects pyramid *ABCD*.

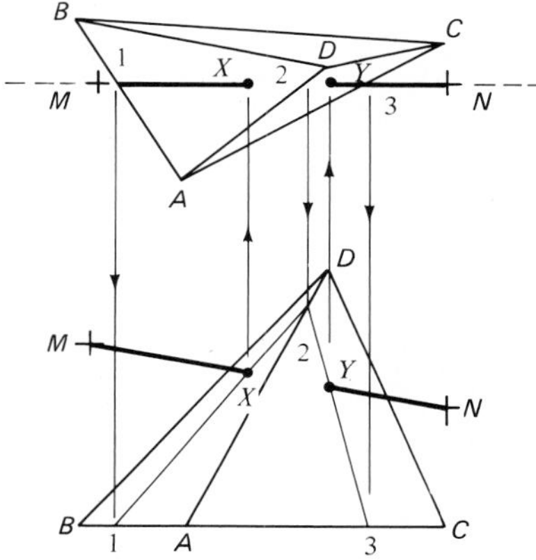

Example 11-5
Intersection of line and pyramid: two-view method.

Solution:

1. Pass a cutting plane perpendicular to the horizontal plane so that it contains line *MN* and shows as an edge in the top view.
2. In the top view, the cutting plane cuts line *AB* at 1, line *AD* at 2, and line *AC* at 3. Project these three points to the front view and draw traces 1-2 and 2-3.
3. In the front view, trace 1-2 intersects line *MN* at piercing point *X* and trace 2-3 intersects line *MN* at piercing point *Y*.
4. Show visibility of line *MN* as it intersects pyramid *ABCD* at piercing points *X* and *Y*. ◀

11-4 INTERSECTION—LINE AND CONE

As in previous two-view methods of solution given in this chapter, the procedure shown in this section involves the use of the cutting plane. Here it is passed through the *vertex* of the cone and includes the line *MN* which intersects the cone. (See Figure 11-2.)

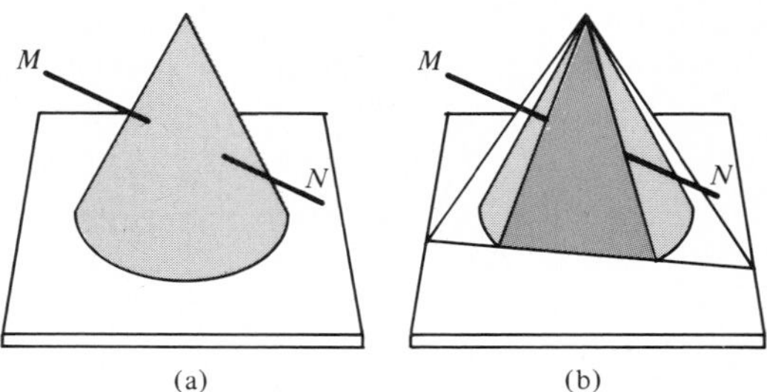

Figure 11-2
Intersection of line and cone.

The principles used in the solution to the following example have been discussed in prior material. It is suggested that the reader study the given solution and develop the required step-by-step procedure.

Example 11-6

Given: Incomplete front and top views of a cone intersected by line *RS*.
Task: Show the piercing points where line *RS* intersects the given cone. Show complete visibility.

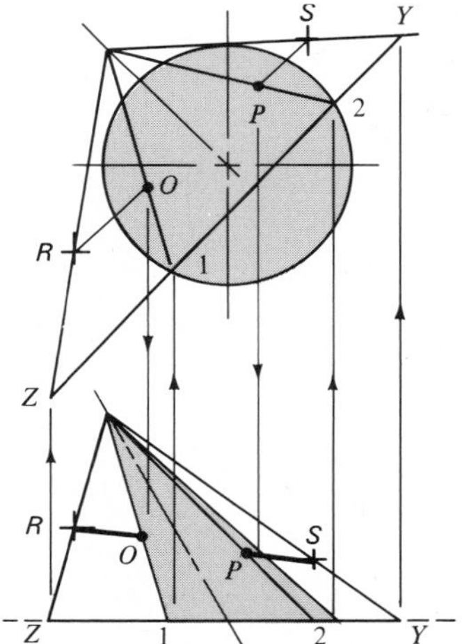

Example 11-6
Intersection of line and cone.

Many other surfaces of general form could be considered in this chapter. These might include such surfaces as spheres, paraboloids, hyperboloids, and warped cones, for example. The reader can find many such surfaces treated in the literature.

The method of computing the intersection between lines and planes can be extended to represent the intersection between lines and solids with flat faces. Figure 11-3 shows the intersection between three lines in

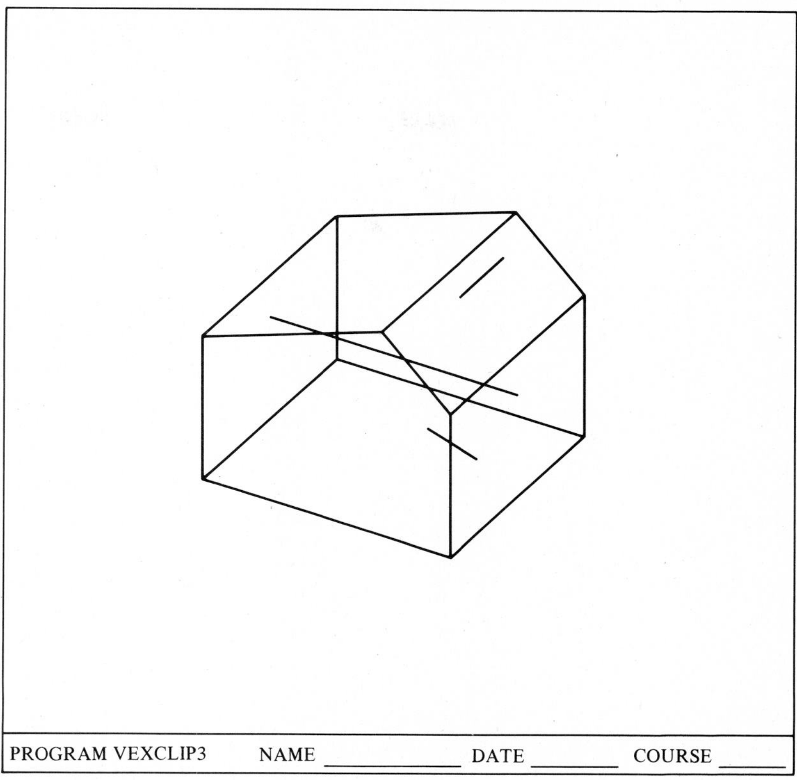

Figure 11-3
Three-dimensional clipping.

space and a polyhedral surface represented by a series of plane facets. Only the portion of the lines inside the object are shown. Figure 11-4 shows three lines intersecting a cone frustum which is approximated by a series of plane facets. This type of computation is called *clipping* in the computer graphics literature. The reader may want to try and implement this type of software. The mathematical theory is given in the following sections, using the method proposed by M. Cyrus and J. Beck, "Generalized Two and Three Dimensional Clipping," *Computers and Graphics,* Vol. 3, 1978, pp 23–28.

11-5 TWO-DIMENSIONAL CLIPPING*

The vector method for calculating the piercing point between a line and a plane used in previous chapters can be generalized by a process known

*Sections 11-5 and 11-6 are for readers who want to know the mathematical details which are the basis for clipping programs such as VEXCLIP3 and XCONE.

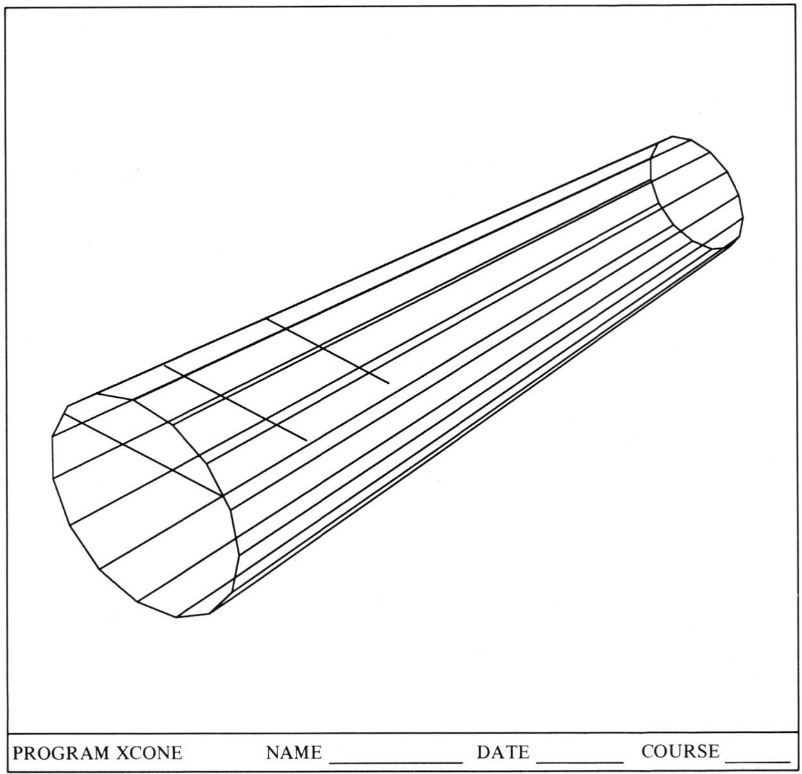

Figure 11-4
Cone clipping.

as *clipping.* Clipping is the process of discarding portions of line segments which do not fall within a defined surface or solid region. In two-dimensional clipping, a plane object is represented by a series of two-dimensional lines which form a closed contour. If the object is convex, then any line lying in the plane of the object will intersect the object in no more than two places. The same is true for a three-dimensional line intersecting a convex three-dimensional solid which is considered in the next section. The following discussion applies only to convex objects.

Consider a two-dimensional planar region whose contour is defined by n-sides of a closed polygon. Let **Q** be *any* point on a line segment between **A** and **B**, where both **A** and **B** lie in the x-y plane. The vector equation for the line is given by

$$\mathbf{Q} = \mathbf{A} + t(\mathbf{B} - \mathbf{A}); \quad 0 \leq t \leq 1.0 \tag{11-1}$$

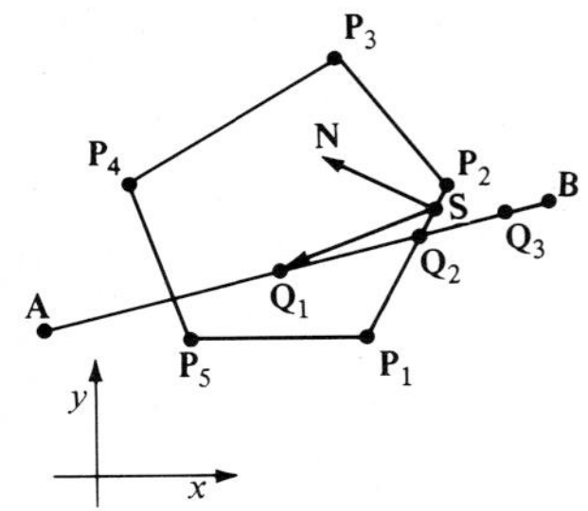

Figure 11-5
Line and polygon.

This line passes through an n-sided polygon as shown in Figure 11-5. Furthermore, let **S** be *any* point on a boundary of the polygon which intersects line **Q**, and let **N** be a vector lying in the *x-y* plane which is *inward* and *normal* to the boundary line on which **S** lies. The dot product $(\mathbf{Q} - \mathbf{S}) \cdot \mathbf{N}$ can be used to classify the position of **Q** relative to the polygon boundary as shown below.

$(\mathbf{Q}_1 - \mathbf{S}) \cdot \mathbf{N} > 0$	$(\mathbf{Q}_2 - \mathbf{S}) \cdot \mathbf{N} = 0$	$(\mathbf{Q}_3 - \mathbf{S}) \cdot \mathbf{N} < 0$
$\mathbf{Q}_1$ is inside boundary	$\mathbf{Q}_2$ is on boundary	$\mathbf{Q}_3$ is outside boundary

Using Equation (11-1) in the first condition gives

$$\{\mathbf{A} + t(\mathbf{B} - \mathbf{A}) - \mathbf{S}\} \cdot \mathbf{N} > 0; \quad 0 \leq t \leq 1.0 \tag{11-2}$$

for $\mathbf{Q}_1$ internal to the polygon. Rearranging gives

$$t(\mathbf{B} - \mathbf{A}) \cdot \mathbf{N} + (\mathbf{A} - \mathbf{S}) \cdot \mathbf{N} > 0; 0 \leq t \leq 1.0 \tag{11-3}$$

Let the directional vector of the given two-dimensional line be $\mathbf{D} = (\mathbf{B} - \mathbf{A})$ and define a weight vector $\mathbf{W}_i = (\mathbf{A} - \mathbf{S}_i)$ for each side of the polygon, $i = 1,2,\ldots\ldots n$. Then, in general, for the ith boundary the condition that $\mathbf{Q}_i$ lies inside the polygon boundary is given by a generalization of Equation (11-3) expressed as

$$t_i\mathbf{D} \cdot \mathbf{N}_i + \mathbf{W}_i \cdot \mathbf{N}_i > 0; i = 1,2,\ldots\ldots,n \tag{11-4}$$

If the *equality* sign is used in Equation (11-4), then n equations for t_i can be written which define the values of the parameter t_i which give the *intersection* points of the two-dimensional line with the n polygon boundaries. Solving for these parameter values gives

$$t_i = \frac{-\mathbf{W}_i \cdot \mathbf{N}_i}{(\mathbf{D} \cdot \mathbf{N}_i)}; \quad i = 1,2,\ldots\ldots,n \tag{11-5}$$

If the points around the contour are labeled in a counterclockwise fashion as in Figure 11-5, and a right-hand coordinate system is used, the *inward* normal vector to each boundary line between $\mathbf{P}_i$ and $\mathbf{P}_{i+1}$ is given by the following vector cross product:

$$\mathbf{N}_i = \mathbf{k} \times (\mathbf{P}_{i+1} - \mathbf{P}_i)$$

$$= \det \begin{vmatrix} \mathbf{i} & \mathbf{j} & \mathbf{k} \\ 0 & 0 & 1 \\ (x_{i+1} - x_i) & (y_{i+1} - y_i) & (z_{i+1} - z_i) \end{vmatrix} \tag{11-6}$$

Expansion gives

$$\mathbf{N}_i = -(y_{i+1} - y_i)\mathbf{i} + (x_{i+1} - x_i)\mathbf{j} \tag{11-7}$$

Since the boundary points $\mathbf{S}_i$ can be chosen to be the same as the vertices $\mathbf{P}_i$, the weight vectors can be written $\mathbf{W}_i = (\mathbf{A} - \mathbf{P}_i)$. Then the set of equations given by Equation (11-5) can be written

$$t_i = \frac{-\{(A_x - P_{i,x})\mathbf{i} + (A_y - P_{i,y})\mathbf{j}\} \cdot \{-(y_{i+1} - y_i)\mathbf{i} + (x_{i+1} - x_i)\mathbf{j}\}}{\{(B_x - A_x)\mathbf{i} + (B_y - A_y)\mathbf{j}\} \cdot \{-(y_{i+1} - y_i)\mathbf{i} + (x_{i+1} - x_i)\mathbf{j}\}}; \quad i = 1,2 \ldots ., n \tag{11-8}$$

Carrying out the dot product gives

$$t_i = \frac{(A_x - P_{i,x})(y_{i+1} - y_i) - (A_y - P_{i,y})(x_{i+1} - x_i)}{-(B_x - A_x)(y_{i+1} - y_i) + (B_y - A_y)(x_{i+1} - x_i)}; \quad i = 1,2, \ldots ., n \tag{11-9}$$

Equation (11-9) gives the same result that would have been obtained if Cramer's Rule had been used to calculate the intersection between two line segments, as discussed in Section 3-14. However, this vector clipping approach can be more easily extended to three-dimensional clipping of convex solids as shown in the following section. Once t_i values are found from Equation (11-9), the actual coordinates of the intersection point for boundaries i are given by

$$\begin{aligned} x_i &= A_x + t_i(B_x - A_x) \\ y_i &= A_y + t_i(B_y - A_y) \end{aligned} \quad i = 1,2, \ldots ., n \tag{11-10}$$

After the parameter values for the piercing points are calculated, a check is made to determine if the corresponding intersections fall between the two endpoints of the clipped, three-dimensional line. When t does not fall between $0 \le t \le 1.0$, the intersection is not of interest. However, Equation (11-5) may give several values of t between $0 \le t \le 1.0$ as illustrated in Figure 11-6. For a convex region, any line can intersect the region at most in only two places. To find which two values of t give these actual intersections, the following logic is used.

a. The calculated values of t are separated into two groups, one with lower values and one with higher values. In Figure 11-3, t_1, t_2, t_3 form the lower group and t_4, t_5 form the higher values.
b. If $\mathbf{D} \cdot \mathbf{N}_i > 0$, then calculated values of t between $0 \le t \le 1.0$ will fall in the lower group. If $\mathbf{D} \cdot \mathbf{N}_i < 0$, the calculated values of t will fall in the higher group.
c. A search is made to find the largest parameter in the lower group, and the smallest parameter in the higher group. In Figure 11-6 these are the parameter values t_3 and t_4.

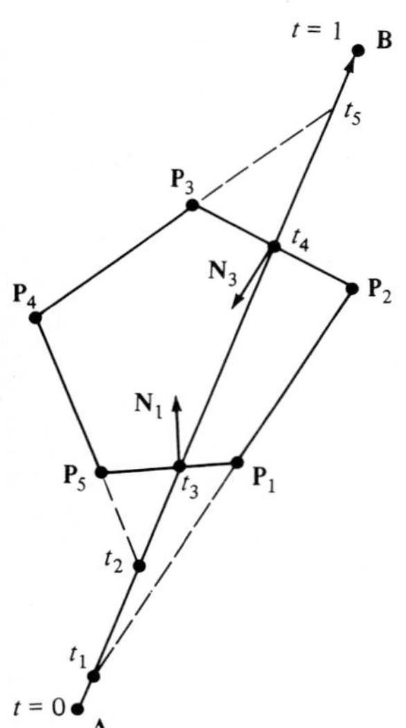

Figure 11-6
Clipping logic.

11-6 THREE-DIMENSIONAL CLIPPING

The method given in the previous section can be extended to three-dimensional clipping. In the literature, one often finds left-hand coordinate sys-

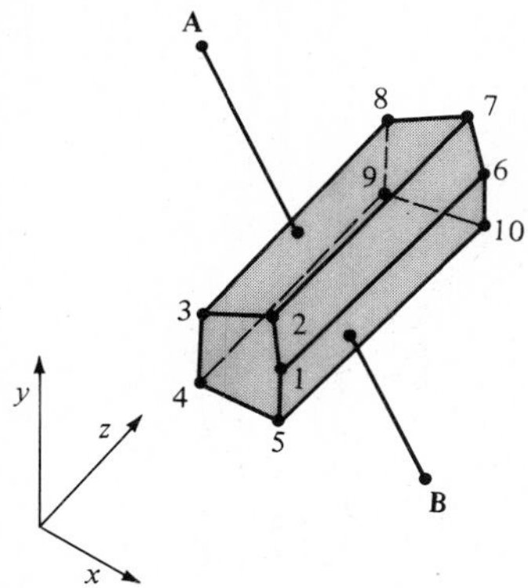

Figure 11-7
Three-dimensional clipping with vertical ends.

tems used for clipping problems since an eye position in front of the x-y plane on the negative z-axis creates a line of sight in the positive z-direction. Recall that in a left-hand coordinate, the direction of the vector $\mathbf{C} = \mathbf{A} \times \mathbf{B}$ is *away* from the observer who sees the rotation of **A** to **B** as counterclockwise. Left-hand coordinates are used in this article.

In three-dimensional clipping, the inward normal vectors on the boundaries become inward surface normals. In addition, inward surface normals for the hither (nearest) and yon (farthest) end planes are needed. If the hither and yon end planes of the solid are both parallel to the x-y plane as shown in Figure 11-7, then the two inward normals on the hither and yon planes are **k** and $-\mathbf{k}$ respectively. In Figure 11-7 the number of vertex pairs is $n = 5$. The complete set of $n + 2$ inward normal vectors are given as follows:

$$
\begin{aligned}
\mathbf{N}_1 &= (\mathbf{P}_6 - \mathbf{P}_1) \times (\mathbf{P}_2 - \mathbf{P}_1) \\
\mathbf{N}_2 &= (\mathbf{P}_7 - \mathbf{P}_2) \times (\mathbf{P}_3 - \mathbf{P}_2) \\
&\cdot \\
&\cdot \\
\mathbf{N}_5 &= (\mathbf{P}_{10} - \mathbf{P}_5) \times (\mathbf{P}_1 - \mathbf{P}_5) \\
\mathbf{N}_6 &= \mathbf{k} \\
\mathbf{N}_7 &= -\mathbf{k}
\end{aligned}
\tag{11-11}
$$

For the solid with nonparallel ends such as shown in Figure 11-8, where $n = 4$, the $n + 2$ surface normals are computed using the vector cross products indicated below, relative to a left-hand coordinate system.

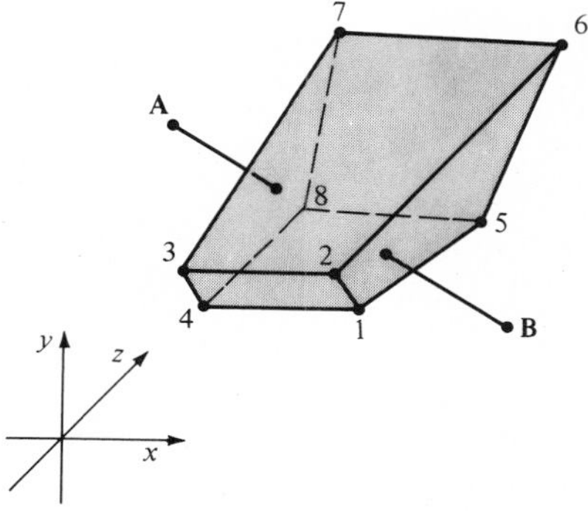

Figure 11-8
Three-dimensional clipping with non-vertical ends.

$$
\begin{aligned}
\mathbf{N}_1 &= (\mathbf{P}_5 - \mathbf{P}_1) \times (\mathbf{P}_2 - \mathbf{P}_1) \\
\mathbf{N}_2 &= (\mathbf{P}_6 - \mathbf{P}_2) \times (\mathbf{P}_3 - \mathbf{P}_2) \\
\mathbf{N}_3 &= (\mathbf{P}_7 - \mathbf{P}_3) \times (\mathbf{P}_4 - \mathbf{P}_3) \\
\mathbf{N}_4 &= (\mathbf{P}_8 - \mathbf{P}_4) \times (\mathbf{P}_1 - \mathbf{P}_4) \\
\mathbf{N}_5 &= (\mathbf{P}_2 - \mathbf{P}_1) \times (\mathbf{P}_4 - \mathbf{P}_1) \\
\mathbf{N}_6 &= (\mathbf{P}_8 - \mathbf{P}_5) \times (\mathbf{P}_6 - \mathbf{P}_5)
\end{aligned} \tag{11-12}
$$

Notice that each of these normal vectors is defined so that the resultant vector will be an *inward* normal vector in a left-hand coordinate system. Figures 11-3 and 11-4 were generated by using the three-dimensional clipping theory presented above. A listing of programs VEXCLIP3 and XCONE is given in Appendix 1.

PROBLEMS: CHAPTER 11

11-1 *Given:* Front and top views of a cone and ends of line segment *MN*.
Task: Show the points at which line *MN* pierces the given cone.

11-2 *Given:* Front and top views of an oblique cone and the ends of line segment *MN*.
Task: Show the points at which line *MN* pierces the given cone.

11-3 *Given:* Front and top views of a prism and line segment end points *MN*.
Task: Show the pierce points of line *MN* and the given prism.

11-4 *Given:* Front and top views of a cylinder and the end points of line segment *MN*.
Task: Show the pierce points of line *MN* and the cylinder.

11-5 *Given:* Front and top views of a tetrahedron and end points of line segment *MN*.
Task: Show the pierce points of line *MN* and the tetrahedron.

11-6 *Given:* Front and top views of a cylinder and end points of line segment *MN*.

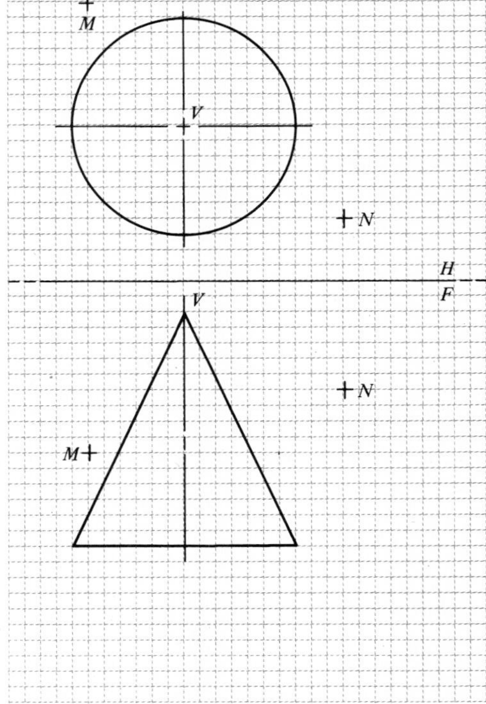

Figure 11-1P

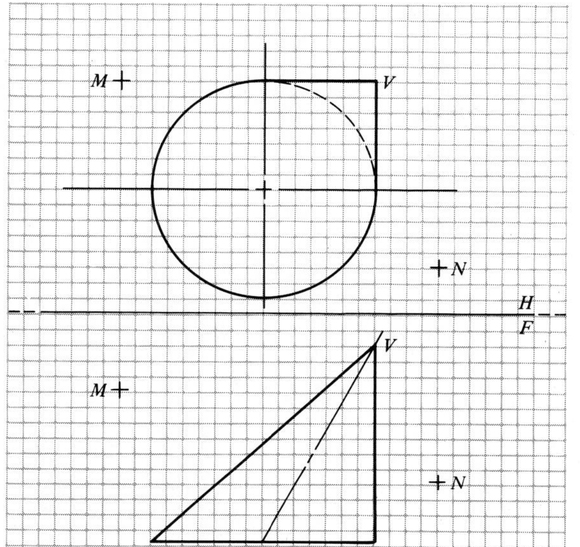

Figure 11-2P

Figure 11-4P

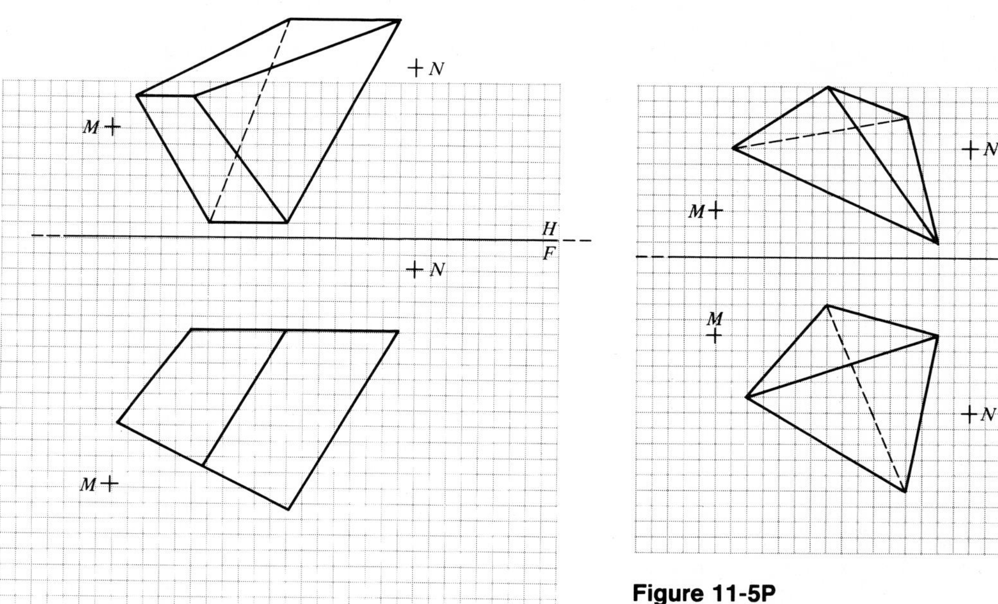

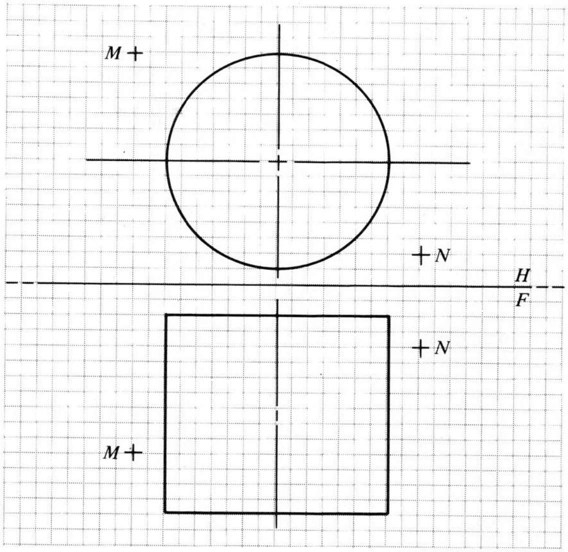

Figure 11-3P

Figure 11-5P

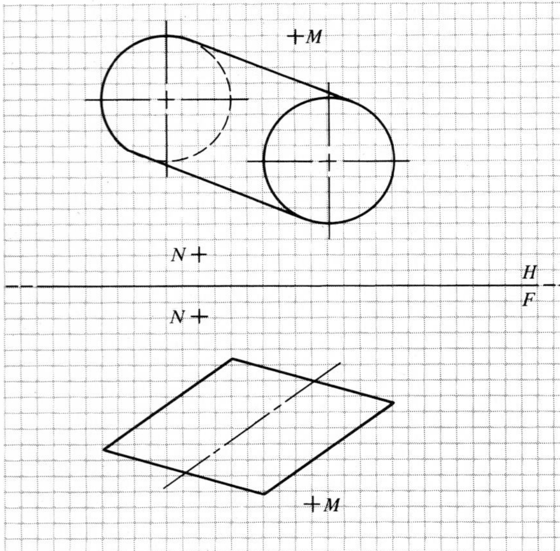

Figure 11-6P

Task: Show the pierce points of the line *MN* and the cylinder.

11-7 *Given:* Incomplete front and top views of a solid which is intersected by line *MN.*
Task: Use only the top and front views to determine where the line *MN* pierces the given solid.

11-8 *Given:* Incomplete front and top views of a sphere and line *MN.*
Task: Determine the points where the line *MN* pierces the given sphere.

11-9 *Given:* Incomplete front view and complete top view of a line *MN* piercing a truncated pyramid.
Task: Determine the points where line *MN* pierces the given solid.

11-10 *Given:* Incomplete front and top views of a line *MN* intersecting a cylinder.
Task: Determine the points at which the line *MN* pierces the given cylinder.

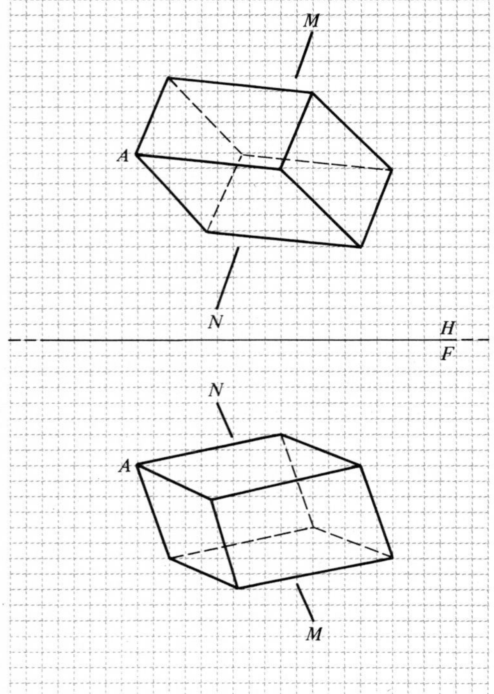

Figure 11-7P

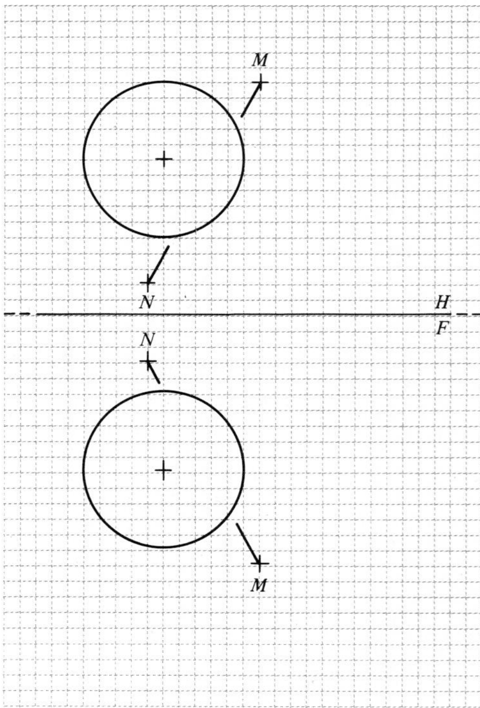

Figure 11-8P

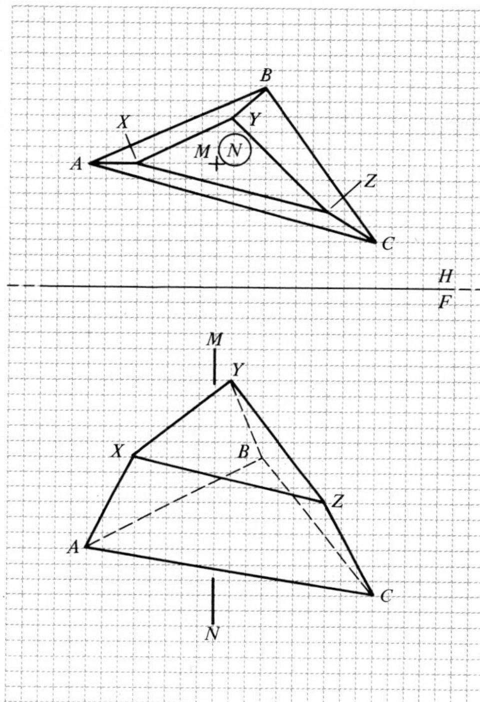

Figure 11-9P

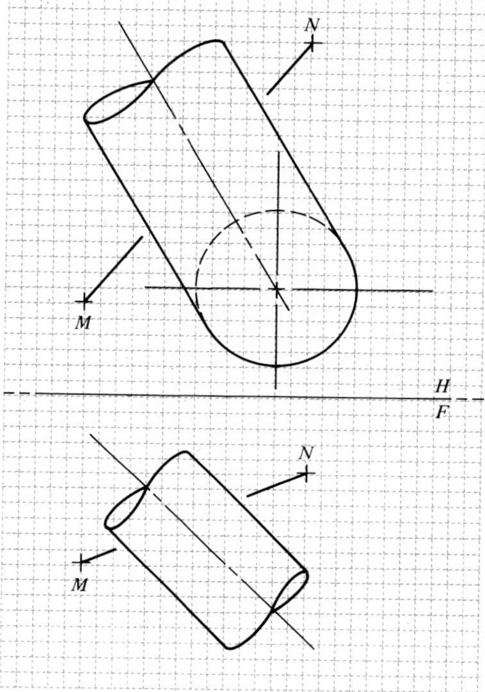

Figure 11-10P

11-11[c] *Given:* Programs XCLIP3 and XCONE.
Task: Create convex volumes such as shown in Figures 11-7 and 11-8 and create computer output to show the clipping of various three-dimensional lines.

[c]Superscript *c* indicates that a computer solution is required.

11-12[c] *Given:* Program VEXCLIP3 and subroutine CLIPSUB.
Task: Modify these programs to obtain a clipping solution for an object such as shown in Figure 11-8.

12

Tangencies

Cones, cylinders, and spheres are used in this chapter to illustrate the descriptive geometry techniques for determining the plane which is tangent to a surface and contains another specified point, or passes parallel to a given line. A second problem considered from a descriptive geometry point of view is to find the plane which contains a given line and makes a specified angle with a principal plane.

12-1 LINE TANGENT TO CIRCLES

Tangent lines may be curved or may involve one that is curved while the second is straight. They are tangent if both are in the same plane and make contact at one point only, with that point being in each line. It should be emphasized here that a line tangent to a curve *must* be in the plane of the curved line and *will* coincide with it for an infinitesimal distance. It must be remembered also that any line tangent to a circle is perpendicular to the radius of that circle when the radius is drawn to the point of tangency. (See Figure 12-1.)

12-2 PLANE TANGENT TO A CONE

Since a plane may be defined by a pair of intersecting lines, a plane may be constructed tangent to a cone by using a lateral element of the cone and a line which intersects it and which is, at the same time, tangent to the circle which forms the base of the cone. (See Figure 12-2.)

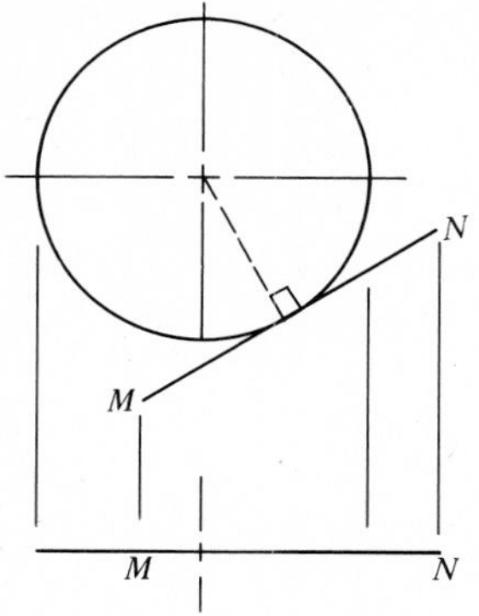

Figure 12-1
Line tangent to a circle.

A. Plane Tangent to Cone and Containing a Given Point on the Cone

Example 12-1

Given: Front and top view of cone. Top view of point *P* which is located on surface of cone.
Task: Draw a plane which is tangent to the given cone and containing the given point *P*.
Solution:

1. In the top view, draw element *VX* from the vertex of the cone through point *P* to the base of the cone.
2. Project the element *VX* to the front view.
3. Project point *P* to the front view where it will be on element *VX*.
4. In the top view, draw intersecting line *MN* perpendicular to element *VX* and tangent to the base of the cone at *X*. Line *MN* is true length since it is in the plane of the cone base and therefore is a horizontal line.
5. Project line *MN* into the front view and form the required plane.
6. *VMN* represents the tangent plane. ◀

Figure 12-2
Plane tangent to a cone.

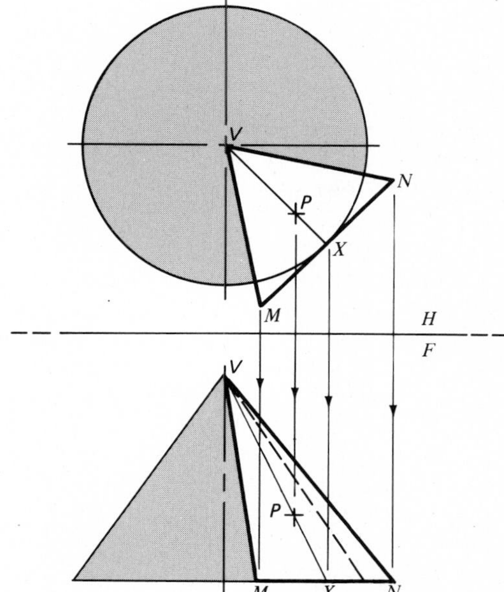

Example 12-1
Plane tangent to a cone and containing a given surface point.

B. Plane Tangent to Cone and Containing Point Outside the Surface of Cone

Example 12-2

Given: Front and top views of oblique cone and point *P* outside its surface.
Task: Draw a plane tangent to the given cone and containing the given point *P*.
Solution:

1. In the front view, draw one edge of the plane from the vertex *V* through point *P* to intersect the base plane of the cone at *M*.
2. In the top view, draw a line from vertex *V* through point *P*.
3. Project point *M* into the top view until it intersects line *VP* extended at *M* to form one edge of the required plane.
4. In the top view, draw a line *MN* tangent to the base of the cone at *X*. In order for this line to be tangent it must be perpendicular to the radius of the base at the point of tangency.
5. Draw the top view of element *VX*. The top view of the plane is now tangent to the cone and contains point *P*.
6. Project all points and lines into the front view.
7. *VMN* represents the plane which is tangent to the cone and contains point *P*.

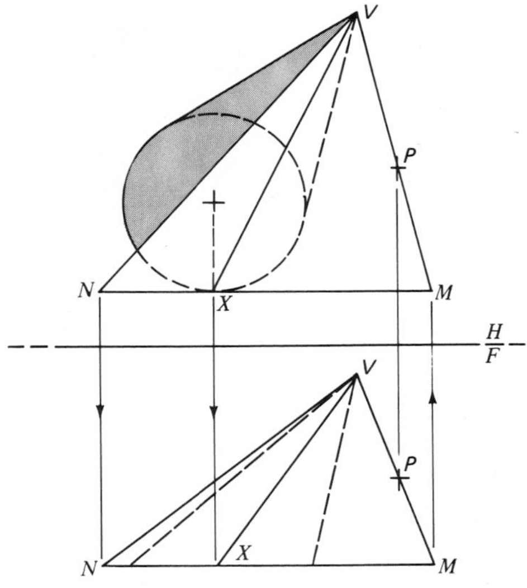

Example 12-2
Plane tangent to a cone and containing a given external point.

Note: A second plane which fulfills the qualifications of the problem could be drawn on the opposite side of the cone. ◀

C. Plane Tangent to Cone and Parallel to a Given Line

A plane is parallel to a line if any line in the plane is parallel to the given line. (See Section 6-3.)

Example 12-3

Given: Front and top views of a cone and line *AB*.
Task: Draw a plane tangent to the given cone and parallel to the given line *AB*.
Solution:

1. In the top view, draw a line from the vertex of the cone parallel to the given line *AB*.
2. In the front view, draw a line from the vertex of the cone parallel to the given line *AB* until it intersects the base plane of the cone at *M*.
3. Project point *M* into the top view until it intersects the line from *V* at *M*. Line *VM* has been drawn parallel to *AB* in both views; therefore, any proposed plane having *VM* as one of its lines will be parallel to given line *AB*.

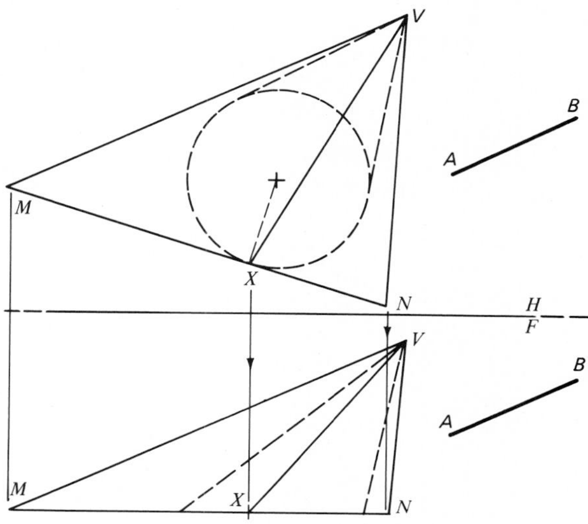

Example 12-3
Plane tangent to a cone and parallel to a given line.

4. In the top view, draw a line *MN* tangent to the base of the cone at *X*. *MN* must be perpendicular to the radius of the circular cone base at *X*.
5. Project line *MN* and its tangent point to the front view. Draw the lateral element in both views.
6. Plane *VMN* is now tangent to the cone along the element *VX* and is parallel to given line *AB* as required. ◀

12-3 PLANE TANGENT TO A CYLINDER

The solutions in this section are based on the theories used in Section 12-2 and on the statement that two intersecting lines form a plane. When one of the two lines is a lateral element of a cylinder and the other is a line tangent to the cylinder base and intersecting the element, the plane is tangent to the cylinder. (See Figure 12-3.)

A. Plane Tangent to Cylinder and Containing a Given Point on the Cylinder

Example 12-4

Given: Front and top views of a cylinder with point *P* located on its surface.
Task: Draw a plane tangent to the cylinder and containing given point *P*.
Solution:

1. In the top view, draw lateral element *YX* parallel to the axis of the cylinder and containing point *P* so that it intersects the cylinder base at *X*.

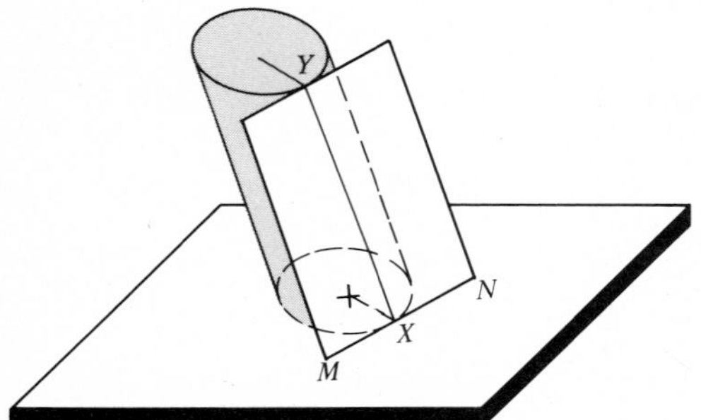

Figure 12-3
Plane tangent to a cylinder.

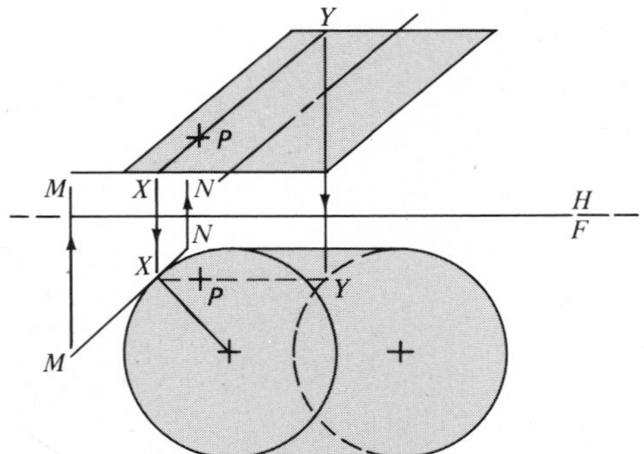

Example 12-4
Plane tangent to a cylinder and containing a given surface point.

2. Project the element *YX* into the front view.
3. In the front view, draw a radius of a circular base to intersect the circle at point *X*.
4. Draw line *MN* tangent to the base of the circle and intersecting the lateral element *YX* at *X*.
5. Project points *M* and *N* into the top view and draw line *MN*. The line must be in the plane of the cylinder base.
6. Plane *MNY* is tangent to the cylinder along *XY* and contains point *P*. ◀

B. Plane Tangent to Cylinder and Containing a Point Outside the Surface of the Cylinder

The method used in the solution of the following problem is similar to that used in Section 12-2-B. It is offered to the student for self-study.

Example 12-5

Given: Front and top views of a cylinder and point *P* which lies outside the surface of the cylinder.
Task: Construct a plane which is tangent to the cylinder and contains point *P*. ◀

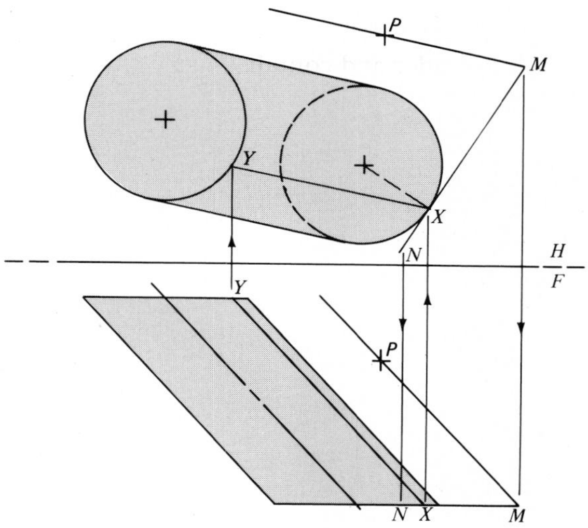

Example 12-5
Plane tangent to a cylinder and containing a given external point.

C. Plane Tangent to a Cylinder and Parallel to a Given Line

Two planes are parallel when two intersecting lines of one plane are parallel respectively to two intersecting lines of another plane. (Reference Section 6-4.)

This problem differs from the previous ones in this section in that it must be solved indirectly by using the rule of descriptive geometry noted above. The solution is found by placing the given line *AB* in a plane which is parallel to the tangent plane. (See Figure 12-4.) If the two planes are parallel, then one plane is parallel to any line in the other plane.

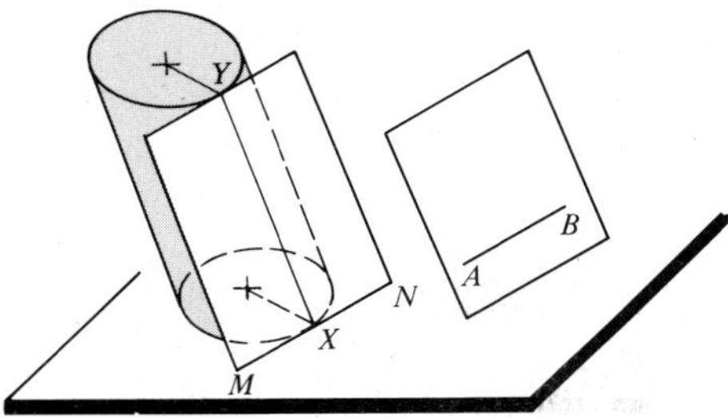

Figure 12-4
Plane tangent to a cylinder and parallel to a given line.

Example 12-6

Given: Front and top views of a cylinder and line *AB*.
Task: Construct a plane tangent to the cylinder and parallel to the given line *AB*.
Solution:

1. In both top and front views, draw line *AC* parallel to the axis of the cylinder.
2. Any line drawn tangent to the base of the cylinder will be in the plane of the base and will therefore be a horizontal line. In order to be parallel to the cylinder base, line *BC* is drawn horizontal in the front view.
3. Project point *C* into the top view and draw line *BC*.
4. Line *MN*, which will be one of the intersecting lines of the tangent plane, is in the top view, drawn parallel to line *BC* and tangent to the cylinder base. Point *X* is located by drawing a perpendicular from the center of the base circle to intersect line *MN* at the tangent point.

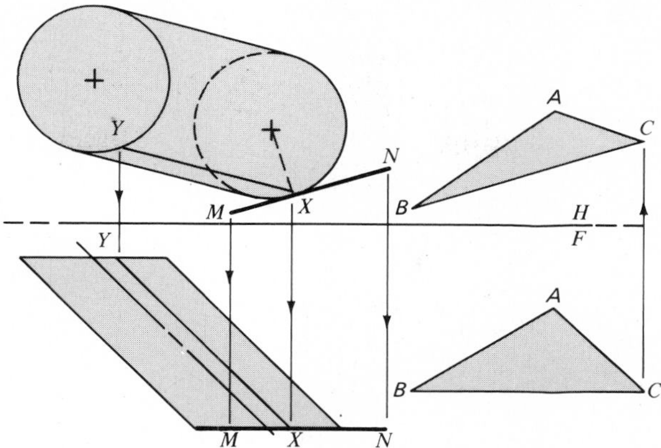

Example 12-6
Plane tangent to a cylinder and parallel to a given line.

5. Draw cylinder element *YX* parallel to the axis to form the second line of the tangent plane.
6. Project all points and lines into the front view.
7. The plane *YMN* is now tangent to the cylinder and parallel to plane *ABC*. It is, therefore, parallel to line *AB* as required. ◀

12-4 PLANE TANGENT TO A SPHERE

If a line is perpendicular to each of two intersecting lines, it is perpendicular to a plane formed by the intersecting lines.

Example 12-7

Given: Front and top views of a sphere.
Task: Place point *P* on the surface of the given sphere. Construct a plane which is tangent to the sphere and contains point *P*.
Solution:

1. In the front view, place point *P* on the sphere in an arbitrary location.
2. Show a horizontal cutting plane on edge in the front view so that it contains point *P*.
3. The horizontal cutting plane will be seen as a circle in the top view. Project point *P* to the top view and locate it on the circle. Show correct visibility.
4. In the top view, draw a true length line *PM* so that it is tangent to the

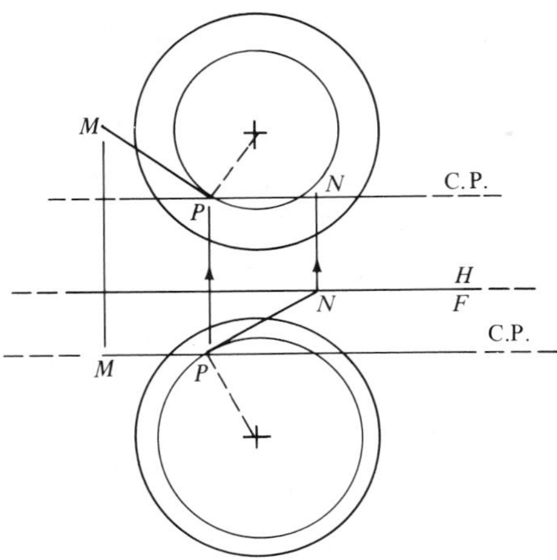

Example 12-7
Plane tangent to a sphere and containing a given surface point.

circle at point *P*. Line *PM* must be perpendicular to the radius at its point of tangency.

5. Since *PM* is true length in the top view and must be in the plane of the circle, it will be seen as a horizontal line in the front view.
6. Repeat the above operations by using a cutting plane which contains point *P*, and shows on edge in the top view. It will appear as a circle in the front view.
7. In the front view, draw true length line *NP* perpendicular to the radius of the circle and tangent to the circle at point *P*.
8. Line *NP* is true length in the front view and must be a frontal line. Project line *NP* into the top view.
9. The radius is perpendicular to two intersecting lines at point *P*. These lines then form a plane which is tangent to the sphere at that point. ◀

12-5 PLANE MAKING A SPECIFIED ANGLE WITH A PRINCIPAL PLANE AND CONTAINING A GIVEN LINE

If one element of a right circular cone is at a specified angle θ with its base plane, then all elements of the cone will form the same angle with that plane. For this reason, a plane tangent to a lateral element of a cone's surface will have a dihedral angle θ with the base plane of the cone. (See Figure 12-5.)

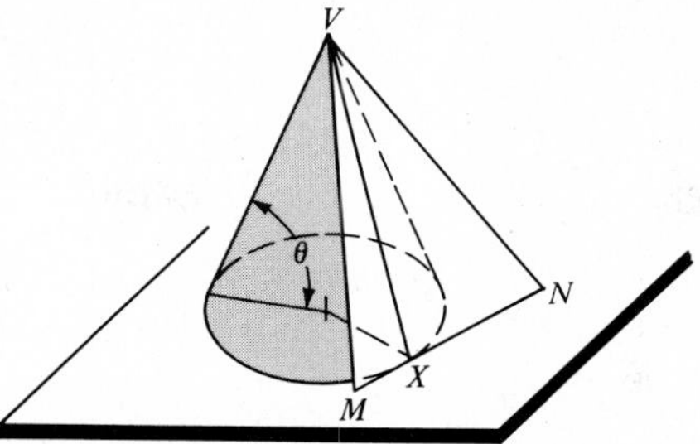

Figure 12-5
Plane tangent to a cylinder and parallel to a given plane.

Example 12-8

Given: Front and top views of line *AB*.
Task: Construct a plane which contains line *AB* and makes an angle of 50 deg with the horizontal plane.
Solution:

1. In the front view, arbitrarily select a point *V* on line *AB*.
2. Using point *V* as the vertex, draw a right circular cone with its axis perpendicular to the horizontal plane and having elements at a 50 deg angle with the horizontal plane.

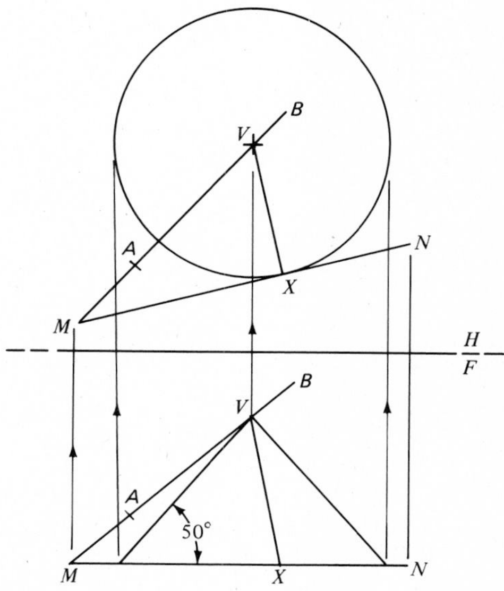

Example 12-8
Plane containing a given line and having a specified angle with a principal plane.

3. In the front view, extend line *AB* until it pierces the base plane (a horizontal plane) at *M*.
4. Project *V* into the top view and draw the circular view of the cone shown in the front view.
5. Project piercing point *M* into the top view until it intersects the given line *AB* at *M*.
6. In the top view, draw line *MN* so that it is tangent to the base circle at *X*. Point *N* is located on line *MX* extended. The radius of the base and the element which coincides with it are perpendicular to line *MN*.
7. Line *MN* is in the base plane and therefore a horizontal line which is drawn in the front view.
8. Line *VX* is the tangent element of the cone and is at a 50 deg angle with the horizontal plane. Since the plane *VMN* is at the specified angle and contains line *AB*, it is the required plane. ◀

A computational approach is not a panacea for every geometry problem. It takes judgment and experience to know when geometric modeling on a computer is a better approach than descriptive geometry on paper, especially when surfaces and solids are considered. Even when a computer is affordable and available, the best approach to a given problem may be by use of a pencil and paper. A well-prepared professional must keep the options for problem solutions open. If solid modeling is to be done on a computer, then many other considerations arise which are related to costs, software availability, types of hardware, ergonomics, and the like. A knowledge of differential calculus is necessary to consider the tangency problem from a mathematical point of view. This topic is discussed in books on advanced geometric modeling and computational geometry.

PROBLEMS: CHAPTER 12

Note: The problems in Chapter 12 are review problems for the first 12 chapters in the text.

12-1 *Given:* Front and top views of a cone and line *MN*.
Task: Construct a plane tangent to the cone and parallel to line *MN*.

12-2 *Given:* Front and top views of a cylinder and line *MN*.
Task: Construct a plane tangent to the cylinder and parallel to line *MN*.

12-3 *Given:* Front and top views of triangle *ABC* and point *M*.
Task: Construct the top and front views of line *MN* which is perpendicular to plane *ABC*.

12-4 *Given:* Top and front views of *MN* and *RS* which represent the centerlines of two pipes of equal diameter to be installed as shown. There must be a minimum clearance of 1.0 unit between the outside surfaces of the pipes.
Task: Determine the diameter (in units) of the pipes which can be used.

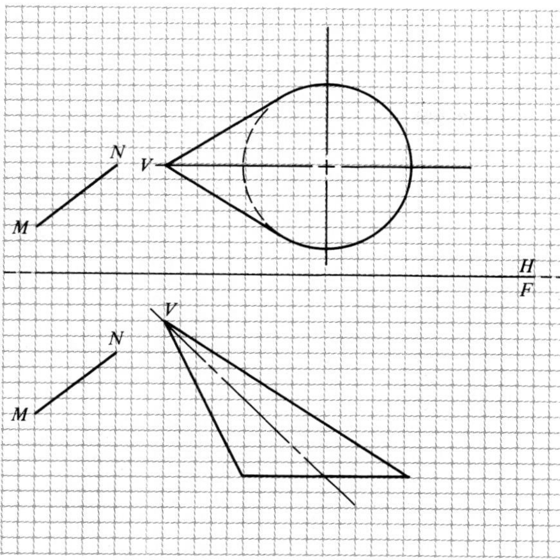

Figure 12-1P

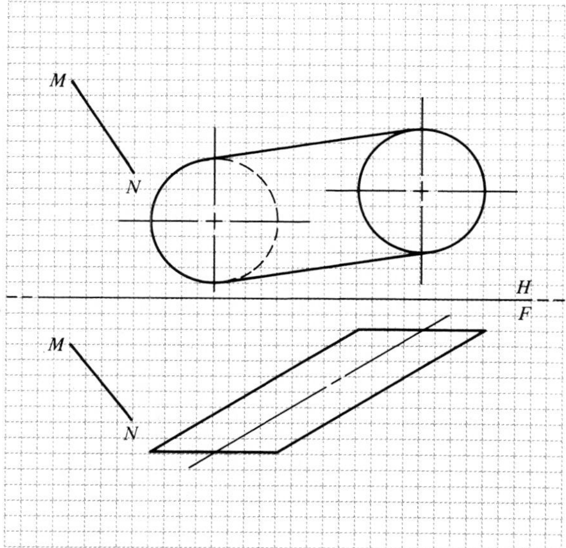

Figure 12-2P

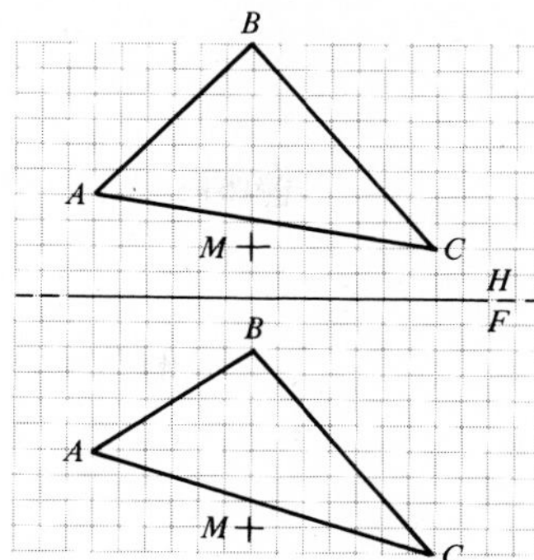

Figure 12-3P

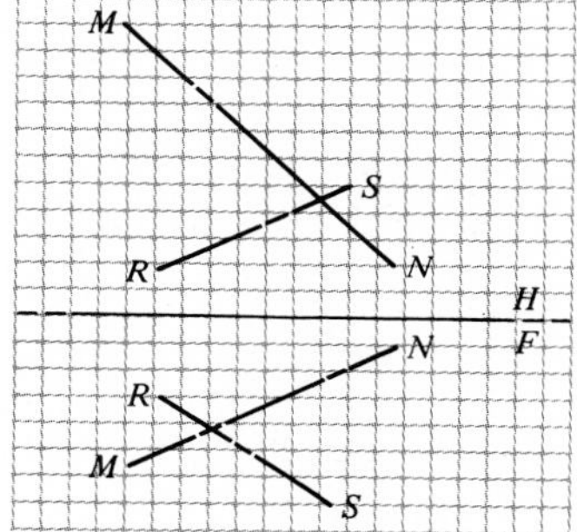

Figure 12-4P

12-5 *Given:* Front and top views of a cylinder and point *P*.
Task: Construct a plane which is tangent to the cylinder and contains point *P*.

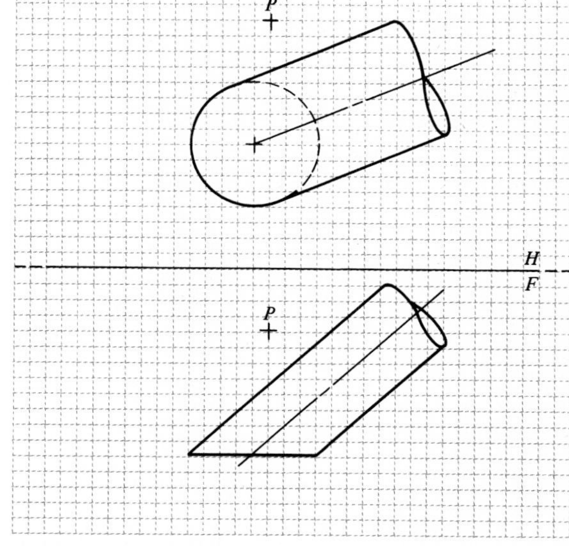

Figure 12-5P

12-6 *Given:* Top and front views of a tetrahedron.
Task: Show the projection and find the shortest length of a line from point *P* to the plane *ABC*.

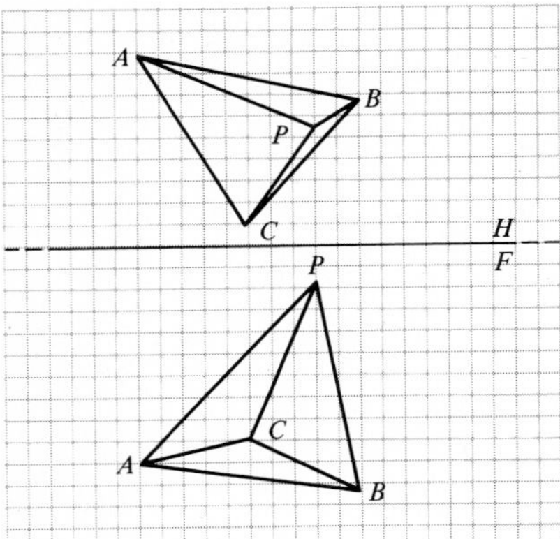

Figure 12-6P

12-7 *Given:* Top and front views of point *P* and triangle *ABC*.
Task: Find the true distance from point *P* to plane *ABC*. Show the point of intersection.

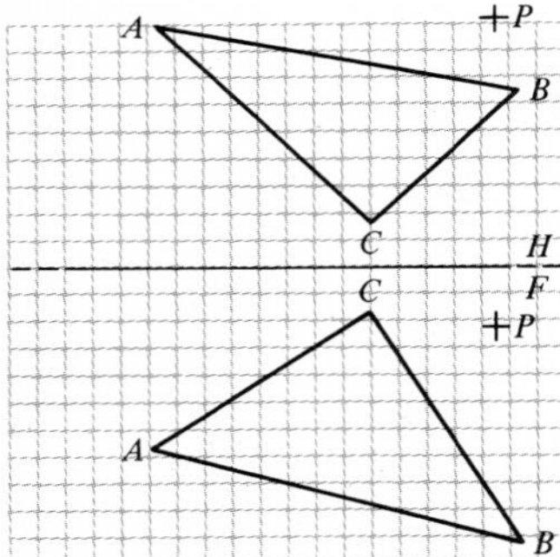

Figure 12-7P

12-8 *Given:* Front and top views of a sphere with point *P* on its surface.
Task: Construct a plane which is tangent to the sphere and contains point *P*.

12-9 *Given:* Front and top views of line *MN*.
Task: Construct a plane which contains line *MN* and has an angle of 55 deg with the frontal plane.

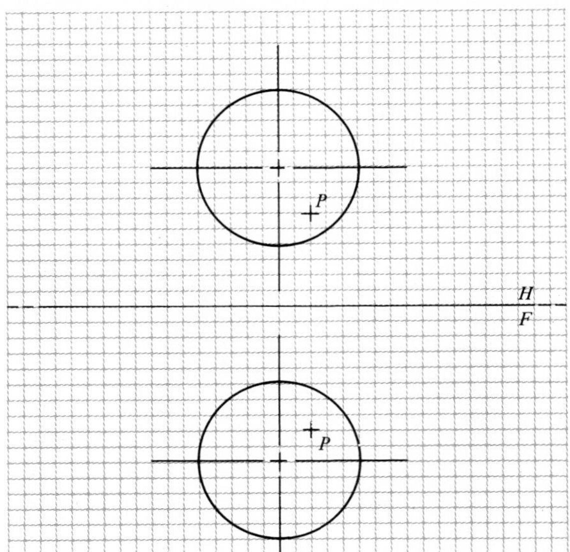

Figure 12-8P

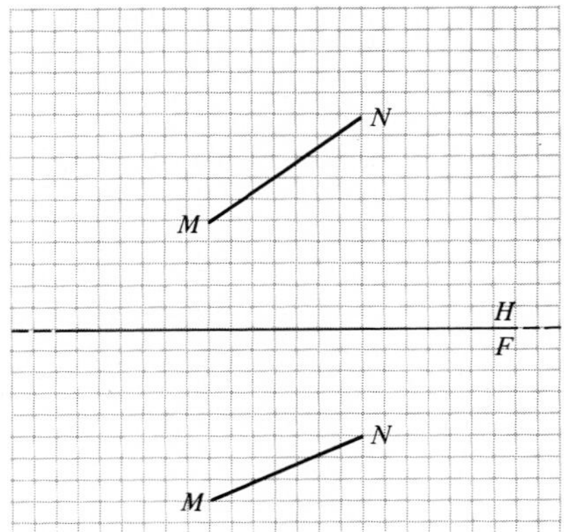

Figure 12-9P

12-10 *Given:* Top and front views of triangle *ABC* and point *O*.
Task: Construct a line through point *O* perpendicular to plane *ABC*. Indicate by point *P* the location where the line pierces the plane. Find the true length of *OP*.

12-11 *Given:* Let **P** be the center of a sphere of radius *r*. An infinite line passes through points **A** and **B**.
Task: Devise a method for determining whether the line intersects the sphere.

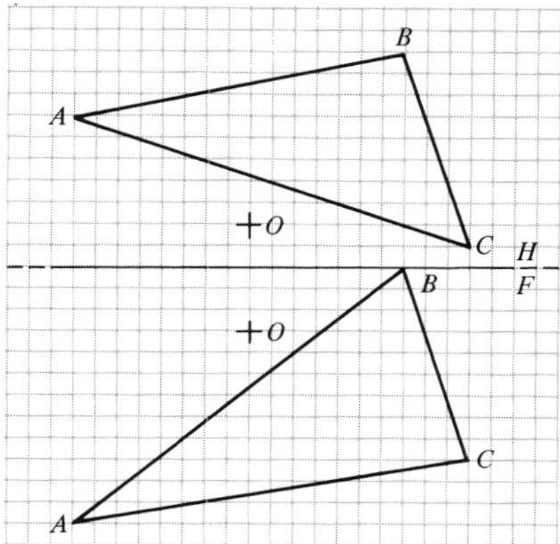

Figure 12-10P

12-12 *Given:* Three planes with unit normal vectors $\mathbf{n}_1$, $\mathbf{n}_2$, $\mathbf{n}_3$ intersect in a point. This point is given by

$$\mathbf{P} = \frac{d_1(\mathbf{n}_2 \times \mathbf{n}_3) + d_2(\mathbf{n}_3 \times \mathbf{n}_1) + d_3(\mathbf{n}_1 \times \mathbf{n}_2)}{\mathbf{n}_1 \cdot \mathbf{n}_2 \times \mathbf{n}_3}$$

where $d_1 = k_1/|\mathbf{N}_1|$, $d_2 = k_2/|\mathbf{N}_2|$, $d_3 = k_3/|\mathbf{N}_3|$.

Task: a. Let $\mathbf{n}_1 = \mathbf{i}$, $\mathbf{n}_2 = \mathbf{j}$, and $\mathbf{n}_3 = \mathbf{k}$. Show that the given equation gives the proper point of intersection.

b. Calculate the point of intersection between the three planes defined in Problem 8-14.

12-13[c] *Given:* A plane is defined by three vectors as follows: $\mathbf{P}_1 = [1.2\ 1.2\ 1.2]$, $\mathbf{P}_2 = [-1.2\ 1.2\ 1.2]$ and $\mathbf{P}_3 = [-1.2\ -1.2\ 0.8]$. Four lines intersect this plane. They are each defined by two points as follows:

Line	**A**	**B**
1	[1 1 −1.8]	[−1 −1 1.8]
2	[−1 1 −1.8]	[−1 −1 1.8]
3	[−1 −1 −1.8]	[−1 −1 1.8]
4	[1 −1 −1.8]	[−1 −1 1.8]

Task: a. Calculate the four piercing points.

b. Create the principal views of the plane and clipped lines.

c. Create an orthographic view as seen from x = 3, y = 4, z = 5.

12-14 *Given:* A plane defined by $\mathbf{P}_1 = [3\ 6\ -2]$, $\mathbf{P}_2 = [5\ 0\ -3]$ and $\mathbf{P}_3 = [7\ 0\ 3]$.

Task: Calculate the point of intersection between the plane and the line perpendicular to the plane and passing through the point $\mathbf{S} = [5\ 8\ 6]$.

12-15[c] *Given:* A triangular, finite plane facet defined by $\mathbf{P}_1 = [1.2\ 1.2\ 1.2]$, $\mathbf{P}_2 = [-1.2\ 1.2\ 1.2]$ and $\mathbf{P}_3 = [-1.2\ -1.2\ 0.8]$. Four line segments defined between given end points are

Line	**A**	**B**
1	[2 2 −1.8]	[.5 .5 −.5]
2	[1 1 −1.8]	[−1 −1 1.8]
3	[−1 1 −1.8]	[−1 −1 1.8]
4	[1 −1 −1.8]	[−1 −1 1.8]

Task: a. Study each finite line segment to determine whether it intersects the finite, triangular plane facet. If so, calculate the piercing point.

b. Create a computer graphics display of the results obtained in part a. in the form of an orthographic view as seen from x = 3, y = 4, z = 5.

12-16[c] *Given:* Any three noncollinear points which define a plane.

Task: Write a computer program which will display the three principal views of the plane along with its normal vector direction. Provide an option which will show the true size of the triangular plane facet in the frontal view.

12-17[c] *Given:* A four sided parallelogram plane facet is defined by the following three vectors.

Vector	*x*	*y*	*z*
1	3.375	3.5	3.0
2	5.375	5.25	3.5
3	6.5	3.5	1.75

The straight path of a laser light beam passes through $\mathbf{A} = [2\ 6\ 1]$ and $\mathbf{B} = [2.75\ 5.75\ 2]$.

Task: a. Does the light path intersect the plane facet?

b. If so, what is the point of intersection? If not, how close to the plane does the light path come? How far is the point of closest distance to the plane from point **A**?

c. Generate principal views which show the solution.

12-18[c] *Given:* Two skew lines in space are defined by two pairs of points as follows:

Line	**A**	**B**
1	[.375 3.375 0]	[3.625 0 3.25]
2	[0 .375 3.375]	[3.625 3.5 1.625]

Task: a. Calculate the shortest distance between the two skew lines.
b. Calculate the points of intersection of the two lines with the common perpendicular.
c. Create the three principal views of the three lines in part b. and an orthographic view as seen from x = 5, y = 5, z = 5.

12-19[c] *Given:* A cable passes through two points given by **A** = [0 1 2] and **B** = [4.5 0 1]. Three planes are defined by three vectors each:

Plane	$\mathbf{P}_1$	$\mathbf{P}_2$	$\mathbf{P}_3$
1	[0 .5 2.5]	[1 0 .5]	[2 1.5 2]
2	[1 0 .5]	[2 1.5 2]	[4 0 3]
3	[1 0 .5]	[3.5 2.5 0]	[4 0 3]

Task: a. Calculate the piercing points between the cable and the planes.
b. Calculate the length from **A** to each piercing point.
c. Create the three principal views and an orthographic view as seen from x = 1, y = 3, z = 5.

12-20[c] *Given:* A hopper consists of four plane surfaces which create four edges. Four corner points for each of the four hopper planes are obtained from drawings, and coordinates are given in the following DATA statements:

DATA 2.625, 0, 2.75, 2.3125, 3.5625, 4, .8125, 3.5625, 1.375, 2, 0, 1.625
DATA 2.3125, 3.5625, 4, 2.625, 0, 2.75, 3.75, 0, 2.125, 4.9375, 3.5625, 2.4375
DATA 3.125, 0, 1, 3.4375, 3.5625, −.3125, 4.9375, 3.5625, 2.4375, 3.75, 0, 2.125
DATA 2, 0, 1.625, .8125, 3.5625, 1.375, 3.4375, 3.5625, −.3125, 3.125, 0, 1

The hopper is to be supported by four beams. The Beam Data File gives a pair of points along each edge of the supporting beams.

Beam Data File

0.625	0.5625	3.375
5.25	0.5625	3.375
0.625	1.5	3.375
5.25	1.5	3.375
0.625	0.5625	2.75
5.25	0.5625	2.75
0.625	1.5	2.75
5.25	1.5	2.75
0.625	1.25	1.125
5.25	1.25	1.125
0.625	2.125	1.125
5.25	2.125	1.125
0.625	1.25	0.5625
5.25	1.25	0.5625
0.625	2.125	0.5625
5.25	2.125	0.5625

Task: Calculate the intersections of interest between the eight edges of the two beams and the four hopper planes. Note: All of the edges may not intersect the hopper. Create the three principal views which show the solution.

12-21[c] *Given:* The axis of a cylinder is to pass through the points $\mathbf{P}_1$ = [0 0 0] and $\mathbf{P}_2$ = [2.6875 2 1.1875]. A point which must lie on the circumference of the cylinder is fixed at **C** = [0.6875 2.25 1.5].
Task: a. Calculate the required diameter of the cylinder.
b. Show the principal views of the cylinder axis, the cylinder cross-section, and the line of shortest distance between **C** and the axis.
c. Create an end view looking along the axis of the cylinder.

12-22 *Given:* A sphere rolls down a V-trough. The bottom of the trough is a line through **M** = [0 1.625 4] and **N** = [2.75 0 1.3125]. The sphere centerline follows a line through **A** = [0.25 2.375 3.25] and **B** = [3 0.75 0.5625].
Task: a. Use descriptive geometry to solve for the angle of the V-trough if the sphere has a radius of 0.5 units.
b.[c] Write a computer program with graphical output which will solve this problem for various values of sphere radius.

13

Intersections—Planes and Solids

Modern society has become so accustomed to mass production that it takes for granted the designs presented by modern production methods. In order to design the many products ranging from the simple gadget to the modern automobile, the engineer must be knowledgeable about principles of intersections of surfaces.

In the material offered in this chapter there will be many times when the problem being discussed will relate to real-life situations and will bring to mind practical problems where such situations could occur. The studies that follow are an extension of work done in previous sections of this text and will use the same relationships as before in order to find the intersections between surfaces. The process of finding points on the intersections usually consists of passing planes that cut elements from both surfaces in order to find where they intersect. The material that follows will use one or a combination of the four general procedures:

1. Edge view method
2. Auxiliary view method
3. Cutting plane (two-view method)
4. Geometric modeling and computer graphics displays

The principles for finding the intersections of lines, planes, and simple surfaces using these procedures were examined in Chapter 8. For this reason, the student should now be familiar with methods of locating points on lines of intersection and the way to determine them. The simple types of illustrations shown in this chapter illustrate general procedures. More advanced examples can be found in available literature.

13-1 PLANE AND PRISM

A. Edge View of Plane Shown

We know from prior study that the intersection of two planes results in a straight line. Since the face of a prism is a plane, the solution to this type of problem is a series of straight lines connecting the piercing points. When the edge view is shown, as it is in Example 13-1, or if it is found by the auxiliary method, it becomes a matter of applying the principles of the cutting-plane method to the particular problem as shown below.

Example 13-1

Given: Top and incomplete front views of a prism intersected by a plane. The plane is defined by four vertices.
Task: Show the intersection of the plane and prism in the top, front, and right-side views.
Solution:

1. Using the principles of orthographic projection draw the right-side view of the intersecting plane and prism. In this view the plane will be shown on edge.
2. In the top view, the faces of the prism are shown on edge and their intersections shown as points. By observation it can be seen that a vertical cutting plane passed through points 1, 2, 3, 4, 5, 6, and 7 in the top view will produce vertical lateral elements on the prism at those points. It will, at these numbered intersections, also create horizontal lateral elements on plane *WXYZ*. These elements are the traces of the cutting plane and can be seen on end in the right-side view.
3. The piercing points and line of intersection can be found by projecting the elements 1, 2, 3, 4, 5, 6, 7 and their respective traces into the front view. Show correct visibility. ◄

B. Cutting-Plane Method

Many of these problems are solved by a combination of methods previously used.

Example 13-2

Given: Incomplete top and front views of intersecting plane and prism.
Task: Find the line of intersection between the given plane and prism.
Solution:

1. Pass a cutting plane so that it includes line *B-B′* and is perpendicular to the horizontal plane. It will show on edge in the top view. In the

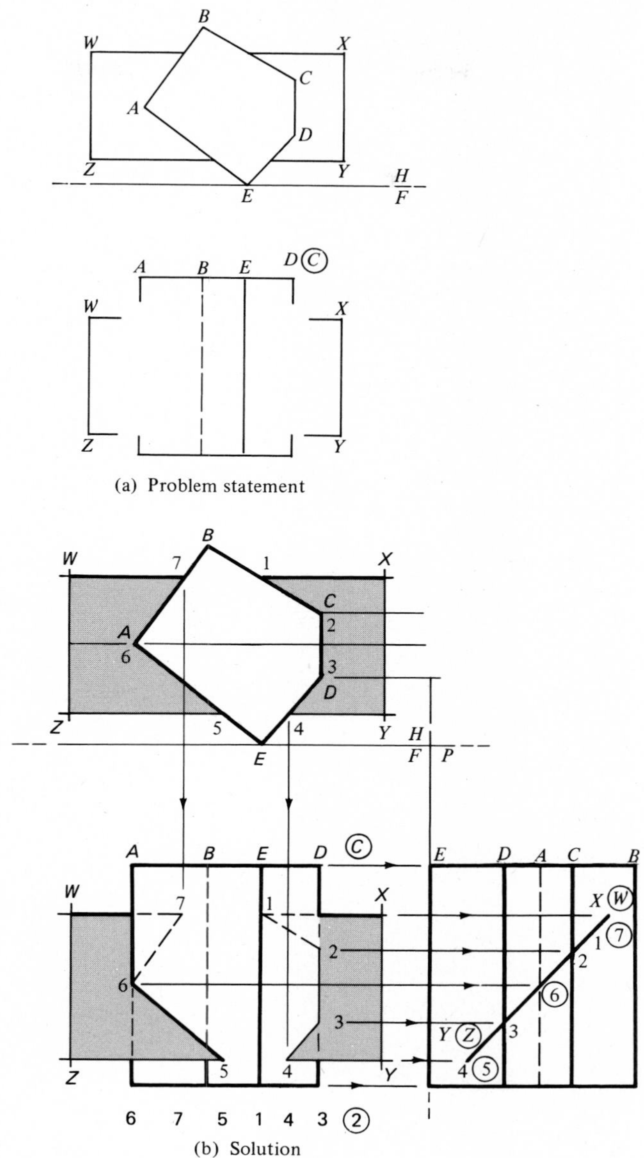

(a) Problem statement

(b) Solution

Example 13-1
Intersection: plane and prism (edge-view method).

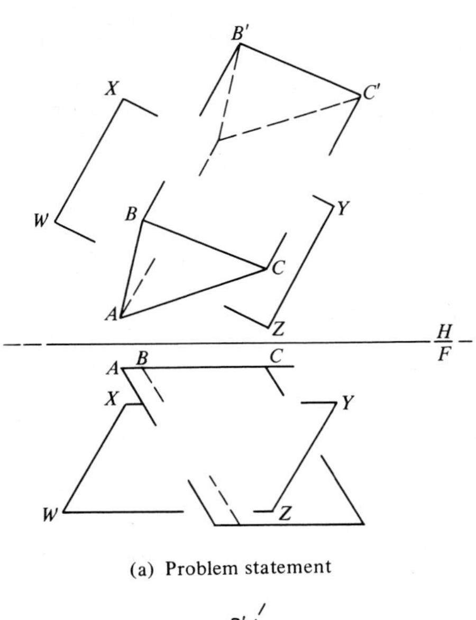

(a) Problem statement

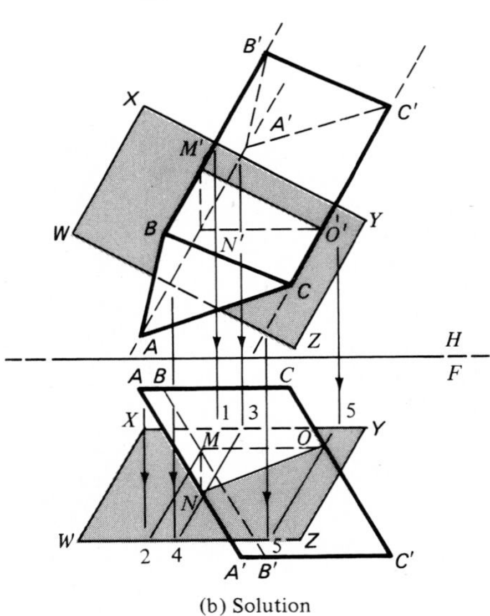

(b) Solution

Example 13-2
Intersection: plane and prism (cutting-plane method).

front view, the cutting plane intersects plane *WXYZ* at points 1 and 2. The trace 1-2 intersects line *B-B′* at point *M* and locates one point of the intersection.

2. Pass a vertical cutting plane to include line *A-A′*. By using the method shown in step 1, the trace 3-4 will be shown and the second point *N* can be found.

3. Using line *C-C′* in the top view, repeat operations of steps 1 and 2 to locate the third point *O* of the intersection.

4. Check visibility of points *M, N,* and *O* and connect them with solid or hidden lines as required to show intersection of plane and prism in the front view.

5. Project the points into the top view to locate *M′, N′,* and *O′*. Check visibility and show the required intersection. ◀

13-2 PLANE AND PYRAMID

A. Auxiliary Method (Edge View of Plane Shown)

Example 13-3

Given: Incomplete top and front views of an intersecting pyramid and plane.
Task: Using the auxiliary method of solution, complete the given top and front views.
The solution to this problem is shown in Example 13-3(b) for study by the student. ◀

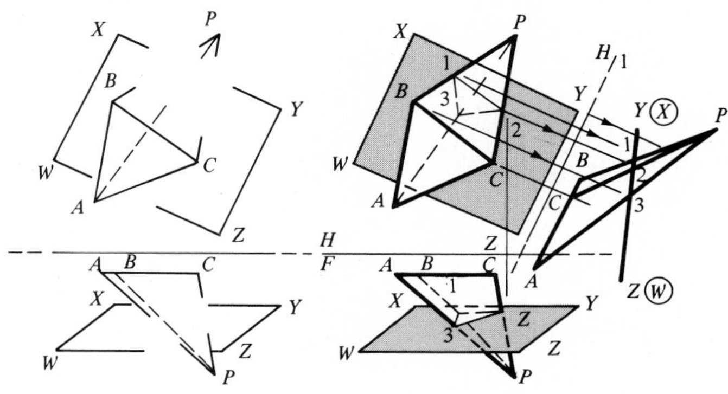

(a) Problem statement (b) Solution

Example 13-3
Intersection: plane and pyramid (auxiliary method).

B. Cutting-Plane Method (Pierce-Point, Two-View)

Example 13-4

Given: Incomplete top and front views of an intersecting pyramid and plane.
Task: Using *only* the front and top views, find the intersection of the plane and pyramid. Complete both views.
Here again, the reader should study the solution shown by Example 13-4(b) and develop a sound step-by-step procedure.

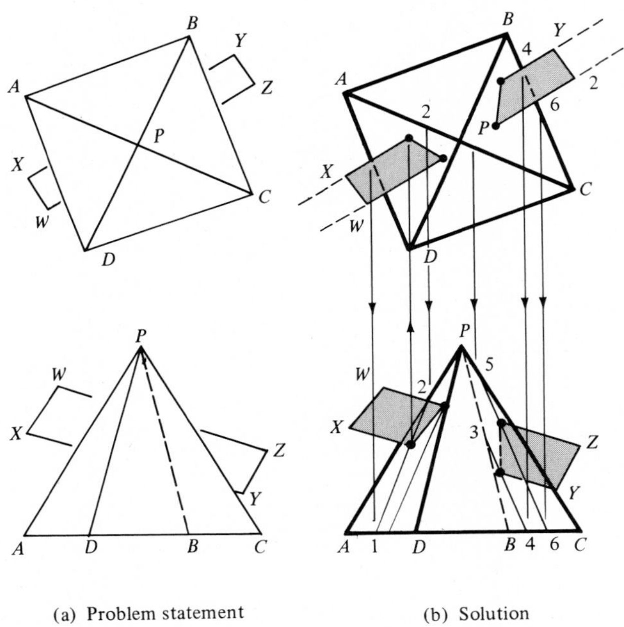

Example 13-4
Intersection: plane and pyramid (two-view method).

13-3 PLANE AND CYLINDER

In all methods of finding the intersection of surfaces where one surface is circular, that surface is treated as if it were a multisided prism. It can be seen that a large number of elements on the cylinder produce an intersection that is more accurate than that produced by a small number of elements; 12 or 16 is generally accepted as an adequate number. By using a 30–60 deg triangle, a spacing of 30 deg and 12 elements may be easily generated.

A. Auxiliary Method—Intersection Plane and Right Cylinder

Example 13-5

Given: Incomplete front and top views of a right cylinder cut by plane *ABC*.
Task: Show complete intersection of plane and cylinder.
Solution:

1. In the top view, use a 30 deg triangle and divide the circular view into 12 equally spaced elements. (Equal spacing is not necessary but will simplify the construction.)
2. Line *B-C* of plane *ABC* is horizontal in the front view and is therefore true length in the top view. Place folding-line *H*/1 perpendicular to the true length line *B-C*. Project lines of sight from elements into the first auxiliary perpendicular to *H*/1. Project plane *ABC* into the first auxiliary. The plane will appear on edge.
3. By inspection, the intersection points of the plane and the elements can be ascertained. Transfer these points to their corresponding elements in the front view.
4. Determine the visibility of the points and, with a French curve, draw the correct curve of intersection in the front view. ◀

B. Auxiliary Method—Intersection Plane and Oblique Cylinder

Example 13-6

Given: Incomplete top and front views of an oblique cylinder cut by a plane.
Task: Find the complete intersection of the given plane and cylinder. The solution is shown for student study. ◀

C. Cutting Plane Method (Pierce-Point, Two-View)

Example 13-7

Given: Incomplete front and top views of an intersecting plane and cylinder.

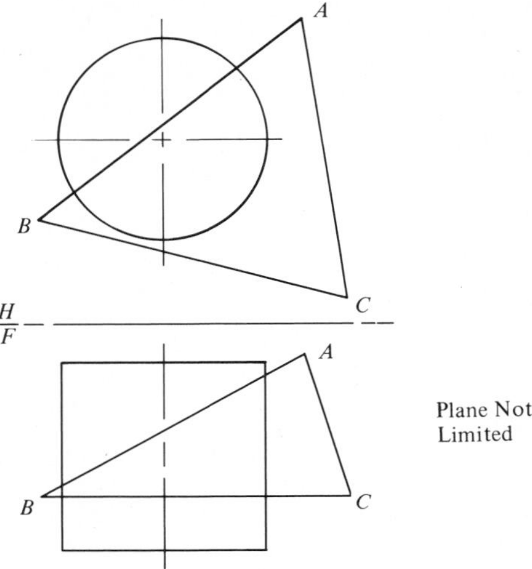

(a) Problem statement

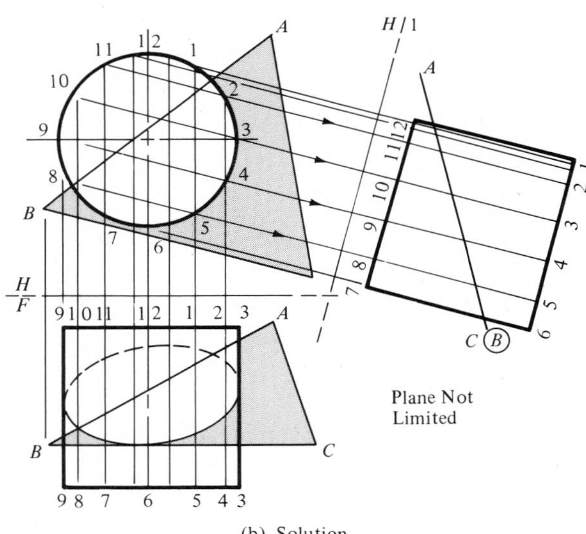

(b) Solution

Example 13-5
Intersection: plane and right cylinder (auxiliary method).

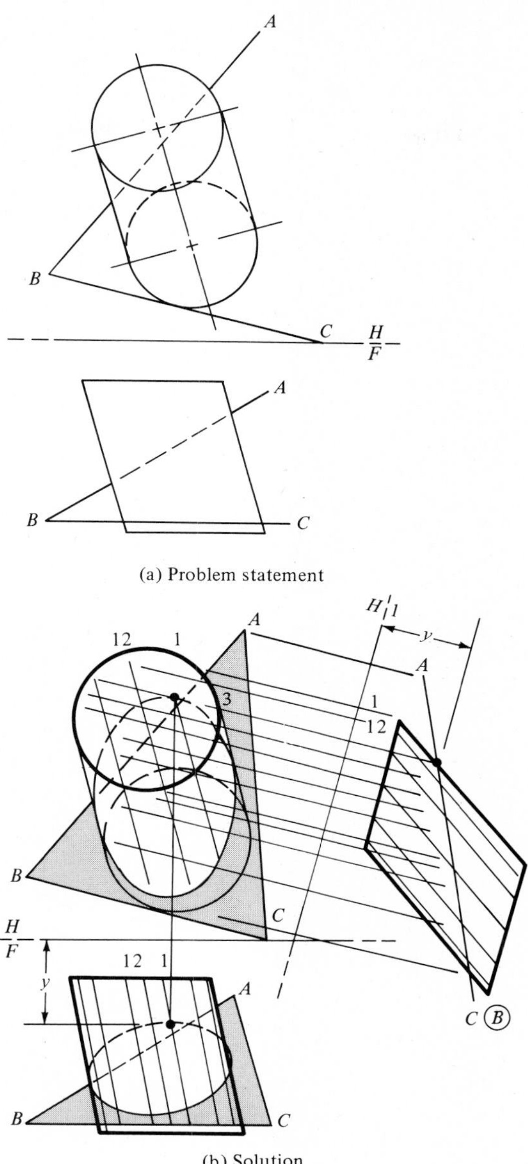

(a) Problem statement

(b) Solution

Example 13-6
Intersection: plane and oblique cylinder (auxiliary method).

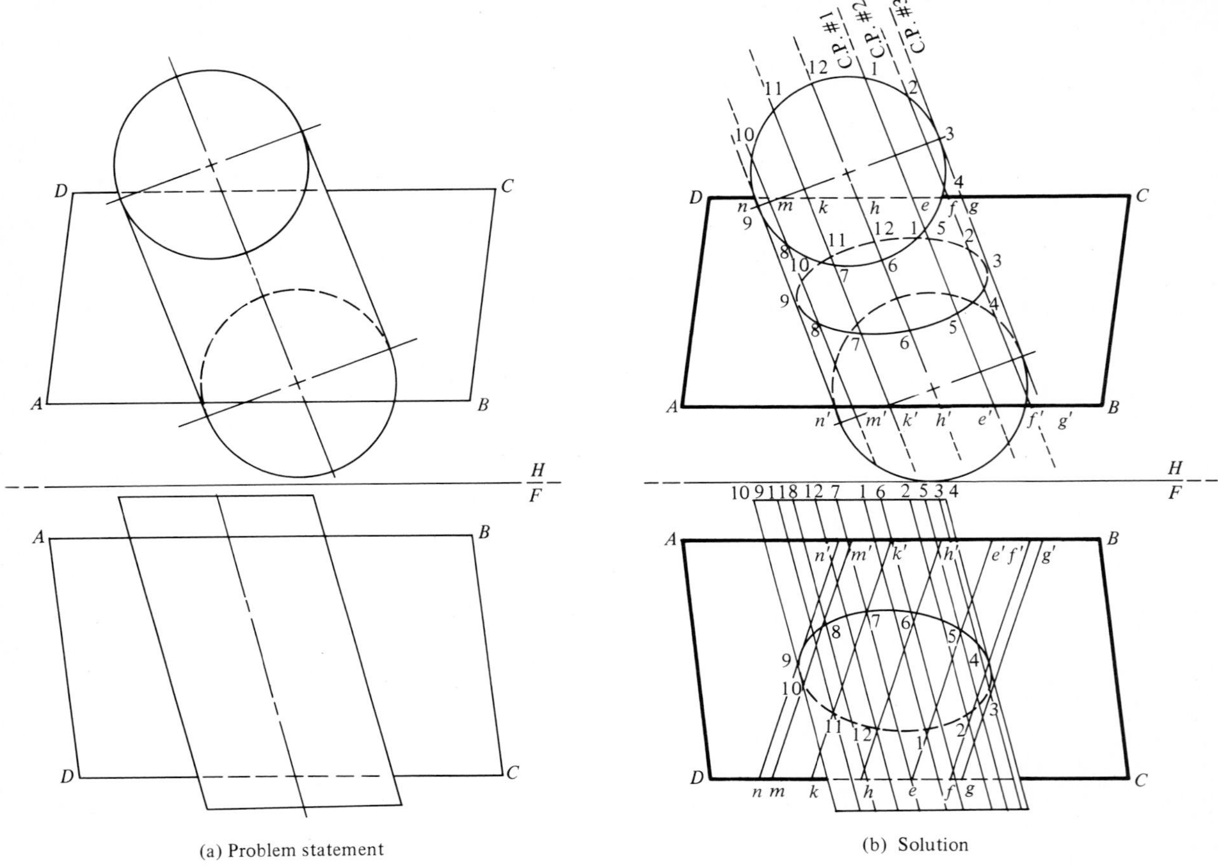

Example 13-7
Intersection: plane and oblique cylinder (two-view method).

Task: Find the complete intersection of the given plane and cylinder.
Solution:

1. In the top view, where the ends of the cylinder project as circles, lay off the locations of the lateral elements of the cylinder.
2. Project these locations 1-12 into the front view.
3. In the top view, pass a vertical cutting-plane (such as C.P. #1) so that it is parallel to the axis of the cylinder and contains elements 1 and 5 on the projected circle. This cutting plane cuts plane *ABCD* at *e'* on line *AB* and at *e* on line *CD* in the top view.
4. Project points *e* and *e'* to their respective locations on plane *ABCD* in the front view and draw trace *e-e'*.

5. The trace *e-e′* intersects line elements 1 and 5 in the front view and locates piercing points 1 and 5 which are both on the curve of intersection between the plane and cylinder.
6. Repeat steps 3, 4, and 5 and pass cutting planes through all elements (i.e., C.P. #2 passes through elements 2 and 4, C.P. #3 passes through element #3, and so on). The cutting-planes cut plane *ABCD* and form traces on the plane in the front view. The intersection of the traces and line elements in the front view determine locations of piercing points on the curve of intersection as shown. Check visibility of the points and connect them with a French curve to show the line of intersection.
7. Project all points into the top view. Check visibility and draw the curve of intersection. ◀

13-4 PLANE AND CONE

The methods and principles for finding the intersections of a plane and a cone in Examples 13-8 and 13-9 are basically the same as those already discussed.

A. Edge View of Plane Shown (Lateral Elements Used)

Example 13-8

Given: Incomplete front and top views of a cone and an intersecting plane.
Task: Show the complete curve of intersection between the given plane and cone.
Solution:

1. In the top view, divide the circular view of the cone into 12 divisions and draw lateral elements from the vertex of the cone to each of the divisions.
2. Project divisions and elements to the front and side views.
3. Determine piercing points in the front view where the plane on edge cuts each lateral element.
4. Project each piercing point to its respective element in the top view and in the side view.
5. Use irregular curve and connect the piercing points to form the line of intersection between plane and cone. ◀

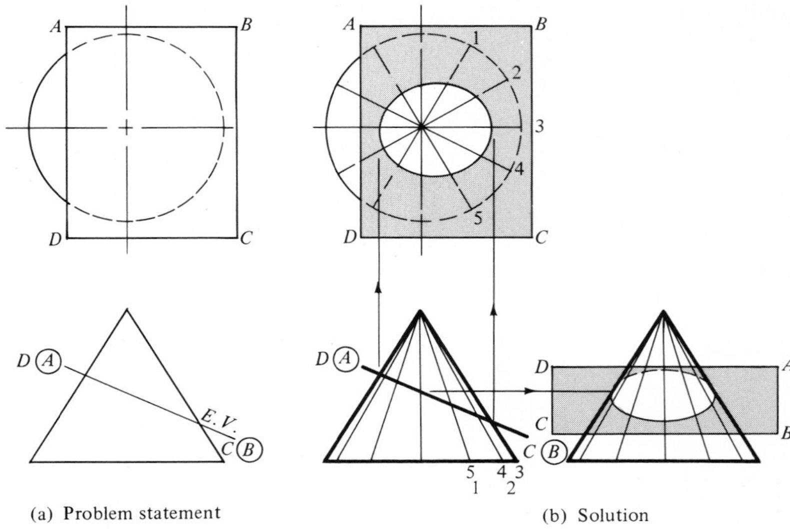

(a) Problem statement

(b) Solution

Example 13-8
Intersection: plane and cone (lateral elements used).

B. Edge View of Plane Shown (Circular Elements Used)

The following example varies slightly from the previous examples by using horizontal cutting planes or elements. These cutting planes eliminate the confusion sometimes caused by the nearly vertical lateral elements of a cone. The student should now be able to establish a regular procedure by studying the solution to Example 13-9.

Example 13-9

Given: Incomplete front and top views of a cone and intersecting plane.
Task: Show the complete curve of intersection between the given cone and plane. ◀

C. Cutting-Plane Method (Piercing-Point, Two-View)

Example 13-10

Given: Incomplete top and front views of a cone and an intersecting plane.
Task: Using only the two given views, show the complete curve of intersection between the given cone and plane.

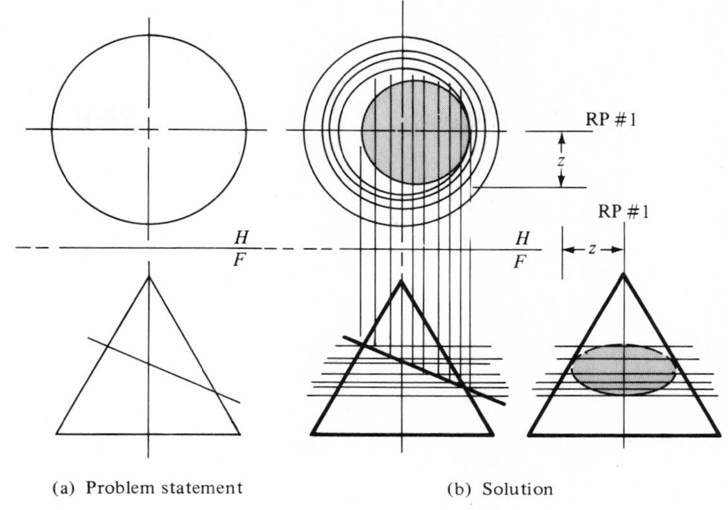

(a) Problem statement (b) Solution

Example 13-9
Intersection: plane and cone (circular elements used).

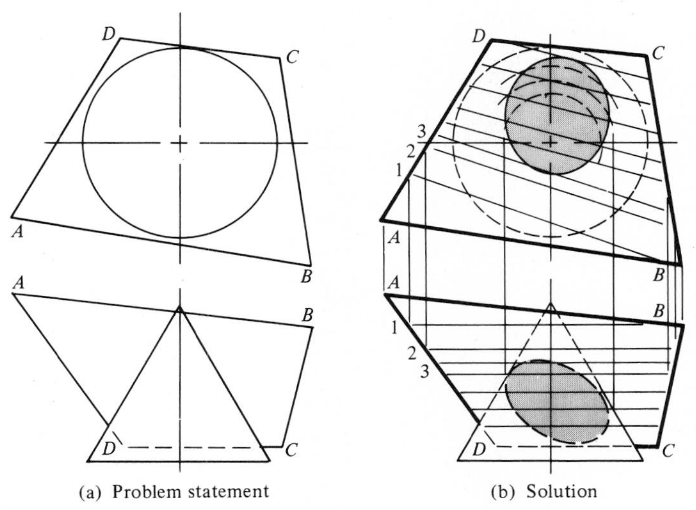

(a) Problem statement (b) Solution

Example 13-10
Intersection: plane and cone (two-view method).

Solution:

1. In the front view, pass arbitrarily spaced cutting planes perpendicular to the frontal plane so that they are seen on edge in the front view. The planes must cut both the cone and plane *ABCD*.
2. Project the points 1, 2, 3, and so on, of the horizontal cutting planes so that their traces are shown in the top view.
3. In the front view, determine the radii of the cone where it is cut by the planes. Using the given radius, draw each respective circular trace in the top view.
4. Locate the points at the intersection of the lateral and circular traces. Join these points with a smooth curve to show the required intersection of the plane and cone.
5. Project the points into the front view and draw the required line of intersection. ◀

13-5 PLANE AND DOUBLE-CURVED SURFACE (SPHERE)

A horizontal plane shown on edge in the front view will show a circle as the line of intersection in the top view. The intersection of the given plane will appear as an ellipse.

Example 13-11

Given: Incomplete front and top views of a sphere and intersecting plane.
Task: Show the intersection of the plane and sphere in both views.
Solution:

1. Pass horizontal cutting planes through the sphere in the front view where they will appear as lines.
2. Establish circular sections in the top view to agree with the sections cut from the front view.
3. Determine the points where the horizontal cutting planes intersect the given plane.
4. Project these points to the circular elements in the top view to determine the points on the line of intersection.
5. Determine correct visibility and draw the required line of intersection.

Note: If a circular view of the intersection is required, it could be obtained by drawing a frontal auxiliary where the line of sight would be perpendicular to the given plane. ◀

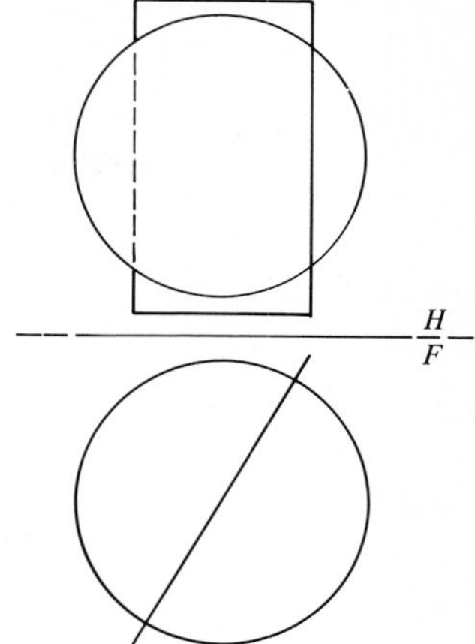

(a) Problem statement

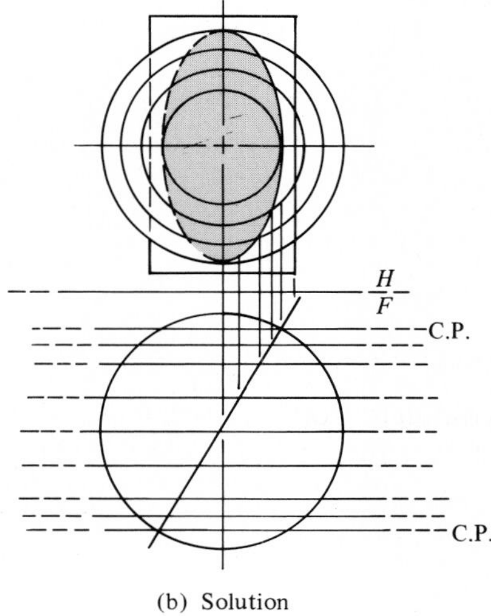

(b) Solution

Example 13-11
Intersection: plane on edge and sphere.

13-6 FAIR SURFACES

The sphere, one of the double-curved surfaces such as the torus and ellipsoid, for example, was used in Section 13-5 to illustrate the method whereby a plane would intersect a standard geometric figure.

In addition to the above, the engineer must be able to define a contoured surface since industry deals with the practical applications of cutting planes in the design of ship hulls, automobile bodies, airplane fuselages, and many other articles of complex curvature. These "shapes" can be envisioned as an envelope encasing the workings of the structure. Since the method of finding contour lines of the ship, automobile, airplane, and the like are basically the same, the line drawings used in this section will adequately illustrate the results obtained when using the principles of descriptive geometry.

To fully describe the complex curvature of a ship's hull, the designer must present a special set of lines which shows the curves formed by the intersection of various planes with the surface of the ship's hull. (See Figure 13-1.)

The student of descriptive geometry who is a stranger to lines drawings will find the arrangement of views to be unusual. Although some of the modern drawings have the conventional descriptive geometry arrangement, the majority tend to follow the traditional naval architecture way to present the views. As shown in the figure, the top and front views are in different positions and appear as they would if first angle projection were used.

The group of horizontal planes includes two planes of primary interest: (1) the *base plane* from which all vertical measurements are taken, and (2) the *designer's waterline* at which the ship is designed to float. Additional horizontal planes are called *waterline planes* and form curves of intersection where they intersect the contour of the ship's hull. The curves are called *waterlines* and are shown in the half-breadth plan which appears in the conventional location of the front view.

Vertical longitudinal planes are called *bow and buttock planes.* They are shown on edge in the half-breadth plan and are perpendicular to the waterline planes. The curves formed by the intersection of these planes with the hull of the ship are called *bow and buttock lines,* or *buttocks,* and show in the *sheer plan.*

Section planes are perpendicular to both the waterline planes and the bow and buttock planes. A station plane intersects the hull in a curve called a *transverse section* or a *station.* Stations are normally spaced at equal intervals over the length of the ship starting at the bow with Station *O.* The stations are shown on the body plan, which shows the front of the ship (the conventional right side). The body plan corresponds to the conventional right-side view except for the fact that the stations of the fore part of the ship are shown on the right side of the center line plane while those of the aft part are shown on the left. The body plan may be placed at any one of several different places on the drawing and is not necessarily aligned with any other view. However, for lines fairing work, the base plane lines for the sheer and body plans should be collinear.

Even though most of the work formerly done by hand in the mold loft

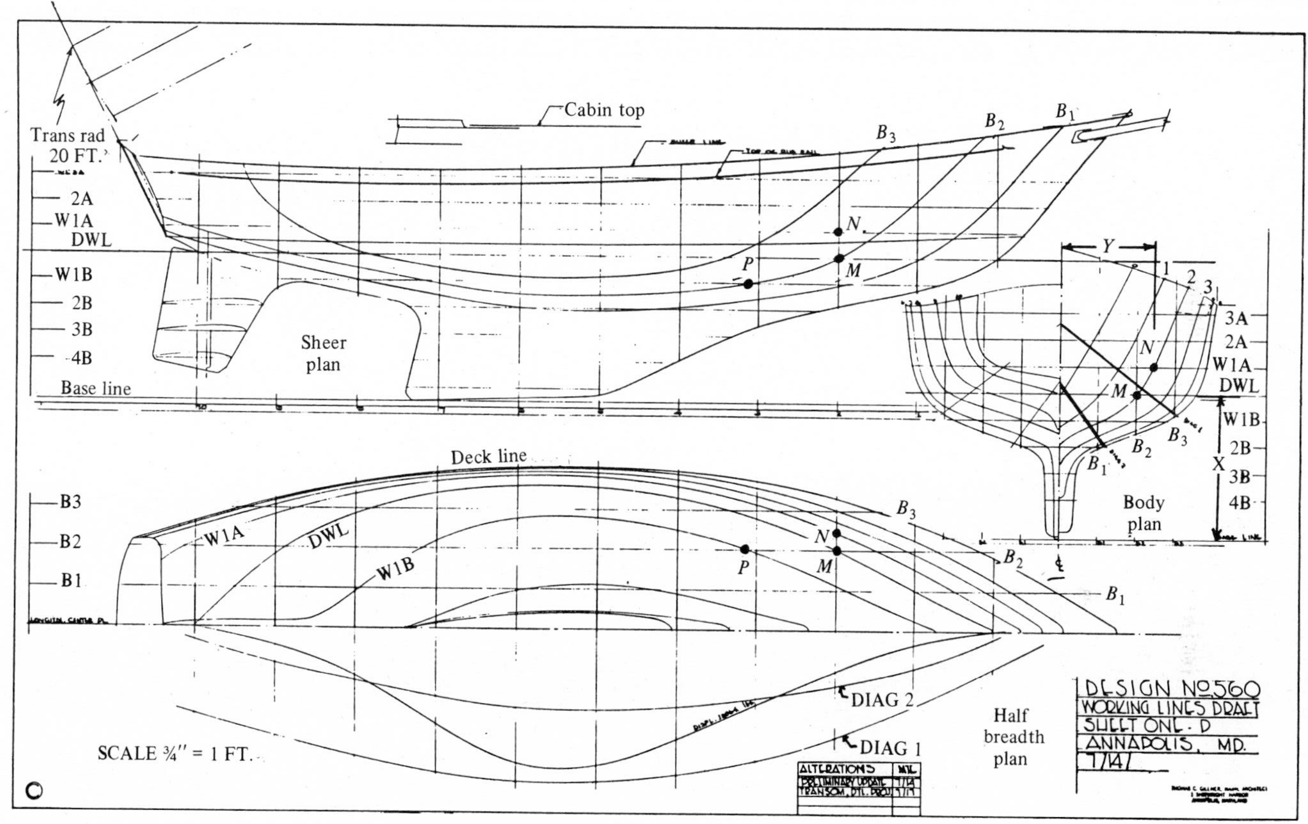

Figure 13-1
Lines drawing: ship's hull. Courtesy of Thomas C. Gillmer, Naval Architect, Annapolis, MD.

is now done by computer, it is imperative that the designer furnish the model maker or the ship builder with a set of plans which has fair and accurate lines. The experience of designers enables them to initiate a set of curves on the body plan and to draw the centerline profile in the sheer plan.

The following illustration and the points noted on the lines drawing of the sailing yacht will show the method used to transfer points on a curve from one view to another. In the body plan, point M is located at the intersection of station 2 and the designer's waterline (DWL) and bow and buttock plane B_2. It is X distance above the base line at station 2 and becomes a point on buttocks B_2.

To draw the waterlines, find the transverse distance Y (from the centerline on the body plan) to the point of intersection between the station line and the waterline. Point N is the intersection of station line 2 and waterline 1A (W1A). It is Y distance from the centerline and is transferred to the half-breadth plan as shown and becomes a point on W1A.

The intersection of a waterline and station line, such as point M, must be a point of the hull surface in all three views for the ships lines to be "fair." The same is true for a point of intersection on a buttock line and a station line. It requires trial and error to draw a complete set of fair lines since the curves must be continuous and smooth.

A further check on the fairness can be made as follows. Point P in the half-breadth plan shows the intersection of waterline W1B and bow and buttock plane B_2. This point will project into the sheer plan and will be a point on the B_2 buttock. The points in all views should be adjusted and changed if necessary until they project between views such that they are located on smooth, fair curves. Fairness requires not only agreement in three views and smoothness, but the far more subjective quality of being aesthetically pleasing to the experienced designer!

A fourth set of planes called *diagonals* is sometimes drawn to check the fairness of the lines of the drawing. Planes inclined to the longitudinal center plane and perpendicular to the stations intersect the hull to form curves called diagnonals. This auxiliary is rotated to a horizontal position so that the curve of intersection is shown in true shape on the half-breadth plan. Two diagonals are shown on the lower half of the half-breadth plan shown in Figure 13-1. These lines appear straight in the body plan where they were initiated.

Computer-aided ship design was one of the first activities in which significant effort was made by several companies to generate and improve software for interactive layout and design of ship hull surfaces. Once the surface is defined the data can be used in other programs to determine drag, stability, volume, and a host of other parameters important to the naval architect. Several advanced methods for geometric modeling have been implemented. Some systems also generate code for the computer control of flame cutters for producing the steel plates that make up the ship hull, or for numerically controlled milling of ship models used in towing-tank testing.

Geometric modeling of complex shapes such as ships, aircraft, and automobiles is a difficult task at best. Often clay or wooden models are first created by hand and then points are digitized from this prototype surface to define points on a series of three-dimensional surface patches which are then defined mathematically. As systems and software improve, designs can be created directly on the computer screen using dials, tablets, a keyboard, and other hardware interfacing devices. The design of the data base which contains the surface information is equally important. Easy data access must be available so that questions of form, shape, size, area, distance, and other parameters can be answered by proper use of the principles of descriptive geometry and geometric modeling.

13-7 INTERSECTING PLANES

Several methods are available to determine the line of intersection between two intersecting planes. Section 8-9 suggested one approach. Program CLIPLANE in Appendix 1 gives an approach which makes use of previously developed subroutines. The three principal subroutines are PLANE, BLINE, and INTLP. Supporting subroutines are EYETOZ, AXIS-ROT, GRAPHICS, and TITLE. The reader should now be familiar with the purpose of each of these routines.

The steps used in CLIPLANE to create output such as shown in Figures 13-2 and 13-3 are as follows:

a. Call EYETOZ to fix the eye position of the observer and create the 3 $\times$ 3 orthographic projection matrix (O).

b. Define three noncollinear points $\mathbf{B}_1$, $\mathbf{B}_2$, $\mathbf{B}_3$ on a base plane using a DATA statement. These points are then stored in arrays R( ), S( ), T( ).

c. Call PLANE to determine the implicit parameters N_1, N_2, N_3, k for the base plane, and to calculate a fourth corner $\mathbf{B}_4$ to define the base plane as a rectangle.

d. Call GRAPHICS to draw a single orthographic (telephoto) view of the base plane.

e. Define a clipping plane which will intersect the base plane specifying three noncollinear points $\mathbf{C}_1$, $\mathbf{C}_2$, $\mathbf{C}_3$ which are then stored in arrays I( ), J( ), K( ).

f. Call PLANE to repeat step c for the clipp˙ g plane.

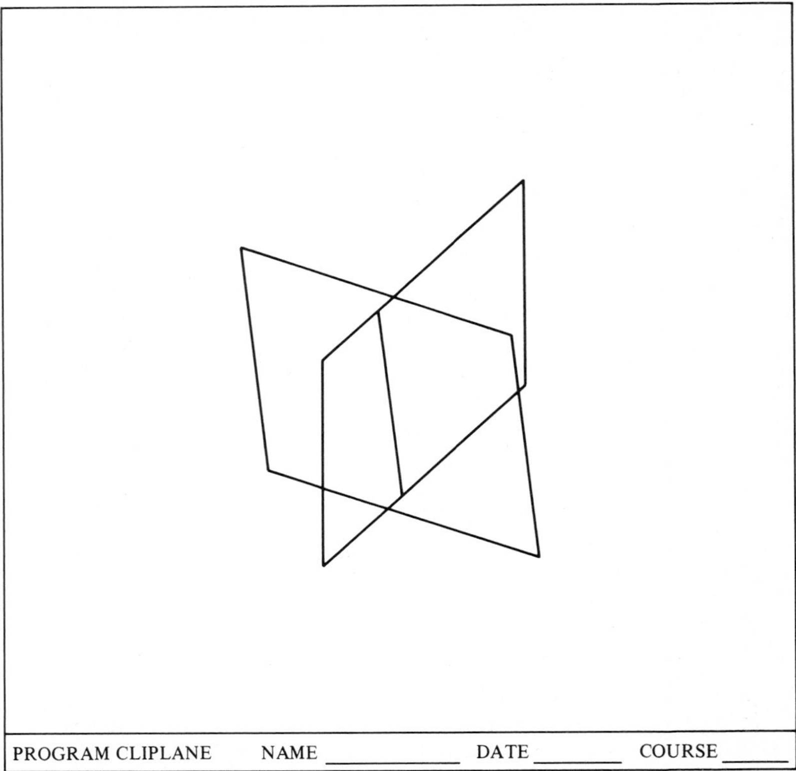

Figure 13-2
Two intersecting planes.

g. Call GRAPHICS to repeat step d for the clipping plane.

h. Define two lines on the clipping plane in parametric form as

$$\mathbf{L}_1 = \mathbf{C}_1 + t(\mathbf{C}_2 - \mathbf{C}_1); \quad 0 \leq t \leq 1.0$$

$$\mathbf{L}_2 = \mathbf{C}_4 + t(\mathbf{C}_3 - \mathbf{C}_4); \quad 0 \leq t \leq 1.0$$

For each line call INTLP to find the piercing point between the line and base plane. Store each piercing point in the arrays U( ), V( ), W( ).

i. Call GRAPHICS to draw the line of intersection between the calculated piercing points.

j. Call TITLE to complete the graphical output.

Figure 13-2 shows an output from CLIPLANE with $\mathbf{B}_1 = [1.2 \;\; 1.2 \;\; 1.2]$, $\mathbf{B}_2 = [-1.2 \;\; 1.2 \;\; 1.2]$, and $\mathbf{B}_3 = [-1.2 \;\; -1.2 \;\; 0.8]$ and with $\mathbf{C}_1 = [0 \;\; -1 \;\; -1]$, $\mathbf{C}_2 = [0 \;\; 1 \;\; -1]$, and $\mathbf{C}_3 = [0 \;\; 1 \;\; 2]$. The line of intersection shown in Figure 13-2 is the correct one for the two finite planes, even

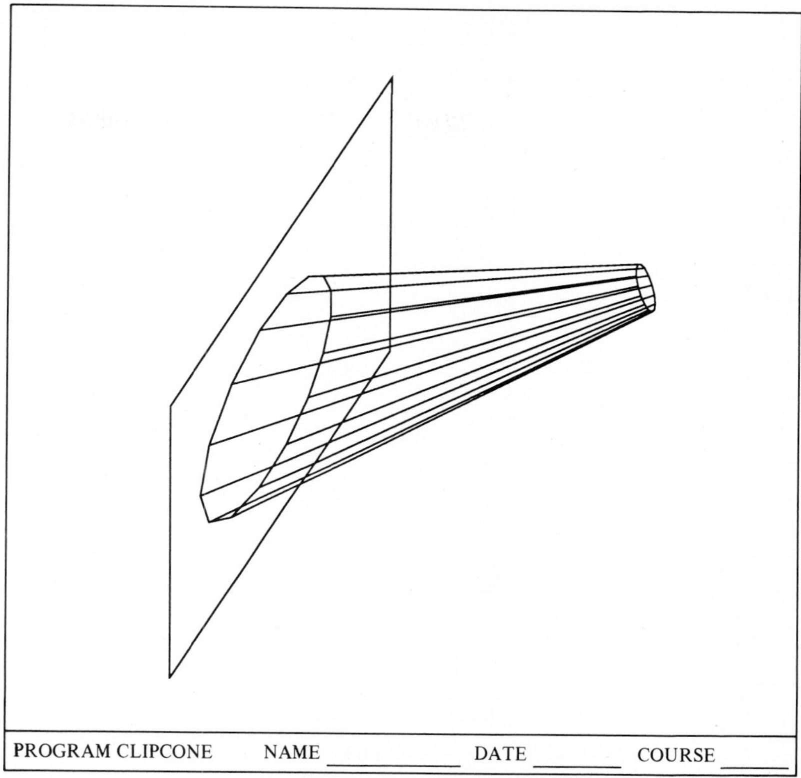

Figure 13-3
Plane-cone intersection.

though the piercing point calculation was based upon the mathematical definition of infinite planes.

Consider the same base plane intersecting with a clipping plane given by $\mathbf{C}_1 = [0 \ \ -2 \ \ -2]$, $\mathbf{C}_2 = [0 \ \ 2 \ \ -2]$, and $\mathbf{C}_3 = [0 \ \ 2 \ \ 2]$. Now the line of intersection displayed by CLIPLANE is not the desired output since it contains points which are not common to both finite planes. This can be corrected by interchanging the base plane and clipping plane, or by using a search technique to determine when a point on the line of intersection lies outside the finite region of one of the planes. A method for doing this was given in Section 4-7.

13-8 INTERSECTING PLANE AND CONE

When a cone can be represented by a series of planar facets, then the lines which form the boundaries of these facets intersect a base plane at piercing points which can be calculated as was done in the previous section. Thus, only a few modifications to the program CLIPLANE are needed to compute the curve of intersection between a plane and a cone. (See Figure 13-3).

Program CLIPCONE in Appendix 1 gives one solution. A clipping cone frustum is defined, rather than a clipping plane. The end radius is $R1$, the base radius is $R2$, and the frustum length is $L1$. The cone is defined using $N = 2\pi / (n - 1)$ facets, where n is arbitrarily chosen to be equal to 16. The endpoints for the lines along the cone surface are calculated using the following equations:

$$x_1 = R_1 \cos \theta \qquad x_2 = R_2 \cos \theta$$

$$y_1 = R_1 \sin \theta \qquad y_2 = R_2 \sin \theta$$

$$z_1 = \frac{-L_1}{2} \qquad z_2 = \frac{L_1}{2}$$

For each value of θ a call to INTPL is made to find the components of the piercing point which is then stored in arrays U( ), V( ), W( ). A call to GRAPHICS produces the computer graphics output. The last part of the program CLIPCONE is discussed in Chapter 15 under the topic of "developable surfaces."

PROBLEMS: CHAPTER 13

13-1 *Given:* Front and top views of a prism and the edge view of an inclined plane.
Task: In the top view, show the line of intersection of the plane and the prism.

13-2 *Given:* Top, front, and right-side views of a cone. The cone is cut by a plane which is shown on edge in the side view.
Task: Show the intersecting line of the plane and the cone in the top and front views.

13-3 *Given:* Top and front views of a cone cut by a plane.
Task: Show the line of intersection of the plane and cone in the top and front views.

13-4 *Given:* Top and front views of a pyramid cut by plane *MNO*.
Task: Show the line of intersection of the plane and pyramid.

13-5 *Given:* Top and front views of pyramid *MNOP* cut by plane *ABCD*.
Task: Show the piercing points and line of intersection of the plane and pyramid.

13-6 *Given:* Front and top views of a sphere cut by a plane shown on edge in the front view.
Task: Show the line of intersection of the plane and sphere.

13-7[c] *Given:* Figure 13-7 shows three principal views and a general orthographic view of a pyramid cut by an inclined plane which is perpendicular to the frontal plane.
Task: a. Define such a plane by one of three methods:
1. Three points in space
2. A normal vector and its base vector
3. An implicit equation

[c]Superscript c indicates a computer solution is required.

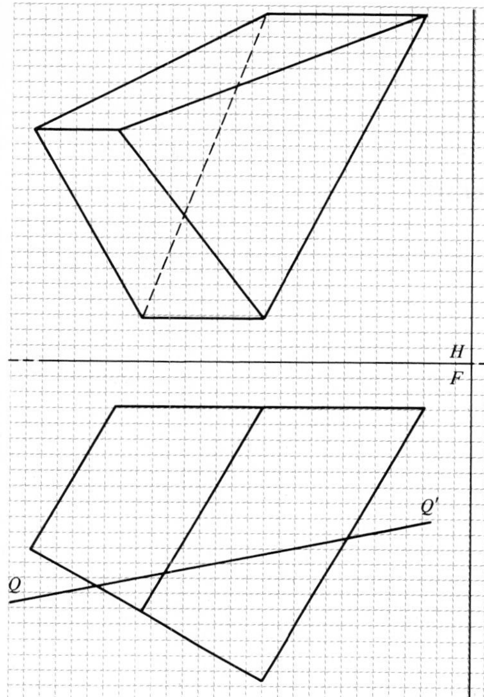

Figure 13-1P

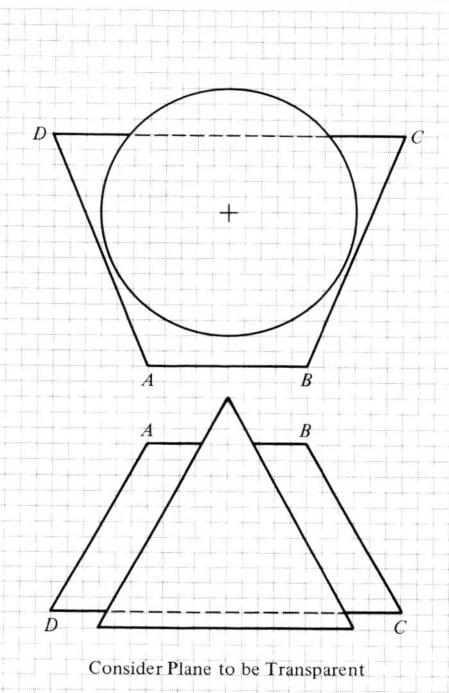

Figure 13-3P

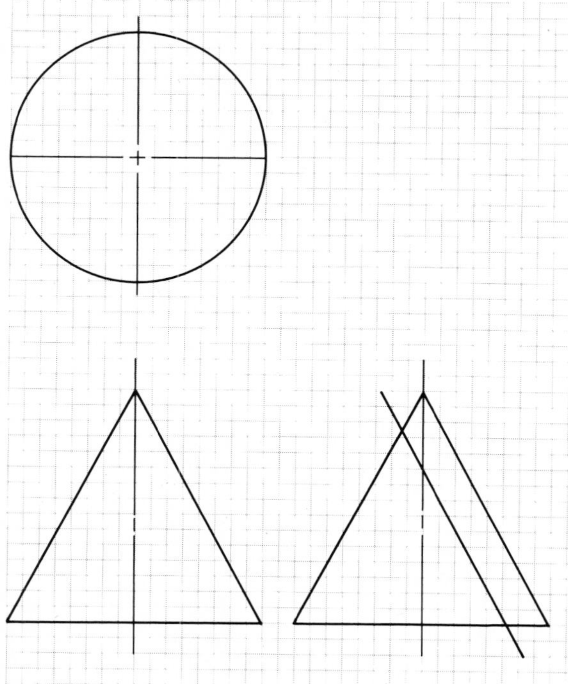

Figure 13-2P

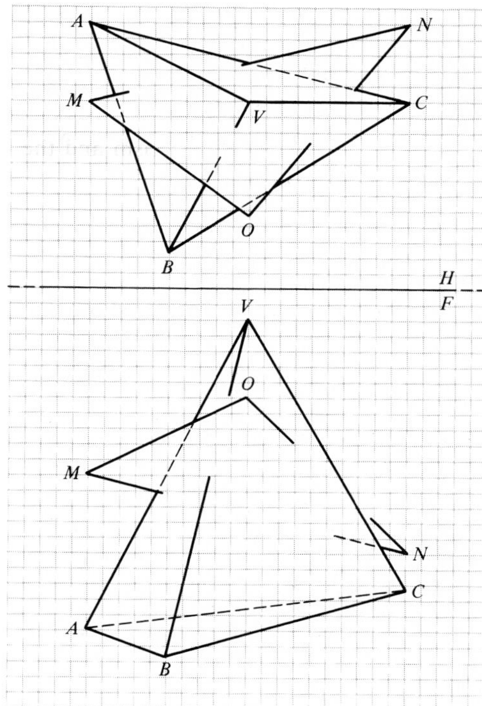

Figure 13-4P

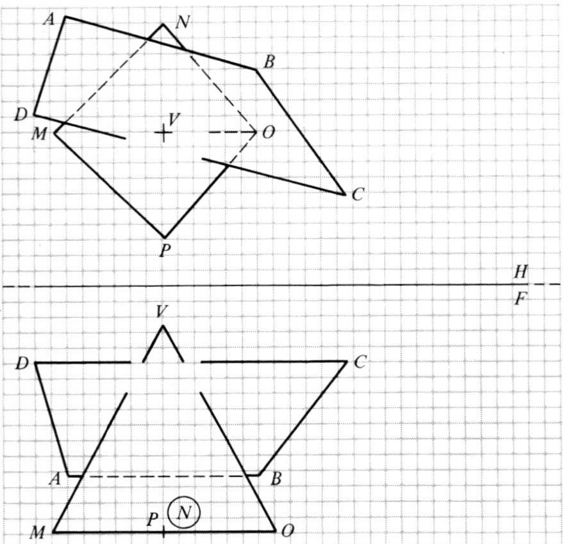

Figure 13-5P

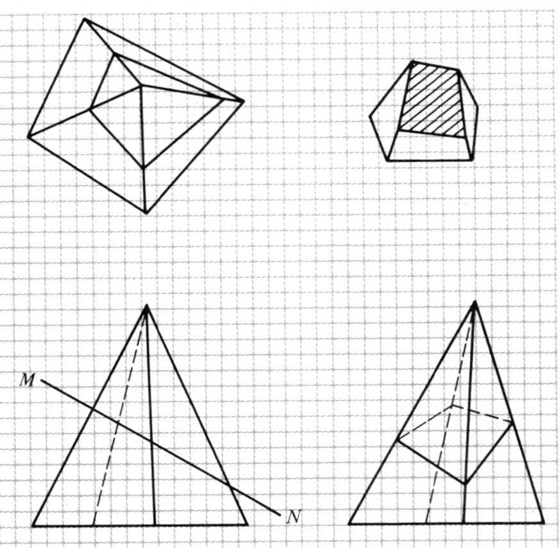

Figure 13-7P
Intersecting pyramid and plane.

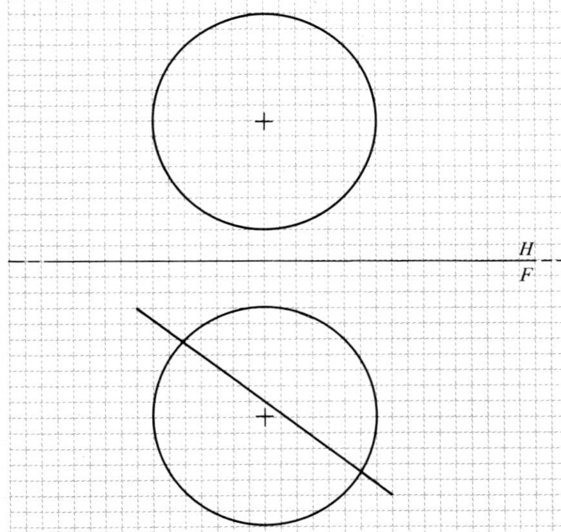

Figure 13-6P

b. Define a pyramid and calculate the points of intersection between the plane and this pyramid.

13-8[c] *Given:* Figure 13-8P shows the principal views and a general orthographic view of a five-sided prism cut by an inclined plane which is perpendicular to the frontal plane.

Task: a. Define such a plane by one of three methods.

1. Three points in space
2. A normal vector and its base vector
3. An implicit equation

b. Define a five-sided prism and calculate the points of intersection between the plane and this pyramid.

13-9[c] *Given:* Figure 13-9 shows views of a triangular prism intersecting a square prism.

Task: a. Define three-dimensional points that can be used to create displays of these two objects which intersect.

b. Calculate the points of intersection and draw the lines of intersection between the two objects.

c. Study the problem for various sizes of square prisms and various orientations of the triangular prism.

---

[c]Superscript *c* indicates a computer solution is required.

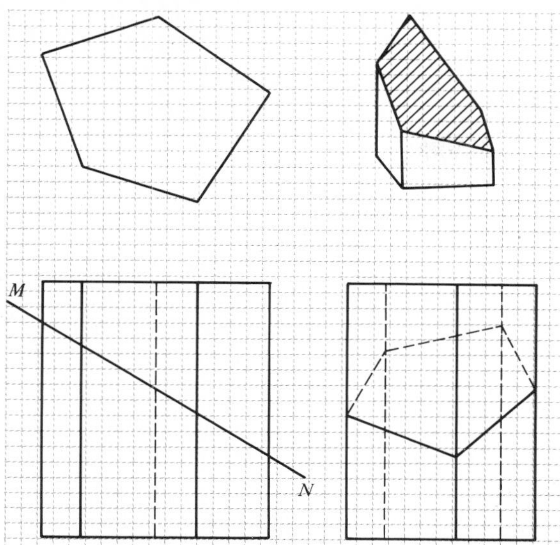

Figure 13-8P
Intersecting prism and plane.

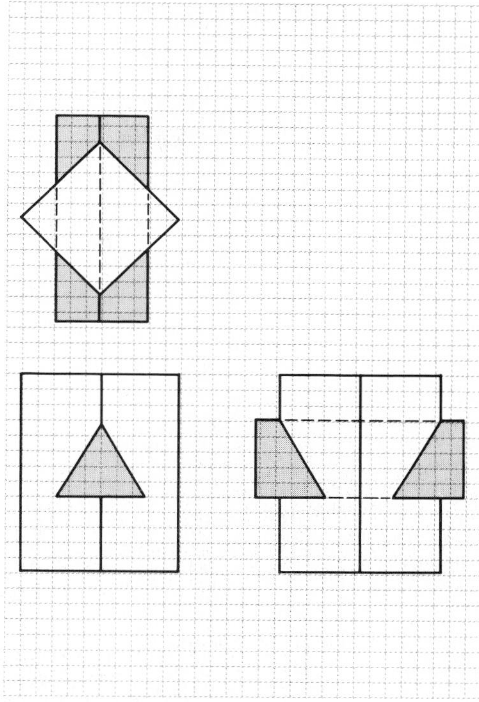

Figure 13-9P
Intersecting prisms.

13-10 *Given:* Incomplete front and top views of plane *ABCD* which intersects a triangular prism.
Task: Using only the top and front views, show the line of intersection of the plane and prism. Consider the plane to be transparent.

13-11 *Given:* Incomplete top and front views of a truncated cone cut by an opaque plane *ABCD*.
Task: Show the line of intersection formed by the plane and truncated cone. Correct visibility is required.

13-12 *Given:* Complete front view and incomplete top view of a duct cut by a plane which is shown on edge in the front view.
Task: In the top view, show the line of intersection formed by the plane and the duct.

13-13 *Given:* Complete front view and incomplete top view of a solid cut by a plane which is shown on edge in the front view.
Task: In the top view, show the line of intersection between the plane and the solid.

13-14[c] *Given:* Figure 13-14P shows a triangular prism intersecting a rectangular base pyramid.
Task: a. Create a data base and a computer graphics display of the three principal views of this situation.
b. Calculate the points of intersection and show the lines of intersection between these two objects.

13-15[c] *Given:* Figure 13-15P shows a triangular prism intersecting a rectangular base pyramid.
Task: a. Create a data base and a computer graphics display of the three principal views of this situation.
b. Calculate the points of intersection and show the lines of intersection between these two objects in each view.

13-16[c] *Given:* Figure 13-16P shows the horizontal and frontal views of a triangular pyramid being pierced by a rectangular prism.
Task: a. Create the profile view using methods of descriptive geometry.

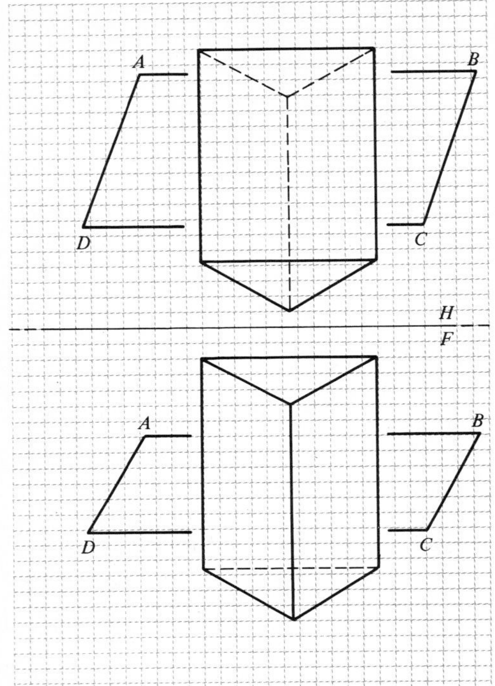

Figure 13-10P

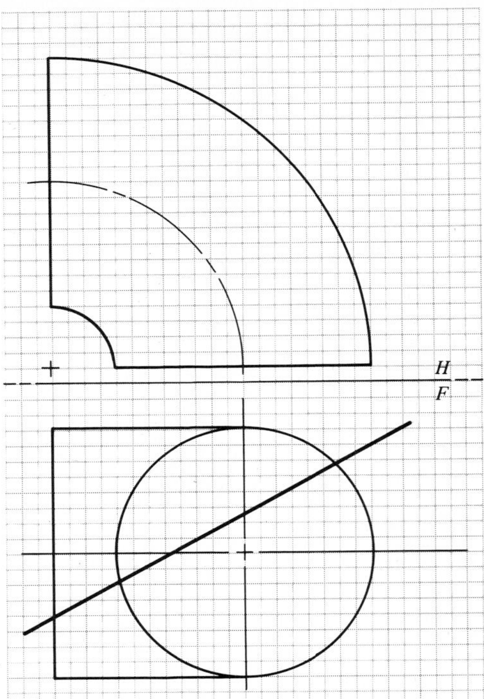

Figure 13-12P

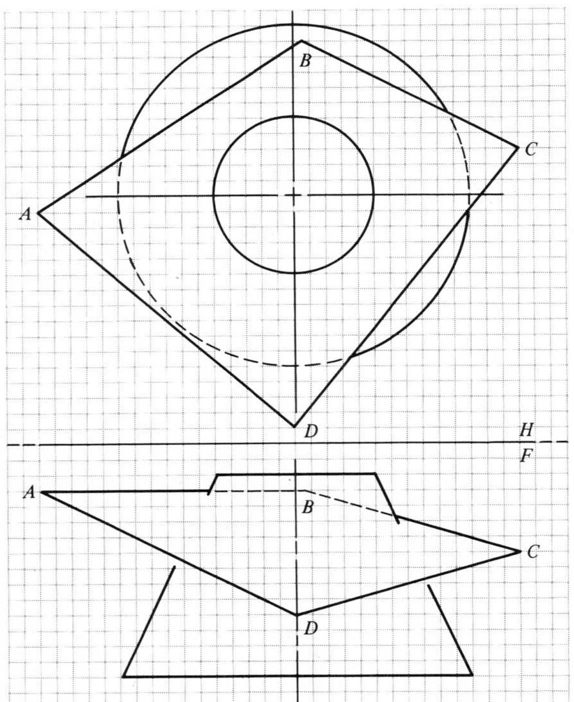

Figure 13-11P

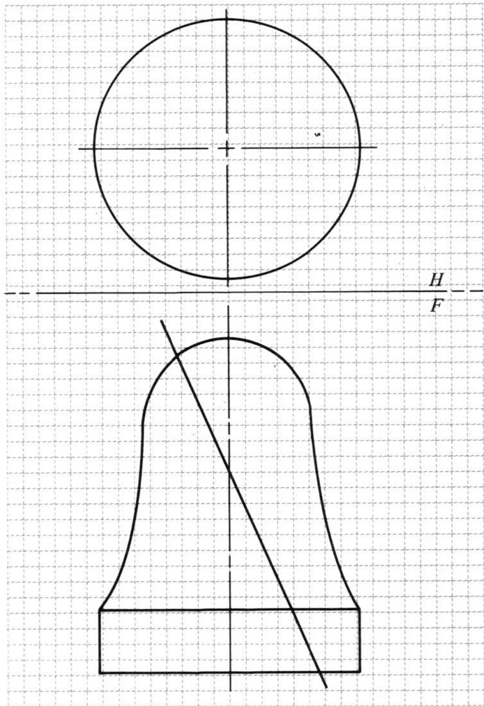

Figure 13-13P

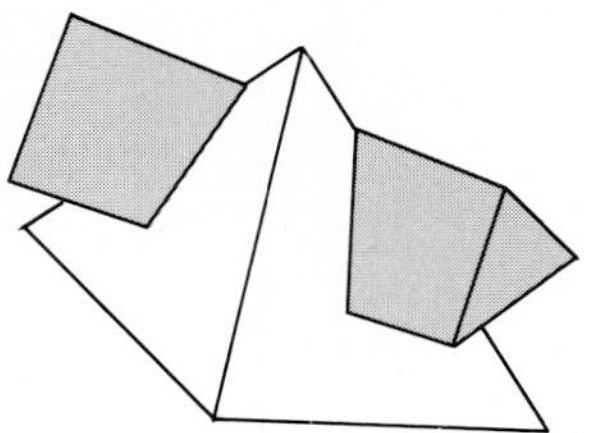

Figure 13-14P
Intersecting prism and pyramid.

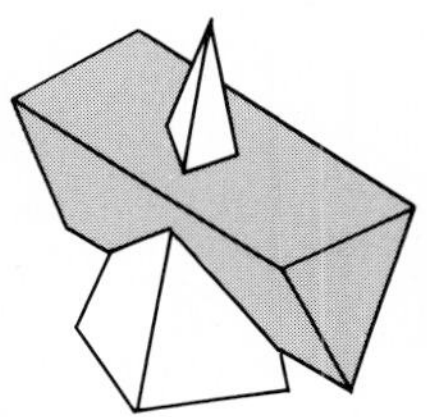

Figure 13-15P
Intersecting pyramid and prism.

b. Calculate the lines of intersection between the two objects using methods of geometric modeling.
c. Create general orthographic projections as seen from any point in space, using methods of computer graphics.

13-17[c] *Given:* Figure 13-17P shows views of a square prism partially intersecting a triangular prism.
Task: a. Calculate the critical points of intersection which can be used to create the lines of intersection between the two objects.
b. Discuss how computer programs might be written to generate proper surfaces for partial intersections between two solids.

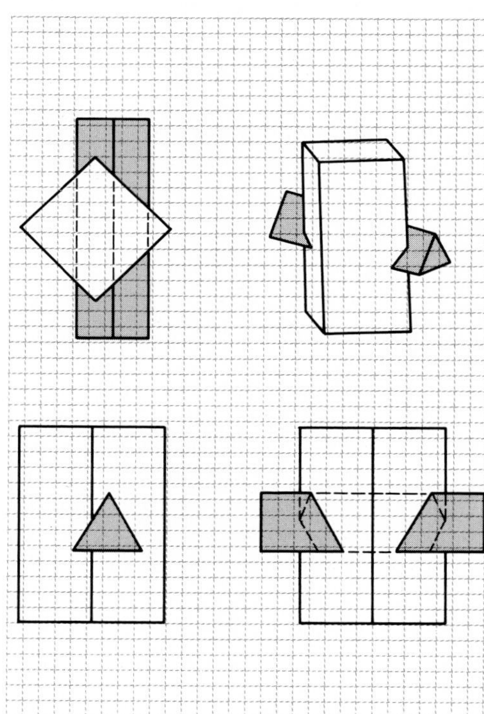

Figure 13-17P
Partial intersection of two solids.

Figure 13-16P
Rectangular prism and triangular pyramid.

14

Intersections—Solids

Surfaces often form contours that meet in lines of intersection which must be determined before fabrication or development takes place. The fundamental principles of intersection will be shown in this chapter using conventional geometric shapes. Having mastered these principles, the student can apply them to practical situations when necessary. A few examples are given to illustrate how clipping algorithms can be used to display the intersection between solid surfaces.

14-1 PRISMS (EDGE-VIEW METHOD AND CUTTING-PLANE METHOD)

Since prisms are composed of planes, the principles used in the previous chapter can be applied to the more complicated intersections found in the following material. Here, as in Chapter 13, a combination of methods may be used to solve the problem under discussion. Some problems may be solved by using edge views only, some may be presented in three views where projection may assist in the solution, while others will require the use of cutting-planes exclusively.

The problem presented in Example 14-1 is solved by using a combination of the edge-view method and the cutting-plane method. In this case, as in all other cases, the lines are terminated when they intersect a surface as they would in sheet metal parts or castings. A note of caution should be entered here in that often hidden lines will be shown.

Example 14-1

Given: Incomplete front and top views of two intersecting prisms.
Task: Using only the two given views, show the intersection of the prisms in the front view.

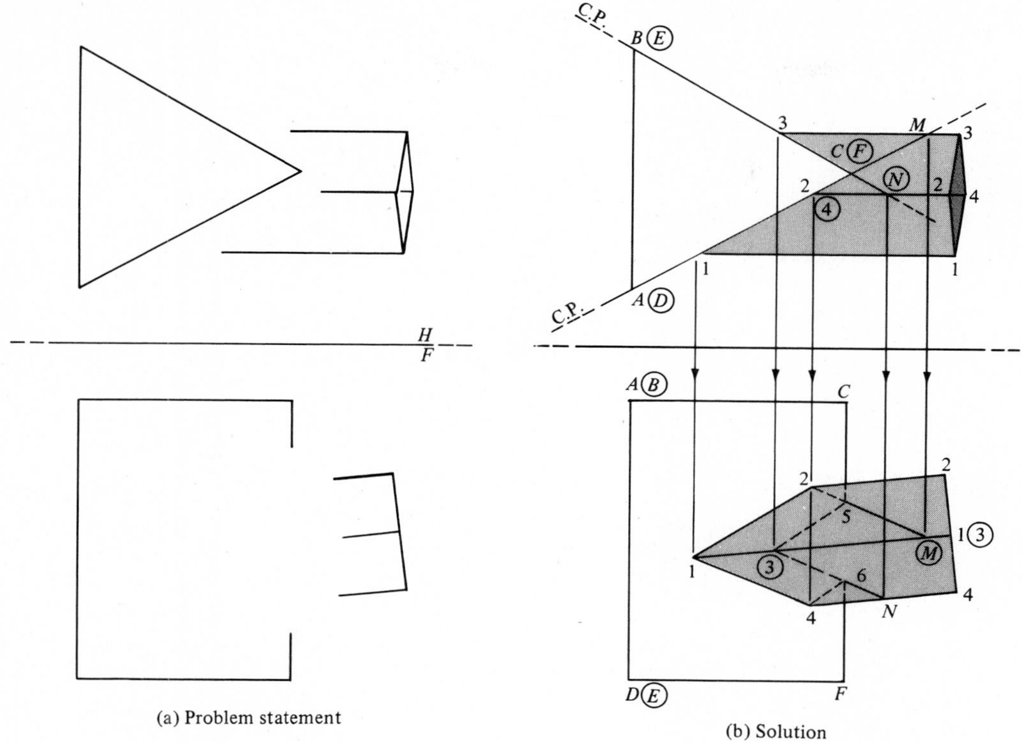

(a) Problem statement

(b) Solution

Example 14-1
Intersection of prisms.

Solution:

1. The lateral surfaces of the triangular prism are shown on edge in the top view. Piercing points 1, 2, 3, and 4 (hidden) are determined by locating the points where the edges of the square prism intersect the lateral surfaces of the triangular prisms.
2. Project the points 1, 2, 3, and 4 into the front view to intersect their respective lateral edges of the square prism.
3. In the top view, pass a vertical cutting plane so that it contains surface *ACFD* and therefore contains line *CF*. The cutting plane cuts the upper rear surface of the square prism at point 2 and point *M*.
4. Project point *M* into the front view. The trace *M*2 in the front view intersects edge *CF* at piercing point 5.
5. In the top view, pass a vertical cutting plane so that it contains surface *BCFE* and also contains line *CF*. It will cut the lower rear surface of the square prism at point 3 and point *N*.

6. Project point *N* into the front view. The trace *N*3 will intersect line *CF* at piercing point 6.
7. Showing correct visibility, connect points 1 through 6 with lines of intersection. ◀

14-2 PRISM AND PYRAMID (CUTTING-PLANE METHOD)

Example 14-2

Given: Incomplete front and top views of a prism intersecting a pyramid.
Task: Using only the two given views, show the intersection of the prism and pyramid.

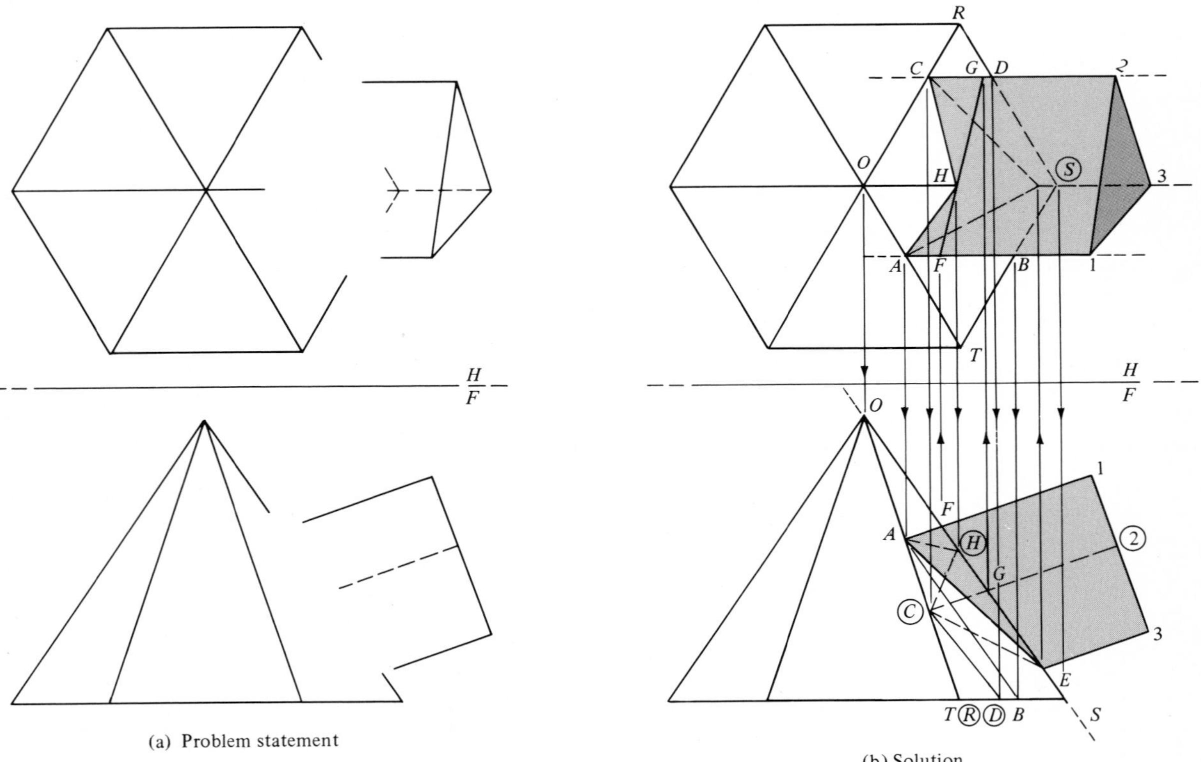

(a) Problem statement

(b) Solution

Example 14-2
Intersection of prism and pyramid.

Solution:

1. In the top view, pass a cutting plane perpendicular to the horizontal plane so that it contains edge #1 of the triangular prism. The cutting plane will cut line *OT* at *A* and base edge *ST* at *B*.
2. Project points *A* and *B* into the front view and draw trace *AB*. This trace intersects edge #1 at *A* which is a point on the line of intersection.
3. In the top view, pass a cutting plane perpendicular to the horizontal plane and containing prism edge #2. The cutting plane will cut the pyramid edge *OR* at *C* and the base of the pyramid at *D*.
4. Project these points into the front view and draw trace *CD* which intersects line #2 at *C* to indicate a point of intersection.
5. In the top view, repeat the process described above by working with the pyramid line *OS*. In the front view, line #3 of the triangular prism intersects the trace at point *E*.
6. In the front view, pass a cutting plane perpendicular to the frontal plane and containing line *OS* of the pyramid. The plane will cut the upper surface of the triangular prism at *F* and *G*.
7. Project *F* and *G* into the top view and draw trace *FG*. The line *OS* and the trace *FG* intersect at *H* and determine the point at which the pyramid edge pierces the triangular prism. Connect all points and show correct visibility for line of intersection in the top view.
8. Project *H* into the front view and connect all points to form the required intersection. Show correct visibility. ◀

14-3 CYLINDERS (CUTTING-PLANE METHOD)

The cutting-plane method (Figure 14-1) shows graphically the theory involved in finding the intersection of two cylinders. Cutting planes are passed so that all resulting traces will be straight line surface elements which are parallel to the axis of the individual cylinder. The intersection of the elements gives points on the line of intersection of the cylinders. Although cylinders intersect in different ways and at different angles, the basic principles of finding their line of intersection are the same.

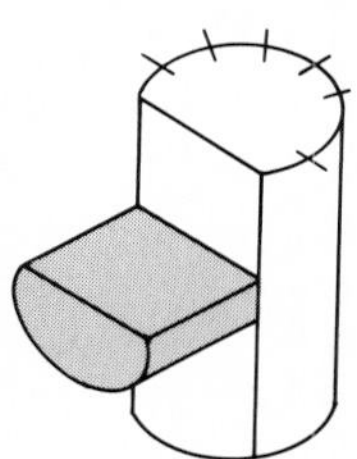

Figure 14-1
Intersecting cylinders.

Example 14-3 (cylinders with two intersecting axes) shows only the visible curve of intersection since the back portion coincides with that in the front. Example 14-4 (cylinders with two nonintersecting axes) dictates that the entire intersection be shown in the front view.

A. Axes Intersecting but Not Perpendicular

Example 14-3

Given: Complete top and incomplete front views of two cylinders whose axes intersect but are not perpendicular.

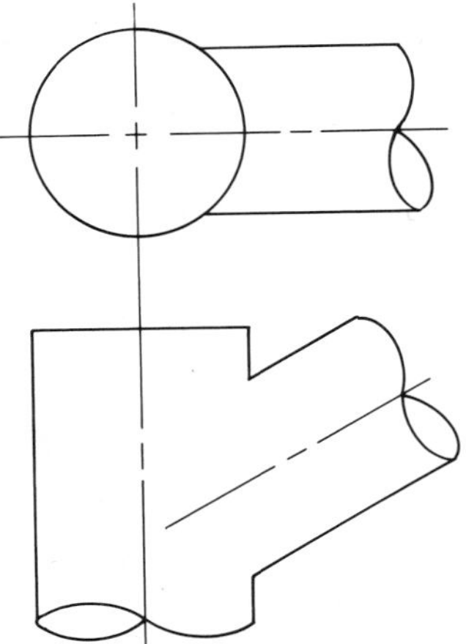

(a) Problem statement

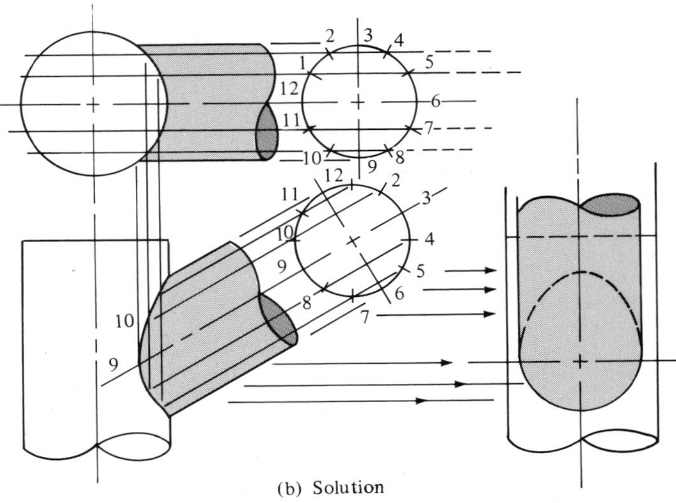

(b) Solution

Example 14-3
Intersecting cylinders with axes intersecting.

Task: Find the curve of intersection between the two cylinders.
Solution:

1. Create a circular view of the inclined cylinder adjacent to the top view as shown.

   Draw a frontal auxiliary of the inclined cylinder. Use the clock method (Section 13-3) and lay off each circle in 12 divisions. *It is very important that the divisions of the circles agree orthographically.*
2. In the top view, pass a cutting-plane through both cylinders at each division thereby creating lateral elements on each cylinder.
3. Show the lateral elements on the inclined cylinder in the front view. It is again emphasized that these elements must be designated so that they agree with those in the top view.
4. The intersection of the respective elements form points which make up the curve of intersection.
5. Draw a complete right-side view. ◀

B. Axes Perpendicular but Nonintersecting

Example 14-4

Given: Incomplete top, front, and complete side views of two cylinders whose axes do not intersect but are perpendicular.
Task: Find the line of intersection of the two cylinders. Show complete visibility.
The solution is shown in Example 14-4(b) for student diagnosis. ◀

C. Two Inclined Cylinders

Example 14-5

Given: Incomplete front, top, and side views of two intersecting, inclined cylinders.
Task: Show the intersection of the cylinders in all views.
Solution:

1. In the top view pass a series of frontal cutting planes which show on edge and pass through both cylinders.
2. Project, to the front views, the points at which the planes cut the edge of the cylinders. From each point draw an element parallel to the axis of the cylinder. The points at which the respective elements meet determine the points on the line of intersection.

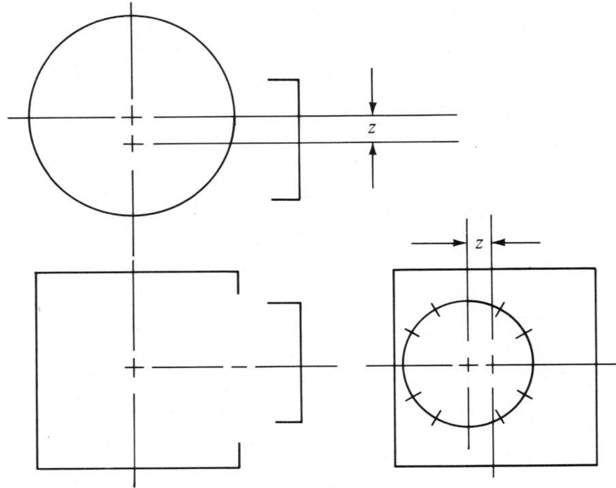

(a) Problem statement

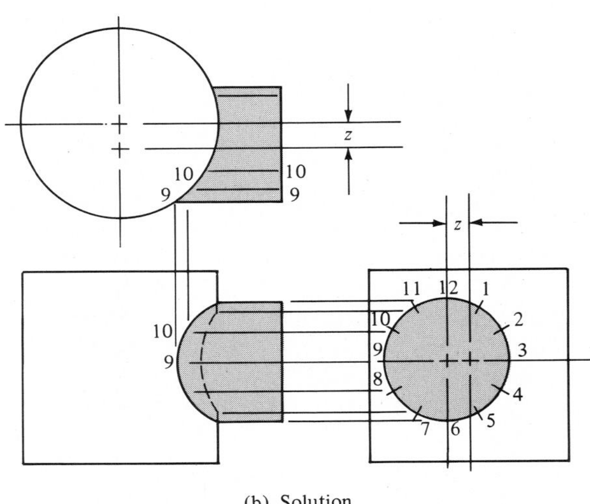

(b) Solution

Example 14-4
Intersecting cylinders with nonintersecting axes.

3. Show the cutting planes in the side view. Project the intersection points from the front view to the side view until they meet their respective cutting planes to form the side view of the intersection.
4. Show both intersections in all views.

 Note: Only two cutting planes have been shown in the side view in order to show the principle involved. Time may be saved by mea-

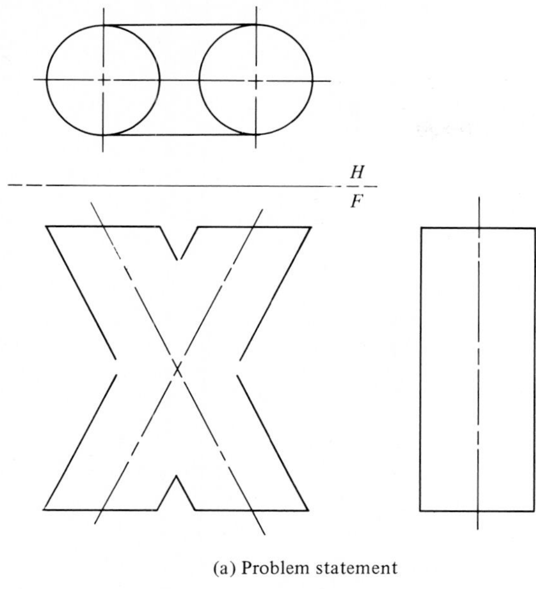

(a) Problem statement

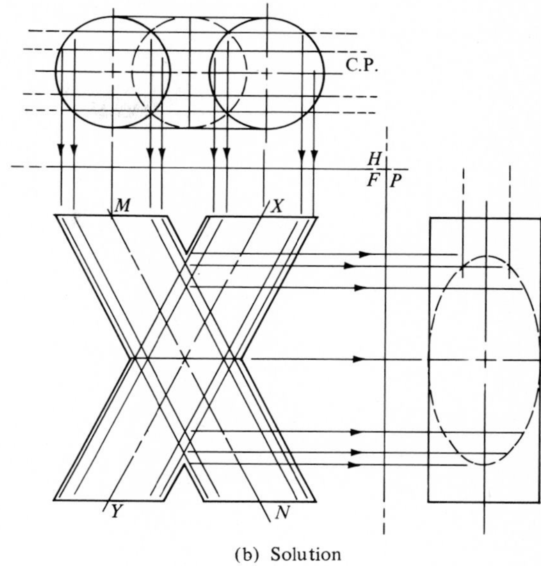

(b) Solution

Example 14-5
Intersection of two inclined cylinders.

suring the required distances in the top view and transferring them to the side view. ◄

14-4 CYLINDER AND CONE

Figures 14-2 and 14-3 show two methods in which cutting planes may be passed in order to find the intersection of a cylinder and a cone.

A. Cone and Upright Cylinder with Parallel Axes—Horizontal Cutting Planes

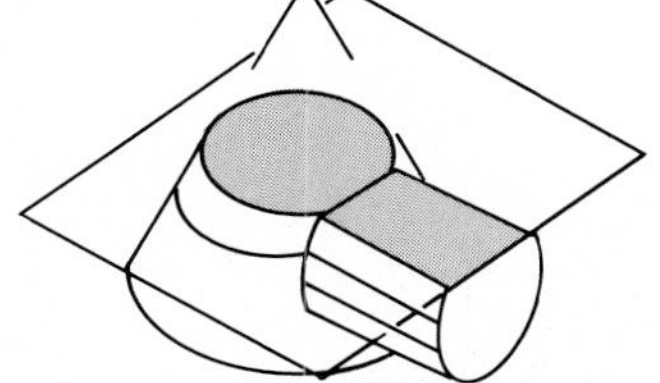

Figure 14-2
Cylinder and cone intersection.

In the following example, a series of horizontal cutting planes are passed so that they produce straight line elements in the front view of the cone and circular elements in the top view. The intersection of the lateral elements of the cylinder and the straight lines of the cone form the series of points which make up the required line of intersection.

Example 14-6

Given: Incomplete front view, incomplete side view, and complete top view of intersecting cylinder and cone.

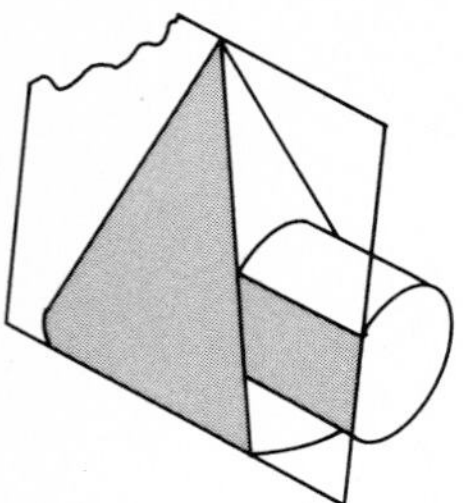

Figure 14-3
Cylinder and cone intersection.

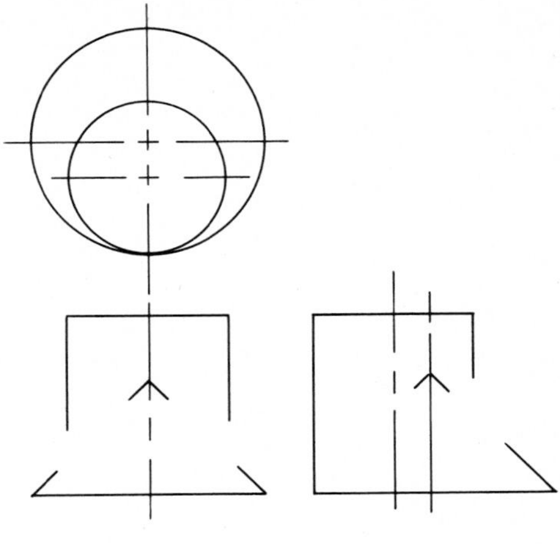

(a) Problem statement

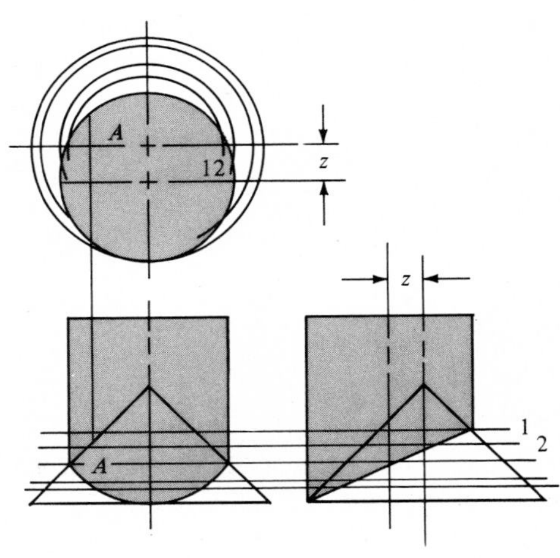

(b) Solution

Example 14-6
Cylinder and cone intersection.

Task: Show the intersection of the cylinder and cone in all views.
Solution:

1. Pass a series of horizontal cutting planes through the cone so that they appear as lines in the front view and as circles in the top view. Be sure each circle has a radius corresponding to its respective cutting plane in the front view as noted by numbers 1 and 2.

2. The top view shows that the circular elements of the cone will be intersected by lateral elements of the cylinder which are parallel to its axis as illustrated at point *A*.
3. Project the intersecting points into the front view and draw the line of intersection as required.
4. Project the points into the side view and draw the line of intersection to complete the solution. ◀

B. Cone and Upright Cylinder with Parallel Axes—Upright Cutting Planes

The following problem was solved by using a vertical cutting-plane which passes through the vertex of the cone and cuts a straight-line element from the cylinder and at the same time cuts a lateral element on the surface of the cone.

Example 14-7

Given: Incomplete front and side views and complete top view of intersecting upright cylinder and cone. Both axes must be parallel.
Task: Show the complete line of intersection of the two given solids.
Solution:

1. Pass a series of vertical cutting planes through the vertex of the cone so they are parallel to both the axes of the cylinder and cone as shown. Each cutting plane will cut a straight-line element on the upright cylinder and a lateral element on the cone at the same time.
2. The elements of the cylinder are shown on end in the top view so that it becomes a simple procedure to determine the location of their intersections with the elements of the cone and to project the required points into all views. ◀

C. Cylinder and Cone Using Upright Cutting Planes

Example 14-8

Given: Incomplete front and top views and complete side view of an intersecting cylinder and cone.
Task: Show the intersection of the given solids.
Solution:

1. Pass cutting planes through the cone from the vertex to the base and parallel to the axis of the cone. These planes will show on edge in the top view and will appear as lateral elements on the cone as shown.

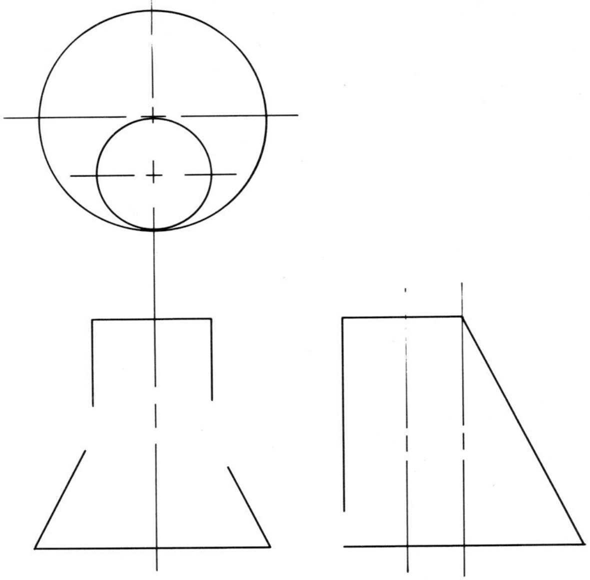

(a) Problem statement

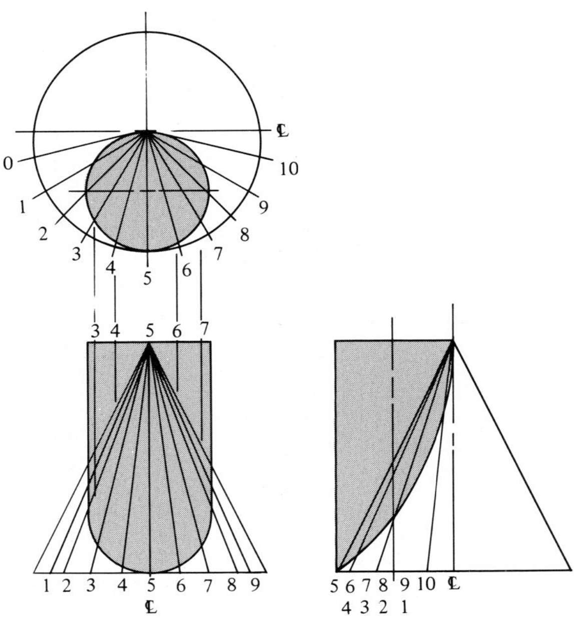

(b) Solution

Example 14-7
Cylinder and cone intersection.

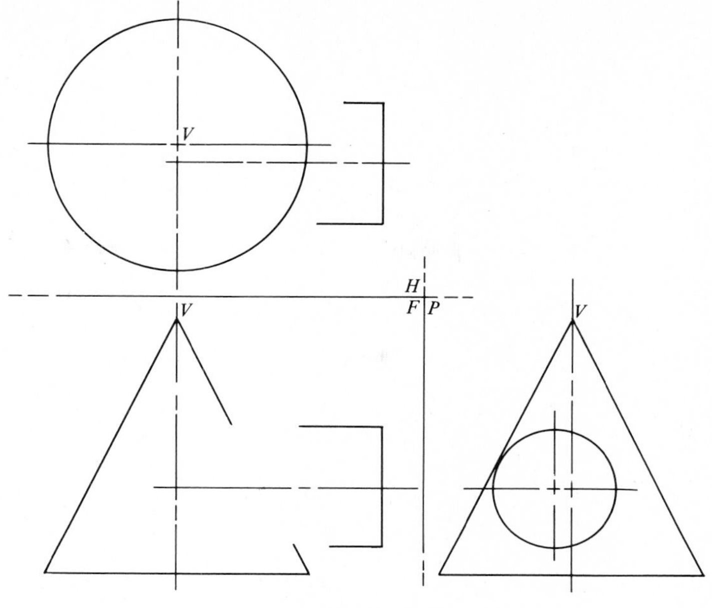

(a) Problem statement

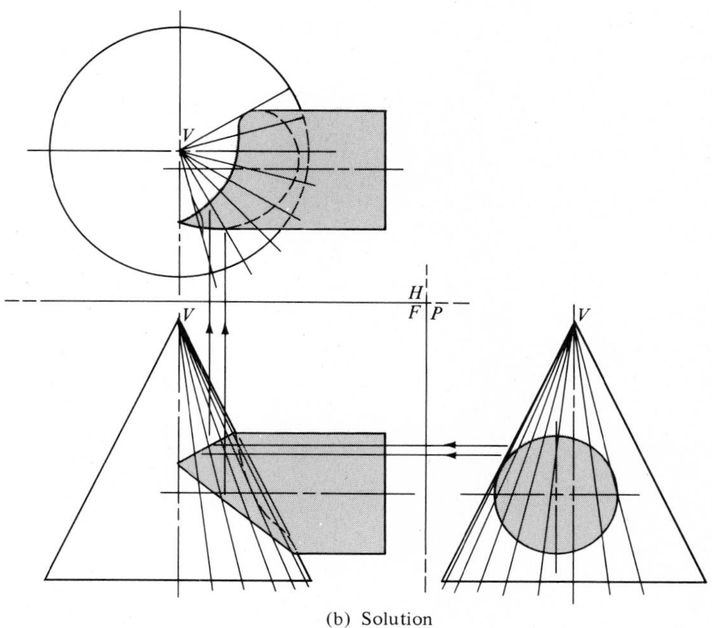

(b) Solution

Example 14-8
Intersection of cylinder and cone.

2. In the side view, the cylinder is shown on edge; therefore, all lateral elements will appear as points. The observer can determine the points at which the lateral elements of the cylinder intersect the lateral elements of the cone.
3. Project the points into the front and top views to form the required line of intersection in each view. ◀

14-5 PRISM AND CONE

Example 14-9

Given: Incomplete front view and complete top view of intersecting prism and cone.
Task: Show the line of intersection between the two given geometric shapes.

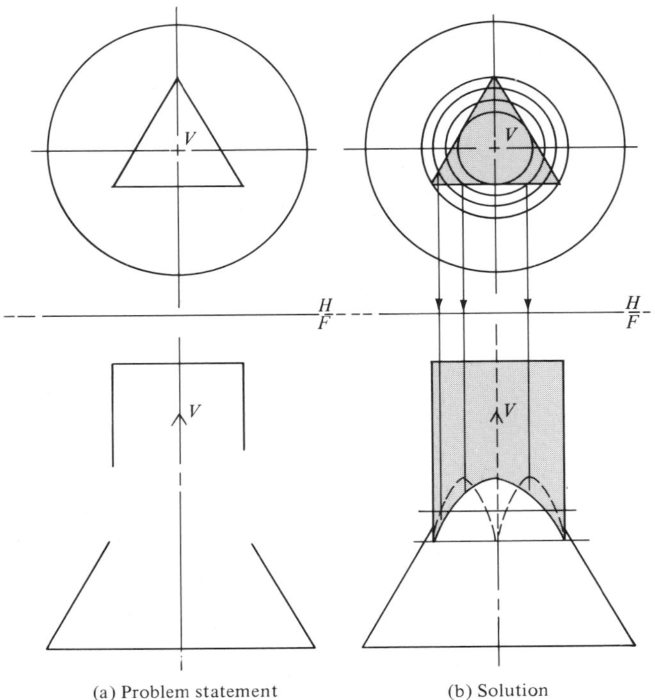

Example 14-9
Intersection of prism and cone.

The methods used in the solution of the above problem have been discussed in prior sections of this chapter. For this reason the student is advised to study the given solution and produce a suitable step-by-step procedure. ◀

14-6 CONES

Example 14-10

Given: Incomplete front, top, and side views of two intersecting cones.
Task: Find the intersection of the two given cones in all views.
Solution:

1. In the front view, pass a series of horizontal cutting planes so that they appear as lines and cut both cones.
2. In the top view, the axis of each cone is used as the center of the circular traces for that particular cone.
3. Locate the points at which the circular traces meet. These points will be located on the curve of intersection.
4. Pass a vertical cutting plane through the two vertices at points P and V. It will cut the base of one cone at S and the base of the other at X. When the traces are drawn to their respective vertices they will intersect at point #1, the highest point on the intersection.
5. Project the points into the front view and draw the intersection of the two cones.
6. From the points of intersection in the front view, project lines of sight into the side view.
7. In the top view, obtain the distances from the points to a frontal reference plane and transfer them to the side view. Draw the required line of intersection. ◀

14-7 CYLINDER AND TORUS

Example 14-11

Given: Incomplete front view and complete top view of intersecting cylinder and torus.
Task: Show the front view of the intersection of the given geometric solids.

The solution of the above problem in shown in Example 14-11(b) for study by the student who is, at this point, familiar with the procedure used. ◀

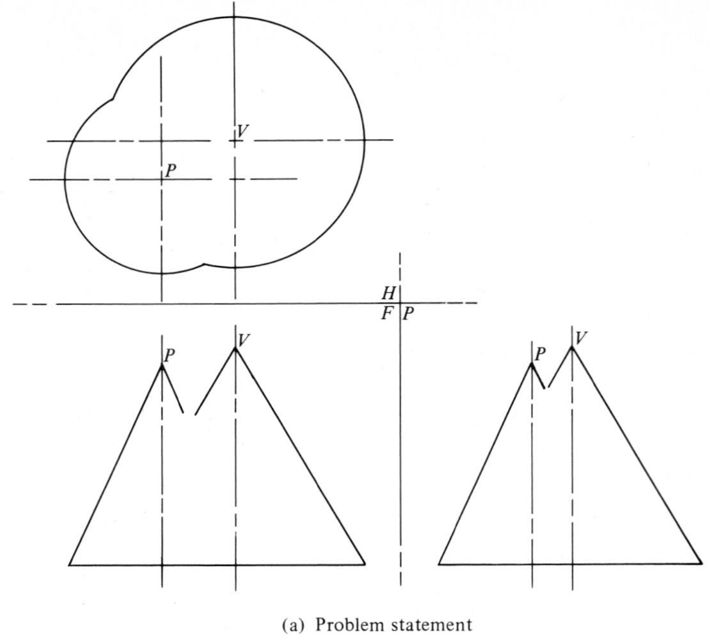

(a) Problem statement

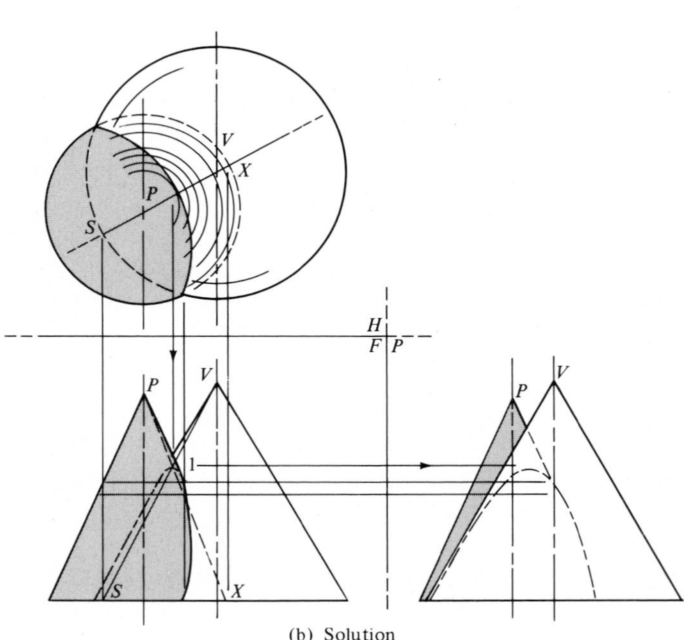

(b) Solution

Example 14-10
Intersection of two cones.

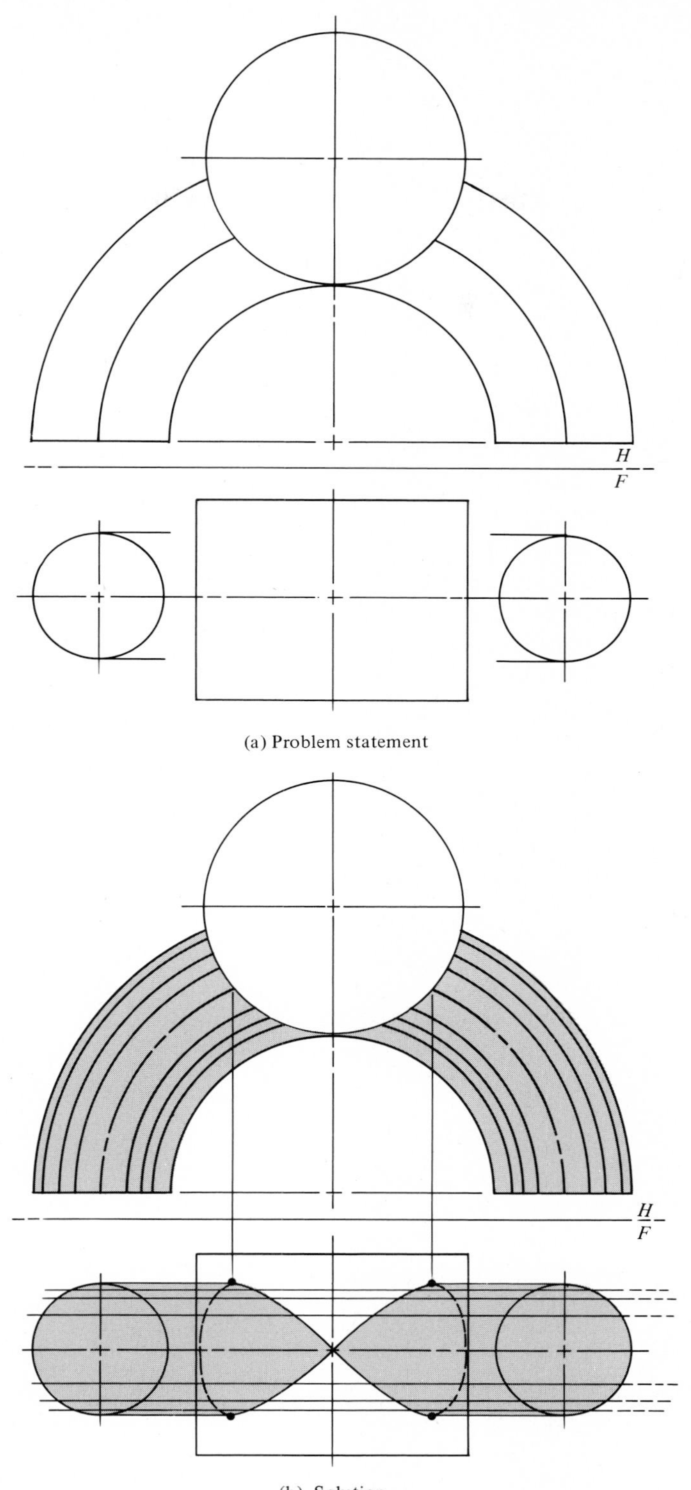

Example 14-11
Intersection of cylinder and torus.

14-8 GEOMETRIC MODELING FOR INTERSECTING SURFACES

Geometric modeling uses the principles of vector analysis to calculate lines of intersection between surfaces. Here we only consider surfaces which are represented by a series of flat faces. The calculation of line-plane intersections can be applied to finding the intersection between solids. Figure 14-4 shows a computer-generated output for the intersection between two triangular prisms. It can be seen that the solution is nothing more than intersections between a series of lines which form the edges of one prism, with a series of plane facets that form the surface of the second prism.

Figure 14-5 extends the technique of line-plane intersections to an intersecting cylinder and prism. Once a computer program exists, then many accurate solutions can be quickly obtained for a variety of config-

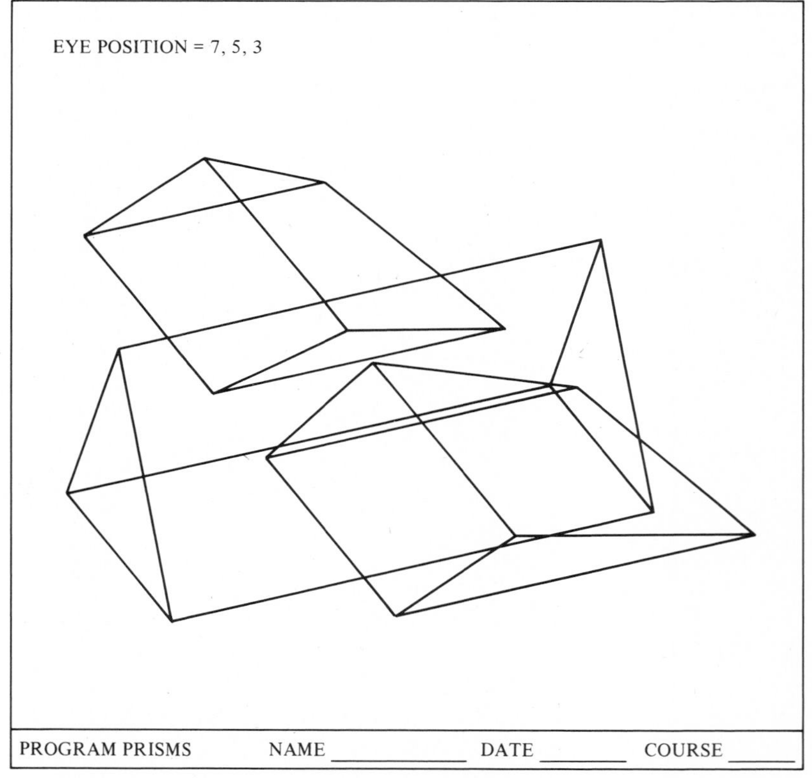

Figure 14-4
Intersecting prisms.

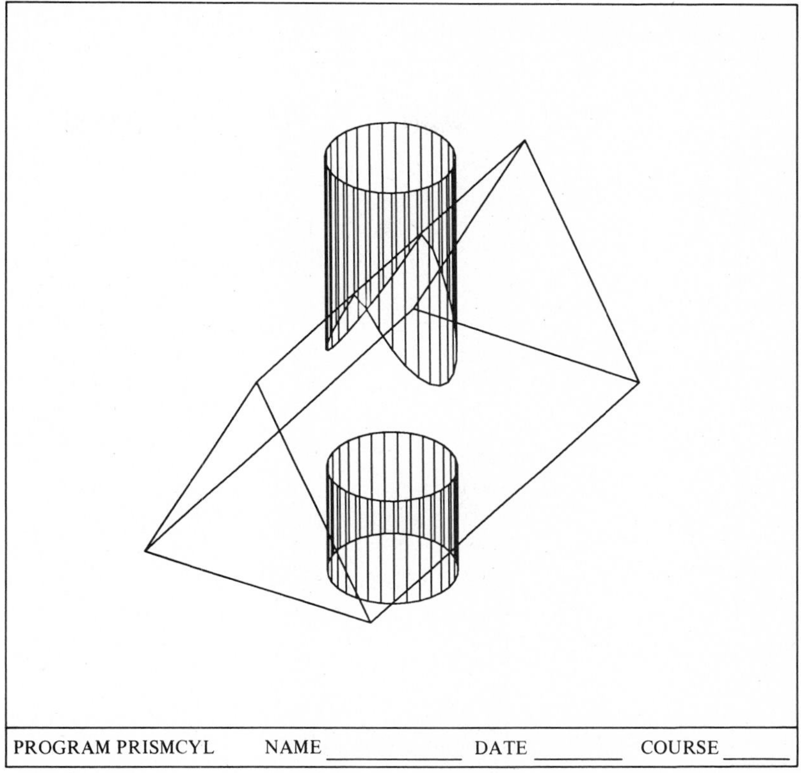

Figure 14-5
Intersecting prism and vertical cylinder.

urations. A final example in Figure 14-6 shows the intersection between a cylinder and a cone. It is easy to increase the number of line segments on the cylinder, or plane facets on the cone, to obtain increased accuracy with a minimum of effort. On the other hand, it takes a considerable amount of effort to obtain high accuracy using traditional descriptive geometry methods.

The program VEXCLIP3 used in Chapter 11 can be modified to calculate and display the intersection between any convex surface and any other surface which can be represented by a series of lines. Two intersecting prisms illustrate a simple example. In program PRISMS, the clipping prism is defined using eight vertices to close the prism as shown in Figure 14-4. When a right-hand coordinate system is used, the vertices must be defined so that the normal surface vectors calculated in subroutine CLIPSUB are inward vectors. The four vertices on the yon plane ($z < 0$) are defined in the first DATA statement (440), and the four vertices on

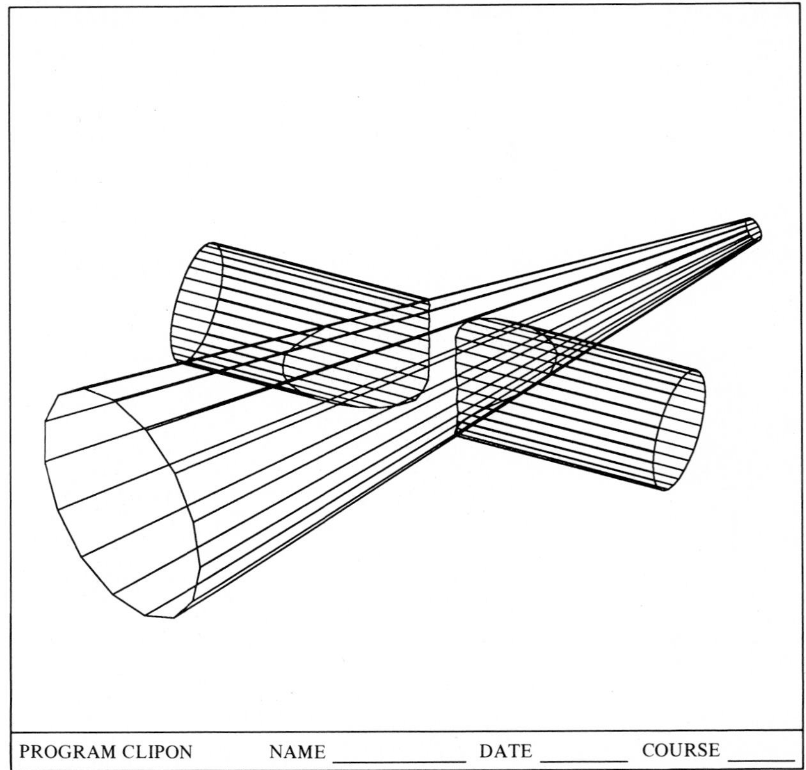

Figure 14-6
Intersecting cone and cylinder: inclined axes.

the hither ($z > 0$) are defined in the second DATA statement (450). These vertices are defined in a clockwise sense on each end of the prism. The inward unit normal vector on the yon plane is **k**, and the inward unit normal vector on the hither plane is $-\mathbf{k}$.

The commands to draw the lines of the clipping prism are given in the third DATA statement (510) and stored in the L(,) array. Then the vertices are transformed in preparation for an orthographic projection, using matrix multiplication with the matrix R(,) which is created in subroutine EYETOZ.

The lines which represent the intersecting prism are defined in the final DATA statement (650) and stored in the X( ), Y( ), Z( ) arrays in program PRISMS. Notice that this occurs in a FOR-NEXT loop where each line is clipped to the prism stored in array P(,) using subroutine CLIPSUB.

The final task is to orthographically transform both the given endpoints of the intersecting prism in X( ), Y( ), Z( ), and the calculated clipped points of intersection in I( ), J( ), K( ). PLOT commands are used to create the computer graphics output.

Once curves of intersection are computed and displayed, the information is normally used for other purposes such as those required for construction or fabrication. Final design decisions may require the calculation of geometric properties such as area, volume, weight, centroids, and mass moment of inertia of solids which are bound by curves of intersection. In structural analysis, the curves of intersection might form the boundary of surface elements used in a finite element model for calculating stresses, deflection, and temperatures under certain operating conditions. The curves of intersection might also be used to define edges of three-dimensional surface patches necessary for further analysis and display. In modern CAD systems, the data is passed to various specialized subroutines to accomplish tasks such as these.

14-9 GEOMETRIC MODELING FOR PRISM-CYLINDER INTERSECTIONS

The clipping prism in program PRISMCYL is defined the same way as in program PRISMS. In addition, a cylinder is defined by use of subroutine CYLSURF. This cylinder, whose axis is along the x-axis, is defined by 32 lines which lie between $-L \leq x \leq L$, where L is the cylinder half-length. These lines on the cylinder are calculated in a clockwise manner, starting at the 6 o'clock position when observing the cylinder from far out on the positive x-axis.

Figure 14-5 shows the intersection between a prism, defined in lines 480–490 in PRISMCYL, and a cylinder of radius 0.5 and length 2.0. These values are passed to the subroutine CYLSURF in line 820. The reader should be able to follow the program logic.

Two features are included in PRISMCYL which allow one to translate the clipping prism (lines 500–570) and rotate the intersecting cylinder (lines 580–620). The 3×3 transformation matrix $[\mathbf{T}]$ is passed to subroutine CYLSURF where the actual rotation takes place. Further study in computer graphics will lead to an ability to use a 4×4 transformation matrix to cause translation as well as rotation; and perspective projection as well as orthographic projection, for any three-dimensional model properly stored in an array. This technique requires the use of homogeneous coordinates and is discussed in several of the references given at the end of the text. (See Reference 21, for example.)

14-10 SOLID MODELING

The geometric modeling techniques given in this text are applied to points, lines, and planes. If solids are to be represented, they can be approximated by a series of plane facets. This procedure produces the same order of accuracy as obtained in the descriptive geometry techniques used to represent cylinders and cones. This is a suitable solid model approximation for many engineering and architectural applications.

Mathematical theory can be used to define some solids analytically. All points of interest on a solid surface, such as the line of intersection with another surface, can then be accurately computed for the exact surface, rather than for a plane facet approximation. Quadric surfaces are solids defined by nonlinear, algebraic equations. The classical quadric surfaces are cones, cylinders, and spheres. Many other types of analytical surfaces exist in the literature. Complex shapes are defined by a series of surface patches. This topic is presented in books on computational geometry and geometric modeling. The analytical approach requires advanced mathematics to evaluate tangents, normals, curvature, twist vectors, and other geometric attributes.

Several software systems are now available for supporting solid modeling. They are being used for computer-aided design and manufacture, technical publications, animated filmmaking, industrial robotics, computer vision schemes, and as a data base for the automatic production of engineering drawings and computer graphics. Three types of solid models are the most common. They are *wireframe, boundary surface representation,* and *solid building block representation.*

The wireframe representation has been used several times in this text. It is suitable for simple objects, but ambiguities can exist in complex objects because it lacks surface and solid or void space information. A wireframe is a *partial* boundary surface representation.

With boundary representation (B-rep), a solid is represented by spatial boundaries enclosing the surface with normal vector directions used to indicate the outside or inside. A complex solid becomes a union of surfaces, bordered by edge curves, which, in turn are defined by vertices. The B-rep computer model contains surface geometry, curve geometry, and point geometry. When plane facets are used to approximate surfaces, and straight lines used as edges, then solid modeling can be done using the elementary modeling theory given in this text. This approach is a special case of boundary representation, often called *polyhedral solids.*

In the *building block systems,* solids are built up using a set of primitive shapes such as blocks, spheres, cones, and cylinders. This building block approach is also called *constructive solid geometry* (CSG). The solid model is stored in computer memory as a list of operations such as addition, subtraction, and intersection. These operations are equivalent to the set operations of union, difference, and intersection studied in Boolean algebra. The stored list of primitive shapes and operations which defines a complex solid is structured like an inverted tree. The root is the solid name, the branches are the Boolean operations, and the leaves are the primitive shapes used to construct an object. The computer is instructed to store, retrieve, and interrogate the model by tracing through the stored list using pointers stored along with each item.

Each solid modeling method has advantages and disadvantages. The building block method is well suited to many mechanical shapes and requires minimal computer storage. On the other hand, B-rep methods have more flexibility since they are not limited to simple primitive surfaces such as quadratics. The B-rep method is frequently used for vehicle design. Hybrid systems are now available which maintain both a B-rep and CSG representation for each solid model so that the best choice can be made to accomplish a particular task. For example, B-rep may be used to create a model for use in structural analysis, whereas a CSG representation might be chosen to produce a color display on a raster graphics device.

CAD/CAM solid modeling is expected to shorten the design cycle, increase productivity, and stimulate creativity. Often the building of expensive prototypes can be eliminated. Every designer can work from the same, up-to-date data base which will eliminate errors and improve communication. Automatic verification of machine tool instructions is used to prevent costly mistakes before manufacturing actually begins. Design systems using solid modeling may eventually lead to a fully automated production environment.

PROBLEMS: CHAPTER 14

14-1 *Given:* Incomplete top and front views of a pyramid and a prism.
Task: Show the intersection line of the two given figures.

14-2 *Given:* Incomplete front and complete views of a pyramid and a triangular prism.
Task: Show the line of intersection of the two figures.

14-3 *Given:* Incomplete views of a cone and cylinder.
Task: Show the line of intersection between the two figures.

14-4 *Given:* Incomplete views of an intersecting cone and cylinder.
Task: Show the line of intersection between the two figures.

14-5 *Given:* Incomplete views of two cylinders.
Task: Show the line of intersection between the two figures.

14-6 *Given:* Complete top view and incomplete front and right side views of a cone intersected by a cylinder.
Task: In the front and side views, show the line of intersection between the cylinder and cone.

14-7 *Given:* Incomplete front and top views of a metal hopper supported by two beams.
Task: Using only the given views, show (in both) where the beams must be cut in order to hold the hopper in a close fit.

14-8 *Given:* Incomplete front and top views of two intersecting ducts.
Task: Show the line of intersection of the ducts.

14-9 *Given:* Complete side view and incomplete front view of an intersecting prism and triangular pyramid.
Task:
1. In the front view, show the line of intersection of the two solids.
2. Draw a complete top view of the prism and pyramid.

14-10 *Given:* Incomplete top and complete side view of an intersecting cylinder and cone.
Task:
1. Solve for the line of intersection in the top view.
2. Draw a complete front view.

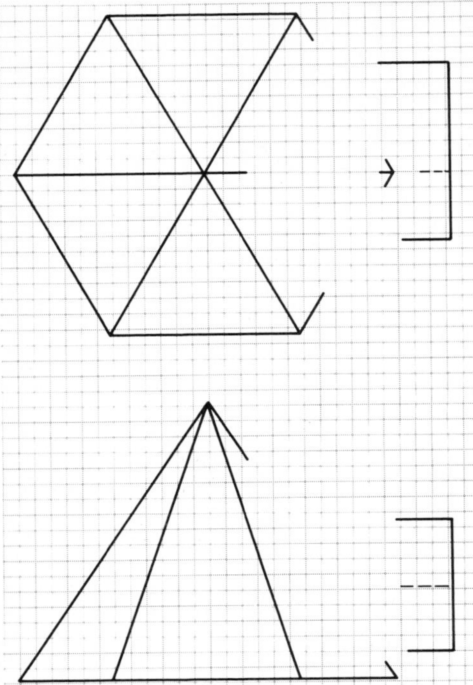

Figure 14-1P

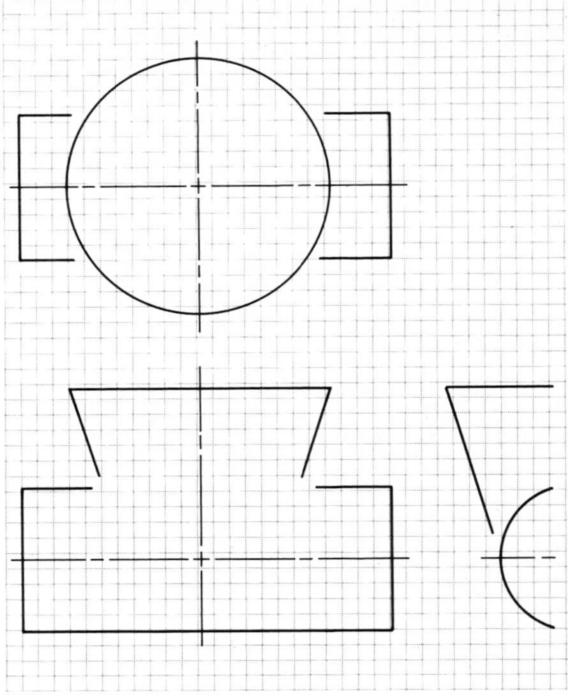

Figure 14-3P

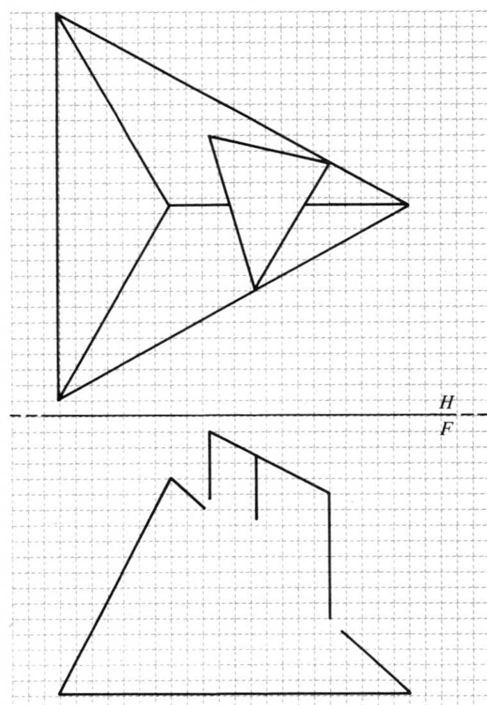

Figure 14-2P

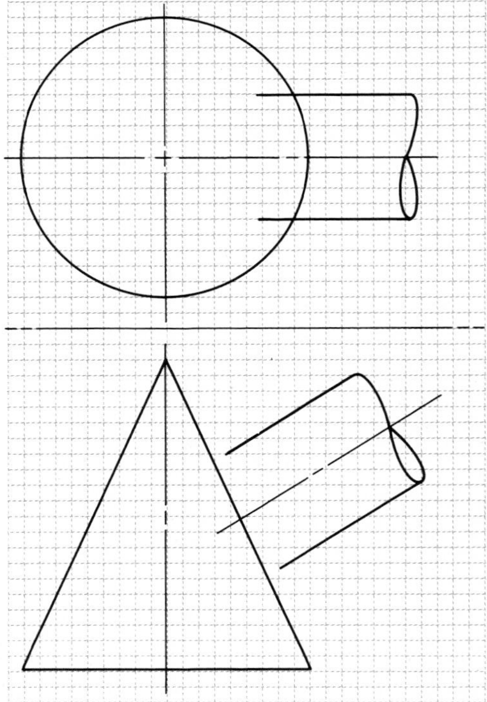

Figure 14-4P

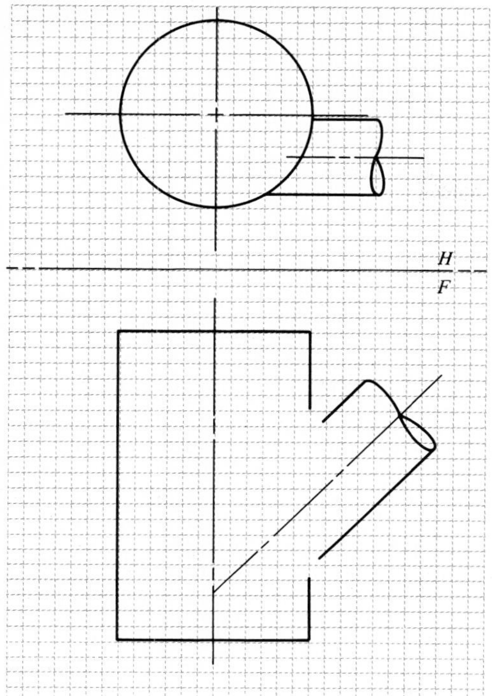

Figure 14-5P

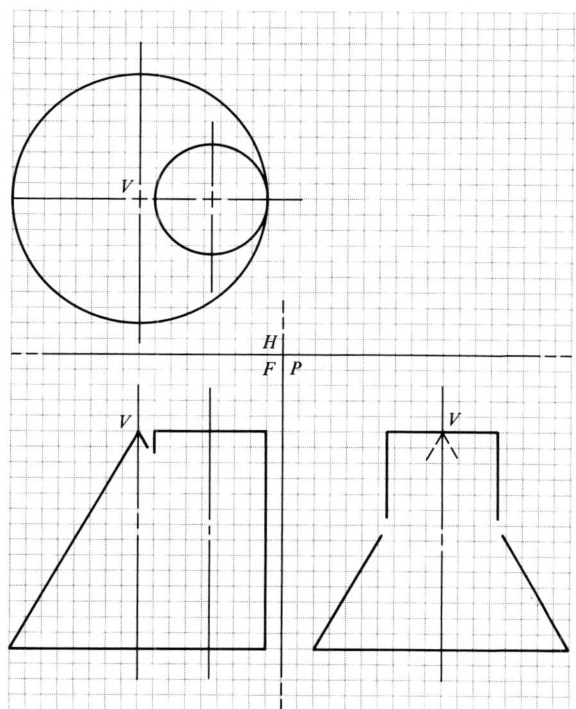

Figure 14-6P

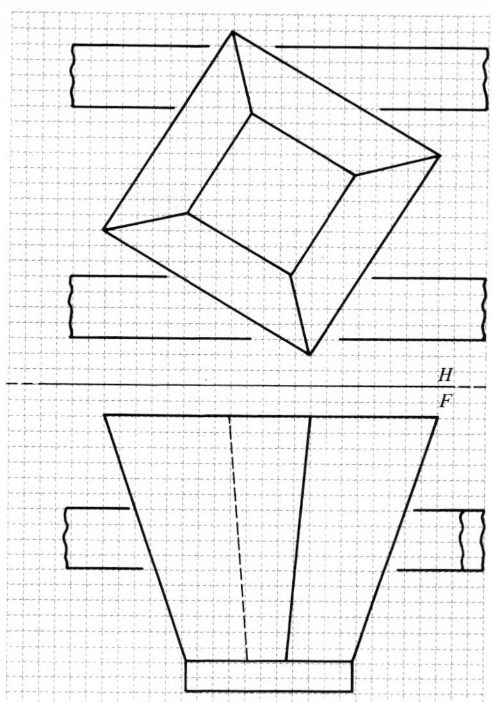

Figure 14-7P

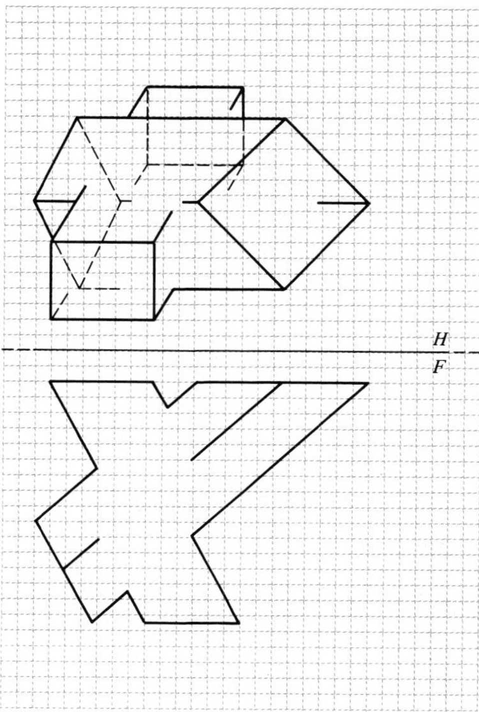

Figure 14-8P

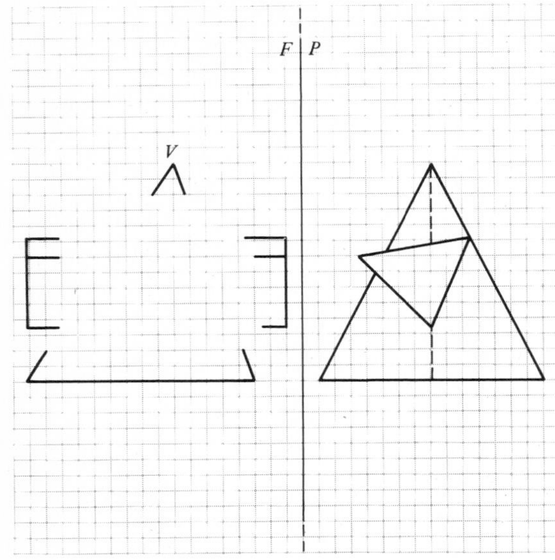

Figure 14-9P

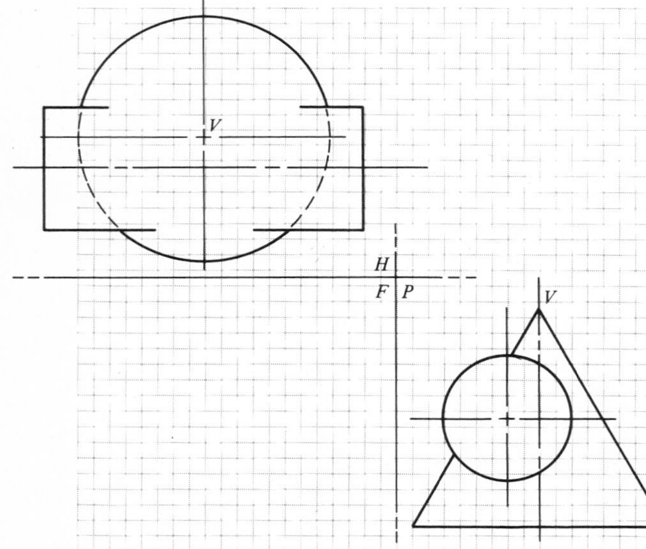

Figure 14-10P

14-11[c] *Given:* Program PRISMCYL and supporting subroutines (Appendix 1).

Task: a. Write a computer program to generate output such as shown in Figure 14-5 for an intersecting prism and cylinder.

b. Replace the prism with an obelisk and generate the proper intersections.

14-12[c] *Given:* Program PRISMS and supporting subroutines (Appendix 1).

Task: Replace the intersecting octagon beam with an intersecting airfoil shape. Obtain the curve of intersection between the airfoil and the base prism.

14-13[c] *Given:* The object defined by the following vertices:

Point	x	y	z
1	−5½	−5	−5
2	5½	−5	−5
3	5½	2	−5
4	0	5	−5
5	−5½	2	−5
6	−5½	−5	−5
7	−5½	−5	5
8	5½	−5	5
9	5½	2	5
10	0	5	5
11	−5½	2	5
12	−5½	−5	5

Task: Replace the base prism in Figure 14-5 with the given object and calculate the intersection with the intersecting cylinder used in program PRISMCYL.

14-14[c] *Given:* Program CLIPON and supporting subroutines (Appendix 1).

Task: Replace the intersecting cylinder with an intersecting cone and display the intersection between two cones.

14-15[c] *Given:* Consider problems in descriptive geometry such as Problems 13-1 through 13-5, 13-14 through 13-16, and 14-1 through 14-3.

Task: Use computational techniques associated with geometric modeling theory to obtain solutions to some of these problems.

[c]Superscript *c* indicates a computer solution is required.

ADDENDUM TO CHAPTERS 13 AND 14

A single, generalized software package based upon three-dimensional clipping as discussed in Chapter 11 can be written to solve problems such as discussed in Chapters 13 and 14. The following sample solutions were generated using such a package, called CSUCLIP, written in the C language for a UNIX environment using the CCAE facilities of Colorado State University. Graphical output was obtained using the "threed" software package developed by the CADIG Group at the U.S. Naval Academy. Commercial software packages that produce similar solutions are available.

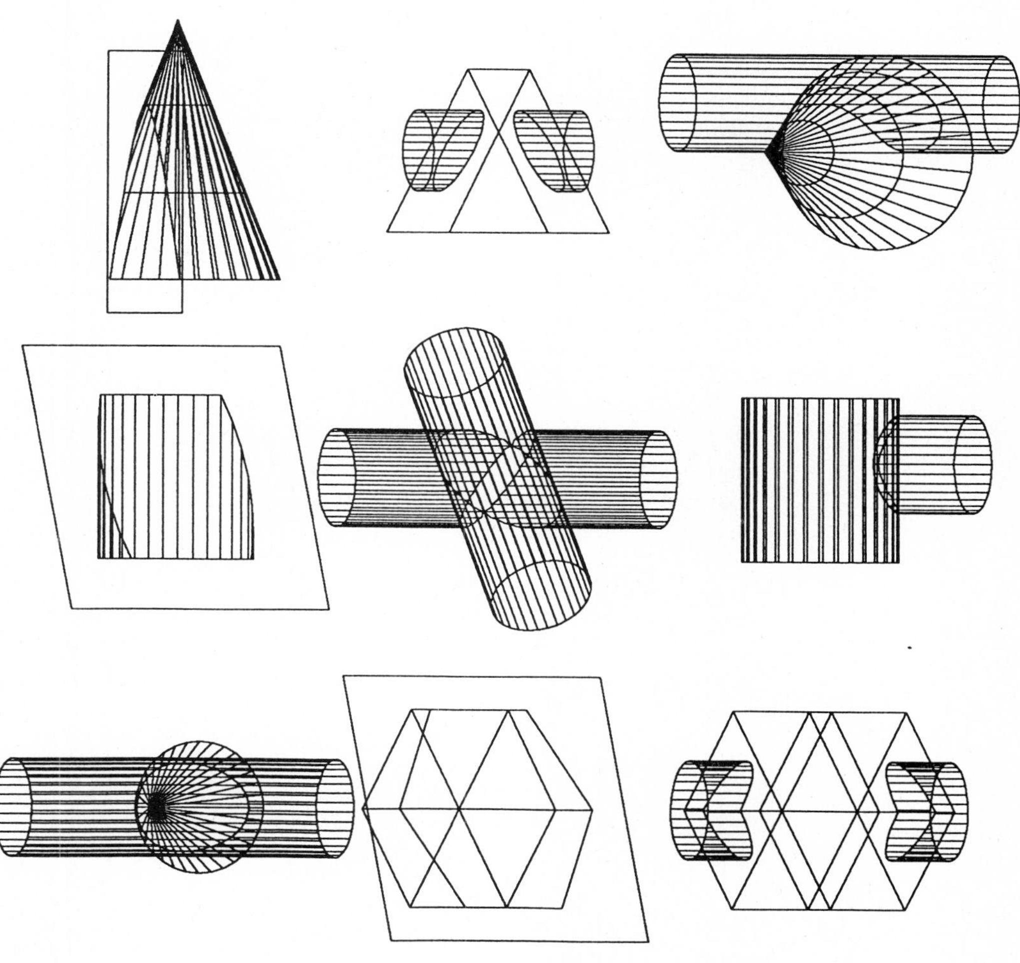

15

Surface Developments

The development of an object is obtained by unfolding or unrolling its surface or surfaces into a plane. By doing this we can obtain a full size pattern, one which shows the actual area of the object and the relationship of the lines and points on that surface. Many areas of industry use developments in the manufacture of such items as air conditioning and heating ducts, containers, aircraft components, furnaces, and similar structures.

Plane or single curved surfaces, called *developable surfaces,* can be developed accurately due to the fact that they can be unrolled into a plane. With this type of development, every line on the actual surface is shown in true length. Since this cannot be done with warped or double-curved surfaces, the latter are called *nondevelopable* and must be laid out approximately. Objects resulting from such approximations can be spun, stamped, formed, and processed if ductile material is used in the manufacturing process.

In this chapter, only developable surfaces are considered. The basic principles involved in the development of surfaces can best be shown using simple geometric objects such as cylinders, prisms, pyramids, and cones. Descriptive geometry methods are followed by examples of computer-generated surface developments based upon geometric modeling theory.

15-1 RIGHT CIRCULAR CYLINDER—PARALLEL LINE DEVELOPMENT

A right cylinder has its axis perpendicular to the base. The development, shown in Figure 15-1, is a rectangle whose width equals the height of the

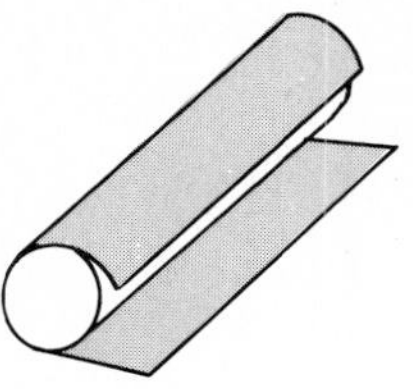

Figure 15-1
Cylinder development.

cylinder, and whose length equals the circumference. An *approximation* of the circumference can be obtained graphically by stepping off the chords of the 12 angles which are established in the circular view. If the *exact* circumference is desired it can be found mathematically and laid off on the stretch-out line with the proper number of elements shown.

Manufacturing details must be considered when designing objects using developable surfaces. Most industrial machines used in bending and forming material require that developments be laid out inside-up. Economy also dictates that the welded or soldered seams be the shortest elements in the development. Extra material must be provided when designing with sheet metal so that patterns may be joined at designated intersections. Further, bend allowances must also be considered since material is deformed by compression and expansion when bending takes place. These topics are not discussed in this text but are reserved for literature related to sheet metal work.

Example 15-1

Given: Top and front views of a truncated, right circular cylinder.
Task: Develop the given truncated cylinder.
Solution:

1. Establish and number at least 12 elements on the circle shown in the top view. For greater accuracy, 24 divisions are shown in Example 15-1.

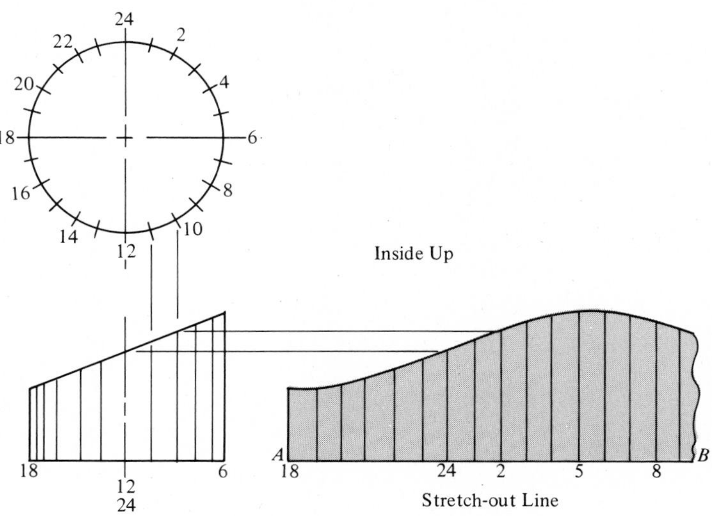

Example 15-1
Development of cylindrical surface.

2. Project the elements into the front view where they appear true length.
3. Obtain chordal distances between elements as shown in the top view and lay off these distances on the stretch-out line *A-B*.
4. Starting with the shortest, number the elements on line *A-B* in a sequence that shows the development inside-up.
5. Transfer the true lengths of the elements from the front view to the corresponding elements shown on the development.
6. Draw a fair curve through the ends of the elements.

15-2 RIGHT PRISM—PARALLEL LINE DEVELOPMENT

Since the base and axis are perpendicular as shown in Figure 15-2, the right prism will roll out to form a rectangular pattern where the length will be the sum of the sides that make up the perimeter of the base, and the width will be determined by the length of the lateral edges. The figure shows the addition of end surfaces on the layout pattern.

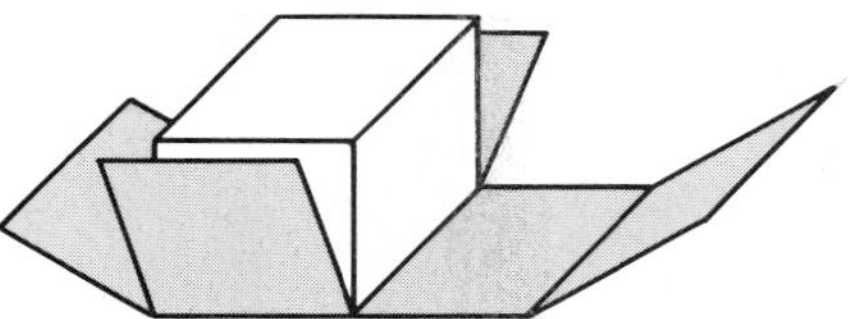

Figure 15-2
Right prism development.

Example 15-2

Given: Top and front views of a truncated right prism.
Task: Develop all surfaces of the truncated prism.
Solution:

1. Identify the lateral edges of the sides of the prism (elements) in the top view and in the front view where they appear in true length.
2. Draw a stretch-out line and starting with the shortest elements lay out (inside-up) the true distances between each element. The true distances are obtained from the top view.
3. Draw each element perpendicular to the stretch-out line.
4. In order to obtain the true length of each element in the development, project each end of the elements horizontally from the front view as shown.

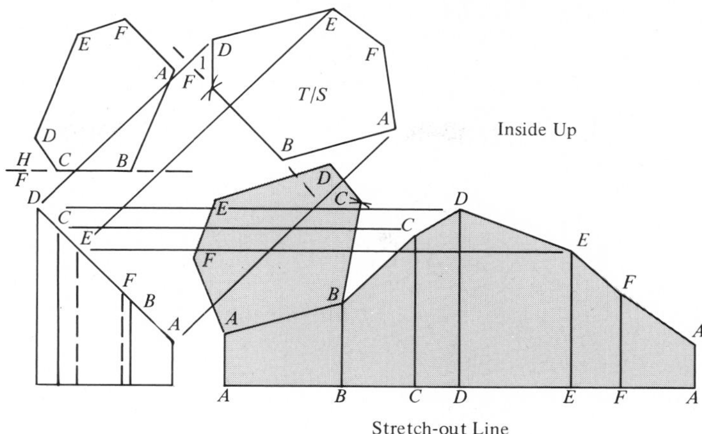

Example 15-2
Development of prism.

5. Connect the endpoints of the lateral elements to complete the development of the sides.
6. In order to complete the pattern of the top truncated surface, draw an auxiliary showing the true size and shape of that surface.
7. Using the triangulation method (Appendix 4), transfer dimensions from the true size and shape of the auxiliary to the existing lay-out so that edges *A-B*, the longest edges, coincide. As shown on the drawing, line *A-B* of the top surface coincides with line *A-B* on the stretch-out. On the development, using point *A* as a center and the true distance *A-C*, scribe an arc. Using *B* as a center and the true length of *B-C*, scribe an arc which cuts arc *A-C* at point *C*. Repeat this process until the entire truncated surface has been attached to the development.
8. Since the base appears as a horizontal line in the front view, it will be true size in the top view. These dimensions may be used to develop the base which will be attached to the pattern so that lines *A-B* coincide. ◀

15-3 RIGHT PYRAMID—RADIAL LINE DEVELOPMENT

The axis of a right pyramid is perpendicular to its base. As shown in Figure 15-3, each face will show as a true size and shape triangle on the development.

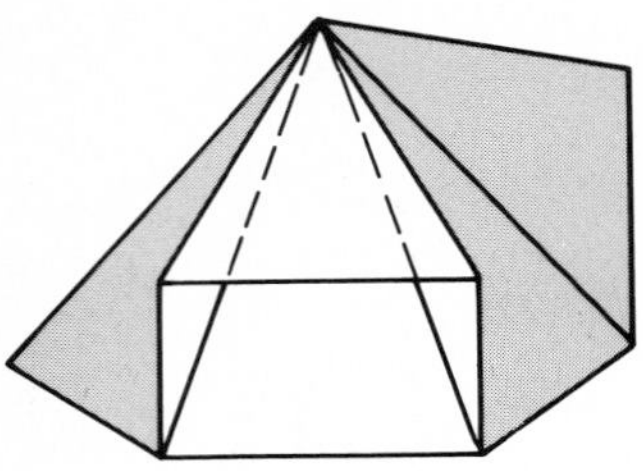

Figure 15-3
Right pyramid development.

Example 15-3

Given: Top and front views of a truncated pyramid.
Task: Develop the lateral surfaces of the pyramid.
Solution:

1. The true length of frontal lines *A*-1 and *C*-3 are shown in the front view.
2. Revolve lines *B*-2 and *D*-4 into the frontal plane and show them true length in the front view. (See Section 10-2.)
3. Since the base of the pyramid is in the horizontal plane, the four edges 1-2, 2-3, and so on, show true length in the top view.
4. Starting with the shortest solid lateral edge (*C*-3) lay out element *O*-*C*-3 on the development. Using 3 as a center and true length 3-2 from the top view, scribe an arc.
5. Using *O* as a center and the true length of element *O*-*B*-2, scribe an arc which cuts arc 3-2 at point 2.
6. Using point 2 as the center and the true length of base edge 2-1, scribe an arc.
7. Again using point *O* as a center and the true length of the lateral element *O*-*A*-1, scribe an arc cutting arc 2-1 at point 1.
8. Using the true lengths of base edges 1-4 and 4-3, and the true lengths of the lateral elements *O*-*D*-4 and *O*-*C*-3 in their respective order, repeat the processes of items 5, 6, 7, and 8.
9. Connect points 3, 2, 1, 4, 3 with straight lines to form the base edges

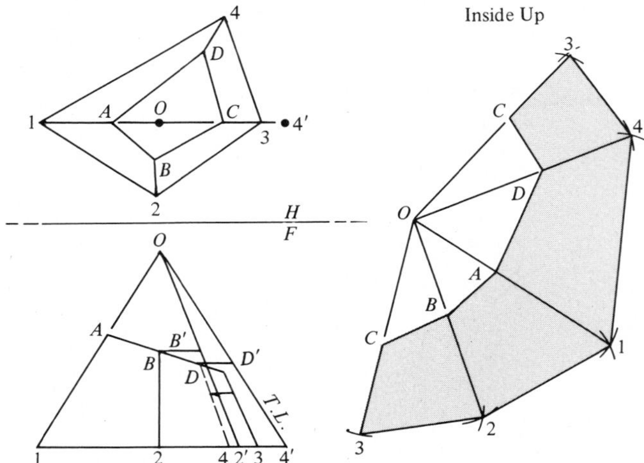

Example 15-3
Development of pyramid.

of the pyramid. Connect points *C, B, A, D, C* to form the truncated portion of the pyramid.

10. In the front view, measure the true distances *O-A, O-B′, O-C,* and *O-D′*. Show these points on the layout and connect them to form the truncated portion of the pyramid. ◀

15-4 RIGHT CIRCULAR CONE—RADIAL LINE DEVELOPMENT

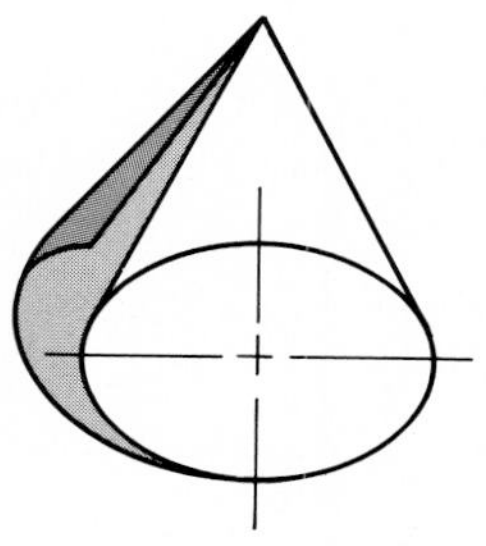

Figure 15-4
Right circular cone development.

Figure 15-4 illustrates the theory of constructing a pattern of a right circular cone where the axis is perpendicular to the base. The student will recognize the fact that the conical surface consists of an infinite number of triangles each having one angle formed at the vertex by two lateral elements and the other angles formed by a lateral element and base. The circumference of the base of the cone can then be found as described in Section 15-1.

Example 15-4

Given: Front and top views of a truncated right circular cone.
Task: Develop the pattern for the given figure.

The solution to the above is shown for analysis by the student. ◀

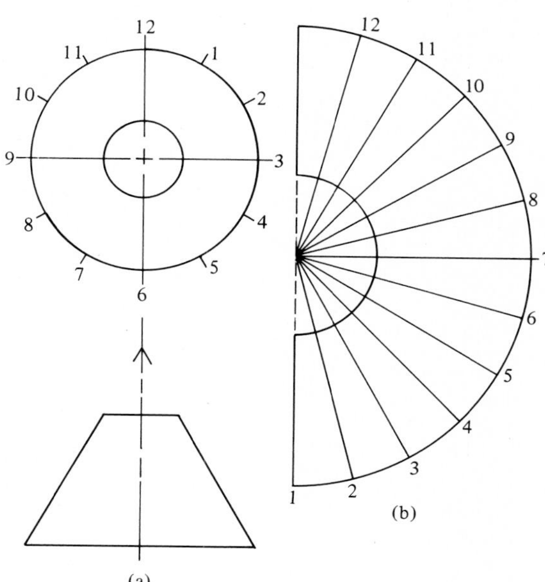

Example 15-4
Development of truncated cone.

15-5 OBLIQUE PRISM—PARALLEL LINE DEVELOPMENT

Example 15-5

Given: Front and top views of oblique prism.
Task: Develop the lateral surfaces of the prism.
Solution:

1. Draw a first auxiliary of the prism showing the true length of the lateral edges.
2. Draw a second auxiliary view showing the surfaces on edge and the lateral edges as points. The true distances between lateral edges can be measured here.

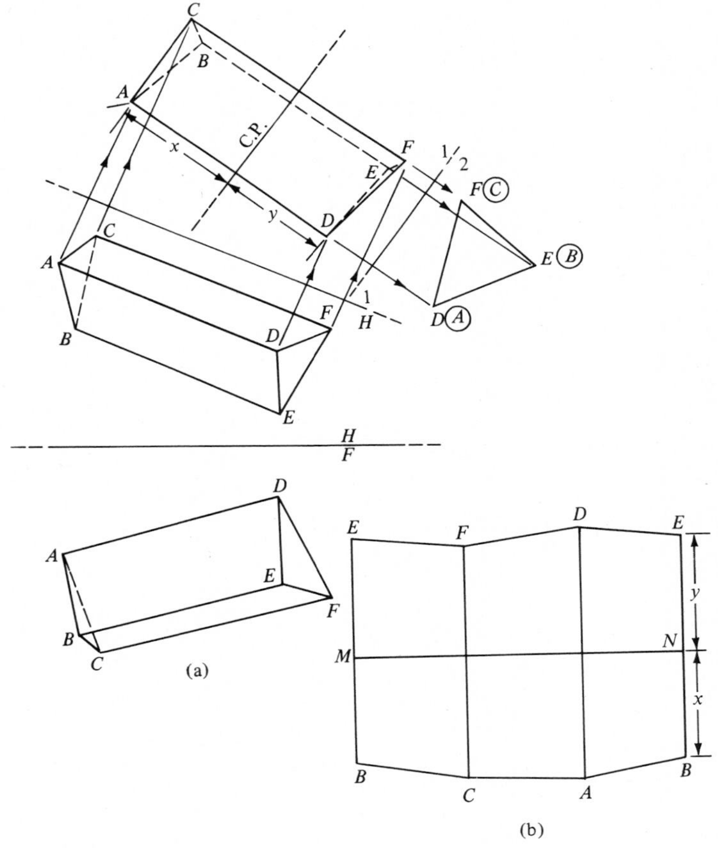

Example 15-5
Development of oblique prism.

3. In the first auxiliary, show the edge view of a cutting plane which is passed perpendicular to the lateral edges. This plane of right section cuts the lateral edges *x* distance from one end and *y* distance from the other.
4. On the stretch-out line *MN* of the development lay out the shortest lateral edge (in this case edge *EB*) so that it is perpendicular to the line. Distance *y* is measured above *MN* and distance *x* is measured below *MN*. This orients line *EB* so that it shows in true length and is in position to develop the pattern inside-up.
5. Obtain the true length distance between lateral edge *EB* and *DA* in the second auxiliary. Transfer this distance onto the stretch-out line *MN* and draw lateral edge *DA*.
6. Repeat the above operations until all surfaces have been shown.
7. Connect the ends of all elements.
8. If the end surfaces of the prism are to be developed, the true size of the ends are found by auxiliary views and transferred to the pattern by triangulation as in Section 15-2. ◀

15-6 OBLIQUE CYLINDER—PARALLEL LINE DEVELOPMENT

With one or two exceptions, the procedure for developing the surface of an oblique cylinder is the same as that used for developing the lateral surface of an oblique prism in Section 15-5.

Example 15-6

Given: Front and top views of an oblique cylinder.
Task: Develop the lateral surface of the oblique cylinder.
Solution:

1. Draw a first auxiliary showing the true length of the axis of the cylinder.
2. In a second auxiliary, the cylinder is shown on edge and appears as a circle. Divide the circle into at least 12 divisions with each point being the end view of a lateral element.
3. Project each point into the first auxiliary view to show the lateral elements parallel to the cylinder axis and in true length.
4. In the first auxiliary view, pass a cutting plane that cuts the cylinder perpendicular to its axis. This plane of right section cuts each element *x* distance from one end and *y* distance from the other.
5. Draw stretch-out line *MN*. In the second auxiliary, obtain the chordal

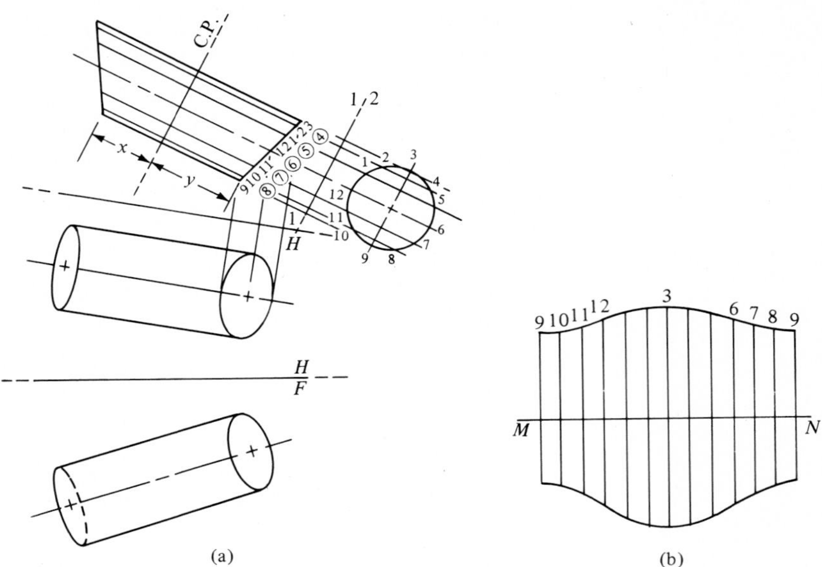

Example 15-6
Development of oblique cylinder.

distances between the lateral elements and transfer them to the stretch-out line *MN*.

6. Starting with the shortest element, locate each so that it is perpendicular to the line *MN* with the *y* distance above the line and the *x* distance below it. Be sure the elements are located so that the pattern is inside-up.
7. Use an irregular curve and draw a fair curve through the endpoints of the elements. ◀

15-7 OBLIQUE CONE—RADIAL LINE DEVELOPMENT

Example 15-7 deals with an oblique cone. The axis is not perpendicular to the base. Therefore, the lateral elements are of different lengths. This means that the cone is made up of a number of different triangles whose sides equal the length of the lateral elements and whose base equals the chords of those elements as shown in the top view.

One method of obtaining the true length of a line is by revolution, a method used in Section 15-3. The other aid used here is known as the *true length diagram.* In drawing this "diagram" a right triangle is formed. The altitude of this triangle is the height of the cone as shown in the front view. The base of the triangle is the projection of the element on the horizontal plane as shown in the top view. The hypotenuse of this right triangle is the true length of the element.

Example 15-7

Given: Top and front views of an oblique cone.
Task: Make one-half of the development required for the given cone. The development is symmetrical about centerline 0-3. For this reason, only one-half of the development is required.
Solution:

1. Establish the height of all triangles in the "true length diagram" by projecting point *O* and the base of the cone horizontally.
2. The base of triangle *OD*9 is found by measuring the distance 0-9 in the top view. Transfer this distance 0-9 to the "diagram" and draw the true length of the element (hypotenuse) 0-9.
3. Repeat this operation and determine the distance of all other elements on the cone.
4. With *O* as the apex and using the triangulation method as demonstrated in Section 15-3, lay out one-half of the development as shown.
5. Remember to start with the shortest element and to view the development inside-up.

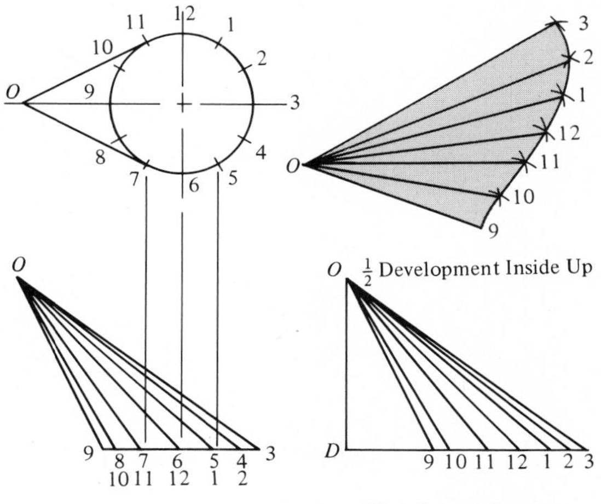

Example 15-7
Oblique cone development.

15-8 TRANSITION PIECE—TRIANGULATION

Many times it is required that a piece be developed which will connect two surfaces of different size and shape. Such a transition piece, shown in Example 15-8, could be used for connecting two ventilating or heating ducts. The conical portion is broken down into several triangular surfaces to form an *approximation* of the required surface. Usually the ductility of the material allows the desired shape to be achieved in the forming process.

Example 15-8

Given: Isometric view of a transition piece.

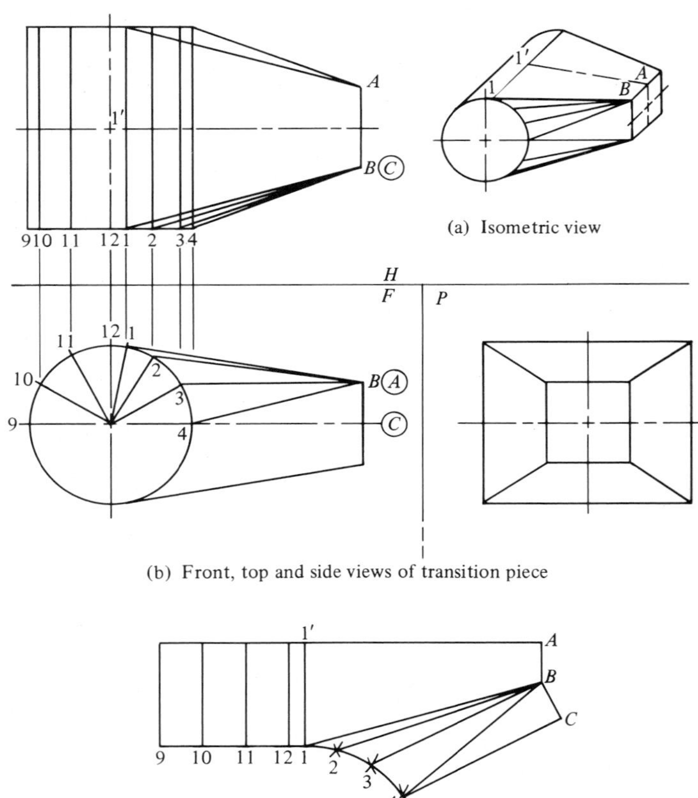

Example 15-8
Transition piece.

Task: Since there are two axes of symmetry, develop only one-fourth of the entire layout.
Solution:

1. Lay off the circular feature in the front view as shown by numbers 9 through 12. Corresponding parallel elements are drawn in the top view. Lay out parallel element portions of the development.
2. By revolution, find the true length of lines necessary to form the pattern 1′*AB*1.
3. Lay off the upper right quadrant of the front view circular feature as shown and form triangles on the conical portion with these points and point *B*.
4. Measure the chords of the triangles in the front view and, by revolution, find the true length of the triangle sides (lateral elements).
5. Using the method of triangulation shown in Example 15-3, construct the triangular areas to form the conical section on the development. (See Example 15-8(c).) ◀

15-9 INTERSECTION AND DEVELOPMENT OF PRISM AND PYRAMID

Example 15-9 shows a prism and a pyramid which intersect. The solution shows the intersection of the solids and the development of the solids after intersection.

Example 15-9

Given: Incomplete top and front views of intersecting prism and pyramid.
Task:

1. Find the line of intersection of the two solids.
2. Make a development of each intersecting solid as shown in Example 15-9(c) and 15-9(d).

Solution: As in prior exercises, the solutions to the two operations are shown, leaving to the student the task of studying the example and establishing the suitable step-by-step procedures. ◀

As illustrated above, the cone and cylinder are two curved surfaces capable of exact development. This means that a sheet of flexible material can be wrapped about the surface without stretching, folding, or other distortion.

The need to work with developable surfaces is very common in the sheet-metal industry. One important application is in the design of elbows and connections for ducts. For example, it is known that the intersection

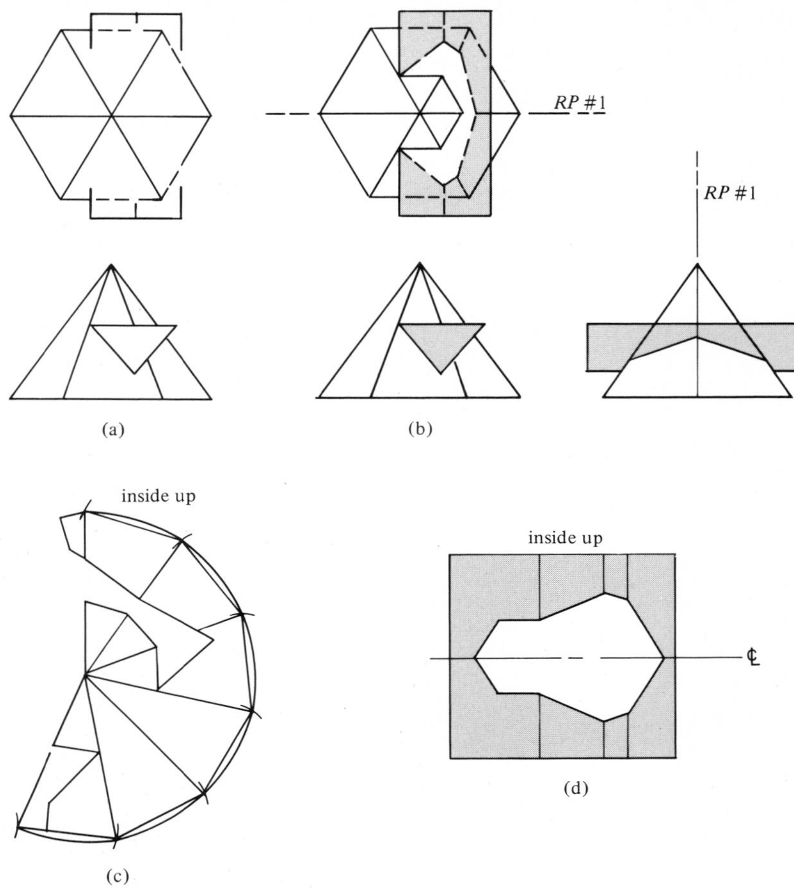

Example 15-9
Intersection of prism and pyramid.

curve between a cone and cylinder which are tangent to the same sphere is always a conic section. In Figure 15-5 the curves of intersection between the cylinder and cone are elliptical. Fabrication is simplified if the intersection curves can be circular and the joints be a plane curve.

15-10 ANALYTICAL DEVELOPMENT OF A CONE

A right cone can be developed analytically using the fact that the base expands into a circular arc whose radius is equal to the slant height of the cone. Figure 15-6 shows the computer-generated intersection between a

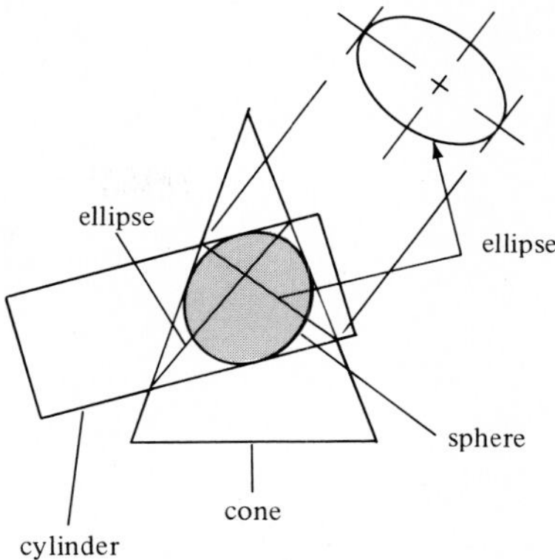

Figure 15-5
Intersection cylinder and cone tangent to the same sphere.

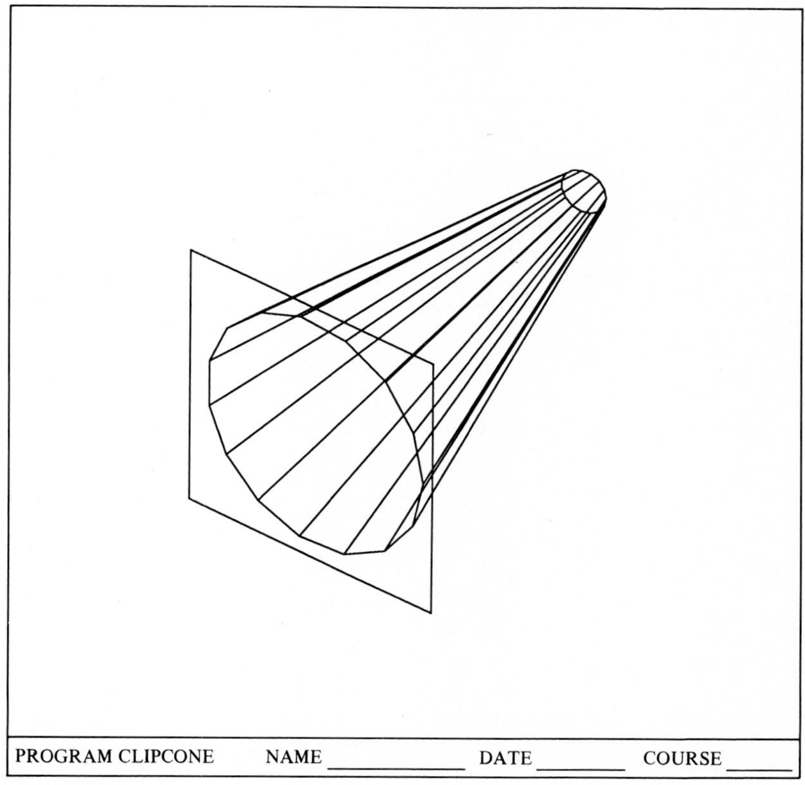

Figure 15-6
Plane-cone intersection.

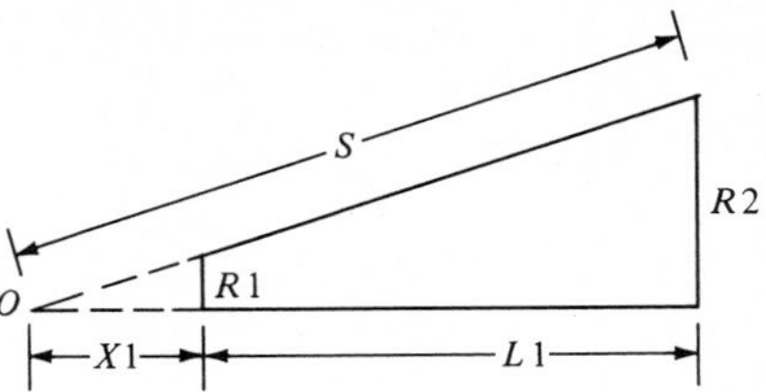

Figure 15-7
Cone slant height.

plane and a cone frustum. Consider a lateral section of the cone frustum as shown in Figure 15-7. By similar triangles one can write

$$\frac{R1}{X1} = \frac{(R2 - R1)}{L1} \tag{15-1}$$

Solving for $X1$ gives

$$X1 = \frac{(R1)(L1)}{(R2 - R1)} \tag{15-2}$$

where $R1$ and $R2$ are the base and top radii, and $L1$ is the length of the cone frustum. The slant height (S) of the full cone, measured from vertex O, is given by

$$S = [R2^2 + (L1 + X1)^2]^{1/2} \tag{15-3}$$

Figure 15-8 shows a full cone development. The sector angle α is given by 2π (base radius/slant height). Using the notation in Figure 15-7, this is

$$\alpha = 2\pi\left(\frac{R2}{S}\right) \tag{15-4}$$

This section angle can be divided into $\alpha/(N1 - 1)$ increments, where $N1$ is the number of lines on the developed surface.

The clipped cone frustum development for the intersection shown in Figure 15-6 is given in Figure 15-9. The lengths of each line on the surface development, from $\alpha = 0$ to $2\pi R2/S$ in increments of $\alpha/(N1 - 1)$, were calculated by finding the magnitude of each surface vector from the base to the top of the clipped cone shown in Figure 15-6.

Figure 15-10 shows the intersection between a cylinder of radius 1.0 unit and length 6.0 units, and the same plane used in Figure 15-6. Here the increment used is $2\pi/(N1 - 1)$ between $0 \leq \alpha \leq 2\pi$. Figure 15-11 shows the corresponding surface development for the intersection cylinder. The surface shown is outside-up but it can easily be "flipped over" to obtain an inside-up surface.

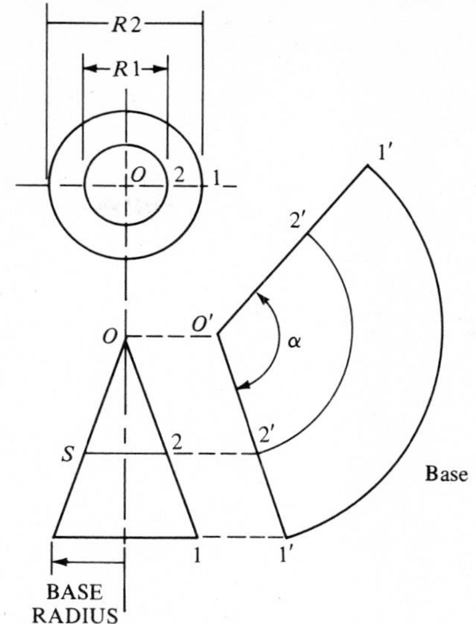

Figure 15-8
Cone development.

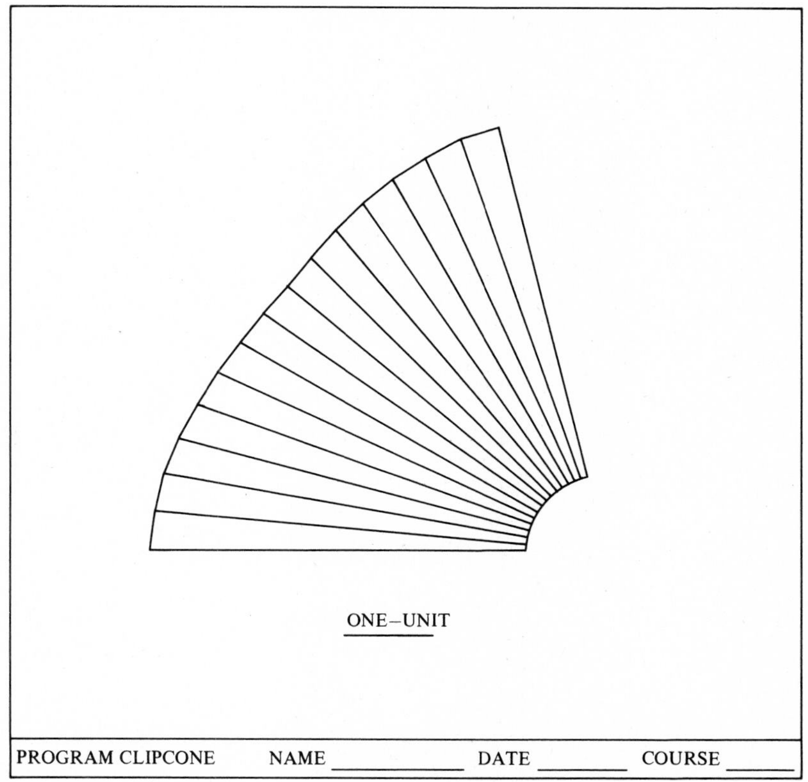

Figure 15-9
Cone development.

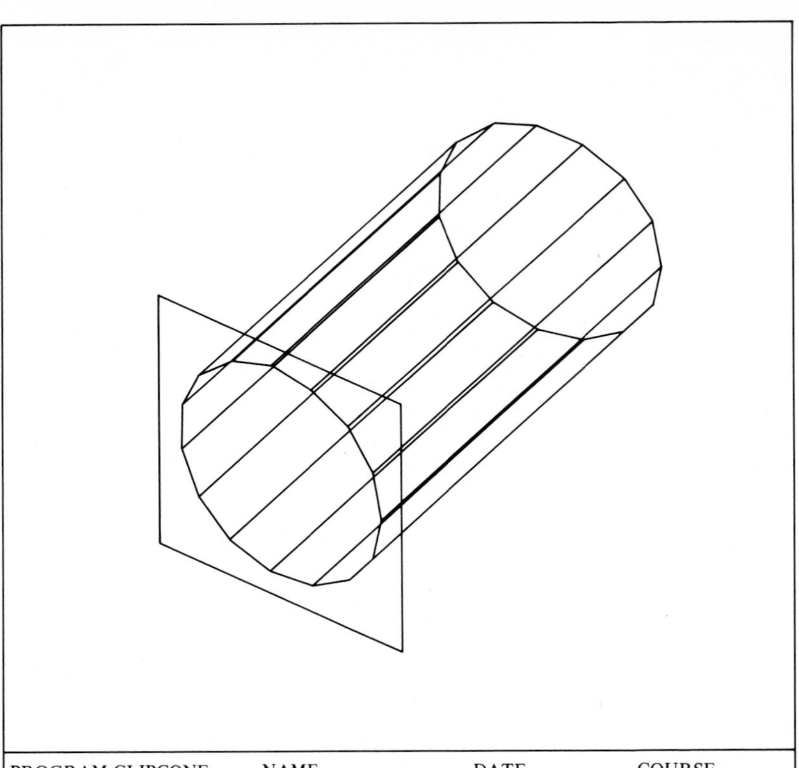

Figure 15-10
Plane-cylinder intersection.

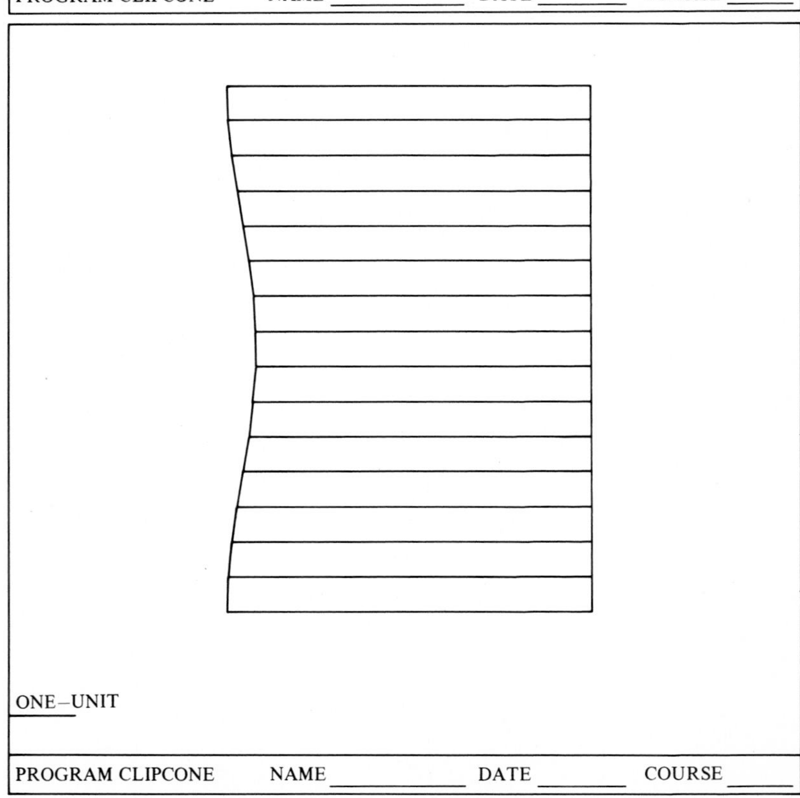

Figure 15-11
Cylinder development.

Figures 15-12 and 15-13 give the two cylinder developments for the cylinder/cone intersection shown in Figure 14-6. Figure 15-14 shows a partial intersection between a cone and cylinder. The orientation is the same as that in Figure 14-6 except that the cylinder has been lowered by 0.2 units. Figure 15-15 shows the resulting clipped cylinder and Figure 15-16 gives the corresponding surface development for the clipped cylinder, outside-up.

Example 15-10

Given: The developable surfaces for the cylinder/cone intersection shown in Figure 14-6 are given in Figures 15-12 and 15-13.
Task: Produce paper models of the three surfaces and connect them.
Solution:

a. The cone has a top radius $R1 = 0.1$, a base radius $R2 = 1.0$, and a length of 6.0. Use the unit distances shown on Figure 15-12 or 15-13

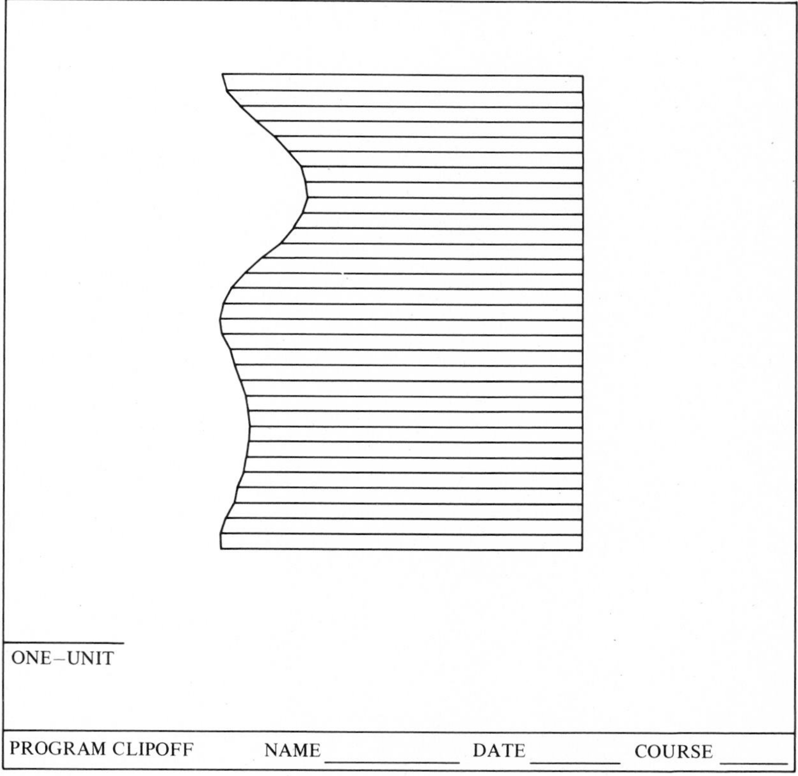

Figure 15-12
Cylinder development.

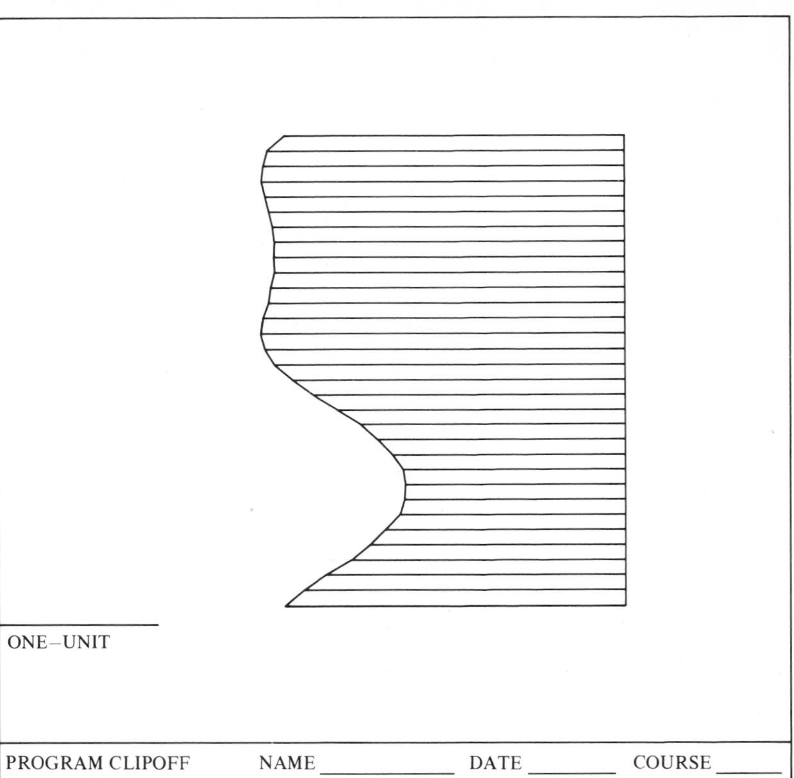

Figure 15-13
Cylinder development.

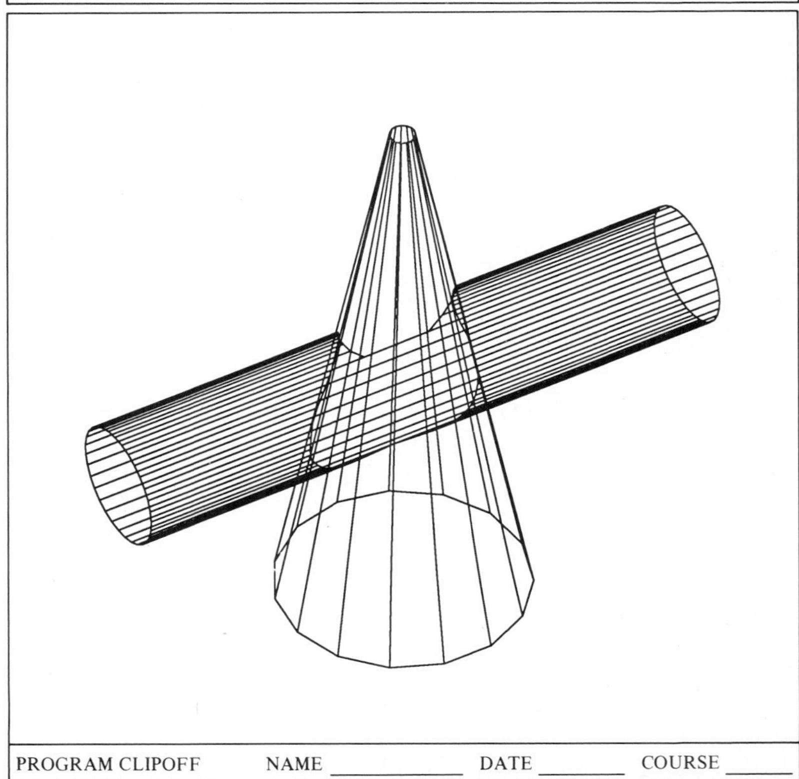

Figure 15-14
Cylinder and cone intersection.

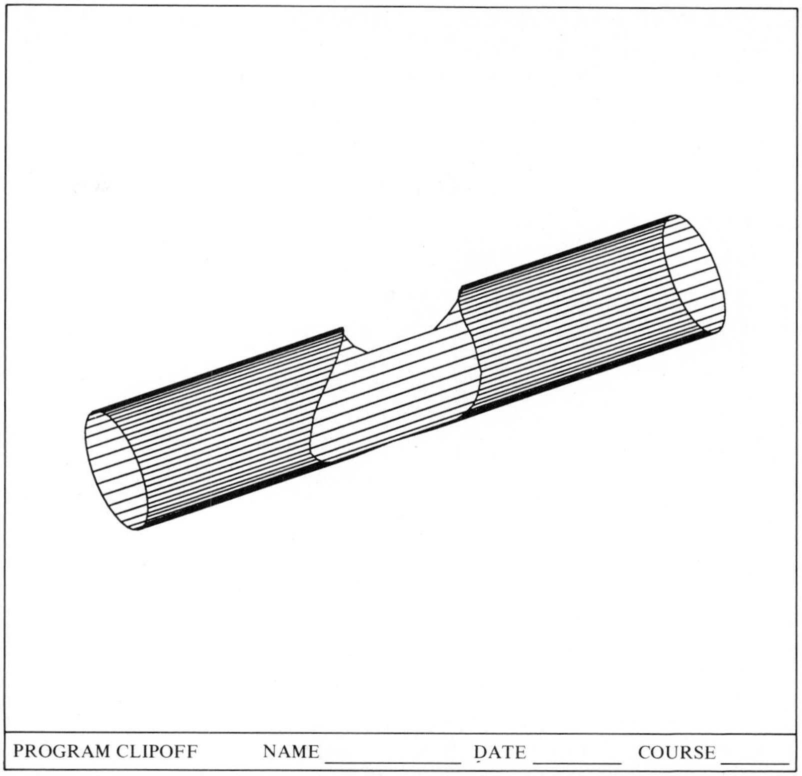

PROGRAM CLIPOFF NAME ________ DATE ________ COURSE ________

Figure 15-15
Clipped cylinder.

PROGRAM CLIPOFF NAME ________ DATE ________ COURSE ________

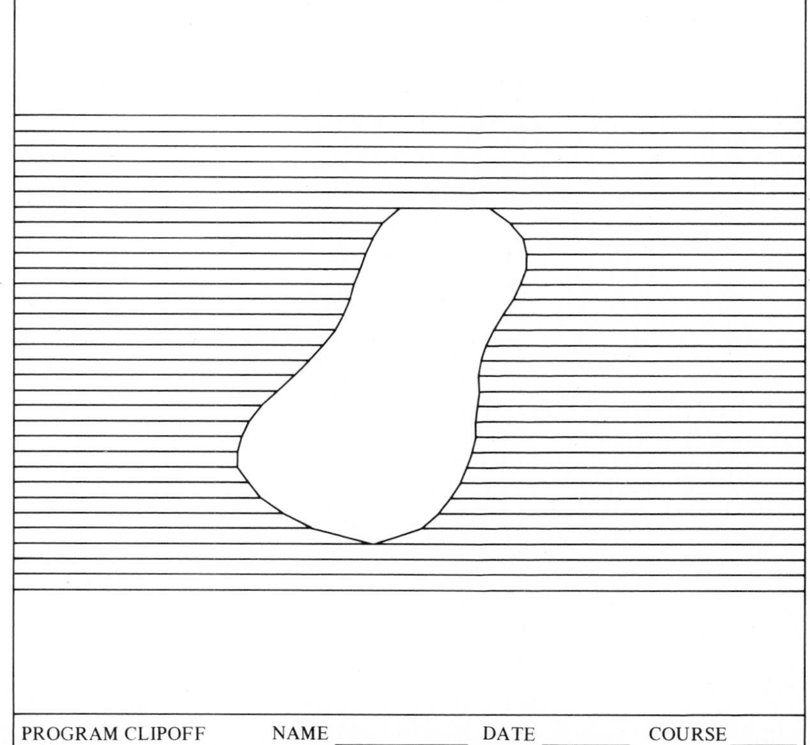

Figure 15-16
Cylinder development.

and lay out the development of the cone as described below. Additional calculations are:

$$X1 = R1\left(\frac{L}{(R2 - R1)}\right) = \frac{0.1(6)}{0.9} = 0.667$$

The slant height is then

$$S = \{R2^2 + (L + X1)^2\}^{1/2} = \{1 + (6.667)^2\}^{1/2} = (45.45)^{1/2} = 6.667$$

The sector angle is calculated to be

$$\alpha = \frac{2\pi\ R2}{S} = \frac{2\pi(1)}{6.667} = 0.942\left(\frac{180}{\pi}\right) = 54 \text{ deg}$$

Thus, the slant height is 6.667 units (in terms of the reference unit), and the sector angle is 54 deg. A layout such as shown in Figure 15-17 is needed. The tabs shown will help to tape the cone together.

b. Cut out the cone development and roll into a cone. It will help to cut some material out of the top as indicated in Figure 15-17.

c. Trace Figure 15-12, cut out, and roll into a cylinder with the lined surface on the outside.

d. Trace Figure 15-13, cut out, and roll into a cylinder with the lined surface on the inside.

e. With care the three paper models should fit together. Keep the seam down on each model. ◀

If the clipping solid is defined to be a cone, then the intersection shown in Figure 14-6 can be obtained. If desired, the clipping cone can be changed to a cylinder by making the end radius equal to the base radius. However, the program CLIPON also requires that there be no partial intersection in order to produce a proper display of the curves of intersection.

When partial intersections occur, a change is required in the graphics output logic. Program CLIPOFF is written to give the curve of intersection between two partially intersecting surfaces when the intersecting cylinder lies below the clipping cone as shown in Figure 15-14. The logic assumes that the first line on a partially intersecting surface will miss the clipping surface. Since the cylinder definition in CYLSURF begins at the 6 o'clock position, this condition is met. If the intersecting cylinder is raised (let LOWER = −0.6 in line 300 in CLIPOFF) then the intersecting cylinder should be defined starting at the 12 o'clock position. This can be done by changing the sign in lines 230, 270, and 280 in CYLSURF.

If the intersecting cylinder is solid, then the interior surface must also be defined. Surface normals can then be calculated and numerically controlled cutting tools can be aligned along these normal vectors and posi-

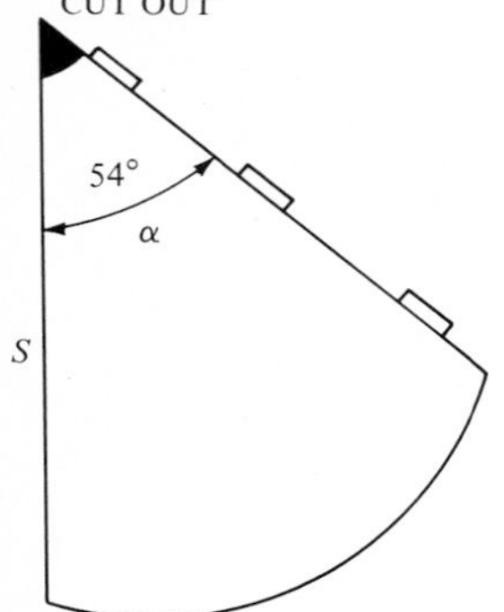

Figure 15-17
Cone mode.

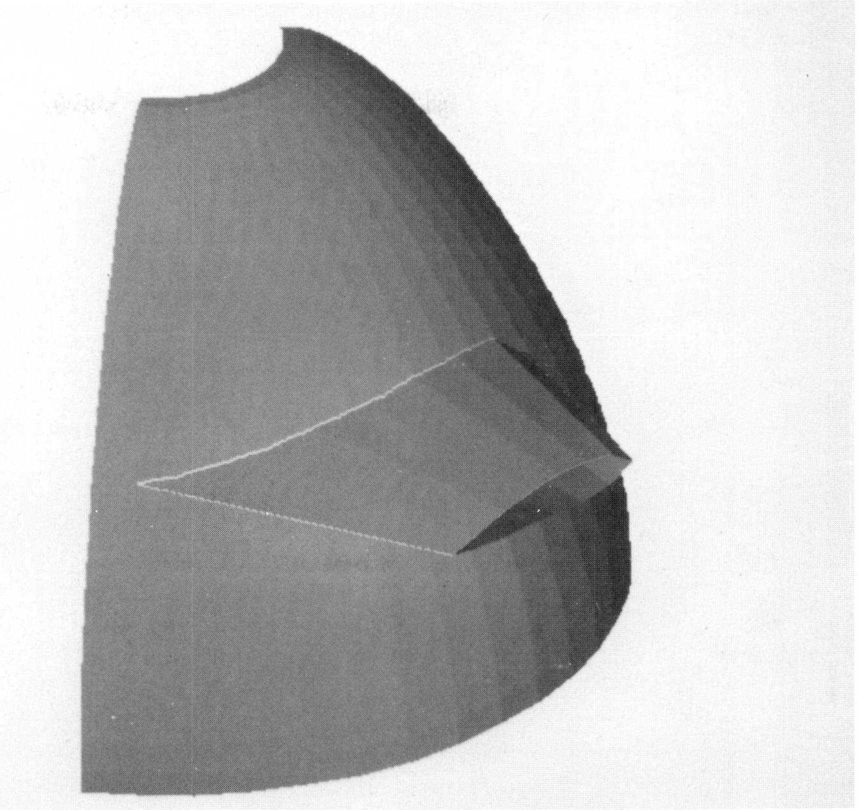

Figure 15-18
Intersecting blade and hub. Courtesy of CADIG Group, U.S. Naval Academy.

tioned to cut out the necessary material to properly connect two intersecting parts of. Software systems for solid modeling are commercially available to provide the necessary support for computer-aided design and manufacture of intersecting surfaces such as shown in Figure 15-18.

PROBLEMS: CHAPTER 15

15-1 *Given:* Top and front views of a hopper.
Task: Develop the given figure.

15-2 *Given:* Top and front views as shown.
Task: Develop one-half of the transition piece.

15-3 *Given:* Top and front views as shown.
Task: Develop the given figure.

15-4 *Given:* Top and front views as shown.
Task: Develop the given figure.

15-5 *Given:* Top and front views as shown.

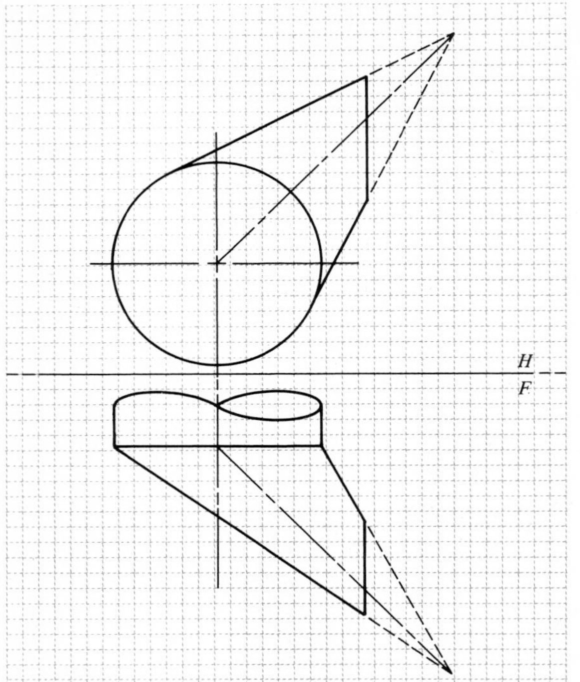

Figure 15-1P

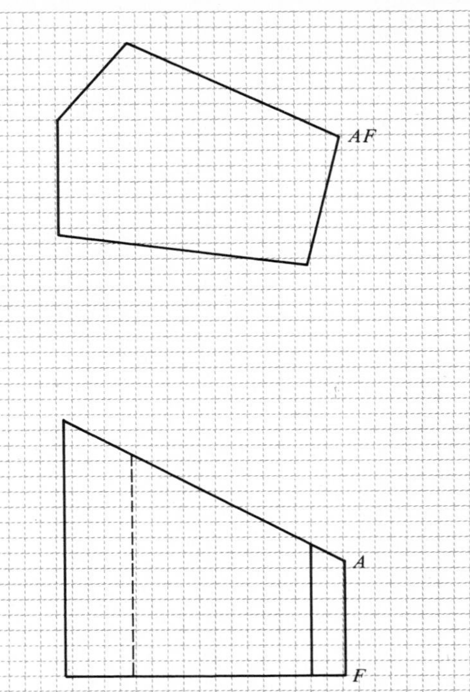

Figure 15-3P

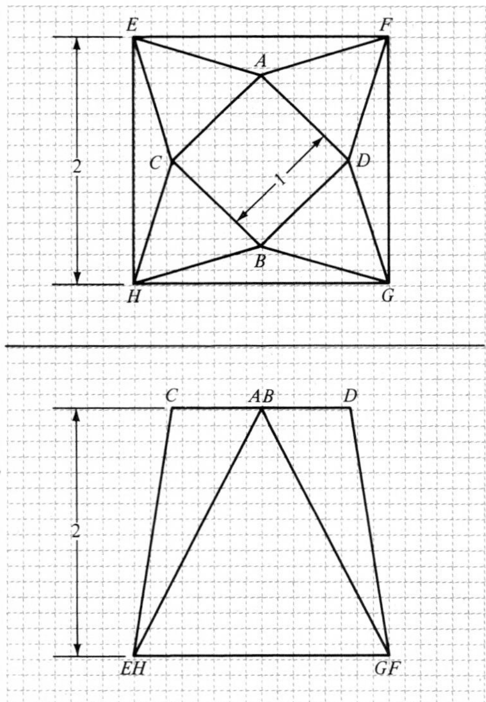

Figure 15-2P

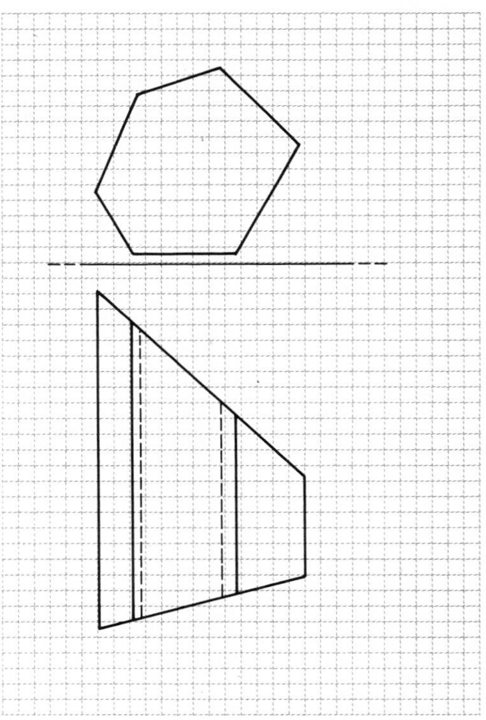

Figure 15-4P

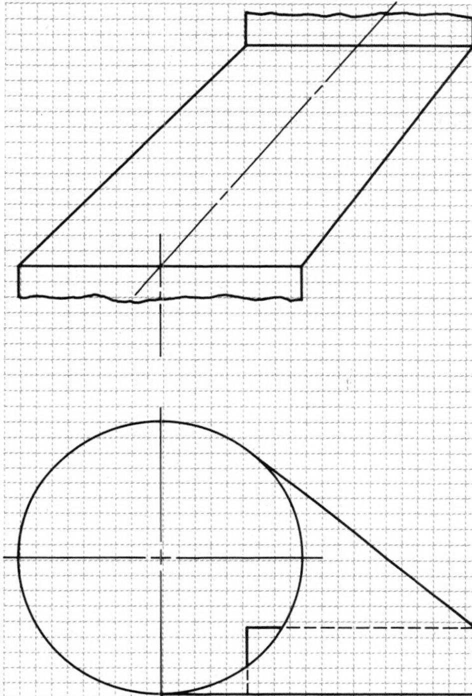

Figure 15-5P

Task: Develop the lateral surfaces of the transition piece.

15-6 *Given:* Front and top views of a round-to-square transition piece.
Task: Develop the transition piece.

15-7 *Given:* Front and top views of a duct in the shape of a truncated pyramid.
Task: Develop the part shown.

15-8 *Given:* Complete side view and incomplete front and top views of an intersecting cylinder and cone.
Task: 1. Show line of intersection in the front and top views.
2. Develop the cone. On the layout show the line of intersection of the cylinder.

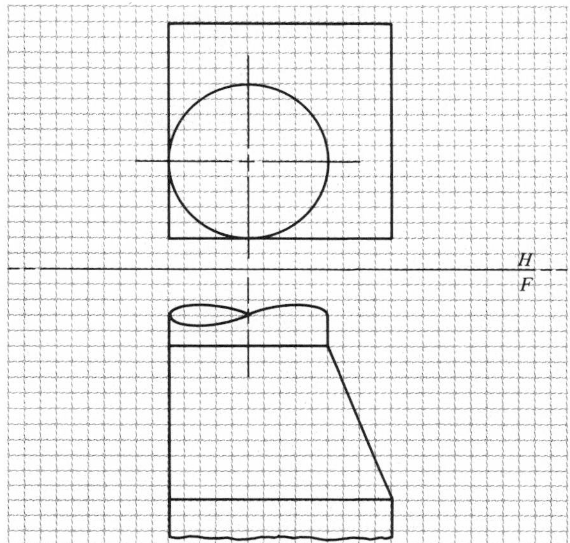

Figure 15-6P

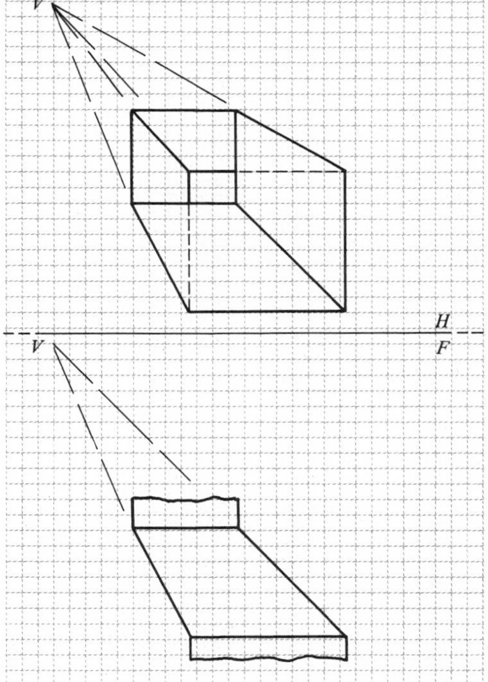

Figure 15-7P

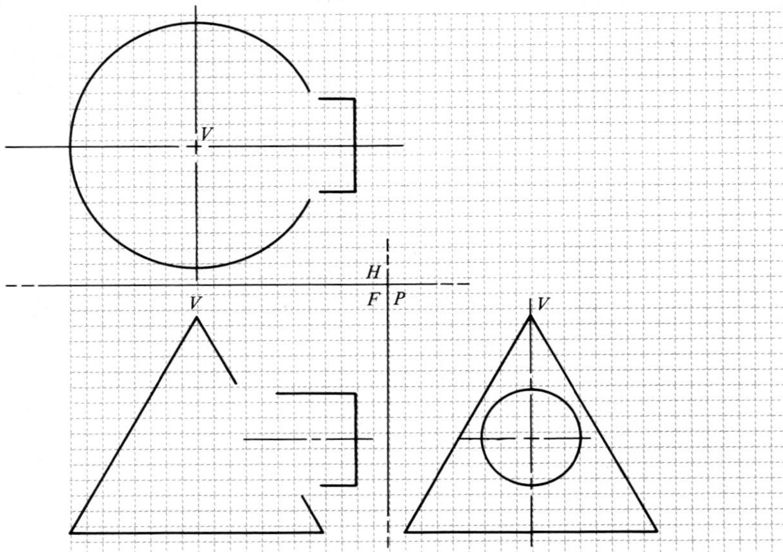

Figure 15-8P

15-9 *Given:* Complete front view and incomplete top view of an anchor block. The dihedral angle between surface *A* and surface *B* is 120 deg.
Task: Complete the top view.

15-10 *Given:* Front and top views of plane *ABC.*
Task: Construct a plane *ACD* so that it intersects plane *ABC* with line *AC* common to both planes. Construct the plane so that the dihedral angle between the planes is 50 deg. Show both planes in all views.

15-11[c] *Given:* Problem 13-11.
Task: Use program CLIPCONE (Appendix 1) and show the conical development of the clipped frustum. Produce output similar to Figure 15-9. Show developments for both inside surface up and outside surface up.

15-12[c] *Given:* Problem 14-3.
Task: Use program CLIPOFF and show the cylindrical development of the two cylinder sections formed by the intersection. Output should be similar to Figure 15-12.

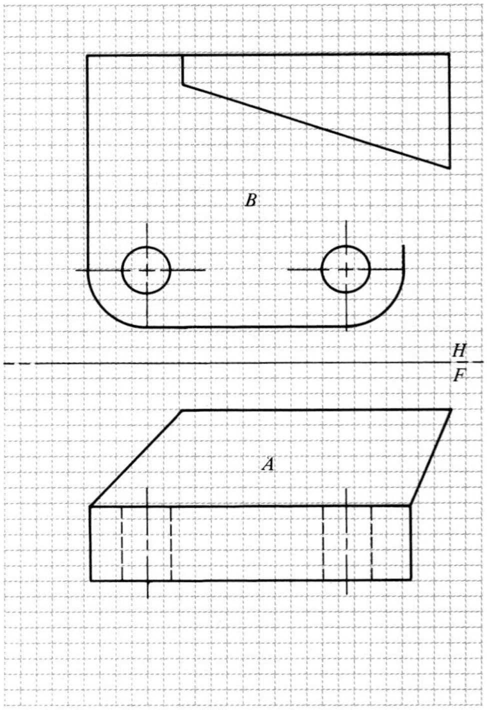

Figure 15-9P

[c]Superscript *c* indicates that a computer solution is required.

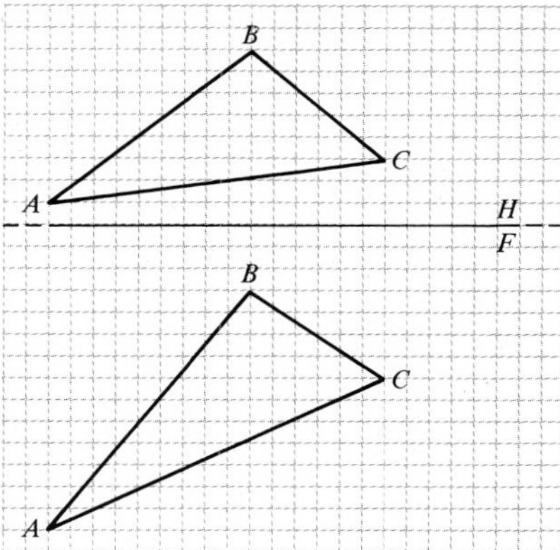

Figure 15-10P

15-13[c] *Given:* Problem 14-4, 14-5, and 14-6.
Task: Obtain solutions to these problems using a computational approach coupled with computer graphics.

15-14[c] *Given:* Problem 14-10.
Task: Modify program CLIPOFF to solve for the intersection. Create output similar to Figures 15-14, 15-15, and 15-16.

15-15[c] *Given:* Problem 15-8.
Task: Solve this problem using geometric modeling and computer graphics.

References

These references can be used to broaden or deepen the subject matter in this text. For study of traditional descriptive geometry and engineering graphics consult References 4, 6, 7, 9, 10, 11, 13, 15, 16, 20, 23, 24, and 25. For supplementary reading on computer graphics and CAD/CAM techniques which give ideas for implementing geometric modeling, use References 1, 2, 3, 5, 8, 12, 17, 18, 19, 21, and 22.

1. Angell, I. O., *A Practical Introduction to Computer Graphics,* John Wiley & Sons, New York, NY, 1981.
2. Angell, I. O., *Advanced Graphics with the IBM Personal Computer,* John Wiley & Sons, New York, NY, 1985.
3. Bower, A. and Woodwark, J., *A Programmer's Geometry,* Butterworths, London, 1983.
4. Earle, J. H., *Engineering Design Graphics,* Addison-Wesley, Reading, MA, 1983.
5. Foley, J. D. and Van Dam, A., *Fundamentals of Interactive Computer Graphics,* Addison-Wesley, Reading, MA, 1982.
6. French, T. E. and Vierck, C. J., *Graphic Science,* McGraw-Hill Book Co., New York, NY, 1958.
7. French, T. E., Vierck, C. J., & Foster, R. J., *Graphic Science and Design,* McGraw-Hill Book Co., New York, NY, 1984.
8. Gasson, P. C., *Geometry of Spatial Forms,* Ellis Horwood, Chichester, England, 1983.
9. Giesecke, F. E., et al., *Engineering Graphics,* 2nd ed., Macmillan, New York, NY, 1975.
10. Goetsch, D. L. and Newson, J. A., *Technical Drawing and Design,* Delmar Publishers, Albany, NY, 1986.
11. Grant, H. E., *Practical Descriptive Geometry,* McGraw-Hill Book Co., New York, NY, 1952.
12. Groover, M. P. and Zimmers, E. W., *CAD/CAM: Computer-Aided Design and Manufacturing,* Prentice-Hall, Englewood Cliffs, NJ, 1984.
13. Hammond, R. H., et al., *Engineering Graphics for Design and Analysis,* Ronald Press, New York, NY, 1964.
14. Hardy, S. E., Editor, *Dartmouth Structured BASIC TM128 User's Guide,* Kiewit Computation Center, Hanover, NH, 1983.
15. Hoelscher, R. P., Springer, C. H., & Dobrovolny, J. S., *Graphics for Engineers,* John Wiley & Sons, New York, NY, 1968.
16. Lamit, L. G., *Descriptive Geometry,* Prentice-Hall, Englewood Cliffs, NJ, 1983.
17. Mufti, A. A., *Elementary Computer Graphics,* Reston Publishing, Reston, VA, 1983.
18. Myers, R. E., *Microcomputer Graphics for the Apple® Computer,* Addison-Wesley, Reading, MA, Fifth Printing, 1983.
19. Newman, W. M. and Sproull, R. F., *Principles of Interactive Computer Graphics,* 2nd ed., McGraw-Hill Book Co., New York, NY, 1979.
20. Paré, E. G., Loving, R. O., & Hill, I. L., *Descriptive Geometry, Metric,* 5th ed., Macmillan, New York, NY, 1977.
21. Rogers, D. F. and Adams J. A., *Mathematical Elements for Computer Graphics,* McGraw-Hill Book Co., New York, NY, 1976.
22. Ryan, Daniel L., *Modern Graphic Communications—A CAD Approach,* Prentice-Hall, Englewood Cliffs, NJ, 1986.
23. Slaby, S. M., *Fundamentals of Three-Dimensional Descriptive Geometry,* Harcourt, Brace & World Inc., New York, NY, 1966.
24. Stewart, S. A., *Applied Descriptive Geometry,* Delmar Publishers, Albany, NY, 1986.
25. Wellman, B. L., *Technical Descriptive Geometry,* McGraw-Hill Book Co., New York, NY, 1957.

APPENDIX 1

Computer Programs

Table A-1
Main Programs and Subroutines Calling Matrix

Main Programs \ *Sub-routines*	*PVIEWS*	*TITLE*	*AXISROT*	*EYETOZ*	*VECPROD*	*AREA*	*INTER2D*	*XCHECK*	*BLINE*	*DOTPROD*	*ORTHOMAT*	*PLANE*	*INTLP*	*AROUND*	*CLIPSUB*	*GRAPHICS*	*DEVELOP*	*CYLSURF*
VIEWS	●	●																
ORTHOEYE		●	●	●														
PLANEDGE		●	●	●	●	●												
2DLINE							●											
3DLINE								●	●									
VECPRODS					●					●								
ANYAUX		●	●		●						●							
PTLINE					●													
PTPLANE					●													
MAIN1									●			●	●					
PIERCE									●			●	●					
CDGPLANE			●				●				●							
CDGSKEW			●				●				●							
CDGHORIZ			●				●				●							
CDGSLOPE			●				●				●							
BODYC														●				
VEXCLIP3		●	●	●	●										●			
XCONE		●	●	●	●										●			
CLIPLANE		●	●	●					●			●	●			●		
CLIPCONE		●	●	●					●			●	●			●	●	
PRISMS		●	●	●	●										●			
PRISMCYL		●	●	●	●										●			●
CLIPON		●	●	●	●										●			●
CLIPOFF		●	●	●	●										●			●

```
100 !*****VIEWS*****
110 !
120 PROGRAM VIEW
130 OPTION NOLET
140 PLOTTER TEK4010
150 CLEAR
160 ! PRINCIPAL VIEWS OF 3-D OBJECT
170 ! DEFINE OBJECT IN FILE #1
180 ! DEFINE LINES IN FILE #2
190 ! POINTS STORED IN X(),Y(),Z()
200 DIM X(1200),Y(1200),Z(1200)
210 DIM L(2,1200)
220 LIBRARY"PVIEWS"
230 LIBRARY"TITLE"
240 PRINT"PT FILE";
250 INPUT PTFILE$
260 OPEN #1:NAME PTFILE$
270 PRINT"LINE FILE";
280 INPUT LNFILE$
290 OPEN #2:NAME LNFILE$
300 PRINT
310 PRINT"XMIN,XMAX";
320 INPUT XMIN,XMAX
330 PRINT"YMIN,YMAX";
340 INPUT YMIN,YMAX
350 PRINT"ZMIN,ZMAX";
360 INPUT ZMIN,ZMAX
370  WIDTH=XMAX-XMIN
380  HEIGHT=YMAX-YMIN
390  DEPTH=ZMAX-ZMIN
400 ! CALC WINDOW LIMITS FOR X-Y SCREEN DISPLAY
410 STRETCH=1.2
420 PRINT"STRETCH=";STRETCH
430 W1=XMIN
440 W2=XMIN+STRETCH*WIDTH+STRETCH*DEPTH
450 W3=YMIN
460 W4=YMIN+STRETCH*HEIGHT+STRETCH*DEPTH
470 SET WINDOW W1,W2,W3,W4
480 LINES=0
490 FOR I=1 TO 1000
500 IF END #1 THEN 530
510     INPUT #1:X(I),Y(I),Z(I)
520 NEXT I
530 FOR I=1 TO 1000
540 IF END #2 THEN 580
550     INPUT #2:L(1,I),L(2,I)
560 LINES=LINES+1
570 NEXT I
580 ! CLEAR
590 CALL PVIEWS(WIDTH,HEIGHT,DEPTH,LINES,L,X,Y,Z)
600 TITLE$="PROGRAM VIEWS"
610 NAME$="NAME___________"
620 DAY$="DATE________"
630 COURSE$="COURSE______"
640 CALL TITLE(W1,W2,W3,W4,TITLE$,NAME$,DAY$,COURSE$)
650 END
```

A1-1
Program VIEWS

```
100 ! *****PVIEWS*****
110 !
120 EXTERNAL
130 SUB PVIEWS(WIDTH,HEIGHT,DEPTH,LINES,L(,),X(),Y(),Z())
140     ! CALLED BY MAIN PROGRAM "VIEWS"
150     ! DRAW 3 PRINCIPAL VIEWS OF OBJECT DEFINED IN X(),Y(),Z()
160     ! WIDTH=WIDTH OF FRONTAL VIEW
170     ! HEIGHT=HEIGHT OF FRONTAL VIEW
180     ! DEPTH=WIDTH OF PROFILE VIEW
190     ! LINES=NO OF LINES DEFINED IN L(,)
200     ! L(,) CONTAINS LINE DEFINITION BETWEEN VERTICES PAIRS
210     FOR Q=1 TO 3
220         FOR I=1 TO LINES
230             LET I1=L(1,I)
240             LET I2=L(2,I)
250             ON Q GO TO 270,300,330
260             ! HORIZONTAL VIEW
270             PLOT X(I1),HEIGHT+DEPTH-Z(I1);X(I2),HEIGHT+DEPTH-Z(I2)
280             GO TO 340
290             ! FRONTAL VIEW
300             PLOT X(I1),Y(I1);X(I2),Y(I2)
310             GO TO 340
320             ! PROFILE VIEW
330             PLOT WIDTH+DEPTH-Z(I1),Y(I1);WIDTH+DEPTH-Z(I2),Y(I2)
340         NEXT I
350     NEXT Q
360 END SUB
```

A1-2
Subroutine PVIEWS

```
100 !*****TITLE*****
110 !
120 EXTERNAL
130 SUB TITLE(W1,W2,W3,W4,TITLE$,NAME$,DAY$,COURSE$)
140 ! CREATE TITLE BLOCK AROUND VIEWPORT
150 ! W1,W2,W3,W3 ARE WINDOW LIMITS
160 ! STRING VARIABLES ARE TITLES DEFINED BY USER
170 OPTION NOLET
180 SET VIEWPORT 30,130,0,90
190 SET WINDOW W1,W2,W3,W4
200 PLOT FRAME
210 SET VIEWPORT 30,130,0,5
220 PLOT FRAME
230 GRAPHIC PRINT W1+0.01*(W2-W1),W3+0.2*(W4-W3):TITLE$
240 GRAPHIC PRINT W1+0.35*(W2-W1),W3+0.2*(W4-W3):NAME$
250 GRAPHIC PRINT W1+0.6*(W2-W1),W3+0.2*(W4-W3):DAY$
260 GRAPHIC PRINT W1+0.8*(W2-W1),W3+0.2*(W4-W3):COURSE$
270 SET VIEWPORT 30,130,0,100
280 END SUB
```

A1-3
Subroutine TITLE

```
100 !*****ORTHOEYE*****
110 !
120 ! GENERAL ORTHO PROJECTION
130 ! STORE 3-D PTS IN FILE #1
140 ! STORE LINE DEF IN FILE #2
150 ! INITIAL PTS IN X(),Y(),Z()
160 ! PROJECTED PTS ARE IN A(),B()
170 DIM X(1200),Y(1200),Z(1200)
180 DIM A(1200),B(1200)
190 DIM R(3,3)
200 DIM L(2,1200)
210 PLOTTER TEK4010
220 OPTION NOLET
230 LIBRARY"AXISROT"
240 LIBRARY"EYETOZ"
250 LIBRARY"TITLE"
260 CLEAR
270 PRINT"WINDOW LIMITS"
280 PRINT
290 PRINT"LEFT,RIGHT";
300 INPUT LEFT,RIGHT
310 PRINT"BOTTOM,TOP";
320 INPUT BOTTOM,TOP
330 SET WINDOW LEFT,RIGHT,BOTTOM,TOP
340 POINTS=0
350 LINES=0
360 PRINT"PT FILE";
370 INPUT PTFILE$
380 OPEN #1:NAME PTFILE$
390 PRINT"LINE FILE";
400 INPUT LNFILE$
410 OPEN #2:NAME LNFILE$
420 CALL EYETOZ(R)
430 FOR I=1 TO 1000
440 IF END#1 THEN 480
450     INPUT #1:X(I),Y(I),Z(I)
460 POINTS=POINTS+1
470 NEXT I
480 FOR I=1 TO 1000
490 IF END#2 THEN 540
500     INPUT #2:L(1,I),L(2,I)
510 LINES=LINES+1
520 NEXT I
530 ! PERFORM POST-OP MATRIX MULTIPLICATION
540 FOR I=1 TO POINTS
550     LET A(I)=X(I)*R(1,1)+Y(I)*R(2,1)+Z(I)*R(3,1)
560     LET B(I)=X(I)*R(1,2)+Y(I)*R(2,2)+Z(I)*R(3,2)
570 NEXT I
580 FOR I=1 TO LINES
590     LET I1=L(1,I)
600     LET I2=L(2,I)
610     PLOT A(I1),B(I1);A(I2),B(I2)
620 NEXT I
630 TITLE$="PROGRAM ORTHOEYE"
640 NAME$="NAME____________"
650 DAY$="DATE________"
660 COURSE$="COURSE______"
670 CALL TITLE(LEFT,RIGHT,BOTTOM,TOP,TITLE$,NAME$,DAY$,COURSE$)
680 END
```

A1-4
Program ORTHOEYE

```
100 ! ****AXISROT****
110 !
120 EXTERNAL
130 SUB AXISROT(AXIS,THETA,P(,))
140     ! CREATE 3X3 TRANSFORMATION MATRIX [P]
150     ! FOR CCW ROTATION ABOUT A PRINCIPAL AXIS
170     ! AS SEEN FROM THE AXIS OF ROTATION LOOKING TOWARD THE ORIGIN
180     ! AXIS=1(X-AXIS)  AXIS=2(Y-AXIS)  AXIS=3(Z-AXIS)
190     ! THETA IS ANGLE OF ROTATION IN DEGREES
200     OPTION NOLET
210      ! CREATE 3X3 IDENTITY MATRIX
220      MAT P=IDN(3,3)
230      ! DEFINE MATRIX TERMS FOR ROTATION
240      M1=MOD(AXIS,3)+1
250      M2=MOD(M1,3)+1
260      C=COS(THETA*PI/180)
270      S=SIN(THETA*PI/180)
280      P(M1,M1)=C
290      P(M2,M2)=C
300      P(M1,M2)=S
310      P(M2,M1)=-S
320 END SUB
```

A1-5
Subroutine AXISROT

```
100 ! *****EYETOZ***
110 !
120 EXTERNAL
130 SUB EYETOZ(R(,))
140 ! CALCULATION OF TRANSFORMATION MATRIX
150 ! CALLED BY  PROGRAM 'ORTHOEYE'
160 ! USE R(,) IN POST-OP SENSE
170 OPTION NOLET
180 DIM D(3,3)
190 DIM E(3,3)
200 PRINT"X,Y,Z OF EYE"
210 INPUT EYE1,EYE2,EYE3
220 THETA=ANGLE(EYE3,EYE1)*180/PI
230 ! ROTATE CW ABOUT Y-AXIS BY THETA
240 CALL AXISROT(2,-THETA,D)
250 DIST=SQR(EYE1*EYE1+EYE3*EYE3)
260 PHI=ANGLE(DIST,EYE2)*180/PI
270 ! ROTATE CCW ABOUT X-AXIS BY PHI
280 CALL AXISROT(1,PHI,E)
290 ! IF MAT FUNCTIONS ARE NOT AVAILABLE DELETE NEXT TWO LINES
300 MAT R=D*E
310 GO TO 430
320 ! IF MAT FUNCTIONS ARE USED, DELETE NEXT 10 LINES
330 !** USE NEXT 9 LINES TO CALCULATE MATRIX DOT PRODUCT R=D*E
340 FOR I=1 TO 3
350 FOR J=1 TO 3
360 SUM=0
370 FOR K=1 TO 3
380 SUM=SUM+D(I,K)*E(K,J)
390 NEXT K
400 R(I,J)=SUM
410 NEXT J
420 NEXT I
430 END SUB
```

A1-6
Subroutine EYETOZ

```
100 !*******PLANEDGE*******
110 !
120 ! DEFINE PLANE USING 3 PTS
130 ! CALCULATE NORMAL TO PLANE
140 ! INITIAL PTS IN X(),Y(),Z()
150 ! PROJECTED PTS ARE IN A(),B()
160 DIM X(3),Y(3),Z(3)
170 DIM A(4),B(4)
180 DIM N(3)
190 DIM R(3,3)
200 DIM L(2,3)
210 PLOTTER TEK4010
220 OPTION NOLET
230 LIBRARY"VECPROD"
240 LIBRARY"AXISROT"
250 LIBRARY"EYETOZ"
260 LIBRARY"AREA"
270 LIBRARY"TITLE"
280 CLEAR
290 PRINT"WINDOW LIMITS"
300 PRINT
310 PRINT"LEFT,RIGHT";
320 INPUT LEFT,RIGHT
330 PRINT"BOTTOM,TOP";
340 INPUT BOTTOM,TOP
350 SET WINDOW LEFT,RIGHT,BOTTOM,TOP
360 PRINT
370 PRINT"DEFINE A PLANE"
380 PRINT"ENTER 3 PTS IN SPACE"
390 FOR I=1 TO 3
400 PRINT"POINT";I;"X,Y,Z";
410 INPUT X(I),Y(I),Z(I)
420 NEXT I
430 ! CALCULATE EDGE VECTORS OF PLANE
440 A(1)=X(2)-X(1)
450 A(2)=Y(2)-Y(1)
460 A(3)=Z(2)-Z(1)
470 B(1)=X(3)-X(1)
480 B(2)=Y(3)-Y(1)
490 B(3)=Z(3)-Z(1)
500 ! CANCLUATE COMPONENTS OF NORMAL VECTOR
510 CALL VECPROD(A,B,N)
520 PRINT
530 PRINT"NORMAL VECTOR=";N(1);N(2);N(3)
540 ! DEFINE MATRIX FOR EYE POSITION
550 CALL EYETOZ(R)
560 ! DEFINE LINE DEFINITIONS
570 L(1,1)=1
580 L(2,1)=2
590 L(1,2)=2
600 L(2,2)=3
610 L(1,3)=3
620 L(2,3)=1
630 POINTS=3
640 LINES=3
650 ! PERFORM POST-OP MATRIX MULTIPLICATION
660 FOR I=1 TO POINTS
670     LET A(I)=X(I)*R(1,1)+Y(I)*R(2,1)+Z(I)*R(3,1)
680     LET B(I)=X(I)*R(1,2)+Y(I)*R(2,2)+Z(I)*R(3,2)
690 NEXT I
700 FOR I=1 TO LINES
710     LET I1=L(1,I)
720     LET I2=L(2,I)
730     PLOT A(I1),B(I1);A(I2),B(I2)
```

```
740 NEXT I
750 CALL AREA(3,A,B,VALUE)
760 VALUE=ABS(VALUE)
770 PRINT"TRUE AREA=";VALUE
780 TITLE$="PROGRAM PLANEDGE"
790 NAME$="NAME___________"
800 DAY$="DATE_______"
810 COURSE$="COURSE_____"
820 CALL TITLE(LEFT,RIGHT,BOTTOM,TOP,TITLE$,NAME$,DAY$,COURSE$)
830 END
```

A1-7
Program PLANEDGE

```
100 ! *****VECPROD*****
110 !
120 EXTERNAL
130 SUB VECPROD (A(),B(),R())
140 ! DETERMINE VECTOR PRODUCT AXB=R
150 ! COMPONENTS OF VECTOR [A] PASSED IN A(1),A(2),A(3)
160 ! COMPONENTS OF VECTOR [B] PASSED IN B(1),B(2),B(3)
170 ! RETURNED COMPONENTS OF VECTOR PRODUCT CONTAINED IN R(1),R(2),R(3)
180 OPTION NOLET
190 FOR I=1 TO 3
200 I1=MOD(I,3)+1
210 I2=MOD(I1,3)+1
220 R(I)=A(I1)*B(I2)-A(I2)*B(I1)
230 NEXT I
240 END SUB
```

A1-8
Subroutine VECPROD

```
70 REM
80 REM
90 REM*****AREA*****
100 EXTERNAL
110 SUB AREA(N,X(),Y(),AREA)
120 REM SUBROUTINE TO CALCULATE AREA OF 2-D POLYGON
130 REM N IS THE NUMBER OF POLYGON VERTICIES
140 REM X() AND Y() CONTAIN THE VERTICES, DEFINED IN THE MAIN PROGRAM
150 REM *** VERTICES MUST BE DEFINED IN CCW MANNER FOR POSIVITE AREA***
160 REM AREA IS RETURNED TO MAIN PROGRAM
170 OPTION NOLET
180 REM CLOSE AREA BY DEFINING LAST VERTEX=FIRST VERTEX
190 LET X(N+1)=X(1)
200 LET Y(N+1)=Y(1)
210 REM INITIALIZE AREA
220 LET AREA=0
230 REM CALCULATE ENCLOSED AREA
240 FOR I=1 TO N
250 AREA=AREA+(X(I+1)-X(I))*(Y(I)+Y(I+1))
260 NEXT I
270 REM CALCULATE ACTUAL AREA
280 AREA=-AREA/2
290 END SUB
```

A1-9
Subroutine AREA

```
100 !*****2DLINE*****
110 !
120 ! INTERSECTING 2-D LINES
130 CLEAR
140 PRINT"PROGRAM 2DLINE"
150 PRINT
160 OPTION NOLET
170 PLOTTER TEK4010
180 LIBRARY"INTER2D"
190 PRINT"GIVE ENDPOINTS OF LINE 1"
200 PRINT"AS X1,Y1,X2,Y2";
210 INPUT X1,Y1,X2,Y2
220 PRINT"GIVE ENDPOINTS OF LINE 2"
230 PRINT"AS X3,Y3,X4,Y4";
240 INPUT X3,Y3,X4,Y4
250 CALL INTER2D(X1,Y1,X2,Y2,X3,Y3,X4,Y4,T,S,X,Y)
260 IF X<99999 THEN 290
270 PRINT"LINES ARE PARALLEL"
280 GO TO 320
290 PRINT
300 PRINT"INTERSECTION AT"
310 PRINT"X=";X;" Y=";Y
320 END
```

A1-10
Program 2DLINE

```
100 ! SUBROUTINE ****INTER2D****
110 EXTERNAL
120 SUB INTER2D(X1,Y1,X2,Y2,X3,Y3,X4,Y4,T,S,X,Y)
130 ! CALLED BY 2DLINES AND CDGXXX PROGRAMS
140 ! INTERSECTION OF TWO 2-D LINES 1-2 AND 3-4
150 ! X1,Y1,X2,Y2 CONTAIN TWO POSITION VECTORS ON LINE 1
160 ! X3,Y3,X4,Y4 CONTAIN TWO POSITION VECTORS ON LINE 2
170 ! CALCULATED PARAMETER VALUES AT INTERSECTION ARE T,S
180 ! POINT OF INTERSECTION, IF ANY, RETURNED IN X,Y
190 ! OTHERWISE X=99999, Y=99999
200 OPTION NOLET
210 DETERM=(X2-X1)*(Y3-Y4)-(X3-X4)*(Y2-Y1)
220 IF ABS(DETERM)>0.00001 THEN 300
230 ! LINES ARE PARALLEL-NO INTERSECTION
240 ! PASS DUMMY VALUES FOR X AND Y TO MAIN PROGRAM
250 X=99999
260 Y=99999
270 PRINT"LINES ARE PARALLEL"
280 GO TO 350
290 ! CALCULATE VALUES OF PARAMETERS T AND S AT INTERSECTION
300 T=((X3-X1)*(Y3-Y4)-(X3-X4)*(Y3-Y1))/DETERM
310 S=((X2-X1)*(Y3-Y1)-(Y2-Y1)*(X3-X1))/DETERM
320 ! CALCULATE X,Y COMPONENTS OF POINT OF INTERSECTION
330 X=X1+T*(X2-X1)
340 Y=Y1+T*(Y2-Y1)
350 END SUB
```

A1-11
Subroutine INTER2D

```
100 !*****3DLINE*****
110 !
120 ! INTERSECTION OF 3-D LINES
130 LIBRARY"XCHECK"
140 LIBRARY"BLINE"
150 OPTION NOLET
160 DIM B(3),C(3),D(3),E(3)
170 DIM X(2),Y(2),Z(2),U(2),V(2),W(2)
180 READ X(1),Y(1),Z(1),X(2),Y(2),Z(2)
190 READ U(1),V(1),W(1),U(2),V(2),W(2)
200 ! DEFINE ENDPOINTS OF LINE 1
210 DATA 2,3,4,3,4,5
220 ! DEFINE ENDPOINTS OF LINE 2
230 DATA -2,-3,-4,-1,-1,-1
240 CALL BLINE (X,Y,Z,B,D)
250 CALL BLINE(U,V,W,C,E)
260 CALL XCHECK(B,C,D,E,I1,J1,K1,A$)
270 PRINT A$
280 IF A$<>"LINES INTERSECT"THEN 300
290 PRINT"INTERSECTION AT";I1;J1;K1
300 PRINT
310 PRINT"LINE 1 ENDPOINTS"
320 PRINT X(1);Y(1);Z(1);X(2);Y(2);Z(2)
330 PRINT
340 PRINT"LINE 2 ENDPOINTS"
350 PRINT U(1);V(1);W(1);U(2);V(2);W(2)
360 END
```

A1-12
Program 3DLINE

```
100 ! *****XCHECK*****
110 !
120 EXTERNAL
130 SUB XCHECK(B(),C(),D(),E(),I1,I2,I3,A$)
140   ! CALLED BY MAIN "3DLINE"
150   ! CHECK FOR INTERSECTION OF TWO 3-D LINES
160   ! VECTOR EQN FOR LINE 1: [B]+T[D] AND FOR LINE 2:[C]+S[E]
170   ! WHERE [B],[C] ARE BASE VECTORS,[D],[E] ARE DIRECTIONAL VECTORS
180   ! POINT OF INTERSECTION(IF ANY) IS [I]=[I1  I2  I3]
190 OPTION NOLET
200 DETERM=-D(1)*E(2)+D(2)*E(1)  !CHECK FOR NON-ZERO DETERMERMINANTS
210 IF ABS(DETERM)>0.0001 THEN 290
220 DETERM=-D(1)*E(3)+D(3)*E(1)
230 IF ABS(DETERM)>0.0001 THEN 330
240 DETERM=-D(2)*E(3)+D(3)*E(2)
250 IF ABS(DETERM)>0.0001 THEN 370
260 A$="LINES ARE PARALLEL"
270 GO TO 470
280   ! *** USE CRAMER'S RULE TO FIND PARAMETERS T AND S
290 T=(-(C(1)-B(1))*E(2)+(C(2)-B(2))*E(1))/DETERM
300 S=(D(1)*(C(2)-B(2))-(C(1)-B(1))*D(2))/DETERM
310 IF ABS((B(3)+T*D(3))-(C(3)+S*E(3)))<0.0001 THEN 430
320 GO TO 400
330 T=(-(C(1)-B(1))*E(3)+(C(3)-B(3))*E(1))/DETERM
340 S=(D(1)*(C(3)-B(3))-(C(1)-B(1))*D(3))/DETERM
350 IF ABS((B(2)+T*D(2))-(C(2)+S*E(2)))<0.0001 THEN 430
360 GO TO 400
370 T=(-(C(2)-B(2))*E(3)+(C(3)-B(3))*E(2))/DETERM
380 S=(D(2)*(C(3)-B(3))-(C(2)-B(2))*D(3))/DETERM
```

```
390 IF ABS((B(1)+T*D(1))-(C(1)+S*E(1)))<0.0001 THEN 430
400 A$="LINES DO NOT INTERSECT"
410 GO TO 470
420 !INTERSECTION PT=[I1  I2  I3]
430 I1=B(1)+T*D(1)
440 I2=B(2)+T*D(2)
450 I3=B(3)+T*D(3)
460 A$="LINES INTERSECT"
470 END SUB
```

A1-13
Subroutine XCHECK

```
90 !           BLINE
100 EXTERNAL
110 SUB BLINE (X(),Y(),Z(),B(),D())
120 ! CALLED BY MAIN PROGRAM PIERCEPT
130 ! CALCULATE BASE VECTOR AND DIRECTIONAL VECTOR
140 ! FOR A 3-D LINE WITH TWO GIVEN ENDPOINTS
150 ! ENDPOINTS OF GIVEN LINE PASSED IN X(1),Y(1),Z(1);X(2),Y(2),Z(2)
160 ! COMPONENTS OF BASE VECTOR RETURNED IN B(1),B(2),B(3)
170 ! COMPONENTS OF DIRECTIONAL VECTOR RETURNED IN D(1),D(2),D(3)
180 OPTION NOLET
190 B(1)=X(1)
200 B(2)=Y(1)
210 B(3)=Z(1)
220 D(1)=X(2)-X(1)
230 D(2)=Y(2)-Y(1)
240 D(3)=Z(2)-Z(1)
250 END SUB
```

A1-14
Subroutine BLINE

```
100 ! *****VECPRODS*****
110 !
120 ! CALL TWO SUBROUTINES "DOTPROD" AND "VECPROD" TO CALCULATE
130 ! DOT PRODUCT [A].[B] AND CROSS PRODUCT [A]X[B]
140 LIBRARY"DOTPROD"
150 LIBRARY"VECPROD"
160 DIM A(3),B(3),C(3)
170 ! INPUT THREE SCALAR COMPONENTS OF [A] INTO A(1),A(2),A(3)
180 PRINT"INPUT 3 ELEMENTS OF VECTOR A";
190 INPUT A(1),A(2),A(3)
200 ! INPUT THREE SCALAR COMPONENTS OF [B] INTO B(1),B(2),B(3)
210 PRINT"INPUT 3 ELEMENTS OF VECTOR B";
220 INPUT B(1),B(2),B(3)
230 PRINT
240 ! CALCULATE DOT PRODUCT [A].[B]=INNER
250 CALL DOTPROD(A,B,INNER)
260 PRINT"DOT PRODUCT IS";INNER
270 ! CALCULATE SCALAR COMPONENTS OF VECTOR PRODUCT AS C(1),C(2),C(3)
280 CALL VECPROD(A,B,C)
290 PRINT"VECTOR PRODUCT ";C(1);C(2);C(3)
300 END
```

A1-15
Program VECPRODS

```
100 ! *****DOTPROD*****
110 !
120 EXTERNAL
130 SUB DOTPROD(A(),B(),INNER)
140 ! DETERMINE VECTOR DOT PRODUCT A.B
150 ! COMPONENTS OF VECTOR [A] PASSED IN A(1),A(2),A(3)
160 ! COMPONENTS OF VECTOR [B] PASSED IN B(1),B(2),B(3)
170 ! CALCULATED AND RETURNED SCALAR VALUE OF DOT PRODUCT IN INNER
180 OPTION NOLET
190 INNER=A(1)*B(1)+A(2)*B(2)+A(3)*B(3)
200 END SUB
```

A1-16
Subroutine DOTPROD

```
100 !*****ANYAUX*****
110 !
120 ! GENERAL AUXILIARY PROJECTION
130 ! LINE OF SIGHT IS NORMAL TO A
140 ! PLANE DEFINED BY 3 POINTS
150 ! STORE 3-D PTS IN FILE #1
160 ! STORE LINE DEF IN FILE #2
170 ! INITIAL PTS IN X(),Y(),Z()
180 ! PROJECTED PTS ARE IN A(),B()
190 DIM X(100),Y(100),Z(100)
200 DIM A(100),B(100)
210 DIM L(2,100)
220 DIM R(3,3)
230 DIM C(3),D(3),N(3)
240 DIM S(3)
250 LIBRARY"AXISROT"
260 LIBRARY"VECPROD"
270 LIBRARY"ORTHOMAT"
280 LIBRARY"TITLE"
290 CLEAR
300 PLOTTER TEK4010
310 OPTION NOLET
320 W1=-10
330 W2=10
340 W3=5
350 W4=25
360 SET WINDOW W1,W2,W3,W4
370 ! CHANGE NEXT 4 LINES
380 VERTICES=5  !NO OF VERTICES IN FILE #1'
390 LINES=4  !NO OF LINES IN FILE #2'
400 OPEN #1: name "PLATE1"
410 OPEN #2: name "PLINE1"
420 FOR I=1 TO VERTICES
430     INPUT #1:X(I),Y(I),Z(I)
440 NEXT I
450 FOR I=1 TO LINES
460     INPUT #2:L(1,I),L(2,I)
470 NEXT I
480 PRINT"INPUT(I) OR CALCULATE(C) EYE POSITION";
490 INPUT E$
500 IF E$="C" THEN 540
510 PRINT"INPUT EYE COORDINATES E1,E2,E3";
520 INPUT E1,E2,E3
530 GO TO 850
540 PRINT"VIEW NORMAL TO PLANE(P) LINE(L) OR TRUE EDGE(E)";
550 INPUT V$
```

```
560 IF V$="L" THEN 700
570 PRINT"INPUT 3 VERTICES TO FORM NORMAL TO PLANE";
580 INPUT P1,P2,P3
590  C(1)=X(P1)-X(P2)
600  C(2)=Y(P1)-Y(P2)
610  C(3)=Z(P1)-Z(P2)
620  D(1)=X(P3)-X(P2)
630  D(2)=Y(P3)-Y(P2)
640  D(3)=Z(P3)-Z(P2)
650 CALL VECPROD (C,D,N)
660  E1=N(1)
670  E2=N(2)
680  E3=N(3)
690 IF V$="P" THEN 840
700 PRINT"INPUT 2 VERTICES TO DEFINE LINE";
710 INPUT P1,P2
720  E1=X(P2)-X(P1)
730  E2=Y(P2)-Y(P1)
740  E3=Z(P2)-Z(P1)
750 IF V$="L" THEN 840
760 ! PLANE EDGE LINE TRUE LENGTH
770  D(1)=E1
780  D(2)=E2
790  D(3)=E3
800 CALL VECPROD (N,D,S)
810  E1=S(1)
820  E2=S(2)
830  E3=S(3)
840 PRINT"E1,E2,E3=";E1;E2;E3
850 CALL ORTHOMAT (E1,E2,E3,R)
860 ! PRINT"ORTHO PROJECTION ON PLANE PERPENDICULAR TO LINE OF SIGHT"
870 FOR I=1 TO VERTICES
880      A(I)=X(I)*R(1,1)+Y(I)*R(2,1)+Z(I)*R(3,1)
890      B(I)=X(I)*R(1,2)+Y(I)*R(2,2)+Z(I)*R(3,2)
900 NEXT I
910 FOR I=1 TO LINES
920       I1=L(1,I)
930       I2=L(2,I)
940 PLOT A(I1),B(I1);A(I2),B(I2)
950 NEXT I
960 LENGTH=SQR((X(2)-X(1))^2+(Y(2)-Y(1))^2+(Z(2)-Z(1))^2)
970 PRINT"LINE LENGTH=";LENGTH
980 IF V$<>"E" THEN 1030
990 ! FIND TRUE ANGLE BETWEEN LINE AND PLANE
1000 ANGLE1=ANGLE(A(1)-A(2),B(1)-B(2))*180/PI
1010 ANGLE2=ANGLE(A(4)-A(3),B(4)-B(3))*180/PI
1020 PRINT"ANGLE=";ANGLE2-ANGLE1
1030 TITLE$="PROGRAM ANYAUX"
1040 NAME$="NAME___________"
1050 DAY$="DATE________"
1060 COURSE$="COURSE_____"
1070 CALL TITLE(W1,W2,W3,W4,TITLE$,NAME$,DAY$,COURSE$)
1080 PRINT"W1=";W1
1090 PRINT"W2=";W2
1100 PRINT"W3=";W3
1110 PRINT"W4=";W4
1120 SET VIEWPORT 0,100,0,100
1130 SET WINDOW W1,W2,W3,W4
1140 PLOT W1,(W3+W4)/2;W1+1,(W3+W4)/2
1150 GRAPHIC PRINT W1,(W3+W4)/3:"ONE-UNIT"
1160 END
```

A1-17
Program ANYAUX

```
100 ! *****ORTHOMAT*****
110 !
120 EXTERNAL
130 SUB ORTHOMAT (EYE1,EYE2,EYE3,R(,))
140 ! CALCULATION OF OBSERVE MATRIX
150 ! CALLED BY MAIN ANYAUX
160 ! SUBROUTINE 'EYETOZ' EXCEPT EYE POSITION EYE1,EYE2,EYE3 IS PASSED
170 !USE RETURNED MATRIX R(,) IN A POST-OP SENSE ON R.H.C ROW DATA MATRIX
180 OPTION NOLET
190 DIM D(3,3)
200 DIM E(3,3)
210 DIM V(3,3)
220 DIM U(3,3)
230 THETA=ANGLE(EYE3,EYE1)*180/PI
240 CALL AXISROT(2,-THETA,D)
250 DIST=SQR(EYE1*EYE1+EYE3*EYE3)
260 PHI=ANGLE(DIST,EYE2)*180/PI
270 CALL AXISROT(1,PHI,E)
280 MAT R=D*E
290 END SUB
```

A1-18
Subroutine ORTHOMAT

```
100 !*****PTLINE*****
110 ! REFERENCE EXAMPLE 7-15
120 ! DEFINE LINE WITH TWO POINTS
130 ! DEFINE A 4TH POINT IN SPACE
140 DIM X(3),Y(3),Z(3)
150 DIM A(2),B(2)
160 DIM N(3),R(3),S(3)
170 PLOTTER TEK4010
180 OPTION NOLET
190 LIBRARY"VECPROD"
200 CLEAR
210 PRINT"PROGRAM PTLINE"
220 PRINT
230 PRINT
240 PRINT"FIND SHORTEST DISTANCE"
250 PRINT"FROM POINT TO A LINE"
260 PRINT
270 PRINT"TO DEFINE A LINE"
280 PRINT"ENTER TWO ENDPOINTS"
290 FOR I=1 TO 2
300 PRINT"PT";I; "X,Y,Z"
310 INPUT X(I),Y(I),Z(I)
320 NEXT I
330 PRINT
340 PRINT"INPUT 3TH POINT IN SPACE"
350 INPUT X(3),Y(3),Z(3)
360 ! FIND DIRECTIONAL VECTOR FOR LINE
370 R(1)=X(2)-X(1)
380 R(2)=Y(2)-Y(1)
390 R(3)=Z(2)-Z(1)
400 RABS=SQR(R(1)^2+R(2)^2+R(3)^2)
410 S(1)=X(3)-X(1)
420 S(2)=Y(3)-Y(1)
430 S(3)=Z(3)-Z(1)
440 SABS=SQR(S(1)^2+S(2)^2+S(3)^2)
450 ! CALCULATE COMPONENTS OF NORMAL VECTOR
```

```
460 CALL VECPROD(S,R,N)
470 PRINT
480 NABS=SQR(N(1)^2+N(2)^2+N(3)^2)
490 DIST=NABS/RABS
500 PRINT"DIST=";DIST
510 PRINT
520 ! CALCULATE PIERCE POINT Q
530 INNER=R(1)*S(1)+R(2)*S(2)+R(3)*S(3)
540 COSA=INNER/(RABS*SABS)
550 E=SABS*COSA
560 T=E/RABS
570 ! CALCULATE INTERSECTION Q ON LINE
580 X(1)=X(1)+T*(X(2)-X(1))
590 Y(1)=Y(1)+T*(Y(2)-Y(1))
600 Z(1)=Z(1)+T*(Z(2)-Z(1))
610 PRINT"PIERCE POINT Q="
620 PRINT X(1);Y(1);Z(1)
630 END
```

A1-19
Program PTLINE

```
100 !*****PTPLANE*****
110 ! REFERENCE EXAMPLE 7-16
120 ! PLANE DEFINED WITH 3 POINTS
130 ! DEFINE A 4TH POINT IN SPACE
140 DIM X(4),Y(4),Z(4)
150 DIM A(3),B(3)
160 DIM N(3),R(3)
170 PLOTTER TEK4010
180 OPTION NOLET
190 LIBRARY"VECPROD"
200 CLEAR
210 PRINT"PROGRAM PTPLANE"
220 PRINT
230 PRINT
240 PRINT"FIND SHORTEST DISTANCE"
250 PRINT"FROM POINT TO A PLANE"
260 PRINT
270 PRINT"TO DEFINE A PLANE"
280 PRINT"ENTER 3 NON-COLLINEAR PTS"
290 FOR I=1 TO 3
300 PRINT"PT";I; "X,Y,Z"
310 INPUT X(I),Y(I),Z(I)
320 NEXT I
330 PRINT
340 PRINT"INPUT 4TH POINT IN SPACE"
350 INPUT X(4),Y(4),Z(4)
360 ! CALCULATE EDGE VECTORS OF PLANE
370 A(1)=X(2)-X(1)
380 A(2)=Y(2)-Y(1)
390 A(3)=Z(2)-Z(1)
400 B(1)=X(3)-X(1)
410 B(2)=Y(3)-Y(1)
420 B(3)=Z(3)-Z(1)
430 ! CALCULATE COMPONENTS OF NORMAL VECTOR
440 CALL VECPROD(A,B,N)
450 PRINT
460 NABS=SQR(N(1)^2+N(2)^2+N(3)^2)
```

```
470 R(1)=X(4)-X(2)
480 R(2)=Y(4)-Y(2)
490 R(3)=Z(4)-Z(2)
500 INNER=R(1)*N(1)+R(2)*N(2)+R(3)*N(3)
510 DIST=INNER/NABS
520 PRINT"DIST=";DIST
530 PRINT
540 ! CALCULATE PIERCE POINT Q
550 X(1)=X(4)-DIST*N(1)/NABS
560 Y(1)=Y(4)-DIST*N(2)/NABS
570 Z(1)=Z(4)-DIST*N(3)/NABS
580 PRINT"PIERCE POINT Q=";
590 PRINT X(1);Y(1);Z(1)
600 END
```

A1-20
Program PTPLANE

```
100 ! *****MAIN1*****
110 !
120 ! INTERSECTION OF A 3-D LINE AND A PLANE
130 ! DEFINE 3-D LINE BY 2 ENDPOINTS
140 LIBRARY"PLANE"
150 LIBRARY"BLINE"
160 LIBRARY"INTLP"
170 DIM R(5),S(5),T(5)
180 DIM X(2),Y(2),Z(2)
190 DIM P(3)
200 ! DEFINE PLANE BY 3 NON-COLLINEAR CORNER POINTS
210 READ R(1),S(1),T(1),R(2),S(2),T(2),R(3),S(3),T(3)
220 DATA 1,4,2.625,2.375,1.75,4,4.5,2.875,1.875
230 ! DEFINE 2 ENDPOINTS OF LINE
240 READ X(1),Y(1),Z(1),X(2),Y(2),Z(2)
250 DATA 0.625,5.5,4.625,4.25,1.875,1
260 CALL INTLP (R,S,T,X,Y,Z,P,L,A)
270 PRINT"INTERSECTION AT X=";P(1); " Y=";P(2); " Z=";P(3)
280 PRINT"LENGTH OF LINE SEGMENT=";L
290 PRINT"ANGLE BETWEEN LINE AND PLANE=";A
300 END
```

A1-21
Program MAIN1

```
100 ! *****PLANE*****
110 !
120 EXTERNAL
130 SUB PLANE (TYPE,R(),S(),T(),N1,N2,N3,K)
140    ! CALLED BY MAIN 'PIERCEPT'
150    ! DEF 3 PTS [R(1) S(1) T(1)], [R(2) S(2) T(2)], [R(3) S(3) T(3)]
160    ! IF TYPE=3 DEFINE 4TH CORNER=1ST CORNER
170    ! IF TYPE=4 CALCULATE 4TH CORNER AND LET 5TH CORNER=1ST CORNER
180    ! CALCULATE IMPLICIT PARAMETERS N1,N2,N3,K FOR PLANE
190 OPTION NOLET
200 LIBRARY"VECPROD"
210 DIM X(3),Y(3),Z(3)
220 X(1)=R(2)-R(1)
230 X(2)=S(2)-S(1)
240 X(3)=T(2)-T(1)
250 Y(1)=R(3)-R(1)
260 Y(2)=S(3)-S(1)
270 Y(3)=T(3)-T(1)
280 CALL VECPROD (X,Y,Z)
```

```
290    ! COMPUTE IMPLICIT COEFFS FOR PLANE
300 N1=Z(1)
310 N2=Z(2)
320 N3=Z(3)
330 K=Z(1)*R(1)+Z(2)*S(1)+Z(3)*T(1)
340 IF TYPE=3 THEN 440
350    ! COMPUTE 4TH CORNER OF PLANE
360 R(4)=R(3)+R(1)-R(2)
370 S(4)=S(3)+S(1)-S(2)
380 T(4)=T(3)+T(1)-T(2)
390    ! DEFINE 5TH CORNER=1ST CORNER
400 R(5)=R(1)
410 S(5)=S(1)
420 T(5)=T(1)
430 GO TO 470
440 R(4)=R(1)
450 S(4)=S(1)
460 T(4)=T(1)
470 END SUB
```

A1-22
Subroutine PLANE

```
100 ! *****INTLP*****
110 !
120 EXTERNAL
130 SUB INTLP (R(),S(),T(),X(),Y(),Z(),P(),LENGTH,A)
140    ! CALLED BY PROGRAM "MAIN1"
150    ! CALCULATE INTERSECTION BETWEEN 3-D LINE AND PLANE
160    ! 3 POINTS ON PLANE DEFINED IN R(),S(),T()
170    ! 2 ENDPOINTS OF A LINE DEFINED IN X(),Y(),Z()
180    ! CALCULATED COMPONENTS OF INTERSECTION POINT RETURNED IN P()
190    ! CALCULATED LENGTH OF LINE FROM X(1),Y(1),Z(1) TO PLANE
200    ! CALCULATED ANGLE BETWEEN LINE AND PLANE IS A
210 OPTION NOLET
220 DIM B(3),D(3)
230    ! CALCULATE IMPLICIT PARAMETERS FOR THE PLANE
240 CALL PLANE (3,R,S,T,N1,N2,N3,K)
250    ! CALCULATE BASE VECTOR AND DIRECTIONAL VECTOR FOR LINE
260 CALL BLINE (X,Y,Z,B,D)
270    ! DOT NORMAL VECTOR WITH DIRECTIONAL VECTOR OF LINE
280 DOTDIR=N1*D(1)+N2*D(2)+N3*D(3)
290 IF ABS(DOTDIR)>0.000001 THEN 330
300 PRINT"LINE IS PARALLEL TO PLANE"
310 GO TO 410
320    ! DOT NORMAL VECTOR WITH BASE VECTOR OF LINE
330 DOTBASE=N1*B(1)+N2*B(2)+N3*B(3)
340    ! SOLVE FOR PARAMETER VALUE AT INTERSECTION
350 PAR=(K-DOTBASE)/DOTDIR
360    ! CALCULATE POINT OF INTERSECTION
370 FOR I=1 TO 3
380 P(I)=B(I)+PAR*D(I)
390 NEXT I
400 LENGTH=SQR((X(1)-P(1))^2+(Y(1)-P(2))^2+(Z(1)-P(3))^2)
410 MAG=SQR(D(1)^2+D(2)^2+D(3)^2)
420 NORMAL=SQR(N1^2+N2^2+N3^2)
430    ! CALC SINE OF THETA AND THEN ANGLE A
440 SINE=DOTDIR/(NORMAL*MAG)
450 A=ATN(SINE/SQR(1-SINE^2))*180/PI
460 END SUB
```

A1-23
Subroutine INTLP

```
100 !*****PIERCE*****
110 !
120 ! LINES AND PLANE INTERSECTIONS
130 ! REFERENCE EXAMPLE 8-10
140 LIBRARY"PLANE"
150 LIBRARY"BLINE"
160 LIBRARY"INTLP"
170 CLEAR
180 PRINT
190 PRINT"PROGRAM PIERCE"
200 PRINT
210 PRINT"FIND INTERSECTIONS BETWEEN"
220 PRINT"LINES AND A PLANE"
230 PRINT
240 PLOTTER TEK4010
250 OPTION NOLET
260 DIM L(4)
270 DIM R(5),S(5),T(5)
280 DIM P(3),B(3),D(3)
290 DIM X(6),Y(6),Z(6)
300 ! DEFINE 3 PTS ON A PLANE
310 READ R(1),S(1),T(1),R(2),S(2),T(2),R(3),S(3),T(3)
320 DATA 1.2,1.2,1.2,-1.2,1.2,1.2,-1.2,-1.2,0.8
330 CALL PLANE(4,R,S,T,N1,N2,N3,K)
340 NO=4 !NO OF LINES IN DATA
350 FOR J=1 TO NO
360 READ X(1),Y(1),Z(1),X(2),Y(2),Z(2)
370 CALL INTLP(R,S,T,X,Y,Z,P,LENGTH,ANG)
380 ! SAVE PTS OF INTERSECTION
390 PRINT"PIERCE POINT ";J
400 PRINT"X=";P(1)
410 PRINT"Y=";P(2)
420 PRINT"Z=";P(3)
430 PRINT
440 L(J)=SQR((X(1)-P(1))^2+(Y(1)-P(2))^2+(Z(1)-P(3))^2)
450 NEXT J
460 DATA 1,1,-1.8,-1,-1,1.8
470 DATA -1,1,-1.8,-1,-1,1.8
480 DATA -1,-1,-1.8,-1,-1,1.8
490 DATA 1,-1,-1.8,-1,-1,1.8
500 PRINT
510 PRINT"LINE SEGMENT LENGTHS ARE"
520 PRINT"SEG ONE=";L(1)
530 PRINT"SEG TWO=";L(2)
540 PRINT"SEG THREE=";L(3)
550 PRINT"SEG FOUR=";L(4)
560 END
```

A1-24
Program PIERCE

```
100 !*****CDGPLANE*****
110 !
120 ! PIERCE POINT BETWEEN A LINE AND A PLANE
130 ! USING COMPUTATIONAL DESCRIPTIVE GEOMETRY
140 ! INITIAL PTS IN X(),Y(),Z()
150 ! PROJECTED PTS ARE IN A(),B()
160 DIM X(5),Y(5),Z(5)
170 DIM A(5),B(5)
180 DIM L(2,100)
190 DIM R(3,3)
200 LIBRARY"AXISROT"
210 LIBRARY"ORTHOMAT"
220 LIBRARY"INTER2D"
230 CLEAR
240 PLOTTER TEK4010
250 OPTION NOLET
260 PRINT"PROGRAM CDGPLANE"
270 VERTICES=5
280 ! USE STD COORDINATES TO DEFINE ENDPOINTS IN FILE #1
290 PRINT
300 PRINT"INPUT DATA"
310 FOR I=1 TO VERTICES
320     READ X(I),Y(I),Z(I)
330 PRINT X(I);Y(I);Z(I)
340 NEXT I
350 PRINT
360 DATA 1.2,1.2,1.2
370 DATA -1.2,1.2,1.2
380 DATA -1.2,-1.2,0.8
390 DATA 1,1,-1.8
400 DATA -1,-1,1.8
410 ! CHOOSE LINE OF SIGHT PARALLEL TO LINE 1-2
420 ! DEFINE EYE COORDINATES
430 E1=X(2)-X(1)
440 E2=Y(2)-Y(1)
450 E3=Z(2)-Z(1)
460 PRINT"E1,E2,E3=";E1;E2;E3
470 CALL ORTHOMAT (E1,E2,E3,R)
480 ! PROJECT VERTICES USING TRANSFORMATION MATRIX R
490 FOR I=1 TO VERTICES
500     A(I)=X(I)*R(1,1)+Y(I)*R(2,1)+Z(I)*R(3,1)
510     B(I)=X(I)*R(1,2)+Y(I)*R(2,2)+Z(I)*R(3,2)
520 NEXT I
530 ! CALC INTERSECTION OF LINES IN LAST AUXILIARY
540 CALL INTER2D(A(2),B(2),A(3),B(3),A(4),B(4),A(5),B(5),T1,S1,P,Q)
550 PRINT"T1=";T1
560 PRINT"S1=";S1
570 ! CALCULATE POINTS OF INTERSECTION ON ORIGINAL LINES
580 X1=X(4)+S1*(X(5)-X(4))
590 Y1=Y(4)+S1*(Y(5)-Y(4))
600 Z1=Z(4)+S1*(Z(5)-Z(4))
610 PRINT"USING STD COORDINATES:"
620 PRINT"PIERCE POINT AT"
630 PRINT X1;Y1;Z1
640 END
```

A1-25
Program CDGPLANE

```
100 !*****CDGSKEW*****
110 !
120 ! SHORTEST DISTANCE BETWEEN TWO SKEW LINES
130 ! USING COMPUTATIONAL DESCRIPTIVE GEOMETRY
140 ! DEFINE TWO LINES USING 4 ENDPOINTS
150 ! STORED IN FILE #1 USING STD COORDS
160 ! INITIAL PTS IN X(),Y(),Z()
170 ! PROJECTED PTS ARE IN A(),B()
180 DIM X(5),Y(5),Z(5)
190 DIM A(5),B(5)
200 DIM L(2,100)
210 DIM C(3),D(3)
220 DIM N(3)
230 DIM R(3,3)
240 LIBRARY"AXISROT"
250 LIBRARY"VECPROD"
260 LIBRARY"ORTHOMAT"
270 LIBRARY"INTER2D"
280 CLEAR
290 PLOTTER TEK4010
300 OPTION NOLET
310 VERTICES=4
320 ! USE STD COORDINATES TO DEFINE ENDPOINTS IN FILE #1
330 FOR I=1 TO VERTICES
340     READ X(I),Y(I),Z(I)
350 NEXT I
360 DATA 1,1.5,2.125
370 DATA 2.6875,2.75,2.5
380 DATA 1.125,2,1.125
390 DATA 0,2.875,2.4375
400 ! CALCULATE COMPONENTS OF VERTEX 5
410 Y(5)=Y(4)  ! TO MAKE LINE 4-5 HORIZONTAL
420 ! CHECK FOR INFINITE SLOPE
430 IF X(2)<>X(1) THEN 460
440 X(5)=X(3)
450 GO TO 500
460 M1=(Y(2)-Y(1))/(X(2)-X(1))   ! SLOPE OF LINE 1-2 IN F PLANE
470 ! SLOPE 3-5 IS PARALLEL TO SLOPE 1-2 IN F-PLANE
480 X(5)=(Y(5)-Y(3))/M1 + X(3)
490 ! CHECK FOR INFINITE SLOPE
500 IF X(2)<>X(1) THEN 530
510 Z(5)=Z(3)
520 GO TO 560
530 M2=(Z(2)-Z(1))/(X(2)-X(1))   ! SLOPE OF LINE 1-2 IN H PLANE
540 ! SLOPE 3-5 IS PARALLEL TO SLOPE 1-2 IN H-PLANE
550 Z(5)=(X(5)-X(3))*M2 + Z(3)
560 PRINT"PROGRAM CDGSKEW"
570 PRINT
580 PRINT"SHORTEST DISTANCE BETWEEN TWO SKEW LINES"
```

```
590 PRINT
600 PRINT"ENDPOINTS FOR LINE 1 ARE"
610 PRINT X(1);Y(1);Z(1);",";X(2);Y(2);Z(2)
620 PRINT"ENDPOINTS FOR LINE 2 ARE"
630 PRINT X(3);Y(3);Z(3);",";X(4);Y(4);Z(4)
640 ! IDENTIFY 3 VERTICES WHICH DEFINE PLANE
650 P1=3
660 P2=4
670 P3=5
680  C(1)=X(P1)-X(P2)
690  C(2)=Y(P1)-Y(P2)
700  C(3)=Z(P1)-Z(P2)
710  D(1)=X(P3)-X(P2)
720  D(2)=Y(P3)-Y(P2)
730  D(3)=Z(P3)-Z(P2)
740 CALL VECPROD (C,D,N)
750  E1=N(1)
760  E2=N(2)
770  E3=N(3)
780 PRINT
790 PRINT "E1,E2,E3=";E1;E2;E3
800 CALL ORTHOMAT (E1,E2,E3,R)
810 ! PRINT"ORTHO PROJECTION ON PLANE PERPENDICULAR TO LINE OF SIGHT"
820 FOR I=1 TO VERTICES+1
830     A(I)=X(I)*R(1,1)+Y(I)*R(2,1)+Z(I)*R(3,1)
840     B(I)=X(I)*R(1,2)+Y(I)*R(2,2)+Z(I)*R(3,2)
850 NEXT I
860 ! CALC INTERSECTION OF LINES IN LAST AUXILIARY
870 CALL INTER2D(A(1),B(1),A(2),B(2),A(3),B(3),A(4),B(4),T1,S1,P,Q)
880 PRINT
890 PRINT"T1=";T1
900 PRINT"S1=";S1
910 ! CALCULATE POINTS OF INTERSECTION ON ORIGINAL LINES
920 X1=X(1)+T1*(X(2)-X(1))
930 Y1=Y(1)+T1*(Y(2)-Y(1))
940 Z1=Z(1)+T1*(Z(2)-Z(1))
950 X2=X(3)+S1*(X(4)-X(3))
960 Y2=Y(3)+S1*(Y(4)-Y(3))
970 Z2=Z(3)+S1*(Z(4)-Z(3))
980 PRINT"USING STD COORDINATES:"
990 PRINT"INTERSECTION ON LINE1=";
1000 PRINT X1;Y1;Z1
1010 PRINT"INTERSECTION ON LINE2=";
1020 PRINT X2;Y2;Z2
1030 DIST=SQR((X2-X1)^2+(Y2-Y1)^2+(Z2-Z1)^2)
1040 PRINT"SHORTEST DISTANCE BETWEEN SKEW LINES=";DIST
1050 PRINT
1060 END
```

A1-26
Program CDGSKEW

```
100 !*****CDGHORIZ*****
110 !
120 ! SHORTEST HORIZONTAL DISTANCE BETWEEN TWO SKEW LINES
130 ! USING COMPUTATIONAL DESCRIPTIVE GEOMETRY
140 ! DEFINE TWO LINES USING 4 ENDPOINTS
150 ! STORED IN FILE #1 USING STD COORDS
160 ! STORE DRAWING INFO IN FILE #2
170 ! INITIAL PTS IN X(),Y(),Z()
180 ! PROJECTED PTS ARE IN A(),B()
190 DIM X(5),Y(5),Z(5)
200 DIM A(5),B(5)
210 DIM L(2,100)
220 DIM C(3),D(3)
230 DIM N(3),E(3)
240 DIM R(3,3)
250 LIBRARY"AXISROT"
260 LIBRARY"VECPROD"
270 LIBRARY"ORTHOMAT"
280 LIBRARY"INTER2D"
290 CLEAR
300 PLOTTER TEK4010
310 OPTION NOLET
320 VERTICES=4
330 ! USE STD COORDINATES TO DEFINE ENDPOINTS IN FILE #1
340 FOR I=1 TO VERTICES
350     READ X(I),Y(I),Z(I)
360 NEXT I
370 DATA 1.5625,1,1.4375
380 DATA 1.375,2.375,2.75
390 DATA .125,1.125,.125
400 DATA 1.25,2.1875,.8125
410 ! CALCULATE COMPONENTS OF VERTEX 5
420 Y(5)=Y(4)  ! TO MAKE LINE 4-5 HORIZONTAL
430 ! CHECK FOR INFINITE SLOPE
440 IF X(2)<>X(1) THEN 470
450 X(5)=X(3)
460 GO TO 510
470 M1=(Y(2)-Y(1))/(X(2)-X(1))   ! SLOPE OF LINE 1-2 IN F PLANE
480 ! SLOPE 3-5 IS PARALLEL TO SLOPE 1-2 IN F-PLANE
490 X(5)=(Y(5)-Y(3))/M1 + X(3)
500 ! CHECK FOR INFINITE SLOPE
510 IF X(2)<>X(1) THEN 540
520 Z(5)=Z(3)
530 GO TO 580
540 M2=(Z(2)-Z(1))/(X(2)-X(1))   ! SLOPE OF LINE 1-2 IN H PLANE
550 ! SLOPE 3-5 IS PARALLEL TO SLOPE 1-2 IN H-PLANE
560 Z(5)=(X(5)-X(3))*M2 + Z(3)
570 ! DEFINE UNIT NORMAL TO HORIZONTAL PLANE
```

```
580 N(1)=0
590 N(2)=1
600 N(3)=0
610 ! INPUT ENDPOINT VERTICES OF TL HORIZONTAL LINE IN H PLANE
620 P1=4
630 P2=5
640  D(1)=X(P2)-X(P1)
650  D(2)=Y(P2)-Y(P1)
660  D(3)=Z(P2)-Z(P1)
670 CALL VECPROD (N,D,E)
680  E1=E(1)
690  E2=E(2)
700  E3=E(3)
710 PRINT"PROGRAM CDGHORIZ"
720 PRINT
730 PRINT"ENDPOINTS FOR LINE 1 ARE"
740 PRINT X(1);Y(1);Z(1);";";X(2);Y(2);Z(2)
750 PRINT"ENDPOINTS FOR LINE 2 ARE"
760 PRINT X(3);Y(3);Z(3);";";X(4);Y(4);Z(4)
770 PRINT
780 PRINT"E1,E2,E3=";E1;E2;E3
790 PRINT
800 CALL ORTHOMAT (E1,E2,E3,R)
810 FOR I=1 TO VERTICES+1
820     A(I)=X(I)*R(1,1)+Y(I)*R(2,1)+Z(I)*R(3,1)
830     B(I)=X(I)*R(1,2)+Y(I)*R(2,2)+Z(I)*R(3,2)
840 NEXT I
850 ! CALC INTERSECTION OF LINES IN LAST AUXILIARY
860 CALL INTER2D(A(1),B(1),A(2),B(2),A(3),B(3),A(4),B(4),T1,S1,P,Q)
870 PRINT"T1=";T1
880 PRINT"S1=";S1
890 ! CALCULATE POINTS OF INTERSECTION ON ORIGINAL LINES
900 X1=X(1)+T1*(X(2)-X(1))
910 Y1=Y(1)+T1*(Y(2)-Y(1))
920 Z1=Z(1)+T1*(Z(2)-Z(1))
930 X2=X(3)+S1*(X(4)-X(3))
940 Y2=Y(3)+S1*(Y(4)-Y(3))
950 Z2=Z(3)+S1*(Z(4)-Z(3))
960 PRINT"USING STD COORDINATES:"
970 PRINT"INTERSECTION ON LINE1=";
980 PRINT X1;Y1;Z1
990 PRINT"INTERSECTION ON LINE2=";
1000 PRINT X2;Y2;Z2
1010 DIST=SQR((X2-X1)^2+(Y2-Y1)^2+(Z2-Z1)^2)
1020 PRINT"SHORTEST HORIZONTAL DISTANCE BETWEEN SKEW LINES=";DIST
1030 PRINT
1040 END
```

A1-27
Program CDGHORIZ

```
100 !*****CDGSLOPE*****
110 !
120 ! SHORTEST LINE OF A GIVEN SLOPE BETWEEN TWO SKEW LINES
130 ! USING COMPUTATIONAL DESCRIPTIVE GEOMETRY
140 ! DEFINE TWO LINES USING 4 ENDPOINTS
150 ! INITIAL PTS IN X(),Y(),Z()
160 ! PROJECTED PTS ARE IN A(),B()
170 DIM X(5),Y(5),Z(5),A(5),B(5)
180 DIM L(2,100)
190 DIM C(3),D(3),N(3),E(3),R(3,3)
200 LIBRARY"AXISROT"
210 LIBRARY"VECPROD"
220 LIBRARY"ORTHOMAT"
230 LIBRARY"INTER2D"
240 CLEAR
250 PLOTTER TEK4010
260 OPTION NOLET
270 VERTICES=4
280 PRINT"PROGRAM CDGSLOPE"
290 PRINT
300 PRINT"SHORTEST LINE OF GIVEN SLOPE(THETA) BETWEEN TWO SKEW LINES"
310 ! DEFINE GIVEN SLOPE
320 THETA=10
330 PRINT"SLOPE=";THETA
340 PRINT
350 THETA=THETA*PI/180
360 PRINT"INPUT DATA IS"
370 FOR I=1 TO VERTICES
380     READ X(I),Y(I),Z(I)
390     PRINT X(I);Y(I);Z(I)
400 NEXT I
410 DATA 0,29,6
420 DATA 29,5,37
430 DATA 3,5,31
440 DATA 23,29,12
450 PRINT
460 ! CALCULATE COMPONENTS OF VERTEX 5
470 Y(5)=Y(4) ! TO MAKE LINE 4-5 HORIZONTAL
480 ! CHECK FOR INFINITE SLOPE
490 IF X(2)<>X(1) THEN 520
500 X(5)=X(3)
510 GO TO 560
520 M1=(Y(2)-Y(1))/(X(2)-X(1))  ! SLOPE OF LINE 1-2 IN F PLANE
530 ! SLOPE 3-5 IS PARALLEL TO SLOPE 1-2 IN F-PLANE
540 X(5)=(Y(5)-Y(3))/M1 + X(3)
550 ! CHECK FOR INFINITE SLOPE
560 IF X(2)<>X(1) THEN 590
570 Z(5)=Z(3)
580 GO TO 630
```

```
590 M2=(Z(2)-Z(1))/(X(2)-X(1))  ! SLOPE OF LINE 1-2 IN H PLANE
600 ! SLOPE 3-5 IS PARALLEL TO SLOPE 1-2 IN H-PLANE
610 Z(5)=(X(5)-X(3))*M2 + Z(3)
620 ! DEFINE UNIT NORMAL TO HORIZONTAL PLANE
630 N(1)=0
640 N(2)=1
650 N(3)=0
660 ! DEFINE VERTICES OF TL HORIZONTAL LINE IN H PLANE
670 P1=4
680 P2=5
690  D(1)=X(P2)-X(P1)
700 P1=4
710 P2=5
720  D(1)=X(P2)-X(P1)
730  D(2)=Y(P2)-Y(P1)
740  D(3)=Z(P2)-Z(P1)
750 CALL VECPROD(N,D,E)
760 ! ROTATE LINE OF SIGHT BY THETA DEGREES ABOUT X-AXIS
770  E1=E(1)
780 E2=E(2)*COS(THETA)-E(3)*SIN(THETA)
790 E3=E(2)*SIN(THETA)+E(3)*COS(THETA)
800 PRINT"E1,E2,E3=";E1;E2;E3
810 CALL ORTHOMAT (E1,E2,E3,R)
820 ! PRINT"ORTHO PROJECTION ON PLANE PERPENDICULAR TO LINE OF SIGHT"
830 FOR I=1 TO VERTICES+1
840      A(I)=X(I)*R(1,1)+Y(I)*R(2,1)+Z(I)*R(3,1)
850      B(I)=X(I)*R(1,2)+Y(I)*R(2,2)+Z(I)*R(3,2)
860 NEXT I
870 ! CALC INTERSECTION OF LINES IN LAST AUXILIARY
880 CALL INTER2D(A(1),B(1),A(2),B(2),A(3),B(3),A(4),B(4),T1,S1,P,Q)
890 PRINT"T1=";T1
900 PRINT"S1=";S1
910 ! CALCULATE POINTS OF INTERSECTION ON ORIGINAL LINES
920 X1=X(1)+T1*(X(2)-X(1))
930 Y1=Y(1)+T1*(Y(2)-Y(1))
940 Z1=Z(1)+T1*(Z(2)-Z(1))
950 X2=X(3)+S1*(X(4)-X(3))
960 Y2=Y(3)+S1*(Y(4)-Y(3))
970 Z2=Z(3)+S1*(Z(4)-Z(3))
980 PRINT"USING STD COORDINATES:"
990 PRINT"INTERSECTION ON LINE1=";
1000 PRINT X1;Y1;Z1
1010 PRINT"INTERSECTION ON LINE2=";
1020 PRINT X2;Y2;Z2
1030 DIST=SQR((X2-X1)^2+(Y2-Y1)^2+(Z2-Z1)^2)
1040 PRINT"SHORTEST REQUIRED DISTANCE BETWEEN SKEW LINES=";DIST
1050 END
```

A1-28
Program CDGSLOPE

```
100 ! *****BODYC*****
110 !
120 ! CREATE A CONE OF REVOLUTION
130 ! BY ROTATING A 2-D LINE
140 ! ABOUT THE Y-AXIS
150 DIM R(3,3)
160 LIBRARY"AXISROT"
170 LIBRARY"AROUND"
180 LIBRARY"EYETOZ"
190 LIBRARY"TITLE"
200 CLEAR
210 PLOTTER TEK4010
220 OPTION NOLET
230 DIM A(100),B(100)
240  W1=-6
250  W2=6
260  W3=-2
270  W4=10
280 SET WINDOW W1,W2,W3,W4
290 CALL EYETOZ(R)
300 PRINT"INPUT NO OF HORIZONTAL DIVISIONS";
310 INPUT HORIZ
320 ! DEFINE ENDPOINTS OF 2-D LINE
330 X1=2
340 Y1=1
350 X2=0.5
360 Y2=8
370 I=1
380 FOR T=0 TO 1
390 A(I)=X1+T*(X2-X1)
400 B(I)=Y1+T*(Y2-Y1)
410 I=I+1
420 NEXT T
430 N=I-1
440 PRINT
450 PRINT"2-D VERTICES"
460 FOR I=1 TO N
470 PRINT A(I);B(I)
480 NEXT I
490 CALL AROUND(A,B,N,HORIZ,R)
500 PRINT
510  TITLE$="PROGRAM BODYC"
520  NAME$="NAME___________"
530  DAY$="DATE_______"
540  COURSE$="COURSE_____"
550 CALL TITLE(W1,W2,W3,W4,TITLE$,NAME$,DAY$,COURSE$)
560 PRINT
570 PRINT"WINDOW LIMITS"
580 PRINT"W1=";W1
590 PRINT"W2=";W2
600 PRINT"W3=";W3
610 PRINT"W4=";W4
620 END
```

A1-29
Program BODYC

```
100 !*****AROUND*****
110 !
120 EXTERNAL
130 SUB AROUND(A(),B(),PTS,HORIZ,R(,))
140 ! CALLED BY MAIN BODY
150 ! DRAW BODY OF REVOLUTION
160 OPTION NOLET
170 DIM X(100),Y(100),U(100),V(100)
180  DELTA=2*PI/HORIZ
190  THETA=0
200 ! TRANSFORM 2-D CURVE
210 FOR I=1 TO PTS
220  U(I)=A(I)*R(1,1)+B(I)*R(2,1)
230  V(I)=A(I)*R(1,2)+B(I)*R(2,2)
240 NEXT I
250 ! SET UP DOUBLE LOOP
260 FOR J=1 TO HORIZ
270 ! DRAW PRESENT VERTICAL SEQUENCE
280 FOR I=1 TO PTS-1
290 PLOT U(I),V(I);U(I+1),V(I+1)
300 NEXT I
310  THETA=THETA+DELTA
320  C1=COS(THETA)
330  S1=SIN(THETA)
340 ! CALCULATE NEXT VERTICAL SEQUENCE OF VERTICES
350 ! BY ROTATING FIRST VERTICAL SEQUENCE ABOUT Y-AXIS BY ANGLE 2*PI/HORIZ=DELTA
360 FOR I=1 TO PTS
370  X1=A(I)*C1
380  Y1=B(I)
390  Z1=A(I)*S1
400 ! TRANSFORM NEW DATA USING MATRIX [R]
410  X(I)=X1*R(1,1)+Y1*R(2,1)+Z1*R(3,1)
420  Y(I)=X1*R(1,2)+Y1*R(2,2)+Z1*R(3,2)
430 ! PLOT HORIZONTAL LINES BETWEEN PRESENT AND NEXT SEQUENCES
440  A0=A(I)
450 IF ABS(A0)<0.0001 THEN 490
460 PLOT U(I),V(I);
470 PLOT X(I),Y(I)
480 ! RESET PRESENT SEQUENCE TO NEXT SEQUENCE
490  U(I)=X(I)
500  V(I)=Y(I)
510 NEXT I
520 NEXT J
530 END SUB
```

A1-30
Subroutine AROUND

```
100 !*****VEXCLIP3*****
110 !
120 ! 3-D CONVEX CLIPPING
130 CLEAR
140 OPTION NOLET
150 PLOTTER TEK4010
160 LIBRARY"CLIPSUB"
170 LIBRARY"EYETOZ"
180 LIBRARY"VECPROD"
190 LIBRARY"AXISROT"
200 LIBRARY"TITLE"
210 NUMBER=12 !TOTAL PTS
220 N1=INT(NUMBER/2)
230 DIM R(3,3)
240 DIM E(3),D(3),K(3)
250 DIM N(50,3)
260 DIM A(50),B(50),C(50)
270 DIM P(12,3)
280 DIM L(2,100)
290 W1=-5
300 W2=25
310 W3=-5
320 W4=25
330 SET WINDOW W1,W2,W3,W4
340 FOR I=1 TO N1
350 READ P(I,1),P(I,2),P(I,3)
360 NEXT I
370 FOR I=N1+1 TO NUMBER
380 READ P(I,1),P(I,2),P(I,3)
390 NEXT I
400 DATA 7,10,0,18,10,0,18,17,0,15,20,0,7,17,0,7,10,0
410 DATA 7,10,10,18,10,10,18,17,10,15,20,10,7,17,10,7,10,10
420 LINES=15
430 FOR I=1 TO LINES
440 READ L(1,I),L(2,I)
450 NEXT I
460 DATA 1,2,2,3,3,4,4,5,5,6
470 DATA 1,7,2,8,3,9,4,10,5,11
480 DATA 7,8,8,9,9,10,10,11,11,12
490 CALL EYETOZ(R)
500 FOR I=1 TO NUMBER
```

```
510 A(I)=P(I,1)*R(1,1)+P(I,2)*R(2,1)+P(I,3)*R(3,1)
520 B(I)=P(I,1)*R(1,2)+P(I,2)*R(2,2)+P(I,3)*R(3,2)
530 NEXT I
540 FOR I=1 TO LINES
550 I1=L(1,I)
560 I2=L(2,I)
570 PLOT A(I1),B(I1);A(I2),B(I2)
580 NEXT I
590 PRINT"NO OF LINES TO BE CLIPPED";
600 INPUT LINES
610 FOR Q=1 TO LINES
620 PRINT"LINE NO";Q
630 PRINT"X1,Y1,Z1,X2,Y2,Z2"
640 INPUT X1,Y1,Z1,X2,Y2,Z2
650 CALL CLIPSUB(N1,P,X1,Y1,Z1,X2,Y2,Z2,X3,Y3,Z3,X4,Y4,Z4,I$)
660 IF I$="INTERSECTION" THEN 690
670 PRINT I$
680 GO TO 830
690 X1=X3*R(1,1)+Y3*R(2,1)+Z3*R(3,1)
700 Y1=X3*R(1,2)+Y3*R(2,2)+Z3*R(3,2)
710 X2=X4*R(1,1)+Y4*R(2,1)+Z4*R(3,1)
720 Y2=X4*R(1,2)+Y4*R(2,2)+Z4*R(3,2)
730 PLOT X1,Y1;X2,Y2
740 PRINT"INTERSECTION ";Q
750 PRINT"X3=";X3
760 PRINT"Y3=";Y3
770 PRINT"Z3=";Z3
780 PRINT
790 PRINT"X4=";X4
800 PRINT"Y4=";Y4
810 PRINT"Z4=";Z4
820 PRINT
830 NEXT Q
840 TITLE$="PROGRAM VEXCLIP3"
850 NAME$="NAME__________ "
860 DAY$="DATE_______ "
870 COURSE$="COURSE_____ "
880 CALL TITLE(W1,W2,W3,W4,TITLE$,NAME$,DAY$,COURSE$)
890 END
```

A1-31
Program VEXCLIP3

```
100 ! *****CLIPSUB*****
110 !
120 EXTERNAL
130 SUB CLIPSUB(N1,P(,),X1,Y1,Z1,X2,Y2,Z2,X3,Y3,Z3,X4,Y4,Z4,I$)
140 ! 2-D OR 3-D CLIPPING SUBROUTINE
150 ! CALLED BY MAIN'VEXCLIP2' AND 'VEXCLIP3'
160 ! N1=NUMBER OF PAIRS IN ARRAY P(,)
170 ! P(,) = CLIPPING REGION (1ST PT=LAST PT)
180 ! INPUT LINE BETWEEN X1,Y1,Z1 AND X2,Y2,Z2
190 ! CLIPPED RETURNED LINE BETWEEN X3,Y3,Z3 AND X4,Y4,Z4
200 ! I$=FLAG IF LINE DOES NOT INTERSECT REGION
210 OPTION NOLET
220 DIM U(3)
230 DIM D(3)
240 DIM E(3)
250 DIM N(52,3)
260 ! CALCULATE NORMALS
270 FOR I1=1 TO N1-1
280 U(1)=P(I1+N1,1)-P(I1,1)
290 U(2)=P(I1+N1,2)-P(I1,2)
300 U(3)=P(I1+N1,3)-P(I1,3)
310 D(1)=P(I1+1,1)-P(I1,1)
320 D(2)=P(I1+1,2)-P(I1,2)
330 D(3)=P(I1+1,3)-P(I1,3)
340 CALL VECPROD(U,D,E)
350 N(I1,1)=E(1)
360 N(I1,2)=E(2)
370 N(I1,3)=E(3)
380 NEXT I1
390 ! DEFINE UNIT VECTOR K
400 N(I1,1)=0
410 N(I1,2)=0
420 N(I1,3)=1
430 ! DEFINE UNIT VECTOR -K
440 N(I1+1,1)=0
450 N(I1+1,2)=0
460 N(I1+1,3)=-1
470 ! CALCULATE INTERSECTION PTS
480 DX=X2-X1
490 DY=Y2-Y1
500 DZ=Z2-Z1
```

```
510 MINT=0
520 MAXT=1
530 ! CALCULATE WEIGHT VECTORS
540 FOR I1=1 TO N1+1
550 WEIGHT1=X1-P(I1,1)
560 WEIGHT2=Y1-P(I1,2)
570 WEIGHT3=Z1-P(I1,3)
580 ! EVALUATE DENOMINATOR OF EQN(11-5)
590 DDOTNORMAL=DX*N(I1,1)+DY*N(I1,2)+DZ*N(I1,3)
600 ! EVALUATE NUMERATOR OF EQN(11-5)
610 WDOTNORMAL=WEIGHT1*N(I1,1)+WEIGHT2*N(I1,2)+WEIGHT3*N(I1,3)
620 ! USE EQN(11-5) TO FIND PARAMETERS
630 IF DDOTNORMAL=0 THEN 710
640 T=-WDOTNORMAL/DDOTNORMAL
650 IF DDOTNORMAL>0 THEN 730
660 ! CHECK TO SEE OF INTERSECTIONS LIE ON LINE SEGMENTS
670 IF T<0 THEN 870
680 IF T>MAXT THEN 700
690 MAXT=T
700 GO TO 760
710 IF WDOTNORMAL<0 THEN 870
720 GO TO 760
730 IF T>1 THEN 870
740 IF T<MINT THEN 760
750 MINT=T
760 NEXT I1
770 IF MINT=>MAXT THEN 870
780 X3=X1+(X2-X1)*MINT
790 Y3=Y1+(Y2-Y1)*MINT
800 Z3=Z1+(Z2-Z1)*MINT
810 X4=X1+(X2-X1)*MAXT
820 Y4=Y1+(Y2-Y1)*MAXT
830 Z4=Z1+(Z2-Z1)*MAXT
840 ! INTERSECTIONS AT X3;Y3;Z3;X4;Y4;Z4
850 I$="INTERSECTION"
860 GO TO 880
870 I$="NO INTERSECTION"
880 END SUB
```

A1-32
Subroutine CLIPSUB

```
100 !*****XCONE*****
110 !
120 ! 3-D CONVEX CLIPPING
130 CLEAR
140 OPTION NOLET
150 PLOTTER TEK4010
160 LIBRARY"CLIPSUB"
170 LIBRARY"EYETOZ"
180 LIBRARY"VECPROD"
190 LIBRARY"AXISROT"
200 LIBRARY"TITLE"
210 NUMBER=32 !TOTAL PTS
220 OPEN #2:NAME"CONE"
230 LINES=46 !LINES IN FILE #2
240 N1=INT(NUMBER/2)
250 DIM R(3,3)
260 DIM E(3),D(3),K(3)
270 DIM N(50,3)
280 DIM A(64),B(64),C(64)
290 DIM P(64,3)
300 DIM L(2,100)
310 W1=-3
320 W2=3
330 W3=-3
340 W4=3
350 SET WINDOW W1,W2,W3,W4
360 R1=0.5 !TIP RADIUS OF CONE
370 NSTEP=2*PI/(N1-1)
380 Q=1
390 FOR ANG=0 TO 2*PI STEP NSTEP
400 P(Q,1)=R1*COS(ANG)
410 P(Q,2)=R1*SIN(ANG)
420 P(Q,3)=-3
430 Q=Q+1
440 NEXT ANG
450 P(Q,1)=P(1,1)
460 P(Q,2)=P(1,2)
470 P(Q,3)=P(1,3)
480 Q=Q+1
490 R2=1 !BASE RADIUS OF CONE
500 FOR ANG=0 TO 2*PI STEP NSTEP
510 P(Q,1)=R2*COS(ANG)
520 P(Q,2)=R2*SIN(ANG)
530 P(Q,3)=3
540 Q=Q+1
550 NEXT ANG
560 P(Q,1)=P(N1+1,1)
570 P(Q,2)=P(N1+1,2)
580 P(Q,3)=P(N1+1,3)
```

```
590 FOR I1=1 TO LINES
600 INPUT #2:L(1,I1),L(2,I1)
610 NEXT I1
620 CALL EYETOZ(R)
630 FOR I=1 TO NUMBER
640 A(I)=P(I,1)*R(1,1)+P(I,2)*R(2,1)+P(I,3)*R(3,1)
650 B(I)=P(I,1)*R(1,2)+P(I,2)*R(2,2)+P(I,3)*R(3,2)
660 NEXT I
670 FOR I=1 TO LINES
680 I1=L(1,I)
690 I2=L(2,I)
700 PLOT A(I1),B(I1);A(I2),B(I2)
710 NEXT I
720 PRINT"NO OF LINES TO BE CLIPPED";
730 INPUT LINES
740 FOR Q=1 TO LINES
750 PRINT"LINE NO";Q
760 PRINT"X1,Y1,Z1,X2,Y2,Z2"
770 INPUT X1,Y1,Z1,X2,Y2,Z2
780 CALL CLIPSUB(N1,P,X1,Y1,Z1,X2,Y2,Z2,X3,Y3,Z3,X4,Y4,Z4,I$)
790 IF I$="INTERSECTION" THEN 820
800 PRINT I$
810 GO TO 960
820 X1=X3*R(1,1)+Y3*R(2,1)+Z3*R(3,1)
830 Y1=X3*R(1,2)+Y3*R(2,2)+Z3*R(3,2)
840 X2=X4*R(1,1)+Y4*R(2,1)+Z4*R(3,1)
850 Y2=X4*R(1,2)+Y4*R(2,2)+Z4*R(3,2)
860 PLOT X1,Y1;X2,Y2
870 PRINT"INTERSECTION ";Q
880 PRINT"X3=";X3
890 PRINT"Y3=";Y3
900 PRINT"Z3=";Z3
910 PRINT
920 PRINT"X4=";X4
930 PRINT"Y4=";Y4
940 PRINT"Z4=";Z4
950 PRINT
960 NEXT Q
970 TITLE$="PROGRAM XCONE"
980 NAME$="NAME___________"
990 DAY$="DATE_______"
1000 COURSE$="COURSE_____"
1010 CALL TITLE(W1,W2,W3,W4,TITLE$,NAME$,DAY$,COURSE$)
1020 END
```

A1-33
Program CONE

```
100 !*****CLIPLANE*****
110 !
120 ! PLANE AND PLANE INTERSECTION
130 LIBRARY"PLANE"
140 LIBRARY"EYETOZ"
150 LIBRARY"AXISROT"
160 LIBRARY"GRAPHICS"
170 LIBRARY"BLINE"
180 LIBRARY"INTLP"
190 LIBRARY"TITLE"
200 CLEAR
210 PLOTTER TEK4010
220 OPTION NOLET
230 SET VIEWPORT 30,130,0,100
240 WIDTH=6
250 HEIGHT=6
260 DEPTH=2
270 PRINT"WIDTH=";WIDTH
280 PRINT"HEIGHT=";HEIGHT
290 PRINT"DEPTH=";DEPTH
300 DIM O(3,3)
310 DIM P(10)
320 DIM R(5),S(5),T(5)
330 DIM B(3),D(3)
340 DIM G(11),H(11)
350 DIM A(2),C(2)
360 DIM X(11),Y(11),Z(11)
370 DIM I(11),J(11),K(11)
380 DIM U(11),V(11),W(11)
390 CALL EYETOZ(O)
400 V$="T" !TELEPHOTO VIEW
410 ! DEFINE 3 PTS ON BASE PLANE
420 READ R(1),S(1),T(1),R(2),S(2),T(2),R(3),S(3),T(3)
430 DATA 1.2,1.2,1.2,-1.2,1.2,1.2,-1.2,-1.2,0.8
440 PRINT"BASE PLANE PTS"
450 PRINT R(1);S(1);T(1)
460 PRINT R(2);S(2);T(2)
470 PRINT R(3);S(3);T(3)
480 PRINT
490 CALL PLANE(4,R,S,T,N1,N2,N3,K1)
500 CALL GRAPHICS(WIDTH,HEIGHT,DEPTH,V$,4,G,H,R,S,T,O)
510 PRINT"INPUT THREE PTS"
520 PRINT"ON CLIPPING PLANE"
530 PRINT"POINT 1 X,Y,Z="
540 INPUT I(1),J(1),K(1)
550 PRINT"POINT 2 X,Y,Z="
560 INPUT I(2),J(2),K(2)
570 PRINT"POINT 3 X,Y,Z="
580 INPUT I(3),J(3),K(3)
590 PRINT
600 CALL PLANE(4,I,J,K,N1,N2,N3,K1)
610 CALL GRAPHICS(WIDTH,HEIGHT,DEPTH,V$,4,G,H,I,J,K,O)
620 LINES=2 !NO OF LINES IN CLIPPING PLANE
630 MU=0 ! SET PARAMETER =0
640 FOR J1=1 TO LINES
650 X(1)=I(1)+MU*(I(2)-I(1))
660 Y(1)=J(1)+MU*(J(2)-J(1))
670 Z(1)=K(1)+MU*(K(2)-K(1))
680 X(2)=I(4)+MU*(I(3)-I(4))
690 Y(2)=J(4)+MU*(J(3)-J(4))
700 Z(2)=K(4)+MU*(K(3)-K(4))
710 MU=MU+1/(LINES-1) !INCREMENT MU
```

```
720 CALL INTLP(R,S,T,X,Y,Z,P,LENGTH,ANG)
730 ! SAVE PTS OF INTERSECTION
740 PRINT"PIERCE POINT ";J1
750 PRINT P(1);P(2);P(3)
760 U(J1)=P(1)
770 V(J1)=P(2)
780 W(J1)=P(3)
790 NEXT J1
800 CALL GRAPHICS(WIDTH,HEIGHT,DEPTH,V$,LINES-1,G,H,U,V,W,O)
810 TITLE$="PROGRAM CLIPLANE"
820 NAME$="NAME__________"
830 DAY$="DATE________"
840 COURSE$="COURSE_____"
850 CALL TITLE(-2,6,-2,6,TITLE$,NAME$,DAY$,COURSE$)
860 END
```

A1-34
Program CLIPLANE

```
100 !*****GRAPHICS*****
110 !
120 EXTERNAL
130 SUB GRAPHICS(WIDTH,HEIGHT,DEPTH,V$,SPANS,G(),H(),X(),Y(),Z(),O(,))
140 ! CREATE FRONT, TOP, SIDE AND ORTHO VIEWS
150 ! CALLED BY MAIN PIERCEPT
160 !WIDTH=WIDTH OF FRONTAL VIEW
170 !HEIGHT=HEIGHT OF FRONTAL VIEW
180 ! V$=N(NORMAL) OR T(TELEPHOTO)
190 !SPANS=NO OF PTS IN X(),Y(),Z()
200 !G(),H() ARE RETURNED TRANSFORMED X,Y VALUES
210 !O(,) IS 3X3 TRANSFORMATION MATRIX
220 OPTION NOLET
230 FOR I=1 TO SPANS+1
240 G(I)=X(I)*O(1,1)+Y(I)*O(2,1)+Z(I)*O(3,1)
250 H(I)=X(I)*O(1,2)+Y(I)*O(2,2)+Z(I)*O(3,2)
260 NEXT I
270 IF V$="N" THEN 300
280  SET WINDOW -WIDTH/2,WIDTH/2,-HEIGHT/2,HEIGHT/2
290 GO TO 340
300 FOR Q=1 TO 4
310 IF Q=4 THEN 330
320 GO TO 340
330 SET VIEWPORT 80,130,50,100
333 IF V$="P" THEN 500
340 S=0
350 FOR I=S+1 TO S+SPANS
360 IF V$="T" THEN 440
370 ON Q GO TO 380,400,420,440
380 PLOT X(I),HEIGHT+DEPTH-Z(I);X(I+1),HEIGHT+DEPTH-Z(I+1)
390 GO TO 450
400 PLOT WIDTH+DEPTH-Z(I),Y(I);WIDTH+DEPTH-Z(I+1),Y(I+1)
410 GO TO 450
420 PLOT X(I),Y(I);X(I+1),Y(I+1)
430 GO TO 450
440 PLOT G(I),H(I);G(I+1),H(I+1)
450 NEXT I
460 S=S+SPANS+1
470 IF S<SPANS+1 THEN 350
480 IF V$="T" THEN 500
490 NEXT Q
500 SET VIEWPORT 30,130,0,100
510 END SUB
```

A1-35
Subroutine GRAPHICS

```
100 !*****CLIPCONE*****
110 !
120 !PLANE AND CONE INTERSECTION
130 LIBRARY"PLANE"
140 LIBRARY"EYETOZ"
150 LIBRARY"AXISROT"
160 LIBRARY"GRAPHICS"
170 LIBRARY"BLINE"
180 LIBRARY"INTLP"
190 LIBRARY"TITLE"
200 LIBRARY"DEVELOP"
210 CLEAR
220 PLOTTER TEK4010
230 OPTION NOLET
240 SET VIEWPORT 30,130,0,100
250 W1=-2
260 W2=4
270 W3=-2
280 W4=4
290 WIDTH=W2-W1
300 HEIGHT=W4-W3
310 DEPTH=0
320 PRINT"WIDTH=";WIDTH
330 PRINT"HEIGHT=";HEIGHT
340 PRINT"DEPTH=";DEPTH
350 DIM P(3),O(3,3)
360 DIM R(5),S(5),T(5)
370 DIM G(32),H(32)
380 DIM E(4),F(4)
390 DIM A(2),B(2),C(2)
400 DIM X(32),Y(32),Z(32)
410 DIM U(32),V(32),W(32)
420 DIM I(32),J(32),K(32)
430 DIM L(32)
440 CALL EYETOZ(O)
450 V$="T" !TELEPHOTO VIEW
460 ! DEFINE 3 PTS ON BASE PLANE
470 READ R(1),S(1),T(1),R(2),S(2),T(2),R(3),S(3),T(3)
480 DATA 1.2,1.2,1.2,-1.2,1.2,0.8,-1.2,-1.2,0.8
490 CALL PLANE(4,R,S,T,N1,N2,N3,K1)
500 PRINT
510 PRINT"PLANE VERTICES"
520 CALL PLANE(4,R,S,T,N1,N2,N3,K1)
530 PRINT R(1);S(1);T(1)
540 PRINT R(2);S(2);T(2)
550 PRINT R(3);S(3);T(3)
560 PRINT R(4);S(4);T(4)
570 CALL GRAPHICS(WIDTH,HEIGHT,DEPTH,V$,4,G,H,R,S,T,O)
580 PRINT"INTERSECTION PTS"
590 ! DEFINE CLIPPING CONE
600 L1=6 !LENGTH OF CONE FRUSTUM
610 R1=0.2 !TIP RADIUS
620 R2=1.5 !BASE RADIUS
630 N1=16 !NO OF FACETS ON CONE
640 N=2*PI/(N1-1)
650 Q=1
660 FOR ANG=0 TO 2*PI+N STEP N
```

```
670 X(1)=R1*COS(ANG)
680 Y(1)=R1*SIN(ANG)
690 Z(1)=-L1/2
700 X(2)=R2*COS(ANG)
710 Y(2)=R2*SIN(ANG)
720 Z(2)=L1/2
730 CALL INTLP(R,S,T,X,Y,Z,P,LENGTH,DIR)
740 L(Q)=LENGTH
750 REM PRINT OUT INTERSECTION PTS
760 PRINT P(1);P(2);P(3)
770 U(Q)=P(1)
780 V(Q)=P(2)
790 W(Q)=P(3)
800 I(Q)=X(1)
810 J(Q)=Y(1)
820 K(Q)=Z(1)
830 A(1)=X(1)
840 B(1)=Y(1)
850 C(1)=Z(1)
860 A(2)=P(1)
870 B(2)=P(2)
880 C(2)=P(3)
890 CALL GRAPHICS(WIDTH,HEIGHT,DEPTH,V$,1,G,H,A,B,C,O)
900 Q=Q+1
910 NEXT ANG
920 CALL GRAPHICS(WIDTH,HEIGHT,DEPTH,V$,N1-1,G,H,U,V,W,O)
930 CALL GRAPHICS(WIDTH,HEIGHT,DEPTH,V$,N1-1,G,H,I,J,K,O)
940 PRINT
950 PRINT"R1=";R1
960 PRINT"R2=";R2
970 PRINT"L1=";L1
980 TITLE$="PROGRAM CLIPCONE"
990 NAME$="NAME__________"
1000 DAY$="DATE________"
1010 COURSE$="COURSE_____"
1020 CALL TITLE(W1,W2,W3,W4,TITLE$,NAME$,DAY$,COURSE$)
1030 PRINT
1040 PRINT"STOP(S) OR CONTINUE(C)";
1050 INPUT G$
1060 IF G$="S" THEN 1170
1070 CLEAR
1080 IF R1=R2 THEN 1110
1090 SET WINDOW -L1,L1/2,-L1/2,L1
1100 GO TO 1120
1110 SET WINDOW -L1/2,L1,-L1/4,1.25*L1
1120 CALL DEVELOP(R1,R2,L1,N1,L)
1130 ! DRAW REFERENCE UNIT
1140 PLOT -L1/2,-L1/6;-L1/2+1,-L1/6
1150 GRAPHIC PRINT -L1/2,-L1/8:"ONE-UNIT"
1160 CALL TITLE(W1,W2,W3,W4,TITLE$,NAME$,DAY$,COURSE$)
1170 END
```

A1-36
Program CLIPCONE

```
100 !*****DEVELOP*****
110 EXTERNAL
120 SUB DEVELOP(R1,R2,L1,N1,L())
130 ! CALLED BY MAIN CLIPCONE
140 ! R1 IS TIP RADIUS OF CONE
150 ! R2 IS BASE RADIUS OF CONE
160 ! L1 IS LENGTH OF CONE OR CYLINDER
170 ! N1=NUMBER OF LINES IN L()
180 ! L() CONTAINS LENGTHS ON SURFACE
190 ! SUBROUTINE TO CREATE AND DRAW A DEVELOPABLE SURFACE
200 OPTION NOLET
210 DIM E(4),F(4)
220 PRINT" LENGTH    ANGLE"
230 Q=1
240 L(N1+1)=L(1)
250 IF R1=R2 THEN 480
260 X1=R1*L1/(R2-R1)
270 SLANT=SQR(R2^2+(L1+X1)^2)
280 ALPHA=2*PI*R2/SLANT
290 INCR=ALPHA/(N1-1)
300 PRINT"SLANT HEIGHT=";SLANT
310 PRINT"LENGTH X1=";X1
320 PRINT"SECTOR ANGLE=";ALPHA*180/PI
330 PRINT
340 FOR ANG=0 TO ALPHA STEP INCR
350 PLOT -X1*COS(ANG),X1*SIN(ANG);-(X1+L(Q))*COS(ANG),(X1+L(Q))*SIN(ANG)
360 IF Q=N1 THEN 450
370 PLOT -X1*COS(ANG),X1*SIN(ANG);-X1*COS(ANG+INCR),X1*SIN(ANG+INCR)
380 FIRSTX=-(X1+L(Q))*COS(ANG)
390 FIRSTY=(X1+L(Q))*SIN(ANG)
400 SECONDX=-(X1+L(Q+1))*COS(ANG+INCR)
410 SECONDY=(X1+L(Q+1))*SIN(ANG+INCR)
420 PLOT FIRSTX,FIRSTY;SECONDX,SECONDY
430 PRINT L(Q);ANG*180/PI
440 Q=Q+1
450 NEXT ANG
460 PRINT L(N1);ALPHA*180/PI
470 GO TO 570
480 N=2*PI/(N1-1)
490 FOR ANG=0 TO 2*PI+N STEP N
500 PRINT L(Q);ANG*180/PI
510 PLOT L1/2-L(Q),R1*ANG;L1/2,R1*ANG
520 IF Q=N1 THEN 560
530 PLOT L1/2,R1*ANG;L1/2,R1*(ANG+N)
540 PLOT L1/2-L(Q),R1*ANG;L1/2-L(Q+1),R1*(ANG+N)
550 Q=Q+1
560 NEXT ANG
570 END SUB
```

A1-37
Subroutine DEVELOP

```
100 !*****PRISMS*****
110 !
120 ! 3-D CONVEX CLIPPING
130 ! PRISM/PRISM INTERSECTION
140 CLEAR
150 OPTION NOLET
160 PLOTTER TEK4010
170 LIBRARY"CLIPSUB"
180 LIBRARY"EYETOZ"
190 LIBRARY"VECPROD"
200 LIBRARY"AXISROT"
210 LIBRARY"TITLE"
220 NUMBER=8 !TOTAL PTS
230 N1=INT(NUMBER/2)
240 DIM R(3,3)
250 DIM X(64),Y(64),Z(64)
260 DIM U(64),V(64),W(64)
270 DIM I(64),J(64),K(64)
280 DIM A(64),B(64)
290 DIM C(64),D(64)
300 DIM P(14,3)
310 DIM L(2,100)
320 W1=-3
330 W2=3
340 W3=-3
350 W4=3
360 SET WINDOW W1,W2,W3,W4
370 FOR Q=1 TO N1
380 READ P(Q,1),P(Q,2),P(Q,3)
390 NEXT Q
400 FOR Q=N1+1 TO NUMBER
410 READ P(Q,1),P(Q,2),P(Q,3)
420 NEXT Q
430 ! DATA TO DEFINE PRISM***CAREFUL-ORDER IS IMPORTANT
440 DATA 1,-1,-2,0,1,-2,-1,-1,-2,1,-1,-2
450 DATA 1,-1,2,0,1,2,-1,-1,2,1,-1,2
460 LINES=9 ! LINES TO DRAW PRISM
470 FOR I1=1 TO LINES
480 READ L(1,I1),L(2,I1)
490 NEXT I1
500 ! DEFINE 9 LINES TO DRAW PRISM
510 DATA 1,2,2,3,3,4,1,5,2,6,3,7,5,6,6,7,7,8
520 CALL EYETOZ(R)
530 FOR Q=1 TO NUMBER
540 A(Q)=P(Q,1)*R(1,1)+P(Q,2)*R(2,1)+P(Q,3)*R(3,1)
550 B(Q)=P(Q,1)*R(1,2)+P(Q,2)*R(2,2)+P(Q,3)*R(3,2)
560 NEXT Q
570 FOR Q=1 TO LINES
580 I1=L(1,Q)
590 I2=L(2,Q)
600 PLOT A(I1),B(I1);A(I2),B(I2)
610 NEXT Q
620 FOR Q=1 TO 7 STEP 2
630 READ X(Q),Y(Q),Z(Q),X(Q+1),Y(Q+1),Z(Q+1)
640 NEXT Q
650 DATA 3,0,1,-3,0,1,3,.5,0,-3,.5,0,3,0,-2,-3,0,-1,3,0,1,-3,0,1
660 FOR Q=1 TO 7 STEP 2
670 E=Q+1
680 CALL CLIPSUB(N1,P,X(Q),Y(Q),Z(Q),X(E),Y(E),Z(E),I(Q),J(Q),K(Q),I(E),J(E),K(E),
690 NEXT Q
```

```
700 FOR Q=1 TO 8
710 A(Q)=X(Q)*R(1,1)+Y(Q)*R(2,1)+Z(Q)*R(3,1)
720 B(Q)=X(Q)*R(1,2)+Y(Q)*R(2,2)+Z(Q)*R(3,2)
730 NEXT Q
740 FOR Q=1 TO 8
750 C(Q)=I(Q)*R(1,1)+J(Q)*R(2,1)+K(Q)*R(3,1)
760 D(Q)=I(Q)*R(1,2)+J(Q)*R(2,2)+K(Q)*R(3,2)
770 NEXT Q
780 FOR Q=1 TO 5 STEP 2
790 PLOT A(Q),B(Q);C(Q),D(Q)
800 PLOT C(Q+1),D(Q+1);A(Q+1),B(Q+1)
810 NEXT Q
820 FOR Q=1 TO 5 STEP 2
830 PLOT A(Q),B(Q);A(Q+2),B(Q+2);
840 NEXT Q
850 FOR Q=1 TO 5 STEP 2
860 PLOT C(Q),D(Q);C(Q+2),D(Q+2);
870 NEXT Q
880 PLOT C(1),D(1)
890 FOR Q=1 TO 5 STEP 2
900 PLOT A(Q+1),B(Q+1);A(Q+3),B(Q+3);
910 NEXT Q
920 FOR Q=1 TO 5 STEP 2
930 PLOT C(Q+1),D(Q+1);C(Q+3),D(Q+3);
940 NEXT Q
950 TITLE$="PROGRAM PRISMS"
960 NAME$="NAME__________"
970 DAY$="DATE_______"
980 COURSE$="COURSE_____"
990 CALL TITLE(W1,W2,W3,W4,TITLE$,NAME$,DAY$,COURSE$)
1000 END
```

A1-38
Program PRISMS

```
100 !*****PRISMCYL*****
110 !
120 ! 3-D CONVEX CLIPPING
130 ! PRISM/CYLINDER INTERSECTION
140 CLEAR
150 OPTION NOLET
160 PLOTTER TEK4010
170 LIBRARY"CLIPSUB"
180 LIBRARY"EYETOZ"
190 LIBRARY"VECPROD"
200 LIBRARY"AXISROT"
210 LIBRARY"CYLSURF"
220 LIBRARY"TITLE"
230 NUMBER=8 !TOTAL PTS
240 N1=INT(NUMBER/2)
250 DIM O(3,3)
260 DIM R(3,3)
270 DIM S(3,3)
280 DIM T(3,3)
290 DIM X(64),Y(64),Z(64)
300 DIM U(64),V(64),W(64)
310 DIM I(64),J(64),K(64)
320 DIM A(64),B(64)
330 DIM C(64),D(64)
340 DIM P(14,3)
350 DIM L(2,100)
360 W1=-3
370 W2=3
380 W3=-3
390 W4=3
400 SET WINDOW W1,W2,W3,W4
410 FOR Q=1 TO N1
```

```
420 READ P(Q,1),P(Q,2),P(Q,3)
430 NEXT Q
440 FOR Q=N1+1 TO NUMBER
450 READ P(Q,1),P(Q,2),P(Q,3)
460 NEXT Q
470 ! DATA TO DEFINE PRISM***CAREFUL-ORDER IS IMPORTANT
480 DATA 1,-1,-2,0,1,-2,-1,-1,-2,1,-1,-2
490 DATA 1,-1,2,0,1,2,-1,-1,2,1,-1,2
500 X0=0 ! X-TRANSLATION OF PRISM
510 Y0=0 ! Y-TRANSLATION OF PRISM
520 Z0=0 ! Z-TRANSLATION OF PRISM
530 FOR Q=1 TO NUMBER
540 P(Q,1)=P(Q,1)+X0
550 P(Q,2)=P(Q,2)+Y0
560 P(Q,3)=P(Q,3)+Z0
570 NEXT Q
580 ABOUTY=0 !ROTATION ABOUT Y-AXIS
590 ABOUTZ=0 !ROTATION ABOUT Z-AXIS
600 CALL AXISROT(2,ABOUTY,O)
610 CALL AXISROT(3,ABOUTZ,S)
620 MAT T=O*S
630 LINES=9 ! LINES TO DRAW PRISM
640 FOR I1=1 TO LINES
650 READ L(1,I1),L(2,I1)
660 NEXT I1
670 ! DEFINE 9 LINES TO DRAW PRISM
680 DATA 1,2,2,3,3,4,1,5,2,6,3,7,5,6,6,7,7,8
690 CALL EYETOZ(R)
700 FOR Q=1 TO NUMBER
710 A(Q)=P(Q,1)*R(1,1)+P(Q,2)*R(2,1)+P(Q,3)*R(3,1)
720 B(Q)=P(Q,1)*R(1,2)+P(Q,2)*R(2,2)+P(Q,3)*R(3,2)
730 NEXT Q
740 FOR Q=1 TO LINES
750 I1=L(1,Q)
760 I2=L(2,Q)
770 PLOT A(I1),B(I1);A(I2),B(I2)
780 NEXT Q
790 NO=32 !NO OF PTS ON CYLINDER
800 RADIUS=0.5 !RADIUS OF INTERSECTING CYLINDER
810 LENGTH=2.0 !LENGTH OF INTERSECTING CYLINDER
820 CALL CYLSURF(NO,.5,2,X,Y,Z,U,V,W,T)
830 FOR Q=1 TO 2*(NO-1) STEP 2
840 E=Q+1
850 CALL CLIPSUB(N1,P,X(Q),Y(Q),Z(Q),X(E),Y(E),Z(E),I(Q),J(Q),K(Q),I(E),J(E),K(E),
860 NEXT Q
870 FOR Q=1 TO 2*(NO-1)
880 A(Q)=X(Q)*R(1,1)+Y(Q)*R(2,1)+Z(Q)*R(3,1)
890 B(Q)=X(Q)*R(1,2)+Y(Q)*R(2,2)+Z(Q)*R(3,2)
900 NEXT Q
910 FOR Q=1 TO 2*(NO-1)
920 C(Q)=I(Q)*R(1,1)+J(Q)*R(2,1)+K(Q)*R(3,1)
930 D(Q)=I(Q)*R(1,2)+J(Q)*R(2,2)+K(Q)*R(3,2)
940 NEXT Q
950 FOR Q=1 TO 2*(NO-1) STEP 2
960 PLOT A(Q),B(Q);C(Q),D(Q)
970 PLOT C(Q+1),D(Q+1);A(Q+1),B(Q+1)
980 NEXT Q
990 FOR Q=1 TO 2*(NO-2) STEP 2
1000 PLOT A(Q),B(Q);A(Q+2),B(Q+2);
1010 NEXT Q
1020 PLOT A(1),B(1)
1030 FOR Q=1 TO 2*(NO-2) STEP 2
1040 PLOT C(Q),D(Q);C(Q+2),D(Q+2);
```

```
1050 NEXT Q
1060 PLOT C(1),D(1)
1070 FOR Q=1 TO 2*(NO-2) STEP 2
1080 PLOT A(Q+1),B(Q+1);A(Q+3),B(Q+3);
1090 NEXT Q
1100 PLOT A(2),B(2)
1110 FOR Q=1 TO 2*(NO-2) STEP 2
1120 PLOT C(Q+1),D(Q+1);C(Q+3),D(Q+3);
1130 NEXT Q
1140 PLOT C(2),D(2)
1150 TITLE$="PROGRAM PRISMCYL"
1160 NAME$="NAME___________"
1170 DAY$="DATE________"
1180 COURSE$="COURSE______"
1190 CALL TITLE(W1,W2,W3,W4,TITLE$,NAME$,DAY$,COURSE$)
1200 PRINT"ABOUT Y=";ABOUTY
1210 PRINT"ABOUT Z=";ABOUTZ
1220 PRINT"X-TRANSLATION=";X0
1230 PRINT"Y-TRANSLATION=";Y0
1240 PRINT"Z-TRANSLATION=";Z0
1250 END
```

A1-39
Program PRISMCYL

```
100 ! *****CYLSURF*****
110 !
120 EXTERNAL
130 SUB CYLSURF(PTS,RADIUS,LENGTH,X(),Y(),Z(),U(),V(),W(),T(,))
140 OPTION NOLET
150 ! CREATE CYLINDRICAL SURFACE AS INTERSECTING SURFACE
160 ! CYLINDER LIES ALONG X-AXIS BETWEEN -LENGTH<=X<=LENGTH
170 ! CALLED BY MAIN XCLIP
180 ! PTS AND COSINES RETURNED IN X(),Y(),Z();U(),V(),W()
190 DIM A(128),B(128),C(128)
200 P9=2*PI/(PTS-1)
210 INCR=P9
220 A(1)=LENGTH
230 B(1)=-RADIUS
240 C(1)=0
250 FOR I=3 TO 2*PTS-2 STEP 2
260 A(I)=LENGTH
270 B(I)=-RADIUS*COS(P9)
280 C(I)=RADIUS*SIN(P9)
290 P9=P9+INCR
300 NEXT I
310 FOR I=2 TO 2*PTS STEP 2
320 A(I)=-A(I-1)
330 B(I)=B(I-1)
340 C(I)=C(I-1)
350 NEXT I
360 FOR I=1 TO 2*PTS
370 X(I)=A(I)*T(1,1)+B(I)*T(2,1)+C(I)*T(3,1)
380 Y(I)=A(I)*T(1,2)+B(I)*T(2,2)+C(I)*T(3,2)
390 Z(I)=A(I)*T(1,3)+B(I)*T(2,3)+C(I)*T(3,3)
400 NEXT I
410 FOR I=1 TO 2*(PTS-1) STEP 2
420 D=SQR((X(I+1)-X(I))^2+(Y(I+1)-Y(I))^2+(Z(I+1)-Z(I))^2)
430 U(I)=(X(I+1)-X(I))/D
440 V(I)=(Y(I+1)-Y(I))/D
450 W(I)=(Z(I+1)-Z(I))/D
460 NEXT I
470 END SUB
```

A1-40
Subroutine CYLSURF

```
100 !****CLIPON****
110 !
120 ! 3-D CONVEX CLIPON
130 ! CONE/CYLINDER INTERSECTION
140 CLEAR
150 OPTION NOLET
160 PLOTTER TEK4010
170 LIBRARY"CLIPSUB"
180 LIBRARY"EYETOZ"
190 LIBRARY"VECPROD"
200 LIBRARY"AXISROT"
210 LIBRARY"CYLSURF"
220 LIBRARY"TITLE"
230 NUMBER=32 !TOTAL PTS ON OBJECT
240 OPEN #2:NAME"CONE"
250 LINES=46 !NO OF LINES IN FILE #2
260 N1=INT(NUMBER/2)
270 DIM O(3,3),R(3,3),S(3,3),T(3,3)
280 DIM X(64),Y(64),Z(64)
290 DIM U(64),V(64),W(64)
300 DIM I(64),J(64),K(64)
310 DIM A(64),B(64),C(64),D(64)
320 DIM P(64,3),L(2,100)
330 DIM L1(33),L2(33)
340 W1=-3
350 W2=3
360 W3=-3
370 W4=3
380 SET WINDOW W1,W2,W3,W4
390 R1=0.1
400 N=2*PI/(N1-1)
410 Q=1
420 FOR ANG=0 TO 2*PI STEP N
430 P(Q,1)=R1*COS(ANG)
440 P(Q,2)=R1*SIN(ANG)
450 P(Q,3)=-3
460 Q=Q+1
470 NEXT ANG
480 P(Q,1)=P(1,1)
490 P(Q,2)=P(1,2)
500 P(Q,3)=P(1,3)
510 Q=Q+1
520 R2=1
530 FOR ANG=0 TO 2*PI STEP N
540 P(Q,1)=R2*COS(ANG)
550 P(Q,2)=R2*SIN(ANG)
560 P(Q,3)=3
570 Q=Q+1
580 NEXT ANG
590 P(Q,1)=P(N1+1,1)
600 P(Q,2)=P(N1+1,2)
610 P(Q,3)=P(N1+1,3)
620 ABOUTY=0 !ROTATE ABOUT Y-AXIS
630 ABOUTZ=0 !ROTATE ABOUT Z-AXIS
640 CALL AXISROT(2,ABOUTY,O)
650 CALL AXISROT(3,ABOUTZ,S)
660 MAT T=O*S
670 FOR I1=1 TO LINES
680 INPUT #2:L(1,I1),L(2,I1)
690 NEXT I1
```

```
700 CALL EYETOZ(R)
710 FOR Q=1 TO NUMBER
720 A(Q)=P(Q,1)*R(1,1)+P(Q,2)*R(2,1)+P(Q,3)*R(3,1)
730 B(Q)=P(Q,1)*R(1,2)+P(Q,2)*R(2,2)+P(Q,3)*R(3,2)
740 NEXT Q
750 FOR Q=1 TO LINES
760 I1=L(1,Q)
770 I2=L(2,Q)
780 PLOT A(I1),B(I1);A(I2),B(I2)
790 NEXT Q
800 NO=32 !NO OF PTS ON CYLINDER
810 LENGTH=2 !LENGTH OF CLIPON CYLINDER
820 RADIUS=0.5 !RADIUS OF CLIPON CYLINDER
830 CALL CYLSURF(NO,RADIUS,LENGTH,X,Y,Z,U,V,W,T)
840 COUNT=1
850 FOR Q=1 TO 2*(NO-1) STEP 2
860 E=Q+1
870 CALL CLIPSUB(N1,P,X(Q),Y(Q),Z(Q),X(E),Y(E),Z(E),I(Q),J(Q),K(Q),I(E),J(E),K(E),
880 L1(COUNT)=SQR((X(Q)-I(Q))^2+(Y(Q)-J(Q))^2+(Z(Q)-K(Q))^2)
890 L2(COUNT)=SQR((X(Q+1)-I(Q+1))^2+(Y(Q+1)-J(Q+1))^2+(Z(Q+1)-K(Q+1))^2)
900 COUNT=COUNT+1
910 NEXT Q
920 L1(COUNT)=L1(1)
930 L2(COUNT)=L2(1)
940 FOR Q=1 TO 2*(NO-1)
950 A(Q)=X(Q)*R(1,1)+Y(Q)*R(2,1)+Z(Q)*R(3,1)
960 B(Q)=X(Q)*R(1,2)+Y(Q)*R(2,2)+Z(Q)*R(3,2)
970 NEXT Q
980 FOR Q=1 TO 2*(NO-1)
990 C(Q)=I(Q)*R(1,1)+J(Q)*R(2,1)+K(Q)*R(3,1)
1000 D(Q)=I(Q)*R(1,2)+J(Q)*R(2,2)+K(Q)*R(3,2)
1010 NEXT Q
1020 FOR Q=1 TO 2*(NO-1) STEP 2
1030 PLOT A(Q),B(Q);C(Q),D(Q)
1040 PLOT C(Q+1),D(Q+1);A(Q+1),B(Q+1)
1050 NEXT Q
1060 PLOT A(1),B(1)
1070 FOR Q=1 TO 2*(NO-2) STEP 2
1080 PLOT C(Q),D(Q);C(Q+2),D(Q+2);
1090 NEXT Q
1100 PLOT C(1),D(1)
1110 FOR Q=1 TO 2*(NO-2) STEP 2
1120 PLOT A(Q+1),B(Q+1);A(Q+3),B(Q+3);
1130 NEXT Q
1140 PLOT A(2),B(2)
1150 FOR Q=1 TO 2*(NO-2) STEP 2
1160 PLOT C(Q+1),D(Q+1);C(Q+3),D(Q+3);
1170 NEXT Q
1180 PLOT C(2),D(2)
1190 TITLE$="PROGRAM CLIPON"
1200 NAME$="NAME___________"
1210 DAY$="DATE________"
1220 COURSE$="COURSE______"
1230 CALL TITLE(W1,W2,W3,W4,TITLE$,NAME$,DAY$,COURSE$)
1240 PRINT"ABOUT Y=";ABOUTY
1250 PRINT"ABOUT Z=";ABOUTZ
1260 END
```

A1-41
Program CLIPON

```
100 !***CLIPOFF***
110 !
120 ! 3-D CONVEX CLIPPING
130 ! CONE/CYLINDER INTERSECTION
140 CLEAR
150 OPTION NOLET
160 PLOTTER TEK4010
170 LIBRARY"CLIPSUB"
180 LIBRARY"EYETOZ"
190 LIBRARY"VECPROD"
200 LIBRARY"AXISROT"
210 LIBRARY"CYLSURF"
220 LIBRARY"TITLE"
230 PRINT"SHOW BOTH CONE/CYLINDER(B)"
240 PRINT"OR ONLY CLIPPED CYLINDER(C)";
250 INPUT S$
260 K1=1
270 NUMBER=32 !TOTAL PTS ON OBJECT
280 OPEN #2:NAME"CONE"
290 LINES=46 !NO OF LINES IN FILE #2
300 LOWER=0.2 !LOWER CYLINDER AXIS
310 N1=INT(NUMBER/2)
320 DIM O(3,3)
330 DIM C$(64)
340 DIM R(3,3)
350 DIM S(3,3)
360 DIM T(3,3)
370 DIM X(64),Y(64),Z(64)
380 DIM U(64),V(64),W(64)
390 DIM I(64),J(64),K(64)
400 DIM A(64),B(64)
410 DIM C(64),D(64)
420 DIM E(64),F(64)
430 DIM P(64,3)
440 DIM L(2,100)
450 DIM L1(33),L2(33)
460 W1=-3
470 W2=3
480 W3=-3
490 W4=3
500 SET WINDOW W1,W2,W3,W4
510 R1=0.1 !TIP RADIUS OF CONE
520 N=2*PI/(N1-1)
530 Q=1
540 FOR ANG=0 TO 2*PI STEP N
550 P(Q,1)=R1*COS(ANG)
560 P(Q,2)=R1*SIN(ANG)
570 P(Q,3)=-2.5
580 Q=Q+1
590 NEXT ANG
600 P(Q,1)=P(1,1)
610 P(Q,2)=P(1,2)
620 P(Q,3)=P(1,3)
630 Q=Q+1
640 R2=1.0 !BASE RADIUS OF CONE
650 FOR ANG=0 TO 2*PI STEP N
660 P(Q,1)=R2*COS(ANG)
```

```
670 P(Q,2)=R2*SIN(ANG)
680 P(Q,3)=2.5
690 Q=Q+1
700 NEXT ANG
710 P(Q,1)=P(N1+1,1)
720 P(Q,2)=P(N1+1,2)
730 P(Q,3)=P(N1+1,3)
740 ABOUTY=30 !ROTATION ABOUT Y-AXIS
750 ABOUTZ=0 !ROTATION ABOUT Z-AXIS
760 CALL AXISROT(2,ABOUTY,O)
770 CALL AXISROT(3,ABOUTZ,S)
780 MAT T=O*S
790 FOR I1=1 TO LINES
800 INPUT #2:L(1,I1),L(2,I1)
810 NEXT I1
820 CALL EYETOZ(R)
830 FOR Q=1 TO NUMBER
840 A(Q)=P(Q,1)*R(1,1)+P(Q,2)*R(2,1)+P(Q,3)*R(3,1)
850 B(Q)=P(Q,1)*R(1,2)+P(Q,2)*R(2,2)+P(Q,3)*R(3,2)
860 NEXT Q
870 IF S$="C" THEN 930
880 FOR Q=1 TO LINES
890 I1=L(1,Q)
900 I2=L(2,Q)
910 PLOT A(I1),B(I1);A(I2),B(I2)
920 NEXT Q
930 NO=32 !NO OF PTS ON CYLINDER
940 LENGTH=2.5 !HALF LENGTH OF CLIPPING CYLINDER
950 RADIUS=0.5 !RADIUS OF CLIPPING CYLINDER
960 CALL CYLSURF(NO,RADIUS,LENGTH,X,Y,Z,U,V,W,T)
970 FOR Q=1 TO 64
980 Y(Q)=Y(Q)-LOWER
990 NEXT Q
1000 COUNT=1
1010 FOR Q=1 TO 2*(NO-1) STEP 2
1020 E1=Q+1
1030 CALL CLIPSUB(N1,P,X(Q),Y(Q),Z(Q),X(E1),Y(E1),Z(E1),I(Q),J(Q),K(Q),I(E1),J(E1)
1040 C$(Q)=I$
1050 IF I$="INTERSECTION" THEN 1090
1060 L1(COUNT)=SQR((X(Q+1)-X(Q))^2+(Y(Q+1)-Y(Q))^2+(Z(Q+1)-Z(Q))^2)/2
1070 L2(COUNT)=L1(COUNT)
1080 GO TO 1110
1090 L1(COUNT)=SQR((X(Q)-I(Q))^2+(Y(Q)-J(Q))^2+(Z(Q)-K(Q))^2)
1100 L2(COUNT)=SQR((X(Q+1)-I(Q+1))^2+(Y(Q+1)-J(Q+1))^2+(Z(Q+1)-K(Q+1))^2)
1110 COUNT=COUNT+1
1120 NEXT Q
1130 L1(COUNT)=L1(1)
1140 L2(COUNT)=L2(1)
1150 FOR Q=1 TO 2*(NO-1)
1160 A(Q)=X(Q)*R(1,1)+Y(Q)*R(2,1)+Z(Q)*R(3,1)
1170 B(Q)=X(Q)*R(1,2)+Y(Q)*R(2,2)+Z(Q)*R(3,2)
1180 NEXT Q
1190 FOR Q=1 TO 2*(NO-1)
1200 C(Q)=I(Q)*R(1,1)+J(Q)*R(2,1)+K(Q)*R(3,1)
1210 D(Q)=I(Q)*R(1,2)+J(Q)*R(2,2)+K(Q)*R(3,2)
1220 NEXT Q
1230 FOR Q=1 TO 2*(NO-1) STEP 2
1240 IF C$(Q)="INTERSECTION" THEN 1280
1250 PLOT A(Q),B(Q);A(Q+1),B(Q+1)
1260 P$="PARTIAL"
1270 GO TO 1350
1280 PLOT A(Q),B(Q);C(Q),D(Q)
1290 E(K1)=C(Q)
```

```
1300 F(K1)=D(Q)
1310 E(K1+1)=C(Q+1)
1320 F(K1+1)=D(Q+1)
1330 PLOT C(Q+1),D(Q+1);A(Q+1),B(Q+1)
1340 K1=K1+2
1350 NEXT Q
1360 ! DRAW CYLINDER END
1370 FOR Q=1 TO 2*(NO-2) STEP 2
1380 PLOT A(Q),B(Q);A(Q+2),B(Q+2);
1390 NEXT Q
1400 PLOT A(1),B(1)
1410 ! DRAW CYLINDER END
1420 FOR Q=1 TO 2*(NO-2) STEP 2
1430 PLOT A(Q+1),B(Q+1);A(Q+3),B(Q+3);
1440 NEXT Q
1450 PLOT A(2),B(2)
1460 IF P$="PARTIAL" THEN 1570
1470 FOR Q=1 TO 2*(NO-2) STEP 2
1480 PLOT C(Q),D(Q);C(Q+2),D(Q+2);
1490 NEXT Q
1500 PLOT C(1),D(1)
1510 PLOT A(2),B(2)
1520 FOR Q=1 TO 2*(NO-2) STEP 2
1530 PLOT C(Q+1),D(Q+1);C(Q+3),D(Q+3);
1540 NEXT Q
1550 PLOT C(2),D(2)
1560 GO TO 1640
1570 ! PLOT PARTIAL INTERSECTION
1580 FOR Q=2 TO K1-3 STEP 2
1590 PLOT E(Q),F(Q);E(Q+2),F(Q+2);
1600 NEXT Q
1610 FOR Q=K1-2 TO 3 STEP -2
1620 PLOT E(Q),F(Q);E(Q-2),F(Q-2);
1630 NEXT Q
1640 TITLE$="PROGRAM CLIPOFF"
1650 NAME$="NAME__________"
1660 DAY$="DATE_______"
1670 COURSE$="COURSE_____"
1680 CALL TITLE(W1,W2,W3,W4,TITLE$,NAME$,DAY$,COURSE$)
1690 PRINT"LOWER AXIS=";LOWER
1700 PRINT"STOP(S) OR CONTINUE(C)";
1710 INPUT G$
1720 IF G$="S" THEN 2390
1730 CLEAR
1740 Q=1
1750 N=2*PI/(NUMBER-1)
1760 IF P$="PARTIAL" THEN 2090
1770 SET WINDOW -LENGTH,LENGTH,-0.25*LENGTH,1.75*LENGTH
1780 PRINT" LENGTH    ANGLE"
1790 FOR ANG=0 TO 2*PI STEP N
1800 PRINT L1(Q);ANG*180/PI
1810 PLOT LENGTH/2-L1(Q),RADIUS*ANG;LENGTH/2,RADIUS*ANG
1820 IF Q=NUMBER THEN 1860
1830 PLOT LENGTH/2,RADIUS*ANG;LENGTH/2,RADIUS*(ANG+N)
1840 PLOT LENGTH/2-L1(Q),RADIUS*ANG;LENGTH/2-L1(Q+1),RADIUS*(ANG+N)
1850 Q=Q+1
1860 NEXT ANG
1870 PLOT -LENGTH,-0.05*LENGTH;-LENGTH+1,-0.05*LENGTH
1880 GRAPHIC PRINT -LENGTH,-0.1*LENGTH:"ONE-UNIT"
1890 CALL TITLE(W1,W2,W3,W4,TITLE$,NAME$,DAY$,COURSE$)
1900 PRINT"STOP(S) OR CONTINUE(C)";
1910 INPUT G$
1920 IF G$="S" THEN 2390
```

```
1930 SET WINDOW -LENGTH,LENGTH,-0.25*LENGTH,1.75*LENGTH
1940 CLEAR
1950 PRINT" LENGTH    ANGLE"
1960 Q=1
1970 FOR ANG=0 TO 2*PI STEP N
1980 PRINT L2(Q);ANG*180/PI
1990 PLOT LENGTH/2-L2(Q),RADIUS*ANG;LENGTH/2,RADIUS*ANG
2000 IF Q=NUMBER THEN 2040
2010 PLOT LENGTH/2,RADIUS*ANG;LENGTH/2,RADIUS*(ANG+N)
2020 PLOT LENGTH/2-L2(Q),RADIUS*ANG;LENGTH/2-L2(Q+1),RADIUS*(ANG+N)
2030 Q=Q+1
2040 NEXT ANG
2050 CALL TITLE(W1,W2,W3,W4,TITLE$,NAME$,DAY$,COURSE$)
2060 PLOT -LENGTH,-0.05*LENGTH;-LENGTH+1,-0.05*LENGTH
2070 GRAPHIC PRINT -LENGTH,-0.1*LENGTH:"ONE-UNIT"
2080 GO TO 2390
2090 PRINT"TOTAL LENGTH=";2*LENGTH
2100 PRINT
2110 PRINT" L1       L2"
2120 PRINT
2130 K1=1 !COUNT INTERSECTIONS
2140 SET WINDOW -LENGTH,LENGTH,-0.25*LENGTH,1.75*LENGTH
2150 FOR ANG=0 TO 2*PI STEP N
2160 IF C$(2*Q-1)="INTERSECTION" THEN 2190
2170 PLOT LENGTH,RADIUS*ANG;-LENGTH,RADIUS*ANG
2180 GO TO 2270
2190 PLOT LENGTH-L1(Q),RADIUS*ANG;LENGTH,RADIUS*ANG
2200 PLOT -LENGTH,RADIUS*ANG;-LENGTH+L2(Q),RADIUS*ANG
2210 PRINT L1(Q);L2(Q)
2220 A(K1)=LENGTH-L1(Q)
2230 B(K1)=RADIUS*ANG
2240 C(K1)=-LENGTH+L2(Q)
2250 D(K1)=B(K1)
2260 K1=K1+1
2270 Q=Q+1
2280 NEXT ANG
2290 PLOT A(1),B(1);
2300 FOR Q=2 TO K1-1
2310 PLOT A(Q),B(Q);
2320 NEXT Q
2330 PLOT C(Q-1),D(Q-1);
2340 FOR Q=K1-1 TO 1 STEP -1
2350 PLOT C(Q),D(Q);
2360 NEXT Q
2370 PLOT A(1),B(1)
2380 CALL TITLE(W1,W2,W3,W4,TITLE$,NAME$,DAY$,COURSE$)
2390 PRINT"ABOUT Y=";ABOUTY
2400 PRINT"ABOUT Z=";ABOUTZ
2410 PRINT"TIP RADIUS=";R1
2420 PRINT"BASE RADIUS=";R2
2430 PRINT"LOWER=";LOWER
2440 END
```

A1-42
Program CLIPOFF

APPENDIX 2

Mathematical Concepts

A. EVALUATION OF ARCTAN FUNCTION

The angular position of a line is given by the value of an angle, say ϕ, measured relative to a chosen coordinate system. When the angular position of the base vectors $\mathbf{P}_1$, $\mathbf{P}_2$, $\mathbf{P}_3$, and $\mathbf{P}_4$ along the circle shown in Figure A2-1 are measured from the positive x-axis in a counterclockwise manner with $0 \leq \phi \leq 2\pi$, the following relationships give the proper values for ϕ.

$$
\begin{aligned}
\phi_1 &= \tan^{-1}\left(\frac{y}{x}\right) \\
\phi_2 &= \pi - \tan^{-1}\left(\frac{y}{x}\right) \\
\phi_3 &= \pi + \tan^{-1}\left(\frac{y}{x}\right) \\
\phi_4 &= 2\pi - \tan^{-1}\left(\frac{y}{x}\right)
\end{aligned}
\tag{A-1}
$$

If $x = 0$, then $\phi = \pm \pi/2$.

Principal values for $\phi = \tan^{-1}(y/x)$ are defined for $-\pi/2 \leq \phi \leq \pi/2$. The ATN function in BASIC is only defined for principal values of ϕ—that is, in the first and fourth quadrants of an x-y coordinate system. In BASIC8 the function ANGLE(X,Y) returns the angle, in radians, for $0 \leq \tan^{-1}(Y/X) \leq \pi$ and $-\pi \leq \tan^{-1}(Y/X) \leq 0$. This ANGLE function is used in the listings given in Appendix 1. An example is

```
250 PHI = ANGLE(X,Y)
```

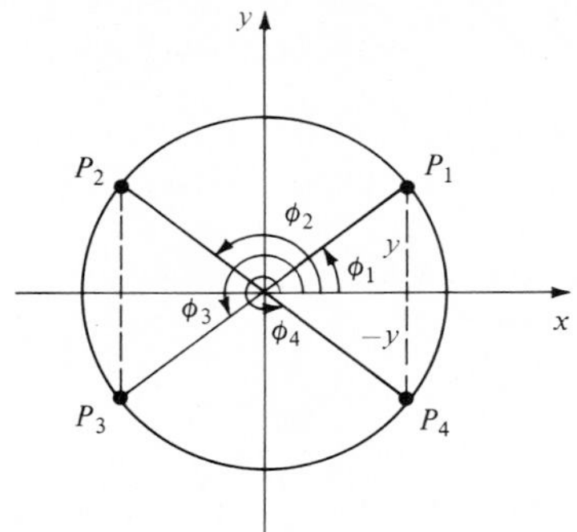

Figure A2-1
Angle measurements for $0 \leq \phi \leq 2\pi$.

If this function is not available, a subroutine such as ATAN can be written and used as follows:

```
100 !*****ATAN*****
110 EXTERNAL
120 SUB ATAN(X,Y,A)
130 ! SUBPROGRAM FOR RETURNING ANGLE A (IN RADIANS) BETWEEN 0 AND 2*PI
140 ! THE ENDPOINT OF THE LINE THROUGH THE ORIGIN DEFINING A IS X,Y
150 OPTION NOLET
160 IF ABS (X)+ABS(Y)=0 THEN 340
170 IF Y<0 THEN 260
180 IF X=0 THEN 220
190 IF X<0 THEN 240
200 A=ATN(Y/X)
210 GOTO 350
220 A=2*ATN(1)
230 GOTO 350
240 A=4*ATN(1)-ATN (ABS (Y/X))
250 GOTO 350
260 IF X=0 THEN 320
270 IF X<0 THEN 300
280 A=8*ATN(1)-ATN(ABS (Y/X))
290 GOTO 350
300 A=ATN(ABS(Y/X))+4*ATN(1)
310 GOTO 350
320 A=6*ATN(1)
330 GOTO 350
340 A=0
350 END SUB
```

A sample call to this subroutine is:

```
250 CALL ATAN(X,Y,PHI)
```

B. MATRIX ALGEBRA CONCEPTS

Matrix algebra is one of several types of algebra which is governed by a special set of mathematical rules. Some of these rules are summarized in this appendix. The rules apply to operations on one or more matrices.

A matrix is a rectangular array of elements. For geometric representations these elements are coordinates of points in space. An $m \times n$ matrix consists of m rows and n columns. For example, consider the four matrices given below.

$$[\mathbf{A}] = \begin{bmatrix} 3 & 2 & 4 \\ 1 & -3 & 2 \end{bmatrix} \qquad [\mathbf{B}] = \begin{bmatrix} 1 & 3 \\ 2 & 4 \end{bmatrix}$$

$$[\mathbf{C}] = [1 \quad 3 \quad 5] \qquad [\mathbf{D}] = \begin{bmatrix} 3 \\ 2 \\ 2 \end{bmatrix}$$

[**A**] is a 2×3 matrix, [**B**] is a 2×2 square matrix, [**C**] is a 1×3 matrix, and [**D**] is a 3×1 matrix.

We can also use subscripts to indicate the row and column of an element in a matrix. For example, for a 2×3 matrix [**A**]:

$$[\mathbf{A}] = \begin{bmatrix} a_{11} & a_{12} & a_{13} \\ a_{21} & a_{22} & a_{23} \end{bmatrix}$$

If this represents the initial [**A**] matrix above, then $a_{11} = 3$, $a_{12} = 2$, $a_{21} = 1$, and so on.

Addition and subtraction of two matrices of the *same size* is defined. The sum or difference of the corresponding elements in each matrix gives the resulting matrix. For example, if

$$[\mathbf{A}] = \begin{bmatrix} 3 & 2 & 4 \\ 1 & -3 & 2 \end{bmatrix} \quad \text{and} \quad [\mathbf{B}] = \begin{bmatrix} 1 & 2 & 3 \\ -3 & 2 & 1 \end{bmatrix}$$

then for $[\mathbf{C}] = [\mathbf{A}] + [\mathbf{B}]$ and $[\mathbf{D}] = [\mathbf{A}] - [\mathbf{B}]$ one obtains

$$[\mathbf{C}] = \begin{bmatrix} 4 & 4 & 7 \\ -2 & -1 & 3 \end{bmatrix} \quad \text{and} \quad [\mathbf{D}] = \begin{bmatrix} 2 & 0 & 1 \\ 4 & -5 & 1 \end{bmatrix}$$

Single three-dimensional vectors can be represented by a single row or single column matrix. If $[\mathbf{R}] = [3\ 5\ 2]$ and $[\mathbf{S}] = [1\ 6\ 4]$, then $[\mathbf{R}] + [\mathbf{S}] = [4\ 11\ 6]$ and $[\mathbf{R}] - [\mathbf{S}] = [\,2\ -1\ -2]$.

In the definition of matrix multiplication, the number of columns of the first matrix must equal the number of rows in the second matrix. Consider the product [**A**][**B**], where [**A**] is a 2×3 and [**B**] is a 3×4 matrix.

$$[\mathbf{A}][\mathbf{B}] = \begin{bmatrix} a_{11} & a_{12} & a_{13} \\ a_{21} & a_{22} & a_{23} \end{bmatrix} \begin{bmatrix} b_{11} & b_{12} & b_{13} & b_{14} \\ b_{21} & b_{22} & b_{23} & b_{24} \\ b_{31} & b_{32} & b_{33} & b_{34} \end{bmatrix} = [\mathbf{C}] \qquad \text{(A-2)}$$

Matrix multiplication is defined and the resulting **[C]** matrix will be a 2 × 4 matrix whose elements are represented by

$$[\mathbf{C}] = \begin{bmatrix} c_{11} & c_{12} & c_{13} & c_{14} \\ c_{21} & c_{22} & c_{23} & c_{24} \end{bmatrix} \tag{A-3}$$

Each element c_{mn} in the product **[C]** is obtained by multiplying the corresponding elements of row m in matrix **[A]** by the corresponding elements in column n in matrix **[B]**, and summing the products. For example:

$$\begin{aligned} c_{12} &= a_{11}b_{12} + a_{12}b_{22} + a_{13}b_{32} \\ c_{13} &= a_{11}b_{13} + a_{12}b_{23} + a_{13}b_{33} \\ c_{22} &= a_{21}b_{12} + a_{22}b_{22} + a_{23}b_{32} \end{aligned} \tag{A-4}$$

In general, element c_{pq} is given by

$$c_{pq} = a_{p1}b_{1q} + a_{p2}b_{2q} + a_{p3}b_{3q} = \sum_{i=1}^{3} a_{pi}b_{iq} \tag{A-5}$$

for the **[A][B]** product indicated above. For a general multiplication of an $m \times n$ matrix **[A]** and an $n \times s$ matrix **[B]**, the result is an $m \times s$ matrix **[C]** where

$$c_{pq} = \sum_{i=1}^{n} a_{pi}b_{iq} \tag{A-6}$$

Matrix multiplication is not commutative, so the order of multiplication is important. That is, in general **[A][B]** ≠ **[B][A]**. This is very important when using matrix multiplication to obtain geometric transformations and vector products. Two vectors represented in matrix form can be multiplied if one is a [1 × 3] and the other is a [3 × 1] matrix. For example:

$$[\mathbf{C}][\mathbf{D}] = [1\ 3\ 5] \begin{bmatrix} 3 \\ 2 \\ 2 \end{bmatrix} = 3 + 6 + 10 = 19$$

This matrix multiplication is equivalent to the vector dot product **C** · **D**.

C. DETERMINANTS

The determinant of a square matrix is used when applying Cramer's Rule to solve a set of linear, algebraic equations, or evaluating a vector cross product. Cramer's Rule is given in the next section. For a 2 × 2 matrix, the determinant **D** is given by

$$\mathbf{D} = \det \begin{vmatrix} a_{11} & a_{12} \\ a_{21} & a_{22} \end{vmatrix} = a_{11}a_{22} - a_{12}a_{21} \qquad \text{(A-7)}$$

The determinant of a 3 × 3 matrix is given by

$$\mathbf{D} = \det \begin{vmatrix} a_{11} & a_{12} & a_{13} \\ a_{21} & a_{22} & a_{23} \\ a_{31} & a_{32} & a_{33} \end{vmatrix}$$

$$= a_{11}(a_{22}a_{33} - a_{23}a_{32}) - a_{12}(a_{21}a_{33} - a_{23}a_{31}) + a_{13}(a_{21}a_{32} - a_{22}a_{31}) \qquad \text{(A-8)}$$

D. CRAMER'S RULE

In this text, Cramer's Rule is used to solve a set of linear, algebraic equations. If these two equations are written in matrix form, with unknowns x and y, the general form is

$$\begin{bmatrix} a_{11} & a_{12} \\ a_{21} & a_{22} \end{bmatrix} \begin{bmatrix} x \\ y \end{bmatrix} = \begin{bmatrix} c_{11} \\ c_{21} \end{bmatrix} \qquad \text{(A-9)}$$

The solution for x and y is found by evaluating the following two expressions:

$$x = \frac{\det \begin{vmatrix} c_{11} & a_{12} \\ c_{21} & a_{22} \end{vmatrix}}{\det \begin{vmatrix} a_{11} & a_{12} \\ a_{21} & a_{22} \end{vmatrix}} \qquad \text{(A-10)}$$

$$y = \frac{\det \begin{vmatrix} a_{11} & c_{11} \\ a_{21} & c_{21} \end{vmatrix}}{\det \begin{vmatrix} a_{11} & a_{12} \\ a_{21} & a_{22} \end{vmatrix}}$$

Cramer's Rule can also be used for more than two equations. However, matrix inversion is usually easier to apply when the number of equations is three or more. Matrix inversion is not needed in this text.

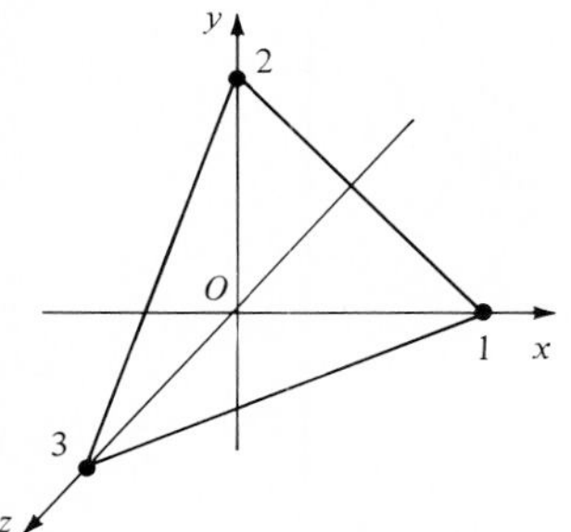

Figure A2-2
Triangular facet.

Example A-1

Given: The three vertices of a triangular facet shown in Figure A2-2 are the unit vectors **i**, **j**, **k**.
Task:

a. Represent these vertices in matrix form and multiply the data matrix by the 3 × 3 transformation matrix given below. Interpret the geometrical transformation achieved.

b. Evaluate the determinant of the given transformation matrix **[R]**.

$$[\mathbf{R}] = \begin{bmatrix} 0.707 & -0.408 & 0.577 \\ 0 & 0.8165 & 0.577 \\ -0.707 & -0.408 & 0.577 \end{bmatrix}$$

Solution:

$$\begin{bmatrix} 1 & 0 & 0 \\ 0 & 1 & 0 \\ 0 & 0 & 1 \end{bmatrix} \begin{bmatrix} 0.707 & -0.408 & 0.577 \\ 0 & 0.8165 & 0.577 \\ -0.707 & -0.408 & 0.577 \end{bmatrix} = \begin{bmatrix} c_{11} & c_{12} & c_{13} \\ c_{21} & c_{22} & c_{23} \\ c_{31} & c_{32} & c_{33} \end{bmatrix}$$

Using the rules of matrix multiplication gives the following elements in the **[C]** matrix:

$$c_{11} = 1(0.707) + 0(0) + 0(-0.707) = 0.707$$

$$c_{12} = 1(-0.408) + 0(0.8165) + 0(-0.408) = -0.408$$

$$c_{13} = 1(0.572) + 0(0.577) + 0(0.577) = 0.577$$

$$c_{21} = 0(0.707) + 1(0) + 0(-0.408) = 0$$

$$c_{22} = 0(-0.408) + 1(0.8165) + 0(-0.408) = 0.8165$$

$$c_{23} = 0(0.577) + 1(0.577) + 0(0.577) = 0.577$$

$$c_{31} = 0(0.707) + 0(0) + 1(-0.707) = -0.707$$

$$c_{32} = 0(-0.408) + 0(0.8165) + 1(-0.408) = -0.408$$

$$c_{33} = 0(0.577) + 0(0.577) + 1(0.577) = 0.577$$

In this case, the resulting 3 × 3 **[C]** matrix is the same as the transformation matrix **[R]**. Figure A2-3 shows the new vertices.

$$[\mathbf{C}] = \begin{bmatrix} 0.707 & -0.408 & 0.577 \\ 0 & 0.8165 & 0.577 \\ -0.707 & -0.408 & 0.577 \end{bmatrix}$$

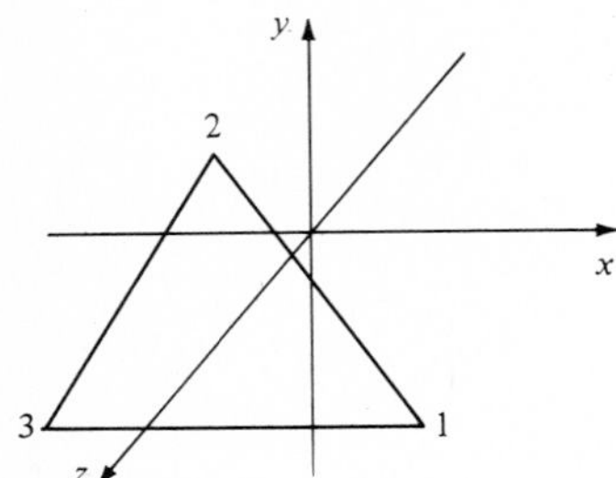

Figure A2-3
Transformed facet.

The determinant of **[R]** is

$$\mathbf{D} = \det \begin{vmatrix} 0.707 & -0.408 & 0.577 \\ 0 & 0.8165 & 0.577 \\ -0.707 & -0.408 & 0.577 \end{vmatrix}$$

$$\begin{aligned} \mathbf{D} = {} & 0.707\,[(0.8165)(0.577) - (0.577)(-0.408)] \\ & -0.408[(0.577)(-0.707) - (0)(0.577)] \\ & + 0.577[(0)(-0.408) - (0.8165)(-0.707)] \end{aligned}$$

$$\mathbf{D} = 0.499 + 0.167 + 0.333 = 1.0$$

Notice that the triangular facet shown in Figure A2-3 has been rotated about the y-axis in a clockwise manner by $\theta = \tan^{-1}(1) = 45$ deg followed by a counterclockwise rotation about the x-axis by $\phi = \tan^{-1}(1/\sqrt{2}) = 35.26$ deg. This places the facet parallel to the x-y plane so that it is seen in true size when viewed orthographically from a position on the z-axis. Rotational transformations are discussed in the following section. Also note that the determinant of the transformation matrix causing this solid body rotation of a surface is unity. This will be true of all such rotational transformations.

E. ROTATION BY MATRIX TRANSFORMATION

Many useful transformations can be created by matrix multiplication. References 1 and 21 give further details (see References). Here we consider only rotations about a principal axis that are necessary to create an orthographic projection from any eye position as discussed in Section 2-9.

The general eye position is specified by the vector $\mathbf{E} = [E_1\ E_2\ E_3]$. The two rotations indicated in Figure A2-4 required to align the line of sight with the z-axis are a clockwise rotation about the y-axis by $\theta = \tan^{-1}(E_1/E_3)$, followed by a counterclockwise rotation about the x-axis by $\phi = \tan^{-1}(E_2/\sqrt{E_1^2 + E_3^2})$. The object as well as the line of sight are rotated about the y- and x-axes in this manner. Then a plot of the transformed vertices in the x-y plane produces the orthographic projection on a plane perpendicular to the line of sight.

The 3×3 transformation matrix which causes a negative (clockwise) rotation about the y-axis is given by

$$[\mathbf{R}_1] = \begin{bmatrix} \cos\theta & 0 & \sin\theta \\ 0 & 1 & 0 \\ -\sin\theta & 0 & \cos\theta \end{bmatrix} \quad \text{(A-11)}$$

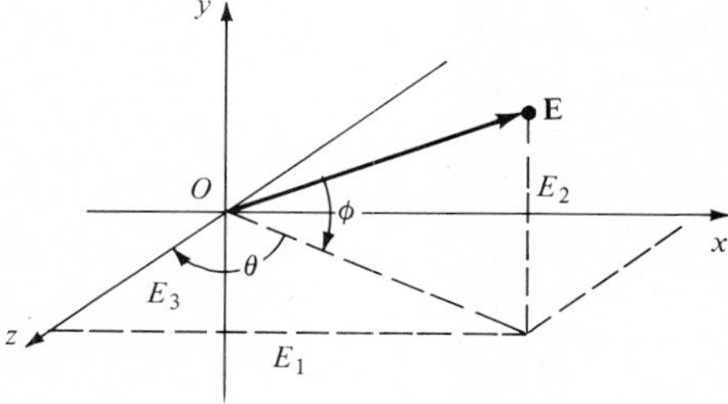

Figure A2-4
Line of sight.

and the 3 × 3 matrix which causes a positive (counterclockwise) rotation about the x-axis is

$$[\mathbf{R}_2] = \begin{bmatrix} 1 & 0 & 0 \\ 0 & \cos\phi & \sin\phi \\ 0 & -\sin\phi & \cos\phi \end{bmatrix} \quad \text{(A-12)}$$

A single 3 × 3 matrix can be created which will perform both rotations in one operation. This is given by the matrix product $[\mathbf{R}] = [\mathbf{R}_1][\mathbf{R}_2]$. Using the rules of matrix multiplication one obtains

$$[\mathbf{R}] = \begin{bmatrix} \cos\theta & -\sin\theta\sin\phi & \sin\theta\cos\phi \\ 0 & \cos\phi & \sin\phi \\ -\sin\theta & -\cos\theta\sin\phi & \cos\theta\cos\phi \end{bmatrix} \quad \text{(A-13)}$$

For completeness, the matrix which produces counterclockwise rotation about the z-axis is

$$[\mathbf{R}_3] = \begin{bmatrix} \cos\gamma & \sin\gamma & 0 \\ -\sin\gamma & \cos\gamma & 0 \\ 0 & 0 & 1 \end{bmatrix} \quad \text{(A-14)}$$

The subroutine AXISROT given in Appendix 1 generates the elements in the matrix $[\mathbf{R}_1]$, $[\mathbf{R}_2]$ or $[\mathbf{R}_3]$, depending upon the calling variables. For example: "CALL AXISROT(1,PHI,E)" returns a 3 × 3 **[E]** matrix which can be used to cause counterclockwise rotation about the x-axis by the angle PHI. The subroutine EYETOZ creates a single 3 × 3 matrix **[R]** which can be used to produce the two rotations discussed above. The lines that create this matrix are reproduced below.

```
240 CALL AXISROT(2,−THETA,D)

280 CALL AXISROT(1,PHI,E)

300 MAT R = D*E
```

Subroutine EYETOZ also contains an algorithm for calculating the matrix product **D*E** without the use of the MAT function.

Subroutine EYETOZ is used in most of the main programs given in Appendix 1 to produce orthographic views from specified eye positions. For example, in the main program ORTHOEYE, the initial data matrix is contained in X( ), Y( ), Z( ) arrays and contains a given number (= POINTS) of three-dimensional vertices. These vertices are then multiplied by the **[R]** matrix returned from the subroutine EYETOZ. In ORTHOEYE this multiplication is carried out in lines 500–530 as given below. Only the x and y values (A(I) and B(I)) are computed since these are plotted in the x-y plane to create the orthographic projection.

```
500 FOR I = 1 TO POINTS
```

```
510 LET A(I) = X(I)*R(1,1) + Y(I)*R(2,1) + Z(I)*R(3,1)
520 LET B(I) = X(I)*R(1,2) + Y(I)*R(2,2) + Z(I)*R(3,2)
530 NEXT I
```

Example A-2

Given: A line in space through **A** = [1 3 2] and **B** = [2 1 −1].
Task: Rotate the line by 30 deg clockwise about the y-axis. Then rotate by 45 deg counterclockwise about the x-axis.
Solution: Using Equation (A-11), a 30 deg rotation about the y-axis gives

$$\begin{bmatrix} 1 & 3 & 2 \\ 2 & 1 & -1 \end{bmatrix} \begin{bmatrix} 0.866 & 0 & 0.5 \\ 0 & 1 & 0 \\ -0.5 & 0 & 0.866 \end{bmatrix} = \begin{bmatrix} -0.134 & 3 & 2.232 \\ 2.232 & 1 & 0.134 \end{bmatrix}$$

Then, from Equation (A-12) a 45 deg counterclockwise rotation about the x-axis gives

$$\begin{bmatrix} -0.134 & 3 & 2.232 \\ 2.232 & 1 & 0.134 \end{bmatrix} \begin{bmatrix} 1 & 0 & 0 \\ 0 & 0.707 & 0.707 \\ 0 & -0.707 & 0.707 \end{bmatrix} = \begin{bmatrix} -0.134 & 0.543 & 3.699 \\ 2.232 & 0.6123 & 0.8017 \end{bmatrix}$$

Example A-3

Given: Same problem as Example A-2.
Task: Perform the combined rotations with a single 3 × 3 matrix.
Solution: First multiply matrix (A-11) and (A-12) to obtain

$$\begin{bmatrix} 0.866 & 0 & 0.5 \\ 0 & 1 & 0 \\ -0.5 & 0 & 0.866 \end{bmatrix} \begin{bmatrix} 1 & 0 & 0 \\ 0 & 0.707 & 0.707 \\ 0 & -0.707 & 0.707 \end{bmatrix} = \begin{bmatrix} 0.866 & -0.3535 & 0.3535 \\ 0 & 0.707 & 0.707 \\ -0.5 & -0.6123 & 0.6123 \end{bmatrix}$$

Using this 3 × 3 as the transformation matrix for the two data points gives

$$\begin{bmatrix} 1 & 3 & 2 \\ 2 & 1 & -1 \end{bmatrix} \begin{bmatrix} 0.866 & -0.3535 & 0.3535 \\ 0 & 0.707 & 0.707 \\ -0.5 & -0.6123 & 0.6123 \end{bmatrix} = \begin{bmatrix} -0.134 & 0.543 & 3.699 \\ 2.232 & 0.6123 & 0.8017 \end{bmatrix}$$

This is the same result obtained in Example A-2. Remember that the order of matrix multiplication is important. A computer solution to this example is given below.◀

```
100 ! PROGRAM A-3
110 !COMPUTER SOLUTION TO EXAMPLE A-3
120 DIM X(2), Y(2), Z(2)
130 DIM E(3,3), F(3,3), R(3,3)
140 DIM A(2), B(2), C(2)
150 LIBRARY "AXISROT"
160 OPTION NOLET
170 ! READ ENDPOINTS OF LINE SEGMENT
180 READ X(1), Y(1), Z(1), X(2), Y(2), Z(2)
190 DATA 1,3,2,2,1,-1
200 ! CREATE MATRIX E FOR ROTATION ABOUT Y
210 CALL AXISROT(2,-30,E)
220 ! CREATE MATRIX F FOR ROTATION ABOUT X
230 CALL AXISROT(1,45,F)
240 ! EVALUATE MATRIX PRODUCT USING MAT FUNCTION
250 MAT R=E*F
260 ! PERFORM MATRIX TRANSFORMATION OF DATA WITHOUT MAT FUNCTION
270 FOR I=1 to 2
280 A(I)=X(I)*R(1,1) + Y(I)*R(2,1) + Z(I)*R(3,1)
290 B(I)=X(I)*R(1,2) + Y(I)*R(2,2) + Z(I)*R(3,2)
300 C(I)=X(I)*R(1,3) + Y(I)*R(2,3) + Z(I)*R(3,3)
310 ! PRINT TRANSFORMED ENDPOINTS
320 PRINT A(I); B(I); C(I)
330 NEXT I
340 END
```

Most computer libraries have routines for evaluating matrix operations such as multiplication and inversion. However, not all software used on personal computers and engineering workstations may have matrix functions. The following program shows how to evaluate the operations of matrix multiplication and inversion without the use of matrix functions, as well as with the use of the BASIC matrix functions. This program can be used as a guide for those who need to replace the MAT functions used in programs in this text with equivalent programming logic. ◀

```
80 ! *****MATRIX*****
90 !
100 ! MATRIX PRODUCT AND INVERSE
110 ! DIM T(3,3), R(3,2), M(3,3), P(3,2)
120 ! [T]=3X3 TRANSFORMATION MATRIX
130 ! FOR ROTATION ABOUT Z-AXIS
140 OPTION NOLET
```

```
150 PLOTTER TEK4010
160 CLEAR
170 PRINT"INPUT ROTATION ANGLE";
180 INPUT A
190 C=COS(A*PI/180)
200 S=SIN(A*PI/180)
210 ! DEFINE ELEMENTS IN [T] MATRIX
220 FOR I=1 to 2
230 T(I,3)=0
240 T(3,I)=0
250 NEXT I
260 T(3,3)=1
270 T(1,1)=C
280 T(2,2)=C
290 T(1,2)=S
300 T(2,1)=-S
310 ! DEFINE TWO 3-D POINTS
320 FOR I=1 TO 2
330 READ P(1,I),P(2,I),P(3,I)
340 NEXT I
350 DATA 1,2,1
360 DATA 2,4,1
370 ! PERFORM MATRIX MULT [R]=[T][P]
380 PRINT"USING MAT FUNCTIONS GIVES"
390 PRINT"MATRIX PRODUCT [T][P]"
400 MAT R=T*P
410 MAT PRINT R;
420 PRINT"MATRIX INVERSE OF [T]"
430 MAT M=INV(T)
440 MAT PRINT M;
450 PRINT
460 ! REPEAT WITHOUT MAT FUNCTIONS
470 PRINT"USING ALGORITHMS GIVES"
480 PRINT
490 PRINT"MATRIX PRODUCT [T][P]"
500 ! COMPUTE [R]=[T][P]
510 FOR I=1 to 3
520 FOR J=1 to 2
530 B=O
540 FOR K=1 to 3
550 B=B+T(I,K)*P(K,J)
560 NEXT K
570 R(I,J)=B
580 NEXT J
590 NEXT I
```

```
600 FOR J=1 TO 2
610 FOR I=1 TO 3
620 PRINT R(I,J)
630 NEXT I
640 PRINT
650 NEXT J
660 PRINT
670 PRINT"MATRIX INVERSE OF [T]"
680 ! FIND INVERSE [M] OF [T] USING ADJOINT METHOD
690 ! [T] MUST BE NONSINGULAR
700 D=T(1,1)*(T(2,2)*T(3,3)-T(2,3)*T(3,2))-T(1,2)*(T(2,1)*T(3,3)-T(2,3)*T(3,1))+T(
710 FOR I=1 to 3
720 ! I-3*INT(I/3)=MOD(I,3) ETC.
730 I1=I-3*INT(I/3)+1
740 I2=Il-3*INT(Il/3)+1
750 FOR J=1 TO 3
760 J1=J-3*INT(J/3)+1
770 J2=J1-3*INT(J1/3)+1
780 M(J,I)=(T(I1,J1)*T(I2,J2)-T(I1,J2)*T(I2,J1))/D
790 NEXT J
800 NEXT I
810 FOR I=1 TO 3
820 FOR J=1 TO 3
830 PRINT M(J,I)
840 NEXT J
850 PRINT
860 NEXT I
870 END
```

APPENDIX 3

Geometric Figures

A3-1 ANGLES

An angle (Figure A3-1) is the opening between two straight lines which, when extended, meet in a point. The straight lines are the sides. The intersection of the sides is the vertex, *B*. The angle in Figure A3-1 is referred to as angle *ABC* or angle *CBA* with the letter at the vertex placed between the other two.

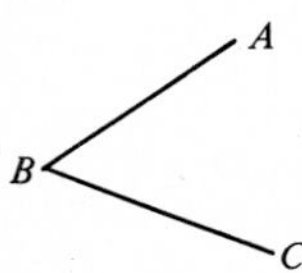

Figure A3-1
Angle.

a. A *right* angle (Figure A3-2) is an angle which measures 90 deg.
b. An *oblique* angle (Figure A3-3) is an angle which is not a right angle.
c. An *acute* angle (Figure A3-4) is one that is less than 90 deg.
d. An *obtuse* angle (Figure A3-5) is greater than 90 deg.
e. A *straight* angle (Figure A3-6) is a straight line and contains 180 deg.
f. A *complementary* angle (Figure A3-7) is one which when added to a given angle equals 90 deg.
g. A *supplementary* angle (Figure A3-8) is one which when added to a given angle equals 180 deg.

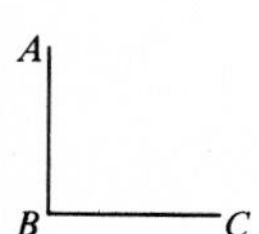

Figure A3-2
Right angle.

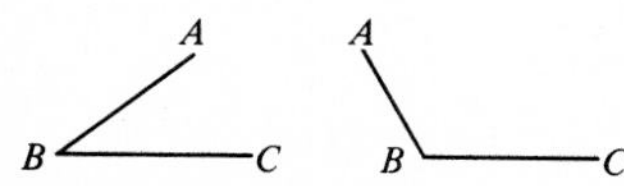

Figure A3-3
Oblique angle.

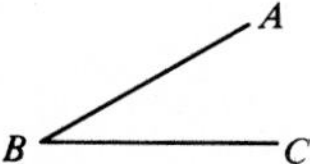

Figure A3-4
Acute angle.

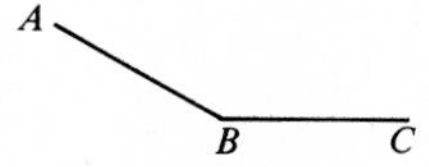

Figure A3-5
Obtuse angle.

Figure A3-6
Straight angle.

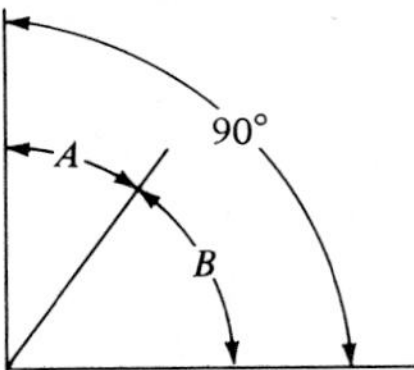

Figure A3-7
Complementary angle.

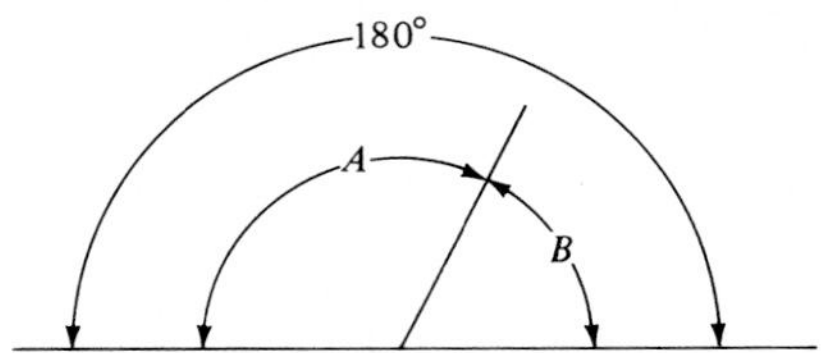

Figure A3-8
Supplementary angle.

A3-2 PLANE FIGURES

A *plane* figure is a plane surface bounded by lines.

a. *Polygons* (Figure A3-9) are plane figures bounded by straight sides. They are named according to the number of sides in the figure. A regular polygon has all sides and all angles equal.

b. A *triangle* is a plane figure having three sides and three angles. They are named according to the number of equal sides in the triangle (Figure A3-10a) and according to their angles (Figure A3-10b).

c. A *parallelogram* (Figure A3-11) is a quadrilateral whose opposite sides are parallel and whose opposite angles are equal.

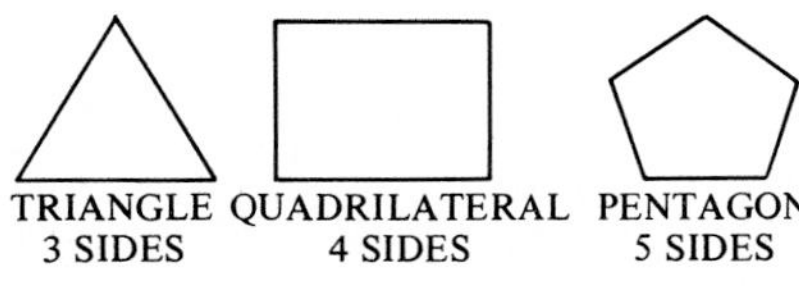

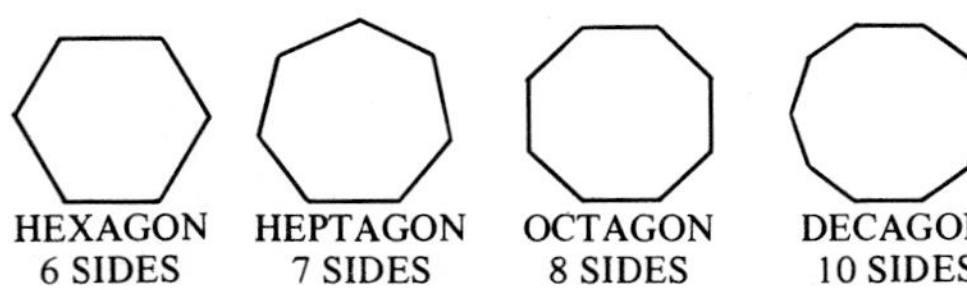

Figure A3-9
Polygons.

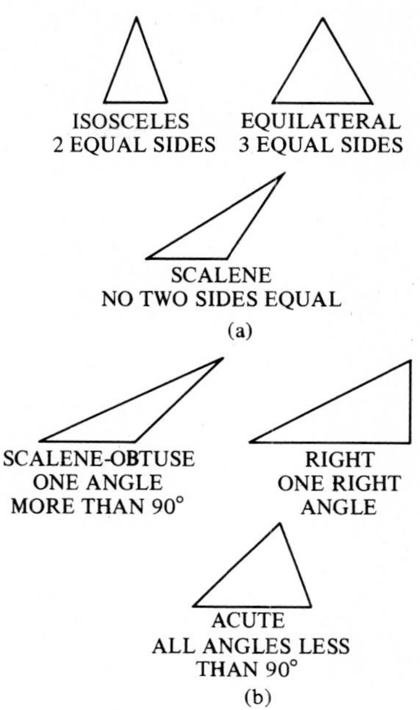

Figure A3-10
Triangles.

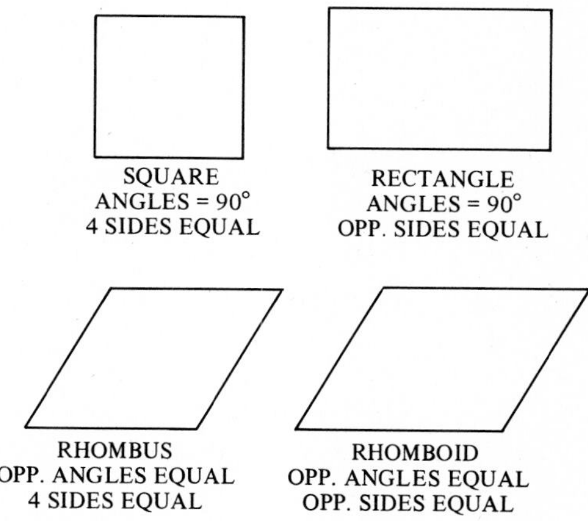

Figure A3-11
Parallelograms.

A3-3 POLYHEDRA

A polyhedron is a three-dimensional figure bounded by plane surfaces or polygons.

A. Pyramids

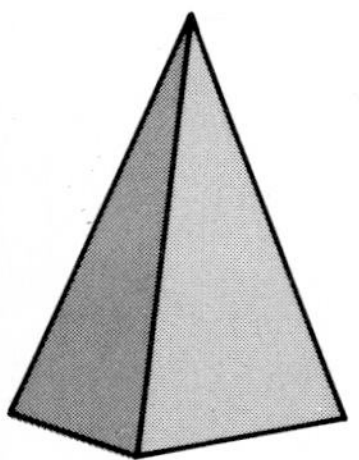

Figure A3-12
Pyramid.

A *pyramid* (Figure A3-12) has a polygon for a base and triangular sides which meet in a common point called the *vertex.*

1. Right pyramid—the axis of the pyramid is perpendicular to its base.
2. Oblique pyramid—the axis of the pyramid is not perpendicular to its base.
3. Triangular pyramid, square pyramid, and the like—named according to the shape of its base.
4. Truncated pyramid—the part remaining after the vertex has been removed by a plane.

B. Prisms

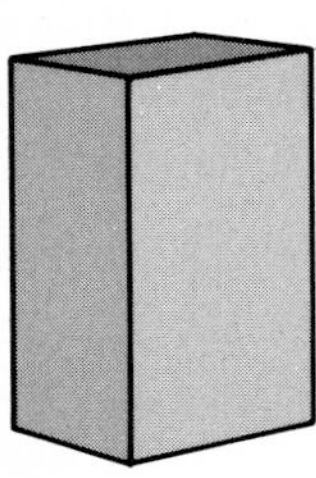

Figure A3-13
Rectangular prism.

A *prism* (Figure A3-13) is a geometric figure whose ends are parallel polygons and whose sides are parallelograms. They take their names from the shapes of their bases. The definition of a right prism, oblique prism, and truncated prism conforms, in general, with those given above for a pyramid.

A3-4 SINGLE-CURVED SURFACES

A. Cylinders

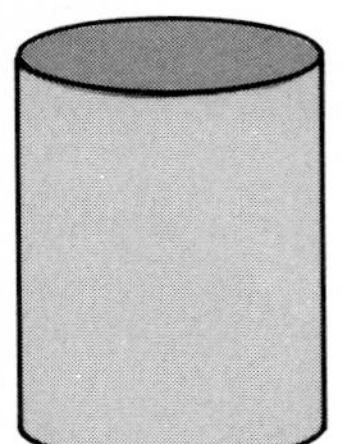

Figure A3-14
Cylinder.

A *cylinder* (Figure A3-14) is formed by a straight line generatrix moving in contact with a curved line directrix which is generally in the shape of a circle or an ellipse. The generatrix is always parallel to its original position and moves to successive positions called *elements.* When the elements are perpendicular to the base, the cylinder is a right cylinder. If the elements are not perpendicular to the base, the cylinder is oblique.

B. Cones

A *cone* (Figure A3-15) is formed by a straight line generatrix moving in contact with a curved line and passing through a fixed point called a *vertex.* When the axis of the cone is perpendicular to the base, the cone is a right cone. If it is not perpendicular to the base, the cone is oblique.

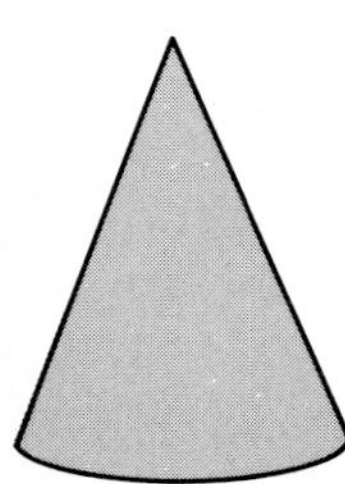

Figure A3-15
Cone.

A3-5 DOUBLE-CURVED SURFACES

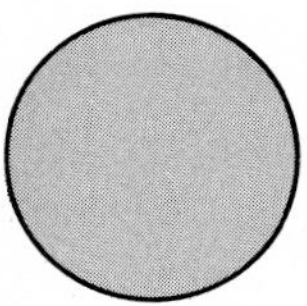

Figure A3-16
Sphere.

A. Spheres

A *sphere* (Figure A3-16) is a solid bounded by a uniformly curved surface. Every point on the surface is equidistant from the center of the sphere.

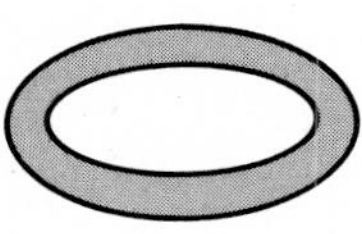

Figure A3-17
Torus.

B. Torus

A *torus* (Figure A3-17) is a surface generated by the revolution of a circle about any axis in its plane other than its diameter.

APPENDIX 4

Geometric Construction

Appendix 4 has been limited to those constructions which may assist in solving the problems presented in the text.

A4-1 LINE DIVISION

Given: Straight line *MN*.
Task: Divide line *MN* into five equal parts.

1. Draw line *MP* at any angle with *MN*.
2. Set off on *MP* (five) equal parts of any dimension.
3. Draw a line from point 5 to *N*.
4. Through points 1, 2, 3, and 4 draw lines parallel to line 5*N* and intersecting line *MN*. These intersections will divide line *MN* into five equal parts.

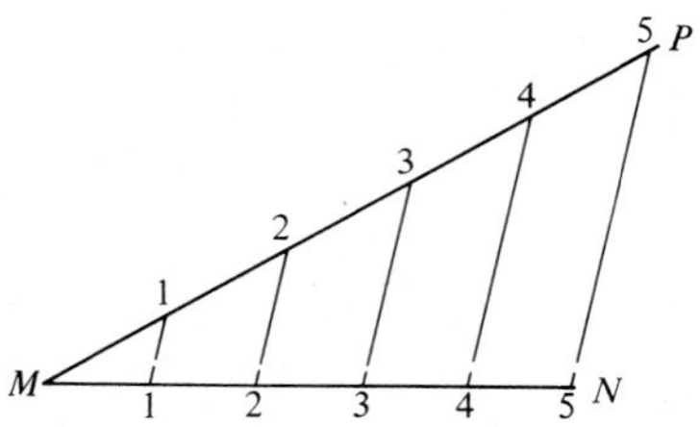

Example A4-1
Line division.

A4-2 TANGENT TO CIRCLE FROM OUTSIDE POINT

Given: Circle and point *P*.
Task: Draw line from point *P* tangent to the given circle.
Solution:

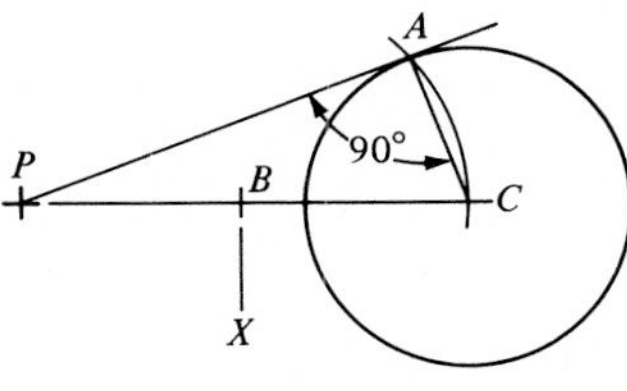

Example A4-2
Tangent to circle.

1. From given point *P* draw line *PC* to center of given circle.
2. Bisect line *PC* at midpoint *B*.
3. Using point *B* as a center and *BC* as a radius, scribe an arc which goes through point *C* and cuts the circle at *A*.
4. Draw a tangent line from point *P* to point of tangency *A*.

NOTE: A line tangent to a circle at any point *must* be perpendicular to a line drawn from the center of the circle to that point and must lie in the plane of the circle.

A4-3 TANGENT TO CIRCLE AT GIVEN POINT ON CIRCLE

Given: Point *P* on circle.
Task: Construct a line which is tangent to the given circle at *P*.
Solution:

Example A4-3
Tangent to circle.

1. Draw line *AB* from center *A* through point of tangency *P*.
2. At point *P* erect a perpendicular on line *AB*. The perpendicular *CD* is the required tangent.

NOTE: A line tangent to a point on a circle must be perpendicular to a line drawn from the center of the circle to the given point.

A4-4 TRIANGULATION

Given: Any plane figure with any number of straight sides (in this case, a five-sided polygon).
Task: Using triangulation, reproduce Example A4-4(b).
Solution:

1. With *AB* as a radius and any point *A′* as a center, scribe arc #1.
2. Draw line *A′B′* from point *A′* at any angle to intersect arc #1 at *B′*.
3. With *AC* as a radius and *A′* as a center, draw arc #2.
4. With *BC* as a radius and *B′* as a center, intersect arc #2 at *C′*.
5. With *AD* as a radius and *A′* as a center, draw arc #3.

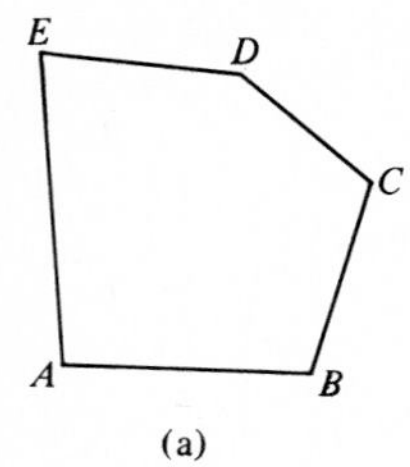

(a)

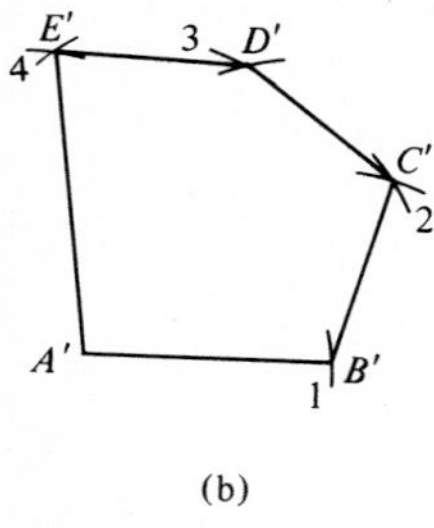

(b)

Example A4-4
Triangulation.

6. With *BD* as a radius and *B′* as a center, intersect arc #3 at *D′*.
7. With *AE* as a radius and *A′* as a center, draw arc #4.
8. With *BE* as a radius and *B′* as a center, intersect arc #4 at *E′*.
9. Draw lines *A′B′*, *B′C′*, *C′D′*, *D′E′*, *E′A′*, to obtain required polygon.

NOTE: To eliminate an accumulation of errors always use the original two points (in this case *A* and *A′* and *B* and *B′*) as the centers of all arcs.

APPENDIX 5

Drafting Techniques

Most drafting offices are equipped with drafting machines or parallel rules. The drafting machine has an adjusting head with a venier attachment so that the blades may be set at desired angles. The parallel rule, either horizontal or vertical, is used in conjunction with protractors and either fixed or adjustable triangles. In busy engineering firms, the above equipment saves a large amount of time and money in drafting operations.

When solving descriptive geometry problems, many parallel and perpendicular lines must be drawn at angles which do not conform to those formed by standard triangles. It is suggested that the paper be allowed to move freely on the board so it can be turned and held at angles suitable for problem solution. The following procedures are used to draw parallel and perpendicular lines at any angle.

A. PARALLEL LINES

Figure A5-1(a), (b), and (c) illustrates the procedure used for drawing parallel lines. The only difference in the three figures is in the positioning of the triangles.

Place one leg of triangle (1) along the initial line *MN*. Place triangle (2) against triangle (1). Hold triangle (2) firmly in place and slide triangle (1) along its edge until it reaches the location of the required parallel line.

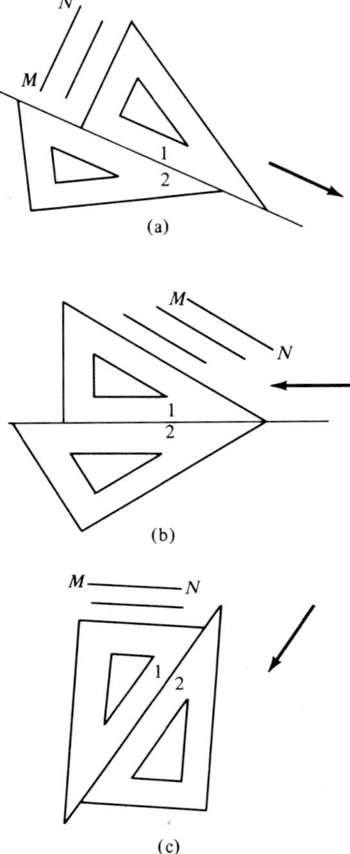

Figure A5-1
Parallel line construction.

B. PERPENDICULAR LINES

I. Rotated Triangle Method

Figure A5-2(a) and (b) shows two positions of the triangles with the same procedure being used. Place the hypotenuse of triangle (1) along the given line *MN*. Position triangle (2) against one of the perpendicular legs of triangle (1). A perpendicular line can be drawn if triangle (1) is rotated 90 deg so that its other perpendicular leg lies along the edge of triangle (2). The required perpendicular can be drawn along the hypotenuse of triangle (1).

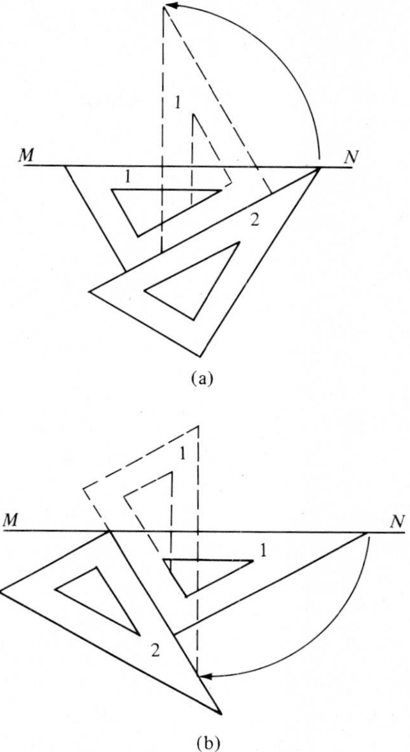

Figure A5-2
Perpendicular line construction.

II. Sliding Triangle Method

Figure A5-3 shows the method which enables the drafter to draw a line perpendicular to a given line.

Align one of the perpendicular sides of triangle (1) with the given line *MN*. Place triangle (2) along the hypotenuse of triangle (1). To draw a perpendicular line, slide triangle (1) to the required location and draw along its second perpendicular side.

The above discussion covers only a few of the many combinations that can be formed and is given to illustrate some of the important techniques used in basic geometric construction. A specific example for the drafting technique for a tangent to a circle at a given point is demonstrated in Example A5-1.

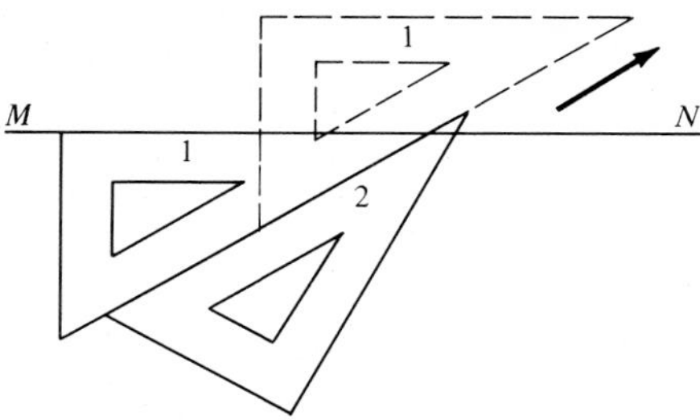

Figure A5-3
Perpendicular line construction using sliding triangles.

Example A5-1

Given: Point *P* on a circle.
Task: Draw a line which is tangent to the given circle at point *P*.
Solution:

1. Place a triangle so that one 90 deg leg passes through the center of the circle (point *A*) and the point of tangency, point *P*.
2. Using two triangles as shown and keeping the 90 deg leg parallel to *AP*, slide the triangle until the other 90 deg leg passes through *P*. Draw the required tangent *CD* through *P* perpendicular to *AP*. ◀

Computer software for automated drafting is beginning to supplement or replace traditional techniques in many current applications. Nonetheless, it is wise not to become completely dependent upon a computer. Keep your options open!

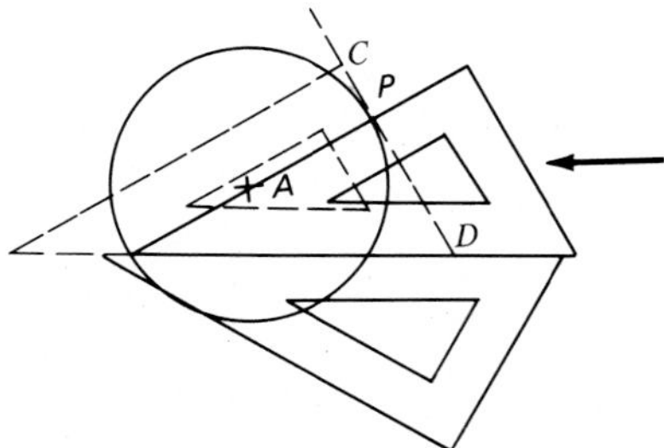

Example A5-1
Tangent to a circle.

APPENDIX 6

Vectors

Geometric modeling normally requires the use of vectors in an analytical manner as demonstrated in this text. Vectors can also be used in a graphical manner, in conjunction with the methods of descriptive geometry. This appendix presents the basics of vector geometry from a graphical point of view. Vectors are quantities which always have both magnitude and direction, and which obey certain laws for addition, subtraction, and multiplication. Often, it may be desirable to use a graphical method for performing vector operations or obtaining vector solutions.

A6-1 DEFINITIONS

1. *Magnitude*—Size; spacial quality; quantity.
2. *Vector Quantity*—A quantity having magnitude, direction, and perhaps position.
3. *Localized Vector*—A line which represents a vector quantity and has definite length, position, and direction in space. The length of the vector is proportional to its magnitude, the line shows position while an arrowhead indicates its direction (sense). A vector is noted in printed material by the use of a capital letter in boldface type.
4. *Free Vector*—A vector which represents direction and magnitude only. It may be moved to another location as long as it remains parallel to its original line of action.
5. *Linear Displacement*—The change of position from one point to another, along a line of action.

6. *Space Diagram*—A drawing (not necessarily to scale) which shows the line of action and the direction of all vectors under consideration.
7. *Vector Diagram*—A drawing (to scale) showing the magnitude and direction of all vectors in a system. The vector diagram generally contains free vectors.
8. *Concurrent Vectors*—Vectors acting through a common point.
9. *Coplanar Vectors*—Vectors which lie in the same plane. The true length of all coplanar vectors can be shown in the same view.
10. *Resultant*—A vector that represents, and can replace, the combination or sum of all vectors in a system. In order to obtain the resultant, the vectors are added geometrically as shown in the next section.
11. *Equilibrant*—A single vector which is equal and opposite to a resultant.
12. *Equality*—To be equal, two vectors must have the same magnitude and direction.

A6-2 CONCURRENT COPLANAR VECTORS

A. Vector System in Equilibrium

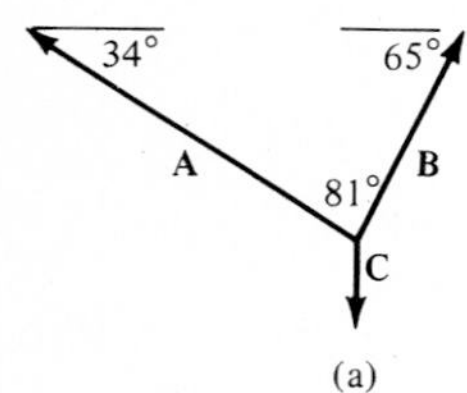

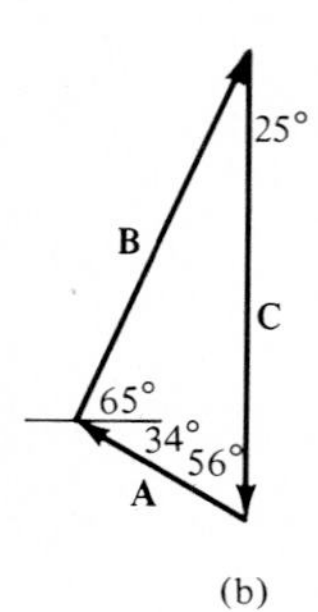

Figure A6-1
Vectors in equilibrium.

Figure A6-1(b) illustrates a system where the vectors, when drawn to scale and laid "tail to head," form a closed polygon. Such a figure denotes a resultant of zero magnitude and a vector system which is in equilibrium.

In order to completely analyze the given system, a space diagram (Figure A6-1(a)) should be drawn to show the line of action and direction of the vectors to be considered. The vector diagram (Figure A6-1(b)) is then drawn to scale and the vectors **A**, **B**, and **C** shown with the same sense and parallel to their counterparts in the space diagram. Since the vectors are treated as free vectors, the original position of the vectors is lost. The magnitude and direction of all the vectors can be found in a view normal to the plane containing the vectors.

B. Vector Addition

1. **Parallelogram Law (Two Vectors).** When the parallelogram method is used to find the sum of two vectors, the resultant is the diagonal of a parallelogram formed by the vectors and lines drawn parallel to them.

   Figure A6-2(a) shows a diagram that gives the position, direction, and length of two vectors. The method used here requires that the vectors be drawn to scale as they appear in the diagram. The parallelogram having the sides of **A** and **B** is then completed as shown in

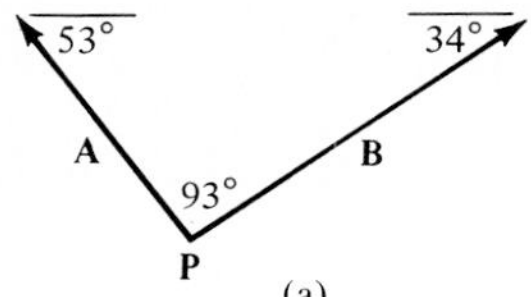

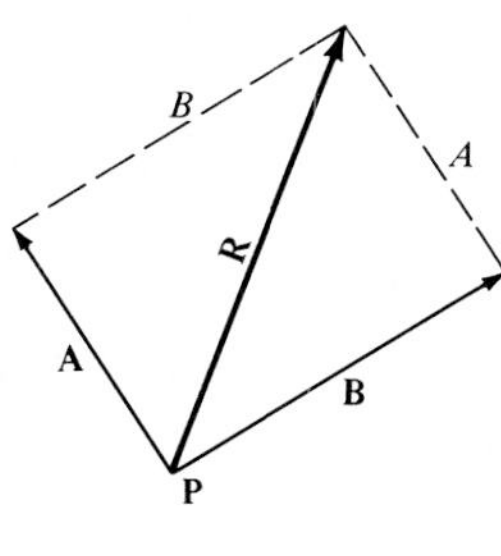

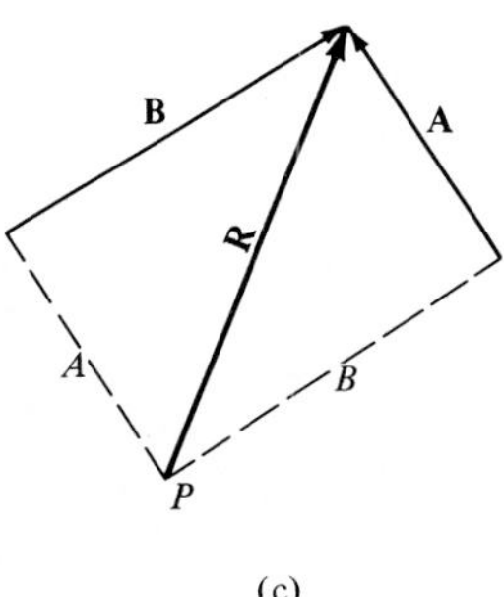

Figure A6-2
Vector addition.

Figure A6-2(b). The resultant of the given vectors showing the magnitude and direction of the vector sum **A**—+→**B** is represented by vector **R** as shown.

Notice that the vectors can be drawn through the point of application either "tail to tail" as in Figure A6-2(b) or "head to head" as in Figure A6-2(c) according to the statement of the problem.

2. **Parallelogram Law (Three Vectors).** The sum of three or more vectors can be found by expanding the parallelogram law to encompass the additional vectors.

   The vectors for this explanation are given in Figure A6-3(a). The resultant of **A**—+→**B** is found by the use of the parallelogram law as shown in Figure A6-3(b). A second parallelogram is drawn where the resultant is added to vector **C** to obtain the final resultant. Figure A6-3(b) and Figure A6-3(c) show that any number of vectors can be added to produce a final resultant.

3. **Triangle Law (Two Vectors).** The triangle law is a simplified version of the parallelogram method and is used to find the resultant of two vectors.

   Figure A6-4(a) shows two vectors which are drawn tail to head as shown in Figure A6-4(b). Vector **B** is drawn parallel to, and in the same direction as, its appearance in the original figure. Vector **A** becomes a free vector and is drawn tail to head with vector **B**. The resultant, a line from the origin of vector **B** to the terminus of vector **A**, is the resultant of **A**—+→**B**.

4. **Polygon Law (Three or More Vectors)**

Three or more vectors can also be added by expanding the rules of the triangle law so they will apply to a polygon of vectors. A vector chain can be formed by drawing the vectors of Figure A6-5(a) tail to head in any order so they are parallel to their original position. A polygon is formed when the closing side, or resultant, is drawn from the origin of the first vector to the terminus of the last (see Figure A6-5(b)).

When it is required to find the sum of three or more vectors the polygon method is used since it requires less construction than the parallelogram method.

A6-3 SPACE VECTORS—TRUE LENGTH

Vectors may be established in space by representing them in at least two views of an orthographic drawing. The principles of descriptive geometry can then be applied to determine the solution to the given problem. Prior material in this text illustrated several methods which can be used to solve vector problems using the rules of vector algebra. A graphical treatment

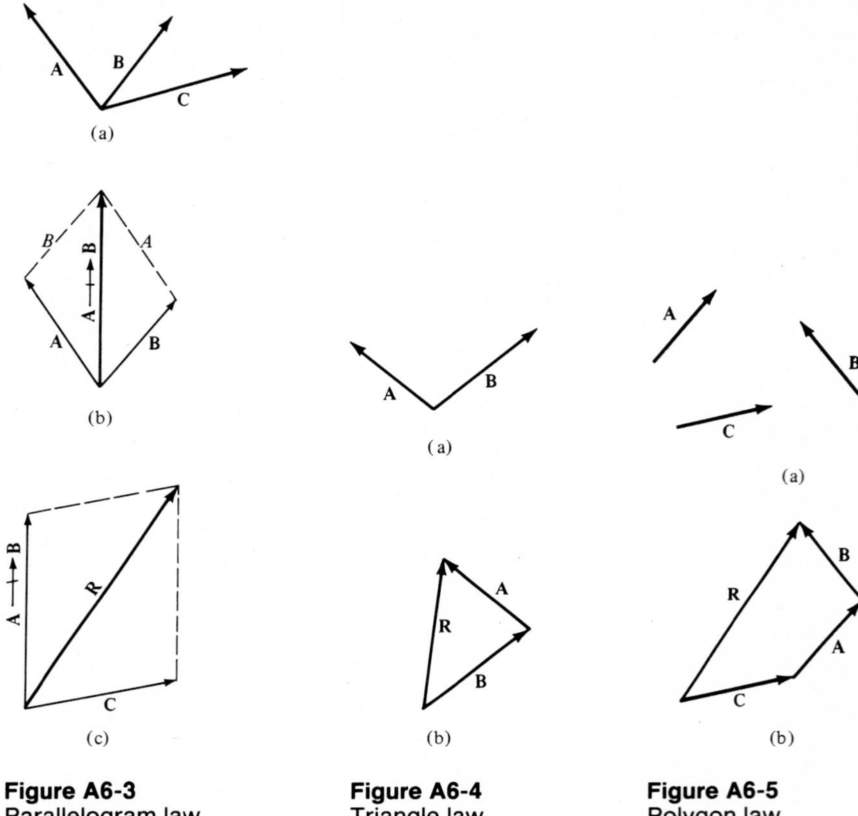

Figure A6-3
Parallelogram law.

Figure A6-4
Triangle law.

Figure A6-5
Polygon law.

depends upon the ability to find the true length of a line which will, in turn, determine the magnitude of a vector quantity. The following examples illustrate ways in which vector lengths can be established.

A. Revolution

Example A6-1

Given: Front and top views of a line segment showing its direction and position only.
Task: Change the given line to a vector having the same direction, position, and a magnitude of five units.
Solution:

1. Extend the line so that it has indefinite length in the top view.
2. Select an arbitrary point *X* on the line and project it into the front view.

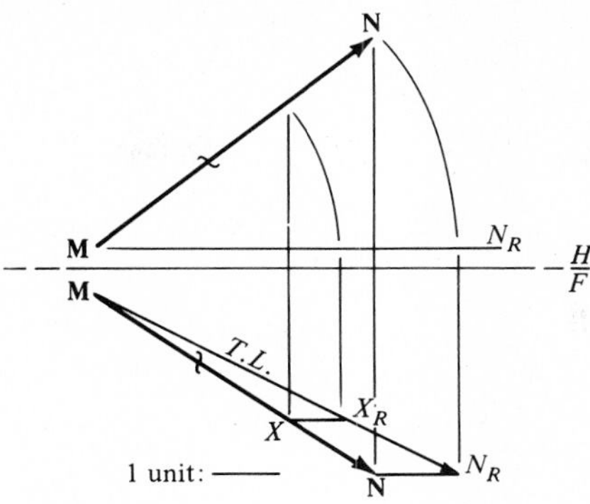

Example A6-1
Vector definition.

3. Using the principles of revolution as shown in Section 3-9, determine the true length of line MX_R in the front view. Here on the true length line, the magnitude of five units can be shown at point N_R. MN_R is the required true length of the vector.
4. By the method of revolution, project point N into both views to complete the top and front views of the vector which now has the required position, direction, and magnitude. ◀

B. Auxiliary Method

Example A6-2

Given: Front and top views of a line.

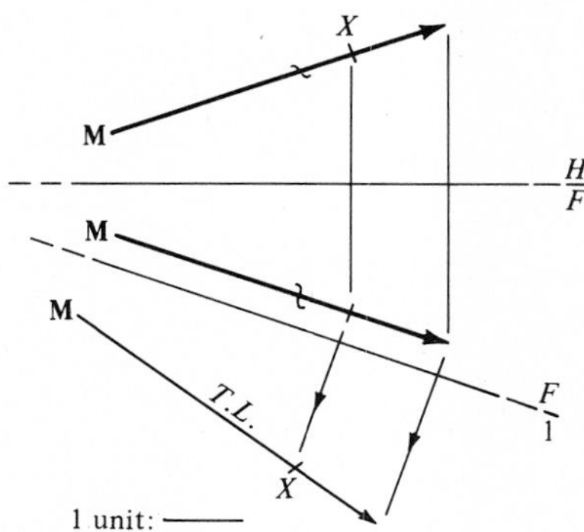

Example A6-2
Vector definition.

Task: Show the front and top views of a vector having the same position and direction as the given line and having a magnitude of five units.
Solution:

1. Extend the line to an indefinite length.
2. Select an arbitrary point *X* on the line in the front view. Project point into top view.
3. Draw a folding line *F*/1 parallel to line *MX* and obtain the true length of the line segment in the first auxiliary. On this true length line (extended if necessary), lay off the required magnitude (five units) of the vector.
4. Project the required vector into all views.

C. Abbreviated Auxiliary Method

Example A6-3

Given: Front and top views of a line segment showing direction and location only.
Task: Establish a vector in both views having the same position and direction as the given line and magnitude of five units.
Solution:

1. In the top view, extend the line to an indefinite length.
2. Select an arbitrary point *X* on the line and project the point into front view.
3. In the front view, find the vertical distance *d* from point *X* to line *MN*.

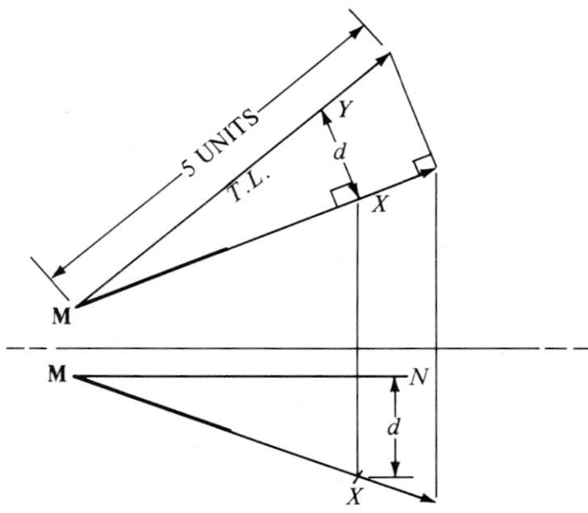

Example A6-3
Vector definition.

4. In the top view, construct a right triangle *MXY* where length *XY* is the distance *d* found in the front view. The hypotenuse *MY* is the true length of line *MX*.
5. Extend true length line, if necessary, and lay off the magnitude of the vector (five units).
6. Project the vector into both views. ◀

A6-4 VECTOR CONSTRUCTION—CONCURRENT NON-COPLANAR VECTORS

A. Revolution

Example A6-4

Given: Front and top views of two concurrent, non-coplanar vectors.
Task: Find the true length of the resultant of the given vectors.
Solution:

1. Using vectors **A** and **B**, construct a parallelogram in the front and top views.
2. Draw the diagonal of the parallelogram shown in each view. The diagonal represents the top and front views of the resultant.
3. Use the principles of revolution to find the true length of the resultant in the front view. ◀

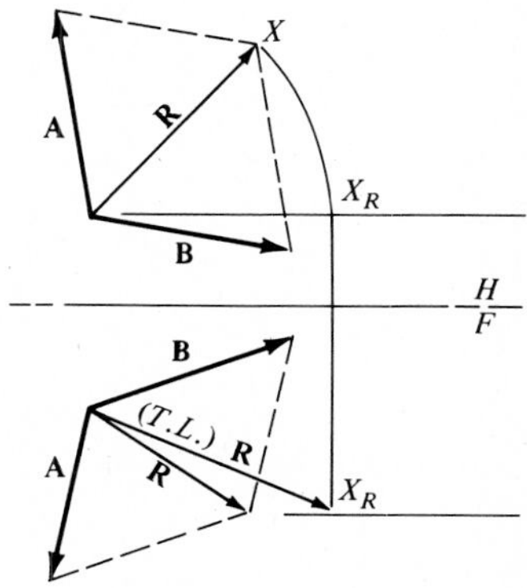

Example A6-4
Vector resultant.

B. Abbreviated Auxiliary Method

Example A6-5

Given: Front and top views of two concurrent non-coplanar vector segments which give the position and direction of two proposed vectors.
Task: Construct both views of two vectors giving the correct magnitude, position, direction, and resultant to replace the given segments. (Vector **A** will equal eight units and vector **B** will equal seven units.)
Solution:

1. Using the abbreviated auxiliary method shown in Example A6-3 and

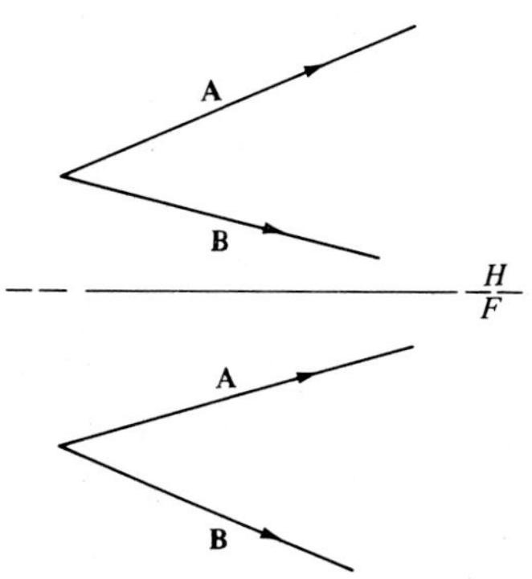

(a) Problem statement

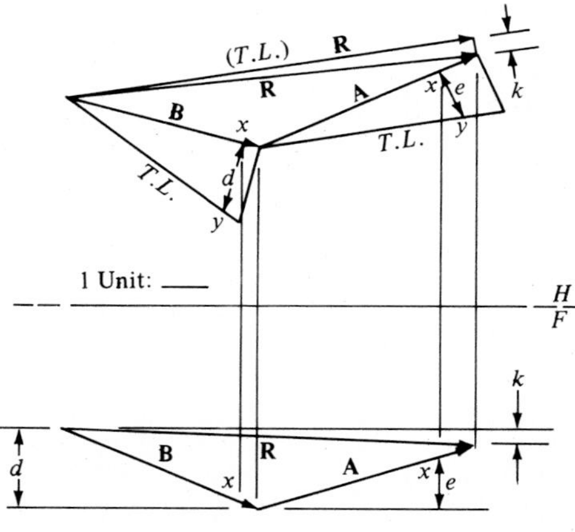

(b) Solution

Example A6-5
Vector resultant.

segment *B*, find the magnitude, direction, and position of vector **B** in both views.

2. Draw segment *A* from the terminus of vector **B** in the direction of vector **A**. By the method used in step 1, find the true length of vector **A** and show its correct representation in both views.
3. Obtain both views of the correct resultant by closing the triangle with a vector **A**⟶**B**.

A6-5 CONCURRENT, COPLANAR COMPONENTS—RESOLUTION OF A KNOWN RESULTANT

If the magnitude, location, and direction of a vector is known it can be resolved into two coplanar vectors by reversing the procedures already described.

Example A6-6

Given: Front and top views of representative segments showing the directions and positions of a proposed resultant and its concurrent, coplanar components.
Task: Show the top and front views of the components of a resultant having a magnitude of 50 units. The components and resultant must have the same position and direction as those given above.

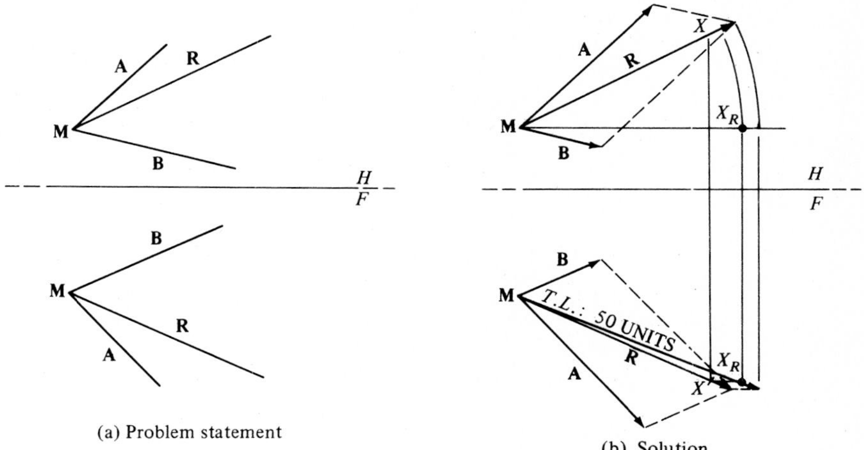

Example A6-6
Vector construction.

Solution:

1. In the front view, place an X at an arbitrary location on line M and project X into the top view.
2. Use revolution to obtain the true length of line MX in the front view.
3. Lay off 50 units on line MX (extended line if necessary).
4. Working in reverse, establish by rotation, the front and top views of vector **R** in space.
5. Establish, in both views, the parallelogram (components) which is drawn through the terminus of vector **R**. ◀

Graphical methods can also be used to obtain solutions for vector multiplications. For example, when multiplying vector **A** by a scalar b, the vector $b\mathbf{A}$ has the same direction as **A** with a magnitude b times the magnitude of **A**. The graphical significance of the dot product and the cross product of two vectors was discussed in the text. Vector products can be solved using graphical techniques if needed. Regardless of which method is used, analytical or graphical, the same result should be obtained, within the limits of accuracy imposed by the skill of the user.

Index

angle
  acute 449
  between line and plane 236–238, 249
  between two intersecting lines 234
  between two planes 242, 249
  between two skew lines 235
  complementary 449
  oblique 449
  obtuse 449
  right 449
  straight 449
  supplementary 449
apparent angle of intersection 248
apparent point 83
area 60, 61, 65
auxiliary 40
  elevation 42, 47
  first 40, 138
  frontal 40, 75
  horizontal 42, 76
  profile 43, 47, 77
  second 140
  successive 140
auxiliary method 113, 162, 164, 186, 188, 189, 203, 206, 208, 260, 280, 313, 315, 465

base plane 324
base vector form 91
bearing 86
bivariate parametric equation 268, 271
boundary representation (B-rep) 356
bow and buttock lines 324
bow and buttock planes 324
building block systems 356

clipping 285–290
clock method 46
complimentary-angle method 240
computational descriptive geometry 216–220
computer graphics 147, 170
cone 452
cone development 374–377
constructive solid geometry (CSG) 356
coordinate system 27, 29
Cramer's Rule 97, 226, 441
cutting plane method 281, 310, 314, 315, 320, 336, 338, 339
cylinder 452
cylinder development 379–382

date files 21
  ASSY3LNS 150
  ASSY3PTS 150
  CONNECT 22, 23
  CORNER 21
  PLATE1 147
  PLINE1 147

determinant 50, 51, 441
developable surfaces 362
diagonals 325
dihedral angle 242
  by revolution 265
direction angles 96
direction components 54, 92
direction cosines 95, 96, 122, 248
direction ratios 167

edge-view method 243, 310, 319, 336
edge views 47, 105
engineering drawing 3
eye position 56, 149

facet 49
  convex 125
  polygon 125
  rectangular 50
  triangular 49
fair surfaces 323
figures
  parallelogram 450
  polygons 450
  triangle 450
folding line 7, 16, 40, 75
frontal plane 4, 7, 10, 138

grade 89

hinge lines 8
horizontal plane 4, 7, 11, 138

intersecting line and plane 212, 213, 215
intersecting lines 82, 97, 100, 108, 164, 246
intersecting planes 208–211, 327
intersection
  cones 349
  cylinder and cone 343–348
  cylinder and torus 349
  cylinders 339–342
  line and cone 284
  line and cylinder 280
  line and prism 277
  line and pyramid 283
  plane and cone 319, 329
  plane and cylinder 314

intersection (*cont.*)
  plane and prism 310
  plane and pyramid 313
  plane and sphere 322
  prism and cone 348
  prism and cylinder 353, 355
  prism and pyramid 338, 373
  prisms 336, 352
irregular plane 45

line 11, 73, 75
  bearing 86
  frontal 12, 13, 78
  grade 89
  horizontal 12, 78
  inclined 12, 13, 236
  infinite 91
  oblique 12, 13, 141, 148, 236
  point view 141
  principal 75
  profile 12, 13, 78
  slope 87
  three-dimensional 92
  two-dimensional 97
  vector equation 91, 197
  vertical 12, 13
line construction 99
line method 181, 238
line of sight 1, 3, 10, 38, 55, 84, 147
line segment 12, 73, 80, 91

magnitude 31, 32, 92
  vector 51, 54
main programs 24
  A-3 446
  ANYAUX 149, 400
  BODYC 269, 414
  CDGHORIZ 219, 410
  CDGSLOPE 219, 412
  CDGPLANE 217, 407
  CDGSKEW 218, 408
  CLIPCONE 330, 424
  CLIPLANE 327, 422
  CLIPOFF 382, 433
  CLIPON 382, 431
  DEFPLANE 124
  FINDAREA 62, 64
  MAIN1 214, 404
  MATRIX 446

main programs (*cont.*)
ORTHOEYE 57, 147, 272, 393, 444
PIERCE 216, 406
PLANEDGE 59, 395
PRISMCYL 355, 428
PRISMS 353, 427
PTLINE 192, 402
PTPLANE 194, 403
SHOWPOLY 128, 130
3DLINE 100, 398
3DLINE 99, 397
VECPRODS 120, 399
VEXCLIP3 290, 353, 416
VIEWS 24, 391
XCONE 290, 420
matrix 20
column 20
multiplication 48, 97, 119, 439
product 48, 118
row 20, 32
subtraction 31
three-dimensional data 21, 48
transformation 58
matrix algebra 439
multiplication 439
rotation 443
transformations 445
multi-planar drawing 3
multi-view drawing 3

nonintersecting lines 83
normal views 38, 45, 47, 74, 143, 145, 234
frontal 38
horizontal 38
profile 38

ortho-box 22, 23, 27, 29
orthogonal drawing 3
orthographic drawing 3, 16
orthographic projection 2, 6, 58
overconstrained 246

parallel line and plane 162, 168
parallel line development 364, 368, 369
parallel lines 110, 158, 161, 162, 166, 457
parallel planes 164, 167
implicit, nonparametric equations 167
parametric equation 91, 97, 126
plane 220
surface 268
perpendicular bisector 100
perpendicular lines 175, 458
perpendicular planes 197
perspective projection 1
picture plane 1, 3, 39, 47, 78
piercing point 202, 212, 216, 329
plane 14, 73
base 324
edge view 105
graphical representation 108
implicit coefficients 122, 197
implicit equation 212
inclined 14, 15
irregular 45, 145
normal 14
normal view 180
oblique 14, 15, 143
parametric equation 126, 220
picture 1, 3, 39, 47
reference 44, 46
scalar equation 121
vector equation 121
vertical 14, 15
plane method 182, 239
plane of projection 3
points 9, 20, 80, 111
in planes 111, 124
noncollinear 110, 122
point view 78, 141
point-view method 179, 181
polygon file 65, 67
polygon test 125
primary auxiliary plane 40
principal views 6, 24, 25, 26, 27, 28, 29
prism 452
profile plane 6
projection 205
line on a plane 205–208, 222
pyramid
oblique 452
right 452